The People Before

The geology, paleoecology and archaeology of Adak Island, Alaska

Dixie West
Virginia Hatfield
Elizabeth Wilmerding
Christine Lefèvre
Lyn Gualtieri

BAR International Series 2322
2012

Published in 2016 by
BAR Publishing, Oxford

BAR International Series 2322

The People Before

ISBN 978 1 4073 0905 7

BAR Publishing is the trading name of British Archaeological Reports (Oxford) Ltd. British Archaeological Reports was first incorporated in 1974 to publish the BAR Series, International and British. In 1992 Hadrian Books Ltd became part of the BAR group. This volume was originally published by Archaeopress in conjunction with British Archaeological Reports (Oxford) Ltd / Hadrian Books Ltd, the Series principal publisher, in 2012. This present volume is published by BAR Publishing, 2016.

Printed in England

BAR
PUBLISHING

BAR titles are available from:

BAR Publishing
122 Banbury Rd, Oxford, OX2 7BP, UK
EMAIL info@barpublishing.com
PHONE +44 (0)1865 310431
FAX +44 (0)1865 316916
www.barpublishing.com

Table of Contents

LIST OF CONTRIBUTORS

Zhanna A. Antipushina: A.N. Severtsov's Institute of Ecology and Evolution, Russian Academy of Sciences, Russia (e-mail: zh.antipushina@gmail.com)

Françoise Bouchet Laboratoire de Paléoparasitologie, CNRS ESA 8045, Université de Reims, France (e-mail: francoise.bouchet@univ-reims.fr)

Joan Brenner Coltrain Dept. of Anthropology, 270 S. 1400 East, University of Utah, Salt Lake City, Utah 84112 USA (e-mail: joan.coltrain@anthro.utah.edu)

Debra Corbett US Fish and Wildlife Service, 1011 E. Tudor Road, Anchorage, Alaska, 99503, USA (e-mail: debbie_corbett@fws.gov

Susan Crockford Pacific Identifications Inc., 6011 Oldfield Rd., Victoria BC V9E 2J4, Canada (e-mail: sjcrock@shaw.ca)

Floyd E. Dowell USDA, ARS, Center for Grain and Animal Health Research, Engineering & Wind Erosion Research Unit, Manhattan, KS 66502, USA (e-mail: floyd.dowell@ars.usda.gov)

Ame Garong Department of Biodiversity Sciences, Graduate School of Social and Cultural Studies, Kyushu University, 744 Motooka, Nishi-Ku, Fukuoka, Fukuoka, 819- 0395, Japan; (e-mail: ame.garong@gmail.com) **Current Addres:** Archaeology Division, National Museum of the Philippines, P. Burgos Street, Manila 1000; (e-mail: ame.garong@gmail.com)

Lyn Gualtieri College of Science and Engineering, Seattle University, 901 12th Ave. P.O. Box 222000, Seattle, Washington 98122-1090, USA (e-mail: gualtieri@seattleu.edu)

Virginia Hatfield Prewitt and Associates, Inc. 3207 29th Lubbock, Texas, USA (e-mail: vlhatfield@hotmail.com)

M. Geoffrey Hayes Division of Endocrinology, Metabolism and Molecular Medicine, The Feinberg School of Medicine, Northwestern University, 303 E. Chicago Ave., Tarry 15-759, Chicago, Illinois, 60611, USA, (e-mail: ghayes@northwestern.edu)

Richard Jeannotte Formerly of Division of Biology, Kansas State University, Manhattan, Kansas 66502, USA; now at Department of Population Health and Reproduction, University of California-Davis, Davis, California 95616, USA (e-mail: rjeannotte@ucdavis.edu)

Taylor Johnson Formerly of Whitman College, Walla Walla, Washington 99362; now at IMPACT Silver Corporation, Suite 1100; 543 Granville Street; Vancouver, BC; Canada V6C 1X8; Now: 202 Peter Koch Tower, Bozeman, Montana 59715

Akihiro Kano Division of Biodiversity Sciences, Department of Environmental Changes, Graduate School of Social and Cultural Studies, Kyushu University, 744 Motooka, Nishi-ku, Fukuoka, Fukuoka, 819-039, Japan

Marvin Kay Department of Anthropology, University of Arkansas-Fayetteville, Arkansas 72701 USA (e-mail: mkay@uark.edu)

Bulat F. Khassanov A.N. Severtsov's Institute of Ecology and Evolution, Russian Academy of Sciences, Russia

Nina K. Kiseleva A.N. Severtsov's Institute of Ecology and Evolution, Russian Academy of Sciences, Russia

Hiroko Koike The Kyushu University Museum, 6-10-1 Hakozaki, Higashi-ku, Fukuoka 812-8581, Japan. (e-mail: koikeroko@gmail.com)

Olga A. Krylovich: A.N. Severtsov's Institute of Ecology and Evolution, Russian Academy of Sciences, Russia (e-mail: constrictor@rambler.ru)

Christine Lefèvre Département Écologie et Gestion de la Biodiversité du Muséum national d'Histoire naturelle and UMR 7209 du Centre National de la Recherche Scientifique, Paris, France, (e-mail: lefevre@mnhn.fr)

R. Game McGimsey Alaska Volcano Observatory, USGS, Anchorage, Alaska 99508-4667, USA

Toshio Nakamura Center for Chronological Research, Nagoya University, Furo-cho, Chikusa-ku, Nagoya 464-8602, Japan

Kirsten Nicolaysen Department of Geology, Whitman College, 345 Boyer Ave. Walla Walla, Washington 99362, USA (e-mail: nicolakp@whitman.edu)

Shin Nishida Division of Biodiversity Sciences, Department of Environmental Changes, Department of Environmental Changes, Graduate School of Social and Cultural Studies, Kyushu University, 744 Motooka, Nishi-Ku, Fukuoka, Fukuoka, 819- 0395, Japan; (e-mail: shin.nsd@nifty.com)

Mitsuru Okuno Department of Earth System Science, Faculty of Science, Fukuoka University, 8-19-1 Nanakuma, Jonan-ku, Fukuoka 814-0180, Japan (e-mail: okuno@fukuoka- u.ac.jp)

Dennis H. O'Rourke Department of Anthropology, 270 S. 1400 East, Rm. 102, Stewart Bldg., University of Utah, Salt Lake City, Utah, USA (e-mail: orourke@anthro.utah.edu)

Andrei M. Pereladov A.N. Severtsov's Institute of Ecology and Evolution, Russian Academy of Sciences, Russia

Brenn Sarata Fugro, Veurse Achterweg 10, 2264 SG Leidschendam, the Netherlands
Arkady B. Savinetsky A.N. Severtsov's Institute of Ecology and Evolution, Russian Academy of Sciences, Russia (e-mail: arkadybs@rambler.ru)
Chizuru Takashima Human Life and Environment Course, Faculty of Culture and Education, Saga University, Honjo-machi 1, Saga, 840-8502, Japan (e-mail: chizurut@cc.saga-u.ac.jp)
Masayuki Torii Kumamoto Gakuen University, 2-5-1 Oe, Kumamoto 862-8680, Japan
Douglas W. Veltre Department of Anthropology, University of Alaska Anchorage, 3211 Providence Drive, Anchorage, Alaska 99508, USA (e-mail: dwveltre@uaa.alaska.edu)
Keiji Wada Department of Earth Science, Asahikawa Campus, Hokkaido University of Education, 9 Hokumon-cho, Asahikawa 070-8621, Japan
Dixie West Natural History Museum and Biodiversity Center, Dyche Hall, 1345 Jayhawk Blvd. University of Kansas, Lawrence, Kansas, USA (e-mail: dlwest@ksu.edu)
Elizabeth Wilmerding Department of Anthropology, Vassar College, Poughkeepsie, New York, USA; (e-mail: ArcticLiz@turbonet.com)

List of Figures

List of Tables

Preface
The People Before: an Introduction to Adak, Alaska

Dixie West

Adak Island, 1928km southwest of Anchorage, Alaska, is located in the Andreanof Island group in the center of the Aleutian volcanic archipelago. The City of Adak, formerly Adak Naval Air Station, is the most occidental municipality in the US. Between 2005 and 2007 the Central Aleutians Archaeological and Paleobiological Project (CAAPP), an international, interdisciplinary scientific team, focused on the paleoenvironment, geology, biology, and anthropology of the central Aleutians using north Adak Island as our research area. Our objective was to determine how prehistoric Native inhabitants interacted with their natural and geological world. The Central Aleutians team wanted to know when the first immigrants arrived in the central Aleutians, and how they impacted, adapted to, and survived the Bering Sea ecosystem. Although maritime biologists study the current ecology of the Bering Sea, archaeologists, geologists, and paleoecologists provide information about the prehistoric Aleutians by excavating and analyzing ancient stone tools, bones, shells, and geological strata.

Brian Garfield's (1969) *The Thousand-Mile War* or Stanley Sandler's (2001) *World War II in the Pacific* document Aleutians campaigns during World War II. And works of anthropologists Lydia Black (1984, *Atka: An Ethnohistory of the Western Aleutians*) and Roza Liapunova (1996, *Essays on the Ethnography of the Aleuts*) contain valuable nuggets of information about the central Aleutians and Adak at the time of Russian contact. However, little published material exists about the prehistoric people who lived in the Andreanof Islands. Nearly all archaeological investigations in the area remain unpublished, and no comprehensive work has focused on the first people who lived on Adak. The Native peoples who inhabit the Aleutian archipelago call themselves Unangan. The words 'Aleutians' and 'Aleut' are Russian terms for the archipelago and the people who lived there during Russian contact. Throughout this book, 'Unangan' and 'Aleut' are terms used for Native Americans who call the Aleutians home. The following chapters provide an initial, albeit partial, account of the prehistoric Unangan people and their central Aleutians environment; this is the story of the people before Russian contact and World War II.

Travelling to Adak requires a three-hour Alaskan Air flight originating from Ted Stevens International Airport in Anchorage. Prior to landing, clearings in the cloud cover over the island reveal color patches dependent on the Aleutian season: Crayola crayon names like 'sepia' and 'antique brass' for autumn through early spring, and 'screamin' green' and 'mountain meadow' for summer.

Adak comprises a foggy or cold-rain terrain of rolling rye grass and tundra-covered hills sprinkled with fresh water lakes and ponds and ocean bays. 'Dreary', 'bleak' or 'desolate' are adjectives that cross the mind of the novice stepping onto the Adak airport tarmac. Part of the Adak Municipal Airport sign is missing, probably sheered off during some past cyclonic storm, and travelers frequently face winds pelting ice crystals. Such is not an unusual greeting for June visitors.

Like all Aleutian Islands, Adak weather frequently includes overcast skies, ubiquitous fog, chilly-to-really-cold temperatures, and omnipresent winds that can escalate to gale force *blows*. During summer, sustained winds sometimes reach 60 to 80km per hour, and winter storms can truly be Homeric. During 1999, a March storm produced winds of 305km per hour before the Adak anemometer broke. During February 2009 gale force winds pummeled Adak, threatening area ships and damaging buildings. This frequent climatic furor has earned Adak, and the Aleutians as a whole, the nickname—*Birthplace of the Winds* (Bank 1956).

The Aleutians are subarctic rather than arctic. Wedged between the cold Bering Sea and milder Pacific Ocean, Adak's weather is maritime with comparatively minor variations in seasonal temperatures. Adak summer temperatures range from mild to chilly, generally averaging between 0° and 10.5° C. However, the recurrent winds often lower summer temperatures down to bone chilling level. The lowest temperature ever recorded on Adak was -16.1° C during January 1963 and February 1964; the highest temperature of 23.9°C was recorded August 1956 (Average-Temperature 2011). Temperature also varies with altitude; cooler temperatures occur on volcanic summits with milder temperatures along the seacoast and south facing slopes. The mean annual precipitation, mostly rain, is 137cm near sea level. June and July are relatively dry months, averaging around 7.6cm of precipitation each.

The heaviest rains, between 17.8 and 20.3cm, occur during November and December, with most of it in the higher elevations (Intellicast.com 2011). Although it varies yearly, snow on Adak can be deep. The average snowfall often exceeds 254cm near the peaks of the two local volcanoes—Moffet and Adagdak (1200m and 650m above sea level respectively) where snow lingers throughout the cool summer months.

The Aleutians comprise the northernmost section of the Ring of Fire, a highly seismic and volcanic zone that encircles the Pacific Ocean basin. Stretching for 40,000km, this geologically unstable region consists of plate movements associated with ocean trenches, earthquakes, and volcanic activity. Eighty percent of the planet's largest earthquakes occur in the Ring of Fire (USGS 2011). The Aleutians are located at the rim of a subduction zone, where the Pacific and North American Plates actively move, and periodically ram each other, resulting in considerable geological volatility. Adak is no exception. Major earthquakes shook the island in 1957, 1964, 1977, 1986, 1996, and 2006 with shock waves measuring between 7.7 and 8.6 on the Richter Scale (Alaska Earthquake Information Center 2011). The United States Geological Survey records smaller earth tremors on a monthly, if not weekly, basis.

Volcanoes formed Adak and the Aleutian Islands. Created some time during the Tertiary—between 65 million and 1.8 million years ago--Adak evolved from a volcanic center on the north side comprised of three volcanoes—Moffet, Adagdak, and Andrew. Moffet and Adagdak are the most recent volcanic domes. The oldest, Andrew, some time in the remote past, exploded creating a basin currently filled with the waters of Lake Andrew and Andrew Bay.

During the 18th century Vitus Bering's ships discovered and explored the Aleutian Islands. Subsequent Russian and Native interactions completely changed the Unangan world. Disease, reshuffling the Unangan people to other islands, and the shift from hunting for household use to large commercial enterprise impacted not only the Native people but also their ecological resource base.

The Imperial Russian Empire was built on the conquest of neighboring territories and tribes. By 1639 the Russians had conquered all of Siberia, and the easternmost limits of the empire lay on the shores of the Chukotka and Kamchatka peninsulas. Peter the Great was determined that his already vast empire be enlarged to include lands to the east. This czar's driving desire was not only for increased territory, but also to locate areas rich in fur bearing mammals that drove the Empire's thriving trade market with China. During the reign of Peter the Great, Russian trappers had considerably reduced the number of valuable fur bearing mammals on the Asiatic mainland. To satisfy the world's glut for animal pelts, Peter the Great determined that the easiest way to expand without conflict would involve sailing to the east to claim land and natural resources on the North American continent (Veltre 1990).

Records of the Russian American Company, church archives, personal journals, and ship logs comprise the earliest historical information about the Andreanof Islands. Roza Lipunova (1996) and Lydia Black (1984) extensively researched these sources, and the summary below is derived from their translations of early Russian records.

Russian exploitation of the Aleutian archipelago commenced from the west and moved east. Finding abundant resources in the Near Islands closer to Kamchatka shores in early exploration years, the Russians felt little need or desire to explore unknown and potentially dangerous waters surrounding the Andreanofs and other island groups farther to the east (Black 1984, 82). Russians with the Vitus Bering expedition first sighted some of the Andreanof Islands during the return voyage to Kamchatka in 1741. Between 1745 and 1747 a few Russian ships, enticed by what appeared to be an unlimited supply of fur seals and sea otters, began investigating the central Aleutians. Sometime during this two-year time period, the first recorded fur hunting ship made landfall in the Andreanofs. In 1749, Captain Ivan Butin, sailing the *Sv. Boris I Sv. Gleb*, reported reaching Atka east of Adak (Black 1984, 74) and by the mid 1750's Russian entrepreneurs knew the bountiful potential of the entire Aleutian chain. Between 1758 and 1764 two Russian ships hunted in the Rat Islands and at least three were in the Andreanof group (Black 1984, 82). In 1753 Captain Petr Bashmakov, sailing the *Sv. Eremiia*, was wrecked on Adak. Bashmakov's crew managed to sail away from the island on a smaller vessel, the *Sv. Petr I Sv Pavel*, built out of wreckage from the mother ship combined with driftwood the crew scavenged on Adak beaches (Black 1984, 74).

Captain Dmitri Pan'kov and a crew of 48 men sailed the *Sv. Vladamir* from Kamchatka in 1758, spent a season in the Commander Islands, and arrived off Atka, July 16, 1759. The captain failed to find a harbor that he considered safe, and he sailed east to Amlia where he established a camp in Pan'kov Bay on the north shore. He remained in the central Aleutians area until the summer of 1761. From his base camp, Pan'kov sent out seasonal hunting parties to nearby islands including Great Sitkin and Adak or Atka (Black 1984, 84).

In the summer of 1760, Unangan villagers killed the Sila Shavyrin hunting party either at Atka or Adak. This uprising spread, involving the base camp of the *Sv. Vladimir* on Amlia. Upon the arrival of the Russian ship *Sv. Gavriil* on Atka September 26, 1760, the conflict ended (Black 1984, 87). The trade relationship between the Russians and central Unangan people was apparently stabilized by the time Andreian Tolstykh arrived on Adak on August 28, 1761; the Unangan people of the Niigugis polity that included the territory or Adak, Atka, Amlia and adjacent islands affably met him on shore. During the time of the Russian contact with the Niigugis political area, the Unangan population was reportedly small and mobile, frequently moving from one village to another and traveling among adjacent islands.

According to the earliest Russian reports, Atka apparently had been previously abandoned and was in the process of being repopulated following a major volcanic explosion. The Russians found only one settlement, inhabited by approximately 60 Natives, on Atka. Apparently a slightly larger Unangan population lived on Adak and Amlia. At the time of initial Russian contact, the Niigugis polity was outnumbered by a polity centered on the island of Kanaga just to the west of Adak (Black 1984, 87). Liapunova (1996, 167), reading the journals of Russian Cossacks M. Lazarev and P. Vasiutinskii in the Andreanof Islands, reported: 'Concerning the largest of the islands, Adak...the number of inhabitants cannot possibly be stated definitely because they move with all their households in large baidaras [sic] from island to island, crossing the straits, and live in localities they like. On Kanaga Island there were approximately 200 inhabitants; on the island of Sitkin (Chetkhina), up to 400 families; on Tagalak Island, up to 400 families; on Atka Island, up to 60 persons; and on Amlia Island up to 600 persons.'

The name *Andreanof Islands* is attributed to Andreian Tolstykh who lived in the island group between 1761 and 1764 (Black 1984, 47; Liapunova 1996, 24-25). In September 1761 Tolstykh established a hunting camp on Adak and he immediately sent out Unangan men in baidarkas to hunt in surrounding islands. He also established friendly contact with the chief of the Kanaga polity to the west. Between 1761 and 1764 Tolstykh 'enjoyed very profitable years, brought home a good cargo, suffered no casualties and produced a detailed description of the islands and their inhabitants' (Black 1984, 90). The next ship to make camp in the area of Adak was the *Sv. Petr I Sv. Pavel* commanded by Ivan Korovin with Griogoril Korenev as foreman. Captain Korovin dropped anchor off Adak on August 3, 1768, and he hunted in the area until 1770 (Black 1984, 91). According to Russian reports, three ships simultaneously hunted in the Andreanof Islands between 1768 and 1774 (Black 1984, 92).

During the late 1770's Grigorii Shelikhov, aggressively seeking complete control of the American fur trade, hunted in the eastern Aleutians. By 1783, Shelikhov owned either outright, or in part, 16 ships hunting in the North Pacific. From his base camp on Kodiak Island, Shelikhov focused on the Alaskan mainland. His largest and most well organized competitors, the Lebedev-Lastochkin Company and the Panov Brothers Company, disputed Shelikhov's claims to the Alaskan Peninsula and the eastern Aleutians. Meanwhile, smaller, independent hunting vessels discovered that it was more profitable to focus their hunting enterprises in the central and western islands. The Unangan people complained about brutalities perpetuated by Russian hunters, and by the late 1770's the Natives suffered the impact of 'systematic economic exploitation and deprivation of their subsistence resources' (Black 1984, 92-93).

In 1797 the Andreanof, Rat and Near Islands were placed under the control of an independent but still Shelikhov-controlled company, the United American Company. In 1799 Shelikhov's heirs received hunting rights for all of the North Pacific controlled by Imperial Russia, along with rights and privileges to Kamchatka, the Kuriles, and the Okhotsk Sea. Overhunting in the region took its toll. By the beginning of the 19th century, with dwindling seal and sea otter populations, few Russian ships patrolled the western Aleutians. A vessel hunted in the west during 1802, 1806, 1809, and 1811, but by 1811, six Russians hunted on Atka and only two on Attu. These Russians, with Unangan families, hunted only for subsistence (Black 1984, 97).

Foreign ships carried not only trade goods and furs but also disease. In 1808 the *Sv. Petr I Sv. Pavel*, sailing from Unalaska to Okhotsk, stopped at Atka to overwinter.

Aboard, sailors carried 'a pestilence, manifest in high fever and chest congestions, which had raged the previous year in Unalaska, causing over 350 deaths' (Black 1984, 97). Hearing word that a ship had arrived in the area, Unangan villagers from various settlements in the Andreanof Island group, 'presumably desirous of news, sociability, and trade goods' gathered at the settlement at Korovin Bay (Black 1984, 98). After contact with the sick Russians, the illness rapidly spread among the locals who possessed little to no natural immunity against Western infections. Apparently, so many Natives died on Atka that not enough people were left alive to bury the dead.

Under the control of the fur trading companies, Unangan people across the archipelago, including the Andreanof Islands, underwent considerable population shifts from their homelands. Because Russian ships did not communicate with remote villages, Natives moved to harbors where the foreign vessels dropped anchor. Later, the forced exile of Natives for the purpose of sea mammal hunting and fox trapping destroyed traditional social order in villages. Furthermore, the removal of Unangan men, the main food suppliers for households, had devastating economic results. In their quest for fur pelts, Russians shuffled Unangan peoples from their native islands to the Pribilofs, Commander Islands, Kamchatka, mainland Alaska and California.

The Russian navigator Ivan Vasil'iv reported that Rat Islanders were settled for a time at Adak and Atka while others were moved to Attu (Black 1984, 42). Sometime after 1768, a 'significant number' of central Unangan people were taken, aboard independent ships, to the Pribilof Islands to hunt; apparently some were returned to their traditional homelands while others were not (Black 1984, 96). And sometime prior to 1790 Ivan Lukanin convinced a group of young Andreanof men and women to sail to the Kenai Peninsula to form a new settlement. The loss of the young people to what were already small settlements would have created a serious setback for the Kanaga/Adak polity. However, at approximately the same time, Unangan groups from Amchitka and surrounding islands were relocated to

Adak and Atka (Black 1984, 93-94). Maksim Lazarev, serving with the last hunting independents and then with Grigorii Shilekhov, relocated the Amchitka Natives to Adak and Atka, apparently to allow the Amchitka sea otter regain in population (Black 1984, 96). However, K.T. Khlebnikov, a Russian chronicler who lived in Russian America between 1818 and 1832, recorded the Amchitka Unangan people were moved east because of poor subsistence on their native island (Black 1984, 96).

This forced exile remains embedded in modern Unangan lore. 'Today, there are still a few remembered traditions concerning this mass movement of villages; some of the Islanders presently on Atka tell of the hardships their ancestors encountered when the Russians forced them to leave Adak and Tanaga to journey eastward to the island of Atka' (Black 1984, 27). According to Jochelson (1925, 27) Unangan individuals remembered: 'Before that time [of the Russians] few Aleut lived on Atka, but Russian hunters transferred to Atka Aleut from the islands of Tanaga, Kanaga, and Adax [sic].'

K.T. Khlebnikov reported that Adak was inhabited at the beginning of the 19th century (Liapunova 1996, 167). However, the death knell for the Andreanof Island population as a whole apparently occurred sometime between 1809 and 18ll, when the population, already largely destroyed by 1808 disease brought by the ship *Sv. Petr I Sv. Pavel*, further declined when three baidarkas of Atkans moved to Attu (Black 1984, 98). Through massive transferrals, 'Eventually, those who traced their roots to Atka and Adak, tended to settle on Bering Island' (Black 1984, 102).

The United States purchased Alaska and the Aleutian Islands from Russian on October 18, 1867. Captain John A. Henriques, commander of the US Revenue Cutter *Lincoln*, was one of the first American officials to visit the central Aleutians. He recorded the tremendous drop in population for the region. Between 1868 and 1871 the Native count dropped from 1075 to 332. Black (1984) proposed that this population decrease could be due to: 1) increased mortality versus birth rate, 2) undercounting by Russian Orthodox priests in the Aleutians and/or, 3) migration from the Andreanof and Near Islands to the eastern Aleutians, Pribilofs, or the American mainland.

By the time America took control of the Aleutians, traditional Unangan settlements had been long abandoned and only few villages were intact. 'By 1881, only Nazan village maintained itself on Atka and one village on Attu' (Black 1984, 105). Under American control the Natives were concentrated near government schools. Like the Russians before, the US lust for furs was keen. By the beginning of the 20th century fur seals and sea otters were near extinction. In 1911, under the International Fur Seal Treaty, the otter was declared a protected species to stop the slaughter until waning sea mammal populations could recover.

On June 3-4, 1942, the Imperial Japanese military attacked Dutch Harbor in the eastern Aleutians. Three days later, on June 7, 1942, the Japanese Northern Army invaded the islands of Kiska (Rat Islands) and Attu (Near Islands) in the western Aleutians. Unangan villagers at Chichagof Harbor were relocated to camps in Hokkaido until 1945. Natives in nine remaining villages on six islands in the Aleutians were evacuated to mainland Alaska (Kohlhoff 1995).

In a matter of months during 1942, the US military built a well-organized installation in the Kuluk Bay area on Adak that allowed American forces to successfully launch attacks against Japanese forces entrenched on Kiska and Attu to the west. The lagoon at Kuluk Bay was filled in to construct runways for transport aircraft and the B-17, B-24, B-25, and B-26 bombers. Amphibious plane facilities were established at Clam Lagoon and Andrew Lake. During 1943, as many as 100,000 troops were stationed on or near Adak. Following the war, the American military built a sizable Naval Air Station on the northeast side of Adak flanking Kuluk Bay. Over 6000 Naval and Coast Guard recruits and their families lived on the island. Despite its isolation, the US military considered Adak a strategic location for scrutinizing the Asiatic mainland during the Cold War. Thus, the US Navy built a enormous complex of buildings and associated infrastructure, including: hundreds of housing units, a college, medical center, two schools, a movie theater, roller skating rink, swimming pool, and a bowling alley.

Until the mid 1990's, Adak was a thriving military base and civilian community. With the end of the Cold War, the Naval Station closed in 1995 under the Base Realignment and Closure Act. In 1997 the Naval Station operationally shut down leaving behind a considerable cacophony of residential units, schools, dormitories and warehouses built to withstand the extreme climate; 26km of paved and gravel roads; and an airport with a control tower. The military legacy on northern Adak also included extensive military destruction, ground and water pollutants, unexploded ordnance, and tons of discarded military debris.

Following the Naval Station closure, the future of Adak was uncertain. Aleutian winds and the generally harsh climate were demolishing Adak facilities, and this natural erosion continues today. Civilians, largely contractor employees, began immigrating to Adak in 1998. A local elementary school soon opened, and by 2001 the City of Adak was incorporated as a second-class city. During 2004 the Aleut Corporation, one of 13 regional Native corporations established during 1971 under the Alaska Native Claims Settlement Act, acquired the north half of Adak Island under a land transfer agreement with the US Department of Interior and the US Navy Department of Defense.

Currently, the Aleut Corporation owns all of the infrastructure as well as most of the land (@186km^2) on the north half of the island. The Alaska Maritime National Wildlife Refuge manages the south side of Adak and along with offshore islets. The US Navy controls sections, containing unexploded ordnance, of north Adak. The decade old Adak community welcomes visitors to camp, hike, deep-sea fish for halibut and cod, and hunt caribou. In the 1950's the military and US Fish and Wildlife Service introduced the original small herd of 23 caribou on Adak to provide sport hunting for the Naval Air Station and as a walking larder for the military base in the event of a national emergency. With no natural predators, the caribou population quickly grew. During 2005 the US Fish and Wildlife Service counted 2751 caribou that Adak residents deem are the largest in the world. Bird watching is a popular summertime tourist attraction, and "birders" travel to Adak to photograph exotic northern and Asian bird species. Bald eagles, a treat for most non-Alaskans, are ubiquitous on Adak. Harbor seals and a remnant population of endangered sea otters are generally seen swimming and feeding at Clam Lagoon. More elusive are Stellers sea lions. Placed on the List of Endangered Species, they haul out on Adak's southern coasts.

At the time of this writing, less than 100 full time residents occupy the island, making a living largely from fuel sales and the local fish plant, and relying on summer tourism. Although businesses come and go, the remote Adak civilian community generally sports a grocery store, several small cafes, a medical clinic, a school, a volunteer fire department, an airport, and a fueling port and dock for Alaskan and Seattle based fishing fleets. Locally owned and commercial Bering Sea fishing vessels from other areas offload their catch at the Adak fishing plant for processing. The Alaskan Maritime Refuge uses the Adak community as a logistic hub during summer months. Scientific teams fly to and from the island to catch the *MV Tiglax*, a research and transportation vessel owned by the US Fish and Wildlife Service that supports scientists in and around remote, uninhabited islands both to the west and east of Adak. The Alaska Volcano Observatory uses the island as a transmitter for monitoring volcanic activity on surrounding islands.

North Adak Island is a dichotomy of brute, natural beauty and a horrific example of what modern humans and war can do to a landscape. The island also contains what up until now has been largely untapped scientific data about the peoples who inhabited the island centuries prior to historic contact. The following scientific papers document the three-year effort to recover, analyze and present that information to the Native American and scientific communities.

Acknowledgements

Funding for the 2005-2007 Adak field campaign and laboratory research was primarily provided by grants from the National Science Foundation's Arctic Social Sciences Program, Award # 0353092 to Lyn Gualtieri and Award # 0353065 to Dixie West. The Aleut Corporation and the US Fish and Wildlife Service provided permits to conduct the research. The generous Adak community welcomed us as their own. We also wish to thank Vernon Byrd, Alaska Maritime National Refuge, Homer, Alaska; Kristi Wallace, Alaska Volcano Observatory, Anchorage, Alaska; Brian Chisholm, University of British Columbia, Canada; Kenneth Pratt, US Bureau of Indian Affairs, Anchorage, Alaska; Ruth Welti, Kansas State University; Joan Coltrain, University of Utah; Jeff Rasic, University of Alaska Museum of the North, Fairbanks; William Banks, CNRS-Bordeaux, France; Kathy Stewart, National Museum, Ottawa, Canada; Lucy Johnson, Vassar College, Géraldine Véron, Département de Systématique et Evolution, Muséum National d'Histoire Naturelle, Paris for either reviewing and/or providing published and/or unpublished literature for many of the following chapters. Their expertise helps keep us on the straight and narrow.

References

Alaska Earthquake Information Center. 2011. http://www.aeic.alaska.edu/quakes/adak_1996.html

Average-Temperature. 2011. (http://www.average-temperature.com/temps/ AK/Adak).

Bank,Ted. 1956. *Birthplace of the Winds.* New York, Thomas Y. Crowell Company.

Black, Lydia. 1984. *Atka: An Ethnohistory of the Western Aleutians*. R. A. Pierce (ed.). Kingston, Ontario, Canada, The Limestone Press.

Garfield, Brian. 1995. *The Thousand-Mile War: World War II in Alaska and the Aleutians*. Fairbanks, Alaska, University of Alaska Press.

Intellicast.com. 2011. http://www.intellicast.com/Local / History.aspx?month=12&location=USAK0002.

Jochelson, Waldemar. 1925. *Archaeological Investigation in the Aleutian Islands.* Publication #267, Washington, DC, Carnegie Institution of Washington.

Kohlhoff D. 1995. *When the Wind Was a River: Aleut Evacuation in World War II.* Seattle: University of Washington Press.

Liapunova, Roza. 1996. *Essays on the Ethnography of the Aleuts* (At the End of the Eighteenth and First Half of the Nineteenth Century). Trans. Jerry Shelest. The Rasmuson Library Historical Translation Series vol. IX. Fairbanks, Alaska, University of Alaska Press.

Sandler, Stanley. 2001. *World War II in the Pacific: An Encyclopedia* (Military History of the United States). New York, Garland Publishing.

USGS. 2011. http://earthquake.usgs.gov/learning/faq.php

Veltre, Douglas. 1990. Perspectives on Aleut Culture Change during the Russian Period. In Barbara Sweetland Smith and Redmond J. Barnett (eds.), *Russian America: The Forgotten Frontier*, 175-183. Tacoma, Washington State Historical Society.

Chapter 1
The Central Aleutians Archaeological and Paleobiological Project

Dixie West, Christine Lefèvre, Elizabeth Wilmerding, Virginia Hatfield, Lyn Gualtieri

Abstract: *During 2005-2007, the Central Aleutians Archaeological and Paleobiological Project (CAAPP) deployed an international and interdisciplinary team of geologists, paleoecologists, biologists and archeologists to: 1) study the human, biotic and environmental history in the central Aleutian Islands, 2) compare the results with those from the western and eastern Aleutian Islands and, 3) provide a more comprehensive understanding of human, natural and geological systems in the Aleutians. This chapter provides an overview of the archaeological history of the Aleutians and previous work on Adak Island, the kinds of research and analyses performed by the Central Aleutians Project members, and a description of the examined archeological sites.*

Keywords: *Adak Island, geology, paleoecology, archaeology, stable isotopes, DNA*

Introduction

By virtue of their geography and history, the Aleutian Islands provide an ideal ecosystem for studying the relationships between humans and their environment during the Holocene. In the Aleutians, research has focused on the eastern and western islands in the archipelago through reconnaissance, recovery and analysis of the archeological, paleobiological, geological, paleoenvironmental, biotic, and historic records. Past Aleutian research reveals a dynamic biocomplexity of changing environments, biotas, and human populations and their cultures. However, this research is neither comprehensive nor synthetic across the Aleutian chain. The missing and potentially unifying ingredient is comparable investigation of the central Aleutian Islands, which have been little explored (see Veltre Chapter 2 this volume).

The Aleutian Islands extend 1800km between North America and Asia and separate the northern Pacific Ocean from the Bering Sea (Figure 1.1). Volcanic in origin and formed during the early Tertiary, the archipelago is comprised of more than 200 islands divided into six groups, separated by ocean passages. The Aleutian chain is a unique biotic bridge between Asia and North America with a rich ecosystem composed of animal and plant species from both continents. Marine vegetation is an important food resource for marine animal life that includes whales, sea lions, sea otters, and numerous species of seabirds, fish and invertebrates. This unique maritime environment was pivotal to the adaptations and survival of the human groups who, in the past, settled the archipelago.

Archaeological evidence suggests that the Aleutians were first settled in the east approximately 9000 years ago (Dumond and Knecht 2001; Laughlin 1963; McCartney 1984). Unangan people gradually migrated westward across the island chain, reaching the central Aleutians (Andreanof Islands) 6000 years ago (O'Leary 1998, 2001; Savinetsky *et al.* Chapter 5 this volume) and the Near Islands to the west approximately 3200 years ago (Corbett *et al.* 2010; Siegel-Causey *et al.* 1994; West *et al.* 1999). Aleutian populations, prior to Russian contact in 1741, are estimated to have numbered 9000-20,000 (Lantis 1970, 1984; Liapunova 1996). These estimates are tentative at best, and it is currently impossible to know prehistoric population levels (Reedy-Maschner 2010).

During the 18th century, the arrival from Russia of seal and sea otter hunters, followed by Russian Orthodox missionaries, drastically impacted Aleutian populations and their maritime hunter-gatherer culture and economy (Black 1984; Reedy-Maschner 2010; Turner 1981; Veltre 1990). Upon contact, the Native population rapidly declined; by 1831 only 16 islands were inhabited by an estimated 2000 Unangan people (Sekora 1973). This initial historical period has provided an often-cited reservoir of data (e.g., Veniaminov 1984) that focuses almost exclusively on Unangan populations in the eastern islands. Black (1984) and Liapunova (1996) filled this void by compiling information on central and western Unangan hunter-gatherers at time of Russian contact. Although useful, historical interpretations are subjective, and biases can influence the interpretation of the observer. Moreover, 18th and 19th century records divulge little about Unangan populations in the remote past.

Scientific study of the archipelago began shortly after Russia sold Alaska to the United States in 1867. Alphonse Pinart (1873, 1875) and William H. Dall (1877, 1899) initiated the first archaeological inquiries, followed by those of Waldemar Jochelson (1925, 1933) and Aleš Hrdlička (1945). Unfortunately, the results of these early collecting expeditions are of little use in understanding

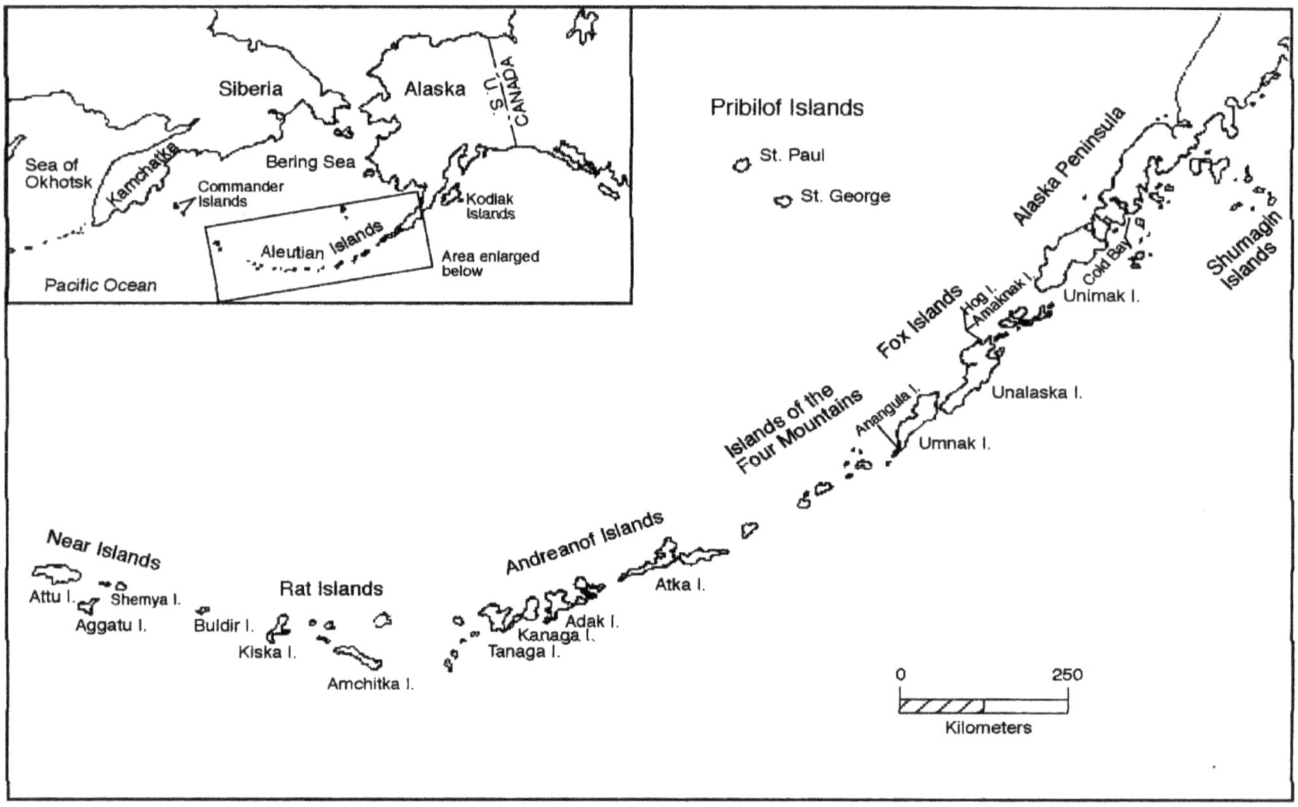

Figure 1.1. The Aleutian archipelago showing Adak in the Andreanof Island group.

prehistoric Unangan-environmental dynamics because the selectively recovered artifacts often lack clear stratigraphic or geochronological contexts.

Following World War II, renewed and concentrated archeology began with Albert Spaulding's (1962) research on Agattu from 1948 to 1951, William Laughlin's (1951, 1958, 1962, 1966) on Umnak Island, and Ted Bank's (1948-1951) throughout the Aleutians. Roger Desautels *et al.* (1970) and John Cook *et al.* (1972) conducted salvage excavation on Amchitka, and archaeologists with the US Bureau of Indian Affairs (USBIA) surveyed the entire archipelago in the 1980's (USBIA 1985, 1989, 1991, 1992a, 1992b). The following compilation, although not comprehensive, documents that much research has focused on the Alaska Peninsula, the eastern and western Aleutians, and the Shumagins:

- *the Alaska Peninsula* (Dumond 1981, 1990, 1998; Jordan and Maschner 2000; Maschner and Jordan 2001; Maschner *et al.* 1997; McCartney 1974; Okada *et al.* 1974, 1976; Workman 1966; Yarborough 1974)
- *the Eastern Aleutians (*Aigner 1966, 1970, 1973, 1976a,1976b, 1977, 1978,1983; Bacon 1977; Davis 2001; Denniston 1966; Dumond and Knecht 2001; Hoffman 1999; Holland 2001; Knecht and Davis 2001; Knecht *et al.* 2001; Laughlin 1951, 1958, 1962, 1963, 1966, 1967, 1974/1975, 1980; Lippold 1966; Turner 1972, 1976; Turner *et al.* 1974; Veltre *et al.* 1984, 1990; Yarborough 1988; Yesner 1988),
- *the Western Aleutians* ((Bouchet *et al.* 1999, 2001; Causey *et al.* 2005; Corbett 1989, 1990, 1991; Corbett *et al.* 1997a, 1997b, 2009, 2010; Funk 2006; Lefèvre and Siegel-Causey 1993; Lefèvre *et al.* 1997, 2001; McCartney 1977; Merritt 1977; Siegel-Causey *et al.* 1983, 1994; West *et al.* 1997, 1998, 1999a, 1999b, 2000, 2001, 2002, 2003, 2007; Wilmerding 1993).
- *the Shumagins* to the south (Johnson 1988, 1992a, 1992b, 2002; Johnson and Winslow 1991; Johnson and Wilmerding 2001; Wilmerding 2005; Winslow and Johnson 1989).

Although the Andreanof Islands are comparatively unstudied, these islands preserve critical archeological and paleobiological links between peoples who inhabited both ends of the Aleutian chain. Integrated archeological, paleobiological and geological study of the central Aleutians is essential in understanding coupled human and natural systems across the Aleutians, and attaining a more comprehensive picture of the biocomplexity of the North Pacific Rim.

Adak Island: Previous Research

Prior to the Central Aleutians Project, little interdisciplinary research had been done on Adak, or in the Andreanof Island group, for that matter (Veltre Chapter 2 this volume). Dall (1877), Jochelson (1925), Hrdlička (1945), Bank (1956) and Laughlin (in Bank 1956) performed minimal investigations in the region, but their work has been poorly documented or remains unpublished. With funding from the Wilderness Studies Division of the US Fish and Wildlife Service, Allen McCartney (1972) conducted a survey of Adak aboard the M/V *Aleutian Tern.* McCartney recorded 36 sites on Adak, with most observations made from aboard ship using binoculars. Bruno Frohlich and David Kopjansky (1975), sponsored by the University of Connecticut and the Aleut Corporation, reported 59 sites on Adak and Kagalaska while circumnavigating the island aboard skiffs. Under the Alaska Native Claims Settlement Act (ANCSA) the Bureau of Indian Affairs surveyed nearly 100 sites on the southern two-thirds of Adak and recorded four sites on the north part of the island (O'Leary 1998), but no choreographed, in depth excavations occurred. Research was limited to mapping and subsurface testing of sites. Douglas Veltre (1997) surveyed and reported nine sites on the north part of Adak for the Aleut Corporation. During 1999, the US Fish and Wildlife Service, under contract to the US Navy, surveyed the coast of Clam Lagoon and parts of Kuluk Bay and Lake Andrew on northeast Adak (Luttrell and Corbett 2000). Nine known sites were evaluated and 28 new sites discovered; 17 new radiocarbon dates (Luttrell and Corbett 2000) revealed a sequence of occupation beginning 6000 years ago extending to the historic period. The same year, due to logistical difficulties that prevented research on Attu in the Near Islands, the Western Aleutians Archaeological and Paleobiological Project (Debra Corbett, Christine Lefèvre, Dixie West) briefly surveyed in the Clam Lagoon and Sweeper Cove areas and excavated test pits at three previously identified archaeological sites (ADK-009, ADK-011, and ADK-171) on northern Adak Island ((Luttrell and Corbett 2000; West 1999). Beginning in 2007 and extending through 2011 Diane Hanson, University of Alaska, Anchorage, is evaluating upland sites on south Adak Island (Hanson *et al.* 2008, 2010). Veltre's (1979) investigations at the Korovinski Site, Atka Island and Matthew O'Leary's (2001) and Diane Hanson's (2010) Adak work are among the few publications to come out of the Andreanof Island group. A history of largely unpublished survey and excavation in the Andreanof Island group leaves a gap in the prehistory of the Aleutian Islands as a whole.

Summary of the 1999 WAAPP Field Season

The Central Aleutians project evolved from the 1999 Western Aleutians Archaeological and Paleobiological Project (WAAPP). During a two-week field season, the Western Aleutians team inspected sites on northern Adak, and opened test pits at ADK-171 and ADK-011 near Clam Lagoon and at ADK-009 at Sweeper Cove south of the present community of Adak (Figure 1.2)

WAAPP archaeologists excavated two test pits at ADK-011 on Zeto Point. Located on a bluff approximately 10m above sea level, ADK-011 is one of the largest and best-preserved prehistoric villages on north Adak. Originally described by Ted Bank as a small site, USBIA archaeologists (O'Leary 1998) recorded at least 36 depressions there. In 1999 WAAPP and USBIA archaeologists recorded at least 101 house pit features ranging in size from 3m to 15m in diameter (Luttrell and Corbett 2000). During 1999 WAAPP archaeologists opened test excavations (Features 1 and 2) in two adjacent house pits. Feature 1, approximately 6m^2 and excavated in two units, represented an abandoned house pit filled with thick accumulations of kitchen debris containing cockles and other invertebrates; mammal, bird, and fish bones; and hundreds of flat, burned basalt slabs believed to be griddle stones. In Unit 1, the kitchen midden measured 35-65cm thick and dated approximately 500 years old, based on radiocarbon dates of terrestrial fauna (Lefèvre *et al.* in press). Below the midden, a series of dark, sandy loam lenses intercalated with charcoal that dated to 2160$\pm$40 years BP (Lefèvre *et al.* in press). In a second unit of Feature 1, archaeologists recognized two distinct layers of cockles, separated by sand layers. Charcoal from the deepest sand layers of Unit 2 dated at 2490$\pm$50 years BP (Lefèvre *et al.* in press). A 2m X 2m test pit (Feature 2) in an adjacent depression to Feature 1 revealed a rich organic layer filled with faunal remains and numerous flat (griddle stone) fragments. Charcoal from this organic layer dated to 220$\pm$50 years BP and 180$\pm$60 years BP (Lefèvre *et al.* in press).

Russian historical ecologists, led by Arkady Savinetsky, cleared and analyzed a profile containing bone and invertebrates at ADK-171. Located on a bluff overlooking the east side of Clam Lagoon, the site was first investigated in 1998 by USBIA archaeologists who identified the eroding shell as a prehistoric kitchen midden (O'Leary 1998; 2001). The USBIA described two distinct cultural layers—a loosely packed cockle layer containing animal bones and human artifacts separated by a thin lens of unidentified volcanic ash from a more densely packed and highly deteriorated shell layer below (Luttrell and Corbett 2000; O'Leary 1998, 2001). Shell from the deposit was radiocarbon dated at approximately 6000 years BP (O'Leary 1998; 2001). During 1999 USFWS archaeologists mapped 55 depressions at the site; these features may represent either prehistoric or 20th century military activity. USFWS archaeologists recorded two bone awls, a needle, and stone flakes for the site (Luttrell and Corbett 2000, 47). O'Leary (2001) described three formal tools, two small stemmed points and a unifacial scraper eroding from the midden. During 1999 the Russian team cleared a 3m thick profile at the eroding shell midden and verified what appeared to be two distinct cultural zones at ADK-171. Both zones were filled with shellfish, and occasional bird, mammal, and fish bones along with stone tools. Savinetsky subsequently dated fish and mammal bones from the two cultural levels at ADK-171 to test the original USBIA date. Dates on bird bones confirm an ancient date for the site (Savinetsky *et al.* Chapter 5 this volume). Currently, site ADK-171 represents the earliest known settlement in the central Aleutians.

Russian scientists also opened a small 1m X 2m test excavation at site ADK-009 near the head of Sweeper Cove. Originally reported by McCartney (1972), Frolich

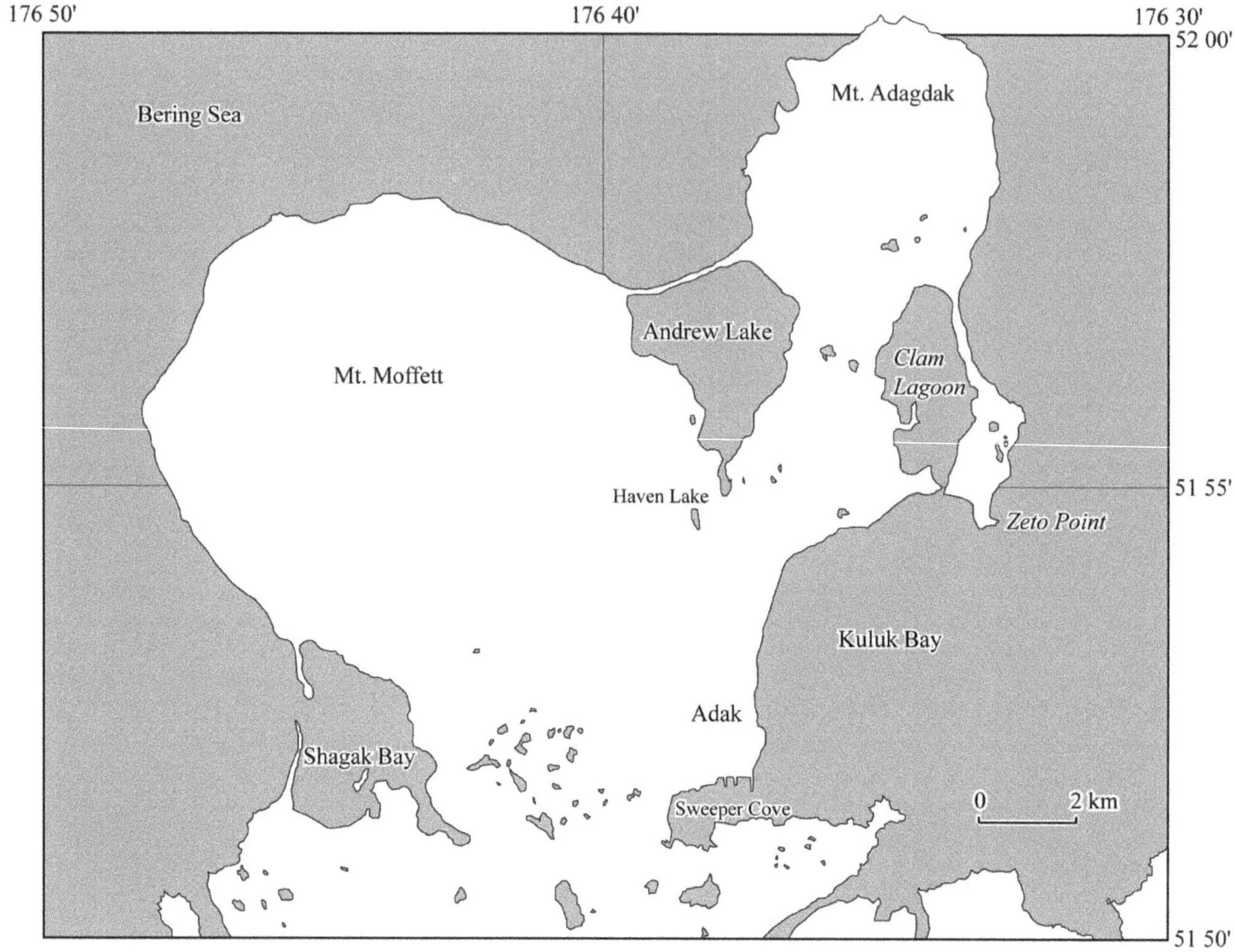

Figure 1.2 The CAAPP research area on North Adak Island, Alaska.

and Kopjanski (1976) mapped the site in 1975. The USFWS (Luttrell and Corbett 2000) mapped 17 features at the site. The 1999 Sweeper Cove test pit measured 198cm deep and contained 166cm of cultural deposit representing five distinct cultural layers filled with invertebrates, fish, bird, and few mammal bones. The cultural deposits accumulated from the beginning of the 6th to the middle of the 14th centuries (2080-970 years ago) (see Savinetsky *et al.* Chapter 5 this volume). Artifacts (N=88) included anvils, choppers, hammer stones, abraders, chipped stone flakes, and bifacially ground ulu knives; bone artifacts included a tooth pendant, awls and harpoon point fragments (Luttrell and Corbett 2000, 17).

Central Aleutians Research Disciplines

Expanding the 1999 research, the Central Aleutians Archaeological and Paleobiological Project team (Appendix 1.A) collected geomorphologic (sea level/glacial), volcanic, paleoenvironmental, and archaeological data around Clam Lagoon and Andrew Bay on Adak Island. Objectives included testing and documenting: 1) impacts of geological events on human activities, 2) long term environmental change during the Holocene, 3) changes in terrestrial and marine animal populations and diversity, and 4) prehistoric inhabitant responses to shifts in the distribution, diversity and abundance of resources, and in turn, human impacts on their resource base. The research disciplines for the CAAP project are briefly described below in the order presented in this volume.

Geomorphology

Published research on the geology of the island and region (e.g., Bradley 1948; Cameron and Stone 1970; Fournelle *et al.* 1994; Thorson and Hamilton 1986; Waythomas 1995) is not tied to the archeological record of Unangan behavior and adaptation. CAAPP integrated geological and archeological investigations to correlate human settlement patterns with documented changes in earth history. Geological data provided a baseline for our understanding of the physical environment prior to and during Unangan occupation. To that end, the project team addressed the following questions. Did specific changes in the physical environment (sea level; glaciation; volcanism; marine, lagoonal or freshwater water sources) affect the initial occupation, continued sustainability or eventual demise of island cultures? What were the local and regional landscapes throughout the Holocene?

Project geologist Lyn Gualtieri (Chapter 3 this volume), led the geological and geomorphologic research component in the field and laboratory. Gualtieri's team tested whether long term geological events, including tectonic uplift and glacial history in the Clam Lagoon region would have impacted Unangan settlement patterns and subsistence activities. She was responsible for geological analyses and determining Holocene landforms, chronology of sea level alterations, and glacial history in each region selected for intensive study. These research components were coordinated with the archaeological record of human habitation to test correlations between changes in the physical environment and the timing and pattern of human adaptations in the central Aleutians.

During 2005, Gualtieri's team mapped terraces and raised beaches on northern Adak, collecting terrestrial material that obtains most of its CO^2 from the atmosphere. Gualtieri sorted datable organic material from a natural peat formation located on the Clam Lagoon barrier beach to date the emergence of this raised landform. Gualtieri also dated organic material found below ash layers at Palisades Lake. During 2006-2007, Gualtieri and colleagues located and collected organic matter from raised, Pleistocene/Holocene beach terraces on southern Adak and on neighboring Kanaga Island to the west where military activity has not disturbed natural stratigraphic layers. Kanaga is geographically close enough to Adak to serve as a useful proxy for geological events on Adak.

Tephrochronology

Volcanism is a significant geological aspect in the Aleutian Islands, particularly on Adak Island (Heusser 1978, 1990). Black (1976a, 287) recognized at least 22 ash bands in peat sections on this island. Black (1976a), Thorson and Hamilton (1986) and Kiriyanov *et al.* (1997) sampled, classified and dated ash-soil sequences on Adak Island. Volcanism has been responsible for altering plant communities on Adak during the Holocene (O'Leary 2001, 218), and volcanism could have had a significant impact on human habitation. Tephra layers found in both natural and archaeological sites also provide a useful, independent dating technique for archaeological sites. O'Leary (2001), recognizing the opportunity, compared Black's (1976a) tephrochronology sequence on north Adak with radiocarbon dated archaeological deposits on the island. CAAPP built on O'Leary's research by concurrently identifying and dating volcanic tephras in both natural deposits and archaeological sites.

Mitsuru Okuno has expertise in radiocarbon dating (AMS) organic material co-deposited with, or draped by, tephras (Okuno *et al.* 2001; Okuno and Nakamura 2003). To test previous chronologies of volcanic eruptions, Okuno *et al.* (Chapter 4 this volume) collected organic debris (mainly charred fragments) in tephra layers from both archaeological sites and from natural contexts across the north half of Adak Island in 2005. He measured the thickness and chemical compositions of volcanic ash layers to determine separate eruption sources. During 2006 and 2007, Okuno, Gualtieri, and colleagues collected volcanic samples from Kanaga Island for comparison with those previously collected on Adak. Okuno identified, dated and sourced volcanic ashes that intercalate north Adak archaeological sites. Okuno determined that the ancient volcanic cones that represent Mt. Moffet and Mt. Adagdak produced none of the ash found in either natural or archaeological deposits on northern Adak during the Holocene. Kanaga volcano produced none of the ash found on Adak, but varying thicknesses of Adak ash deposits suggest that a yet to be identified volcano produced some of the Adak ash.

Paleoenvironmental Analysis

Paleoenvironmental reconstruction was the third critical and integrative component of the central Aleutians project. Project historical ecologist Arkady Savinetsky (Chapter 5 this volume) led the paleoenvironmental research component for the CAAP project. Savinetsky has extensive experience in paleoenvironmental analyses in both the Asiatic North Pacific Rim (Dinesman *et al.* 1999) and the western Aleutians (Savinetsky *et al.* 2004; Savinetsky *et al.* 2010; Siegel-Causey *et al.* 2005; West *et al.* 2007).

In the central Aleutians, Savinetsky's team provided the CAAP project with chronological, diatom and sedimentary analyses of peat bog deposits at Haven Lake, and analyses of fauna recovered from ADK-009 and ADK-171. Savinetsky and colleagues confirmed that ADK-171 is the oldest Unangan settlement found thus far in the central Aleutians. The discovery of saffron cod, a cold loving species at ADK-171 compared with the ratio of boreal and north-alpine diatoms recovered at Haven Lake infers that Unangan immigrants colonized Adak during an extremely cold and highly bioproductive period. Comparing fauna from a very old (ADK-171) and a comparatively young (ADK-009) archaeological site, historical ecologists infer that the Bering Sea faunal composition (invertebrates and vertebrates) of the central Aleutians area has not dramatically changed over the last six millennia.

Faunal Analysis

Paleoenvironmental reconstruction and changing subsistence strategies are generally derived from comparative analyses of paleofauna recovered from different prehistoric time periods (e.g., Causey *et al.* 2005; Crockford and Frederick 2000; Davis 2001; Lefèvre and Siegel-Causey 1993; Lefèvre *et al.* 1997; Siegel-Causey *et al.* 1991; Yesner 1977). CAAPP investigated changing central Unangan subsistence strategies and prey composition at five archaeological locales on north Adak: 1) ADK-171 (c. 6000 years BP) represents the oldest period; 2) ADK-011-Component 1 (2590-2390 years BP), ADK-

012 (1900 to 2500 years BP) and ADK-009 (1900-800 years BP) represent the middle period; 3) ADK-011, Component 2 and ADK-084 (520-290 years BP) represent the late period. Arkady Savinetsky and colleagues (Chapter 5 this volume) identified fish and mammal remains from ADK-171. Susan Crockford (Chapter 6 this volume) identified fish, bird and mammal remains from ADK-012, ADK-011 and ADK-084, as well as fish remains from ADK-009. Crockford then compared available data from all Adak sites (ADK-171, ADK-012, ADK-011, ADK-009, ADK-084) with data from sites of comparable ages to the east and west, essentially a pan-Aleutian comparison of prehistoric vertebrate fauna. Shin Nishida and colleagues (Chapter 7 this volume) analyzed sea otter bones and teeth from ADK-171, ADK-012, and ADK-011 to compare ancient central Aleutian mitochondrial DNA haplotypes. Amee Garong and colleagues (Chapter 8 this volume) analyzed oxygen and carbon stable isotopes of archaeological sea otters to determine the prehistoric diet of otters over time at north Adak. Hiroko Koike and colleagues (Chapter 9 this volume) compared growth rings with oxygen and nitrogen isotope analyses of cockles (*Clinocardium*) to establish ages, growth rates and collection seasonality during the most recent prehistoric occupation (ADK-011-Component 2) on north Adak. Dixie West and colleagues (Chapter 10 this volume) identified invertebrates from ADK-171, ADK-009, ADK-012 and ADK-011 for comparison with invertebrate data from both the eastern and western Aleutian Islands

Stone Artifact Analyses

Reconstructing past technological strategies and other human behavior depends on understanding relationships among source material procurement, tool manufacture, hafting, use, maintenance, and discard (Sellet 1993, 107).

CAAPP archaeologists recovered small obsidian fragments at archaeological sites ADK-171, ADK-012, and ADK-011. The only known geological sources of obsidian in the Aleutians are Okmok Caldera on Umnak Island and Akutan Island (Knecht, Davis and Carver 2001), approximately 650km and 800km east of Adak respectively. Obsidian artifacts can be traced to their volcanic origins based on trace element chemical signatures. Kirsten Nicolaysen and colleagues (Chapter 11 this volume) compared five Adak samples believed to be obsidian with known obsidian samples from Akutan and Okmok. Using inductively coupled plasma mass spectrometry (ICP-MS) at the Pacific Centre for Isotope and Geochemical Research (PCIGR), University of British Columbia, and principle components analysis of these results, Nicolaysen and colleagues suggest that the Adak obsidian samples most closely match the source obsidian from Okmok volcano. The occurrence of obsidian on Adak Island, 650km from its source, suggests prehistoric long-distance trade, raiding, or considerable time investment and safety risk in traveling great distances to procure this material.

Stone tool analysis focuses on the tools and production debris to determine the methods and strategies used in stone tool manufacture, maintenance, and discard. Wilmerding and Hatfield (Chapter 12 this volume) analyzed stone tools and debitage from archaeological sites ADK-171, ADK-012, and ADK-011 on north Adak Island. The authors compare lithic tools and debitage over time and place the Adak stone assemblages within a generalized framework of Aleutian technology stemming from the phases defined by Knecht and Davis (2001). This sequence of technology suggests that the earliest technologies are blade/microblade and irregular flake technologies, as well as ground stone technology, followed by the appearance of bifacial technology around 6000 years BP. In the western and central Aleutians, only bifacial and irregular flake technologies, as well as some evidence for ground stone technology occur, but little to no evidence of blade/microblade technologies is documented.

A tool loses its technological meaning as soon as it is removed from its behavioral context (Schlanger 1990, 20). 'The only methodologies that allow us to take a *chaîne opératoire* study to its natural completion are those of use-wear analysis' (Banks 2004, 5). When a stone tool contacts bone, other stone, wood, or plant materials during use, striations, wear patterns, polish, and residues accumulate on the working and hafting surfaces. These marks suggest whether the tool was used for cutting, scraping, sawing, or piercing and what materials the tool contacted. By examining the working surfaces and edges of stone tools using a high-powered microscope, the marks and residues found on prehistoric stone tools can be compared with marks replicated from different activities on modern, manufactured tools of similar size and shape. Use-wear analysis helps identify tools in the lithic assemblage that would 'most likely go unrecognized with only macroscopic examination and typological studies' (Banks 2004, 6). Marvin Kay (Chapter 13 this volume) performed a *chaîne opératoire* study and use-wear analyses on selected stone tools and debitage from ADK-171, ADK-283, ADK-012, ADK-011 and a blow out site at ADK-187. Kay collected basalt and andesitic cobbles and tabular stones in the general region of Clam Lagoon to determine prehistoric Unangan choices in source materials for specific tool types. Kay proposes that prehistoric groups on Adak were conservative in their approach to stone tool manufacture and that Unangan people frequently selected locally available, medium and coarse-grained stones for tool production and maintenance.

At north Adak archaeological sites, archaeologists frequently recovered flat, andesite slabs that Jochelson (1925) called griddle stones. The source of the material was probably the adjacent outcrops of intermediate volcanic material located around Clam Lagoon. In an initial qualitative analysis, Kirsten Nicolaysen (see Jeannotte Chapter 14 this volume) determined that over 90% of the recovered griddle stones are composed of hornblende-plagioclase andesite, like much of the source material found in the Clam Lagoon vicinity. Griddle stone fragments from the nearly 2m deep stratigraphic spread of ADK-011 were compared with potential natural stone sources. Nicolaysen performed petrographic analysis of the samples using standard rock thin sections. A platy andesitic flow exposed in beach cliffs approximately 5km from Clam Lagoon has the same mineralogical and textural characteristics as most of the griddle stones collected from archaeological site ADK-011. Petrographic analysis shows that it is possible to trace human movements between local stone sources and places of use and discard.

Biochemical Analysis on Stones

Soil layers found in Aleutian archaeological sites contain organic residues in different stages of decomposition that potentially reflect the prehistoric Unangan diet. Organic debris prehistorically introduced to a site sometimes does not leave a visible signal for typical analytical methods because acidic soil conditions, microbial activity, water leeching and other natural processes degrade organics. It is potentially possible to recover chemical signatures from archeological sites even from degraded samples. Organic matter discarded by ancient people may be degraded, but the biochemical signatures potentially remain and should be identifiable. By identifying these biomarkers it is possible to trace human activities that cannot be derived from macroscopic analyses of bones and shells. Richard Jeannotte and colleagues (Chapter 14 this volume) initiated investigations on the biological origins of organic residues coating one or more surfaces of flat andesitic slabs (griddle stones) recovered from archaeological site ADK-011. The biochemical analysis of ADK-011 charred griddle stones is preliminary, but the study demonstrates that it is possible to recover chemical signatures from Aleutian archeological sites.

Bone Tool Analysis

Bone tools comprise an important component of Unangan technology. Dixie West and Virginia Hatfield (Chapter 15, this volume) analyzed bone artifacts recovered from archaeological sites ADK-171, ADK-012, and ADK-011.

Bone tools and ornaments are rare at the three sites. Mammal bone chunks, slivers and 'pinwheels' from well preserved strata at ADK-011 Component 2 represent a simple *chaîne opératoire* for bone tool manufacture.

Human Morphology, Genetics and Stable Isotopes

West and colleagues (Chapter 16 this volume) analyzed the skeleton of a six to seven year old Unangan child from the Andreanof Islands group. The analysis provides new information about the genetics, general health, diet, and mortuary practices surrounding youngsters in the prehistoric Aleutian world. Stable isotope analysis reveals that the Andreanof child ate a diet high in shellfish and probably contributed to his/her personal subsistence by foraging in reefs.

Adak Archaeological Research Area: Site Descriptions

Excavation and Sampling

US Fish and Wildlife archaeologists recorded 37 archaeological sites on north Adak (Luttrell and Corbett 2000). Of the 37 known sites, CAAPP 2005-2007 archaeological research targeted cultural sites ADK-171, ADK-012, ADK-011, and ADK-283 in the area of Clam Lagoon and Zeto Point. Archaeologists also excavated small test pits at ADK-193 and ADK-013, overlooking Sitkin Sound, and at ADK-084, located in a small cove at Andrew Bay. WAAPP 1999 work by Savinetsky's team at ADK-009 is also included (Chapter 5 this volume). Archaeologists also collected surface scatters at the blowout site at ADK-187. Locations of specific village sites are not shown in this volume to protect the sites from looting. Site locations can be obtained by contacting the Aleut Corporation (Anchorage, Alaska), owners of north Adak. Biological sampling occurred in Andrew Bay, Sitkin Sound, Clam Lagoon, Kuluk Bay, and Sweeper Cove.

Archaeological fieldwork emphasized: (1) excavations and sampling at sites of different ages to document changes over time in stone tool manufacture and utilization, (2) economic patterns and, (3) social organization. Of particular interest is the time period between 6000 and 7000 years ago when the earliest archaeologically identified immigrants to the central Aleutians settled at Zeto Point (ADK-171). Circa 2590 and 2300 years BP Unangan people inhabited a protected cove overlooking Sitkin Sound at ADK-011 (Component 1). Between 2510 and 1515 years BP occupation occurred on a barrier beach separating Clam Lagoon from Sitkin Sound (ADK-012) and at ADK-283 on a small island in Clam Lagoon. Between 400 and 170 years BP people again inhabited the protected cove at Zeto Point (ADK-011 Component 2) and also took up residence in a cove at Andrew Bay (ADK-084).

ADK-171, The Tutiakoff Site, Pits 1 and 2

Because of its age, Russian historical ecologists with the Central Aleutians Project returned to ADK-171 to evaluate: (1) the dietary breadth of the earliest migrants in the central Aleutians, and (2) what recovered prehistoric fauna disclose about the ecosystem and environment of Adak Island 6000-7000 years ago (Savinetsky *et al.* Chapter 5 this volume). Additionally, CAAPP wanted to know how the stone and bone technologies of earliest central Aleutian residents compare with technologies found at later sites on north Adak and with the Aleutians as a whole (West and Hatfield Chapter 15 this volume; Wilmerding and Hatfield Chapter 12 this volume).

In 2005 researchers excavated a 2m X 1m unit (Pits 1 and 2), approximately 7m upslope from the 1999 Russian profile.

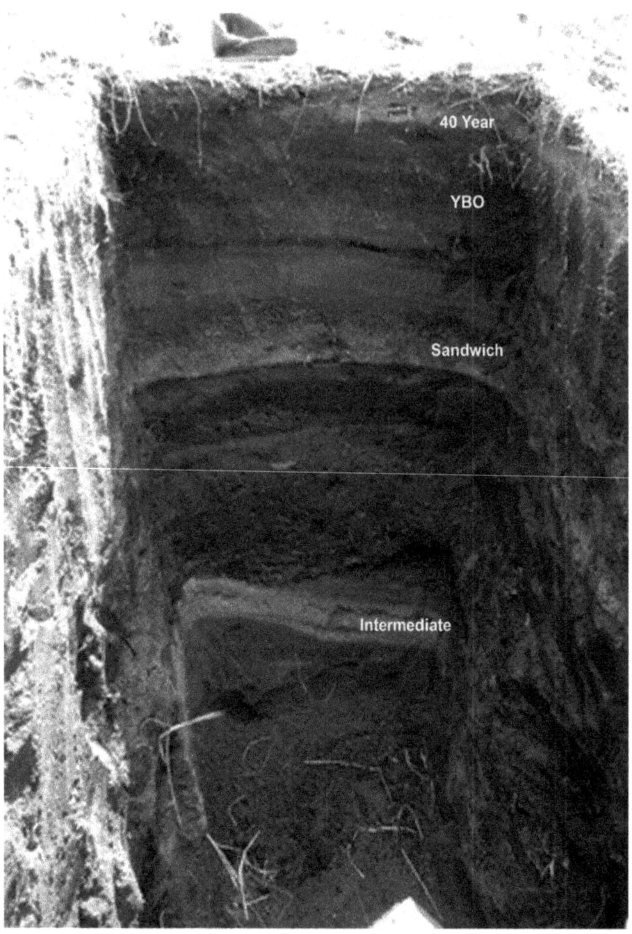

Figure 1.3. South profile of ADK-171, Pits 1 and 2, showing 40 Year, YBO, Sandwich and Intermediate volcanic ash layers.

Researchers recognized a series of well-defined volcanic ash levels corresponding to Black's (1976a) original Haven Lake stratigraphic profile (Figure 1.3). A pair of cultural charcoal bands occurred 40cm to 60cm below ground surface (Layer 4) and a single, ephemeral charcoal lens occurred in Layer 6, 80cm–90cm below ground surface. No artifacts or specific features were clearly associated with the charcoal lens. Layer 6 charcoal dated at 6005±20 years BP (UCIAMS-22151) and compares favorably with dates from the 1999 midden profile (O'Leary 1998; 2001). Excavators recovered no faunal materials in these pits. A single stone artifact represents an isolated find probably associated with Layer 6 (see Wilmerding and Hatfield Chapter 12 this volume and Kay Chapter 13 this volume).

ADK-171, The Tutiakoff Site, Pit 3 and Pit 3 Annex, Clinocardium Midden

Failing to find faunal remains in Pits 1 and 2, CAAPP excavated Pit 3, a 1m X 1m excavation unit, adjacent to the 1999 Russian profile and 2.5m west of the datum point (SE corner) of Pit 2. A second small .5m X 1m excavation unit (Pit 3 Annex) was opened adjacent to Pit 3 to increase the statistical faunal sample. Archaeologists opted not to open a large unit because the midden is a small remnant of what was once a larger area now eroded down a bluff. CAAPP chose to preserve the midden for future archaeological research by others. Layer 1 in both units represented soil development and Sandwich Ash lenses of variable thickness (Black 1976a; O'Leary 2001). Layer 2 comprised a loosely packed cultural lens containing whole and large *Clinocardium*, invertebrates, animal bones, and stone artifacts (Figures 1.4 and 1.5). A thin, gray-brown band (Layer 3) of redeposited Intermediate Ash (identified by Mitsuru Okuno) separated Layer 2 from a thin deposit of compressed and crushed mussel shells and animal bones (Layer 4). Layer 4 was located above 6cm of gray-brown sand (Layer 5) and at least 12cm of Intermediate Ash (Layer 6).

Charcoal from Layers 2 and 4 dated to 5760±20 years BP (UCIAMS-22153) and 5735±30 years BP (UCIAMS-22154) respectively. The chronological proximity of these dated strata suggests that Unangan people did not abandon ADK-171 during the geological or climatic event that redeposited the Intermediate Ash (Layer 3). However, prey species somewhat changed over time at the site (Savinetsky *et al.* Chapter 5 this volume). Cultural

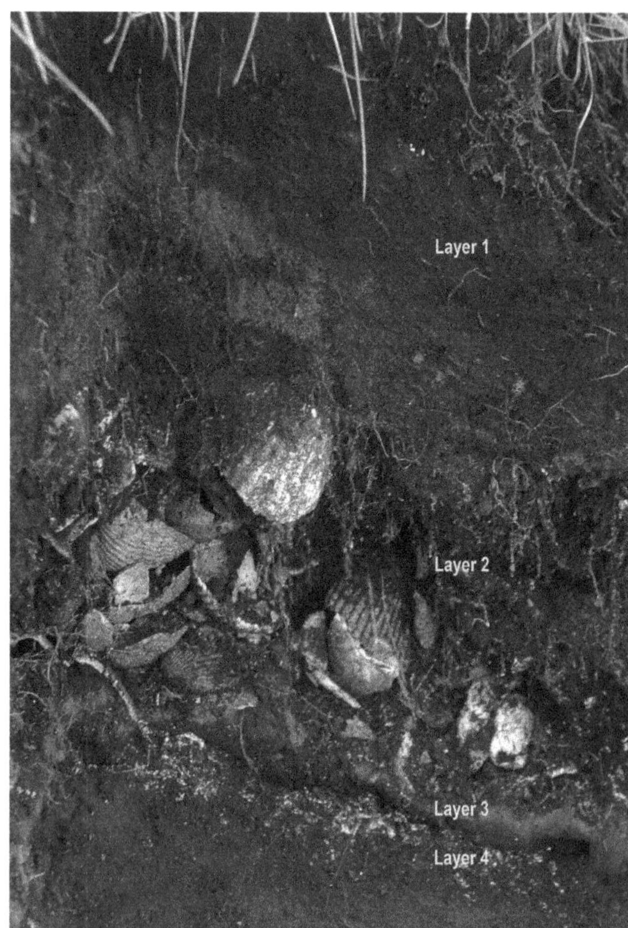

Figure 1.4. ADK-171, Pit 3, Adak Island, Alaska. An ephemeral band of redeposited Intermediate Ash (Layer 3) separates Layer 2 (filled with large cockles) from Layer 4 (a thin stratum of crushed mussel shells).

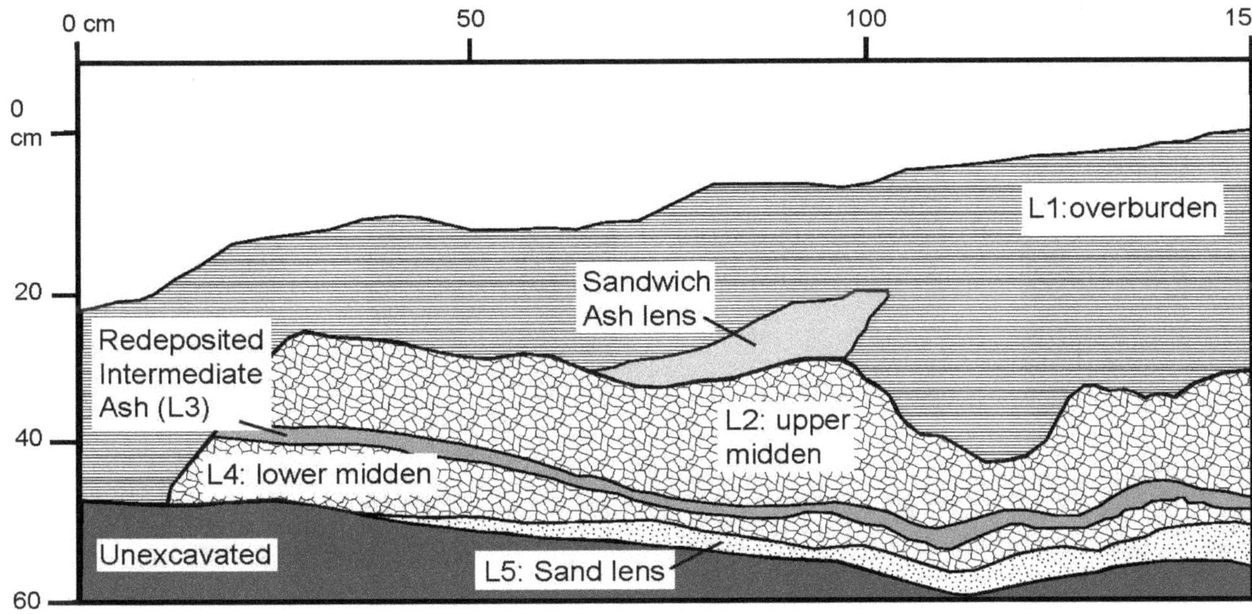

Figure 1.5. Profile of the Clinocardium midden at ADK-171. (Drawn by Elizabeth Wilmerding).

material located immediately above Intermediate Ash at ADK-171 is worth mentioning. Lydia Black (1981, 315, 317) argued that volcanism had a short-term impact on local economies as well as a powerful psychological effect on prehistoric Unangan people. Sizeable eruptions and seismic events potentially disrupted food sources by killing fish and ground nesting bird colonies and impacting shellfish beds and sea mammal rookeries. The 2008 eruption of Kasatochi volcano is a proxy for past volcanic impacts on humans and the ecosystem. The August 2008 Kasatochi explosion, with a plume that rose 45,000 to 50,000 feet, killed several thousand unfledged bird chicks, destroyed the breeding colonies of over 100,000 ground nesting birds, and blanketed the Kasatochi coast with meters of ash. The associated pyroclastic flow created a new coastline approximately 400m further into the sea. (AVO 2011). Cultural layers immediately above the Intermediate Ash suggest that the volcanic explosion that resulted in the ash did not dissuade Unangan people from living at ADK-171 (see Okuno *et al.* Chapter 4 this volume).

ADK-171, The Tutiakoff Site Pit 55-1

Archaeologists excavated a 1.3m X 1m unit test excavation (Test Pit 55-1) within, and adjacent to, Feature 55, a depression mapped in 1999 by the USFWS (Luttrell and Corbett 2000). This 7m X 4m depression flanks the western margin of a low swale at ADK-171. CAAPP attempted to locate a house structure edge or intercept a midden deposit to recover suitable organic material for radiocarbon dating to place the bluff depression within a chronological framework. Archaeologists found no house walls or midden in the 140cm deep excavation. Possibly the US military dug the surface depression or disturbed the upper layers of the site. World War II photographs show that Quonset huts once dotted the bluff at Zeto Point; these huts have since been demolished. The surface pit feature does not appear to correlate with deeply buried cultural strata below.

In Test pit 55-1 Layer 2, ephemeral charcoal lines of possible volcanic origin ran diagonally between 50cm and 70cm below ground surface (Figure 1.6). No cultural debris was associated with these charcoal lenses. However, a charcoal lens 120-140cm deep and situated immediately above a layer of Intermediate Ash (Layer 7), produced stone tools and debitage (Nicolaysen *et al.* Chapter 14 this volume; Wilmerding and Hatfield Chapter 12 this volume). Charcoal from the 120-140cm deep cultural lens dated to 6040±30 years BP (NUTA2-10967), a date roughly contemporaneous with cultural layers in the ADK-171 *Clinocardium* midden.

ADK-012, The Dozered Site

Site ADK-012 is located at the northern end of a barrier beach separating Clam Lagoon from the open water of Sitkin Sound (Luttrell and Corbett 2000). As its name implies, during 1943 the American military pushed the prehistoric site into a berm while building a road around the lagoon. Once a minefield, military contractors swept and cleared the north end of the site prior to the 2005 CAAPP excavation. Ted Bank first reported and sketched a map of the site in 1949 (Luttrell and Corbett 2000). US Fish and Wildlife archaeologists relocated and mapped the site remnant, describing it as an 80m by 20m mound on a 'badly disturbed linear ridge of disturbance vegetation with indistinct features' running parallel to the coastline (Luttrell and Corbett 2000, 32-34). USFWS archaeologists recovered cobble tools, flakes, a flat basalt fragment, and animal bones from a test pit at ADK-012.

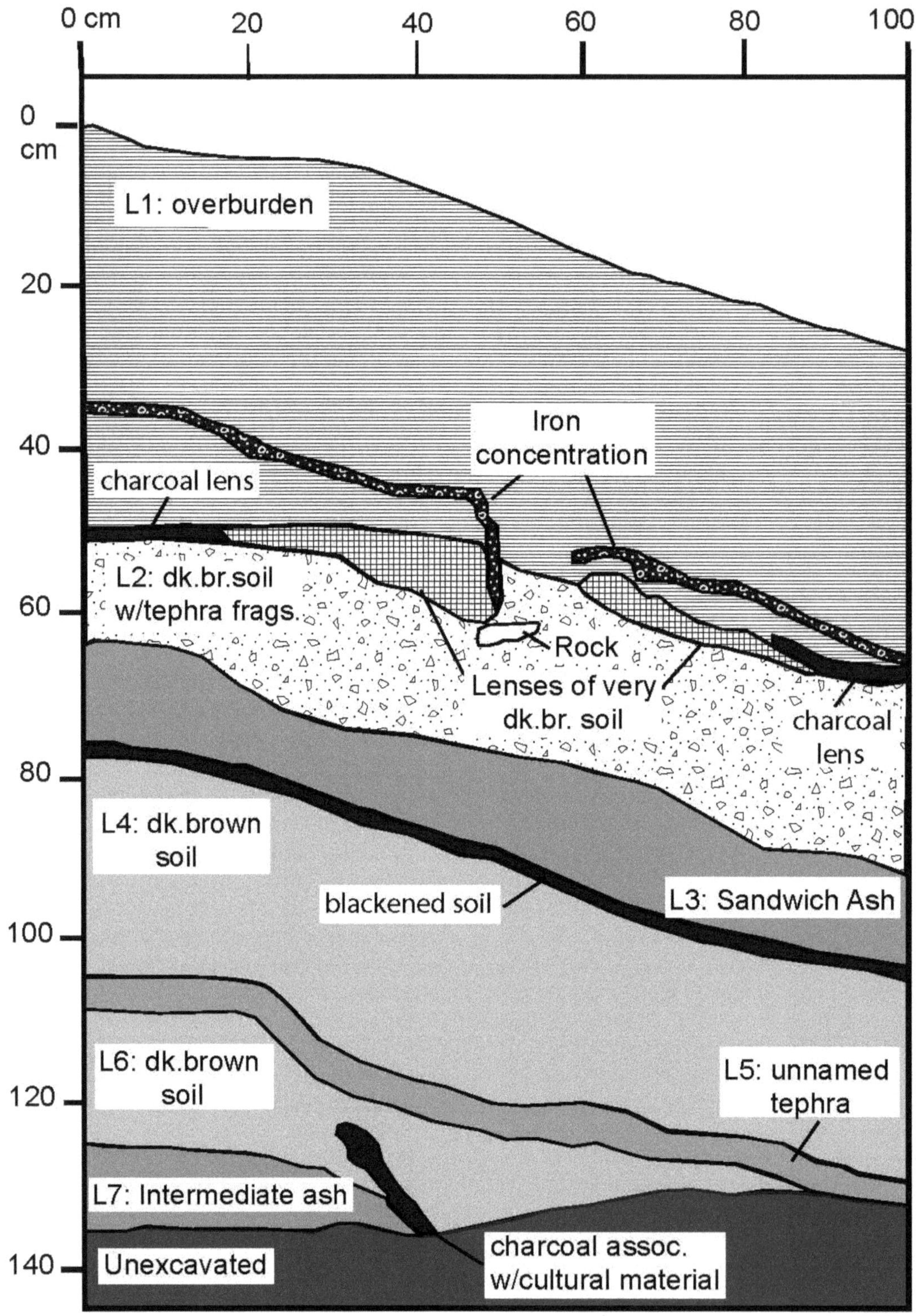

Figure 1.6 West Profile of ADK-171, Pit 55-1, Adak Island, Alaska. (Drawn by Elizabeth Wilmerding).

In 2005 CAAPP selected the highly disturbed ADK-012 for excavation because it is the only identified prehistoric site on the Clam Lagoon barrier beach (Figure 1.7). Archaeologists believed that when the barrier beach developed, it formed Clam Lagoon and connected Zeto Point to the rest of the Adak Island. Lagoon formation could have changed local ecological dynamics, influenced settlement patterns, and provided a new microenvironment for humans as well as harbor seals and sea otters—remnant populations that currently use the lagoon as a refuge from the open sea. CAAPP anticipated that dated organic materials from ADK-012 could help define the timing of the barrier beach formation. CAAPP also wanted to know when and why ancient peoples used such a low lying and

Figure 1.7. Aerial photograph of the barrier beach (upper right) that connects Zeto Point (lower right) to Adak Island. (Photo courtesy of Mitsuru Okuno).

exposed location (see Crockford Chapter 6 this volume). Convenient access to both the lagoon and Sitkin Sound open ocean could explain its location.

CAAPP archaeologists shovel tested the disturbed site until they found artifacts and animal bones in what appeared to be an undisturbed mound near the northern edge of the site. Archaeologists cleared a 110cm thick profile containing disturbed midden, sand lenses, dark loamy soil, tephra, and bulldozed boulders (Figures 1.8 and 1.9). Approximately 70-80cm below datum, a milled board ('Wd' in Figure 1.9) jutted out of the profile. This confirmed that military activities displaced all cultural remains above 70cm. This disturbed, cultural deposit (Layer 2 upper midden and Layer 2 lower midden) contained a rich inventory of animal bones, shell, tools and tool making debris. Mitsuru Okuno identified a distinct band (Layer 4) of 40 Year Ash 70cm below datum, and dark brown, peaty stratum beneath the volcanic ash appeared to be intact. Archaeologists opened a 1m X 1m test pit immediately in front of the profile, and excavations continued for 30cm below the 40 Year Ash layer. Cultural materials were rare in the undisturbed stratum (Layer 5). The site remnant essentially contained a 70cm thick deposit of cultural and natural lenses bulldozed out of context covering a 30cm thick, intact stratum (Layer 5 and below) largely devoid of cultural material.

Radiocarbon dates for ADK-012 are presented in Appendix 1.B. Charcoal from the lower, undisturbed zone dated at 1515±15 years BP (UCIAMS #22158). All other dates were taken on organics from the disturbed midden zone. A thrush leg bone recovered from Layer 2 of the upper zone dated to 1865±15 years BP (UCIAMS #22161). Dates from the site taken on other bones range between 2085±15 (UCIAMS #22159) and 2470±15 years BP (UCIAMS #22273). Radiocarbon dates on terrestrial bird (thrush), marine fish (rockfish) and plant charcoal imply that dates on rockfish bones (2455 years BP and 2470 years BP) might be artificially old. Thrush bone (1865 years BP) and terrestrial plant material (1515 years BP) dates probably represent a more accurate timeframe for ADK-012. Based on the thrush bone, a terrestrial based organism, the barrier

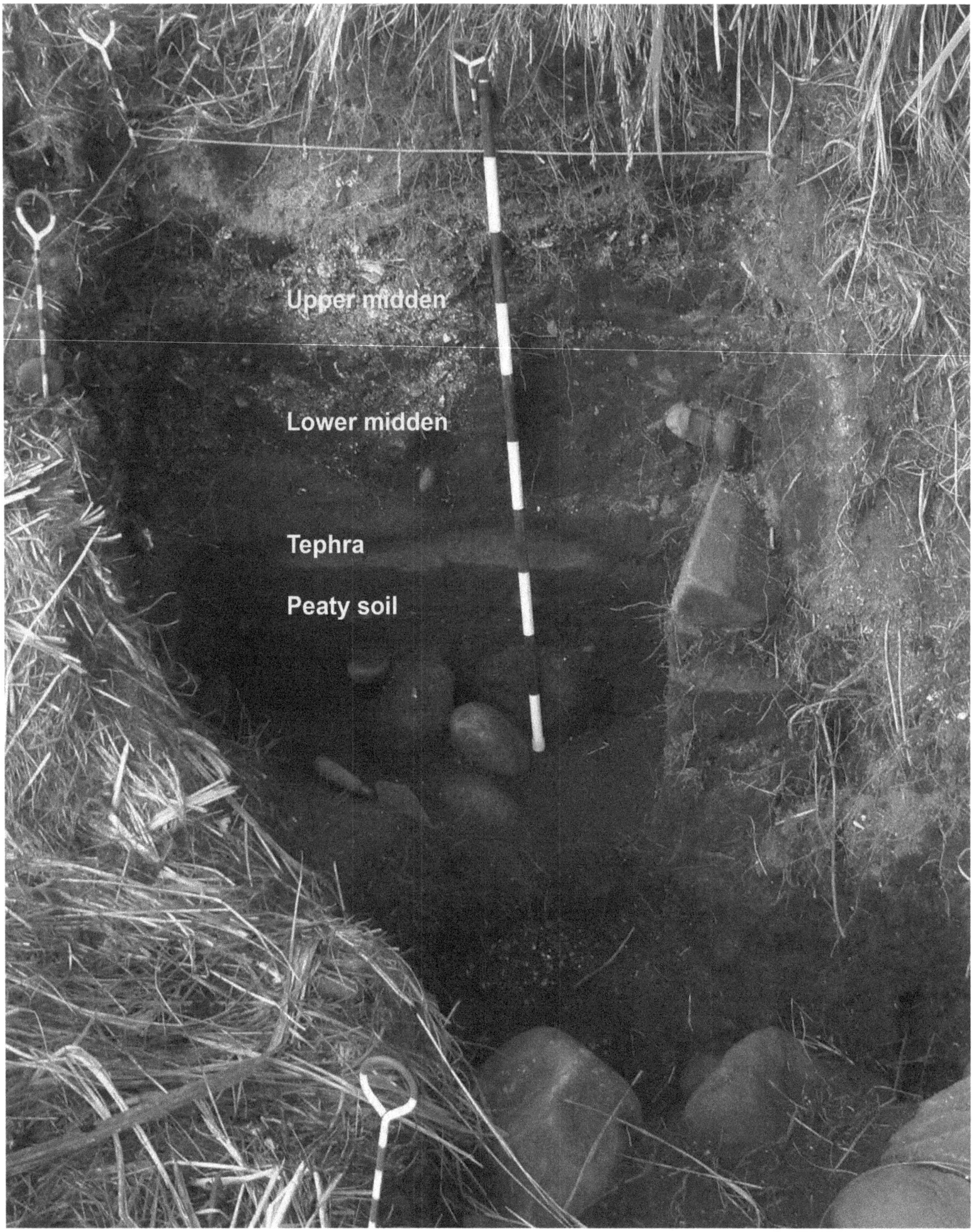

Figure 1.8. Photo of west profile of ADK-012 test excavation, Adak Island, Alaska.

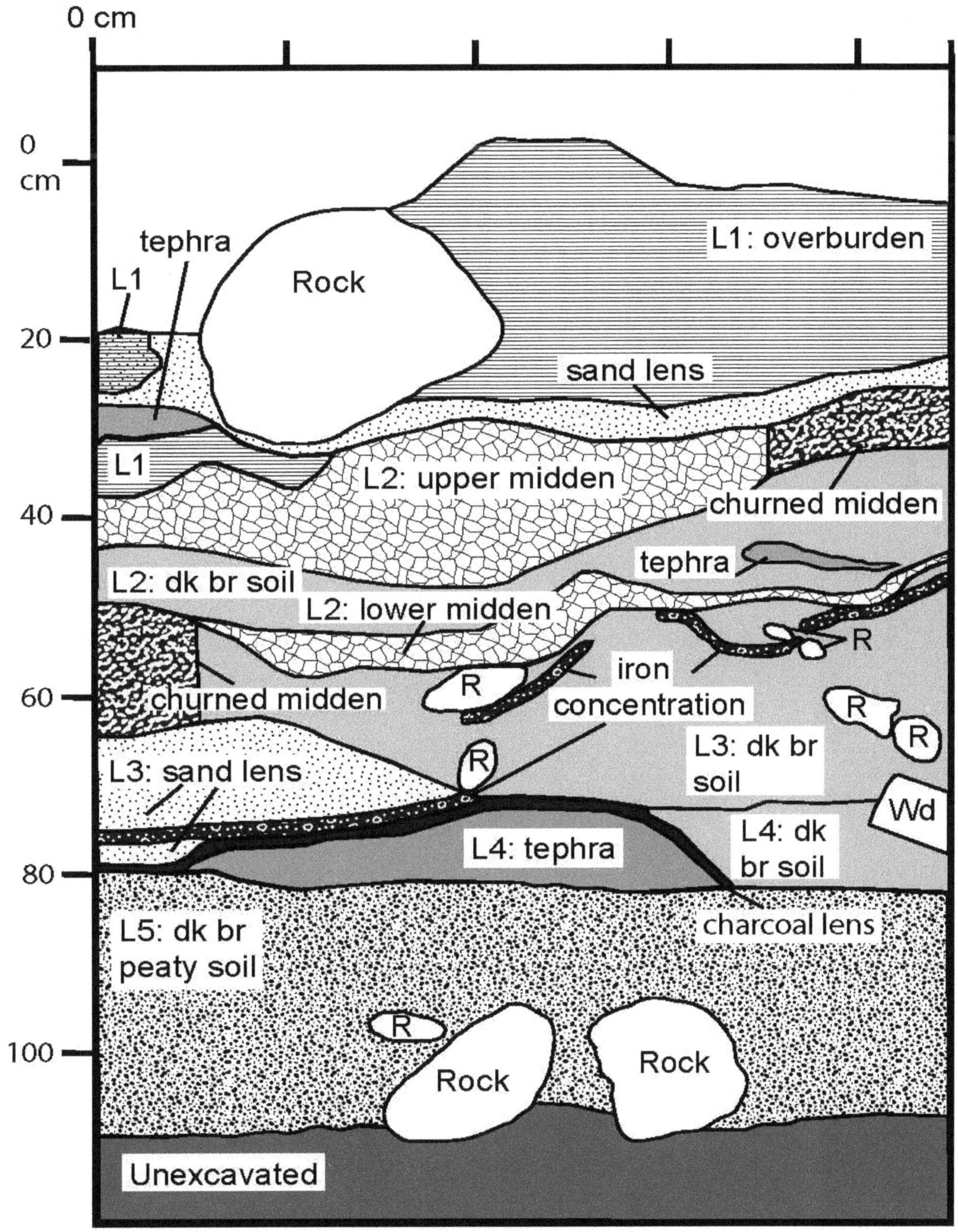

Figure 1.9 West profile of the ADK-012, Adak Island, Alaska. (Drawn by Elizabeth Wilmerding).

beach in the vicinity of ADK-012 lay above sea level *at least* by 1900 years ago, but the barrier beach is probably much older (see Savinetsky *et al.* Chapter 5 this volume). Okuno and colleagues (Chapter 4 this volume) dated the 40 Year Ash to 400 years BP. The absence of YBO Ash, (dated at 3600 years BP by Okuno *et al.* Chapter 4, this volume) at ADK-012 might suggest that: (1) the beach terrace had not formed when the YBO Ash was deposited, (2) YBO ash did not disperse here or washed away, or (3) CAAPP archaeologists did not excavate the unit deeply enough to encounter this ash. Dated terrestrial organic materials from both the disturbed and intact archaeological layers are roughly concurrent, suggesting ADK-012 is a single component site.

ADK-283, The Swetzoff Site

In 2005, CAAPP located four depressions on a small island situated on the east side of Clam Lagoon. The east half of the island is *Elymus* grass covered bedrock. However, soil development on the west end of the small island allowed prehistoric Unangan people to dig four semi-subterranean pits. The Swetzoff Site, although small, is intriguing because of its unique location on a small, unprotected knoll in Clam Lagoon. In 2005 CAAPP excavated a 0.5m X 0.5m test pit in one of the depressions and recovered a stone lamp fragment 50cm below ground surface. Burned residue from the lamp interior dated to 3835 $\pm$ 20 years BP (UCIAMS-22149). A sample believed to be charcoal filled soil from the 50cm level dated to 890 $\pm$ 25 years BP (UCIAMS-22150). Archaeologists also recovered a stone point.

In 2007 CAAPP opened five 1m X 1m units (B, C, D, E, and F) oriented on magnetic north in the depression tested during 2005. Archaeologists excavated the site in natural layers with 10cm arbitrary levels. Units B and C, to the south, were excavated to 20cm and backfilled when archaeologists detected military activity. Archaeologists excavated Units D and E to just below 90cm. Unit F, the northernmost unit that incorporated the 2005 test, was excavated to 120cm. To save dirt for backfilling, excavation matrix was dry screened on tarps using 1/4 inch screens. A 30cm X 30cm column was established in the NW corner of Unit E for water screening with 1/8inch mesh.

The upper 20cm of the site contained bullet casings and windowpane glass. Clam Lagoon was used as a floatplane base during World War II so this modern debris can be expected. Deeper archaeological levels were churned throughout. Rodent burrows, filled with black, organic, greasy matrix, permeated all levels of the site. A very large tunnel (32cm X 22cm) in the northeast corner of Unit F appeared to be an abandoned arctic fox den. Artifacts, found at angles rather than horizontal, suggest rodents, foxes and/or 20th century human activity displaced them from their original positions. Military activity is frequently cited as damaging archaeological sites on Adak Island, but rodent burrowing is extensive and ongoing. It occurs in all north Adak sites and undoubtedly on all Aleutian islands infested with rats.

Archaeologists found a few modern bird bones and invertebrate fragments in Layer 1, but recovered no prehistoric shell or bones. The vertebra of a small whale was lodged in the west wall of Unit E (31-35cm below datum). Archaeologists recovered a small, eyeless needle in Unit F (Level 7). CAAPP recovered five complete or broken stone points made from low to medium grade basalt haphazardly scattered in layers 3, 6, 7, 9, and 10. The lithic

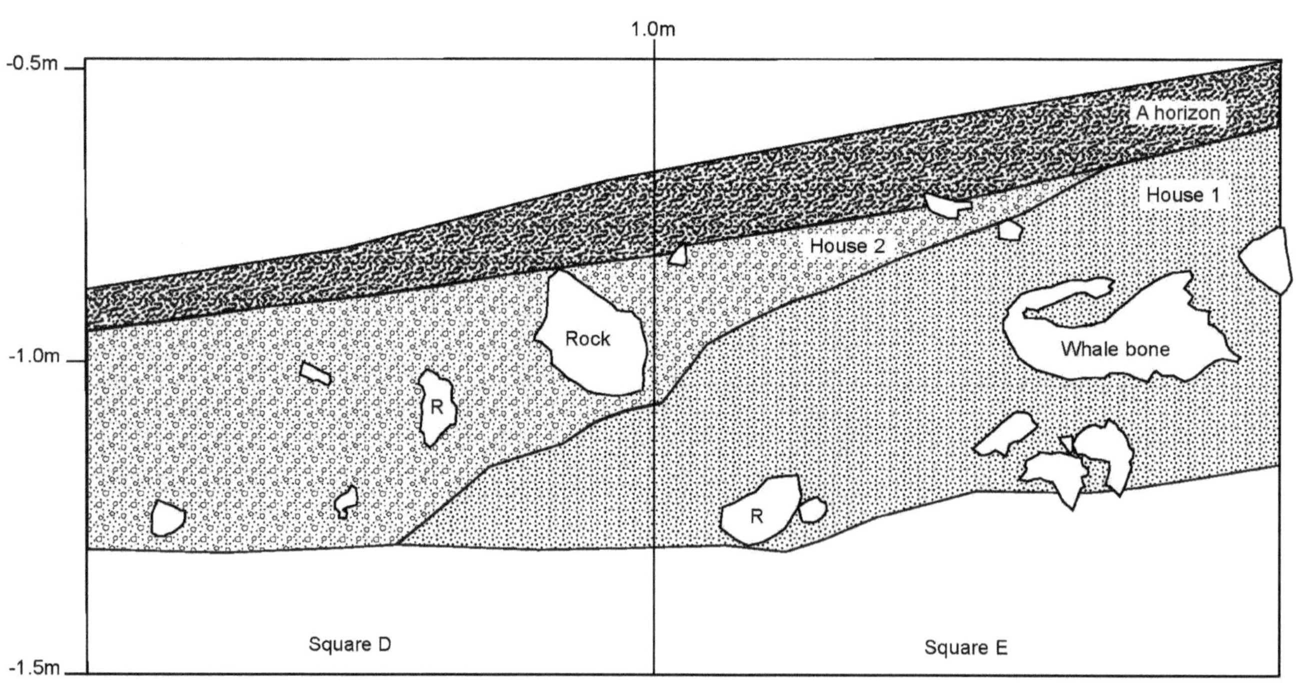

Figure 1.10. West profile of the Swetzoff site excavations showing the locations of House 1 and House 2 (Drawn by Marvin Kay and Elizabeth Wilmerding).

assemblage largely consisted of cobbles from the Clam Lagoon beach, a relatively short wading distance. These rounded cobbles do not naturally occur on the small island's shore. Scant debitage suggests minimal tool manufacture or retooling at the site. The bone needle and broken lamp fragment were the only indications of domestic activity. Hunting equipment (stone points) coupled with sparse debitage and household debris suggest the site may have been an occasional hunting camp or military outpost, rather than a seasonal occupation. Lack of fresh water may have limited daily household activities.

Two architectural features are of interest at the Swetzoff Site. At the contact of Units E and F at 92cm below datum CAAPP excavated a hearth represented by an oxidized pile of stone rubble and burned soil 50cm in circumference. At the same depth a trench ran diagonally from the south to the west side of Unit E into Unit D. The northeast corner of Unit E represents the edge of a prehistoric structure that intruded into older strata located in the rest of Unit E. (Figure 1.10). A charcoal sample from Unit E, Level 3 dated to 1735± 20 years BP (UCIAMS-42405). A second charcoal sample, associated with the hearth feature from Unit E, Level 7 dated to 1755± 20 years BP (UCIAMS-42404). These dates correspond to those from the ADK-012 occupation.

ADK-011: the Zeto Point Site

Located 1.75km north of Zeto Point, Ted Bank reported ADK-011 as a small shell heap (Luttrell and Corbett 2000). During 1999, WAAPP and USFWS archaeologists mapped 101 features at the site. In 2006 CAAPP excavated at ADK-011 because it is one of the largest, richest and best-preserved multicomponent sites in the Clam Lagoon vicinity with dates ranging between 2490±50 years BP and 180±60 years BP (Appendix 1.B).

During 2006, CAAPP laid out an 8m long trench of 1m X 1m units (B-I) oriented on magnetic north to transect the long axis of an oval house depression (G-I) and an adjacent kitchen midden (B-F). Two additional 1m X 1m units (J and K) were opened to the east adjacent to Units G and H to trace a 2m long pocket of tan matrix and to expose more of the house interior. A dense shell and fish bone deposit, intercalated with charcoal, sand, and mottled soil, and containing many stone and bone artifacts dominated the midden area (Figures 1.11 and 1.12). The house area consisted of two shallow bowl-shaped depressions just below a 40-year ash deposit and separated by a blackened ash layer 1-8cm thick. These depressions contained a small quantity of cultural material on their surfaces and were interpreted as house floors (Figures 1.13 and 1.14). Units were excavated following natural stratigraphic levels with additional arbitrary levels no more than 10cm thick where necessary. Matrix from units B-F and I were water screened with 1/8-inch mesh. Units G and H were excavated to provide stratigraphic control to find the bottom depth of the cultural deposits; these units were water screened with 1/4-inch mesh. CAAPP excavated four archaeological layers that span from 170-2500 years BP (Appendix 1.B). Mitsuru Okuno (Chapter 4 this volume) identified the tan matrix (Layer 2=tan lens) as 40 Year Ash that dates to 400 years BP. At ADK-011 archaeological materials dating 170-395 years BP occur above the tan ash pocket. O'Leary (2001, 228) noted: 'Virtually all of the aboriginal sites on Adak are covered by 40 Year Ash and are thus protohistoric or older....Only at Igalax (ADK-018), at the south entrance to Bay of Islands, and ADK-015 at Shagak Bay, do Unangan cultural remains overlie 40 Year Ash (US BIA 1985:26).... Sites capped by 40 Year Ash may be assigned minimally to the protohistoric period before AD 1750-1800.' CAAPP volcanic ash dates suggest the sites reported by the USBIA and ADK-011 Component 2 cultural layer above 40 Year Ash are no older than 400 years BP.

Radiocarbon dates indicated the ADK-011 kitchen midden and Layers 1-3 house floors formed between 170 and 400 years BP (Layer 1: 170 years BP; Layer 2, level 2: 285 years BP; Layer 2, levels 3, 4,5, 6,and 7: 400 years BP). In the house feature, Layers 1-3 date to 395-325 years BP. The age of the midden in Layer 2 (levels 3, 4, 5, 6, and 7) overlaps with the younger layers of the house. A series of charcoal lenses in Layer 4 in the house date to approximately 2500 years BP (Appendix 1.B). CAAPP's ADK-011 housepit represents two Unangan occupations separated by a 2000-year hiatus.

Exploratory Excavations

CAAPP briefly investigated sites in the Clam Lagoon (ADK-187), Sitkin Sound (ADK-013; ADK-193), and Andrew Bay (ADK-084) vicinities. A single day was devoted to each site.

ADK-187

Deeply buried ADK-187 is associated with a blow out site near Candlestick Bridge. Although CAAPP excavated a test pit for approximately 1.5m at what was believed to be ADK-187, we found no intact cultural debris. Perhaps ADK-187 might be as deeply buried as ADK-171 deposits. Although CAAPP did not find intact stratigraphic deposits archaeologists returned to ADK-187 during 2005-2007 to collect eroded lithics associated with the site. Kay (Chapter 13 this volume) conducted use-wear analyses on some of these artifacts.

ADK-013, Ocean View Site /ADK-193-test excavations

These two sites, in proximity to each other, are located 1km northeast of the barrier beach that separates Clam Lagoon from Sitkin Sound. Ted Bank partially excavated ADK-013 in 1949 (Luttrell and Corbett 2000). US Fish and Wildlife archaeologists relocated the site in 1999 (Luttrell and Corbett 2000). ADK-013 consists of pit features overgrown with dense, broad-leafed vegetation near beach level. USBIA archaeologists dated two organic material samples from ADK-013 that range between

Figure 1.11. Photo of the ADK-011 shell midden, Adak Island, Alaska. (Photo courtesy of Virginia Hatfield.)

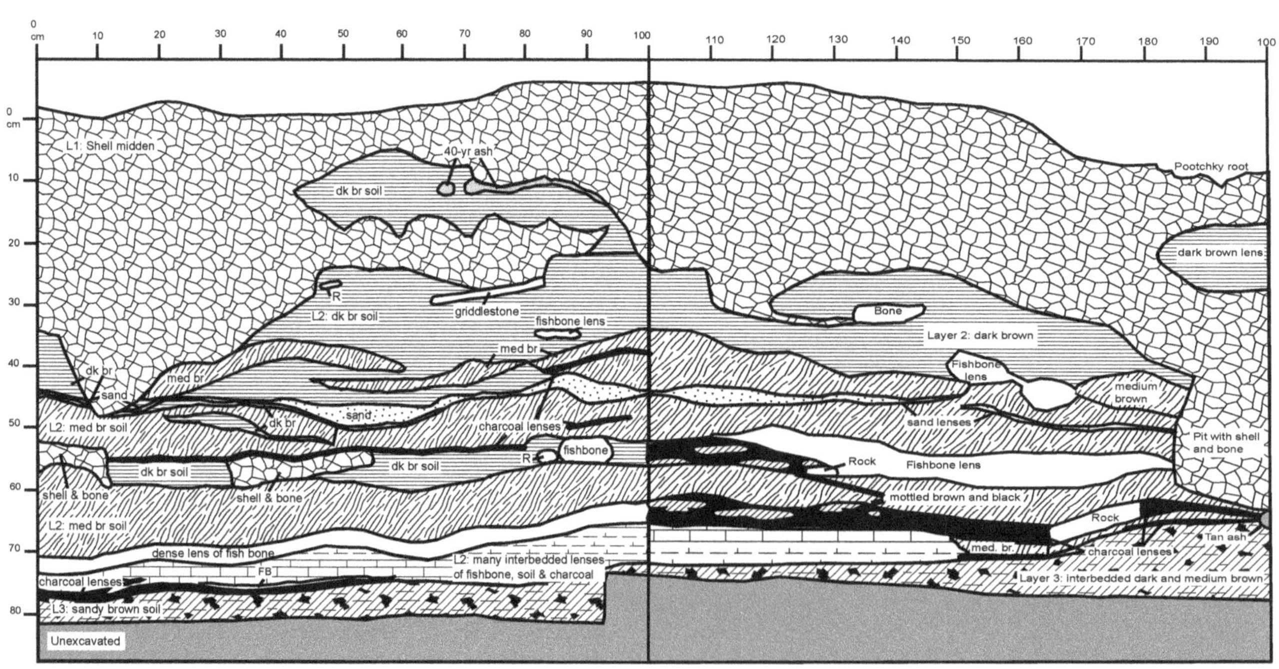

Figure 1.12. West profile of the ADK-011 shell midden, Adak Island, Alaska. (Drawn by Elizabeth Wilmerding).

Figure 1.13. Photo of the ADK-011 house layers, Adak Island, Alaska. Light colored stratum is 40 Year Ash. (Photo courtesy of Virginia Hatfield).

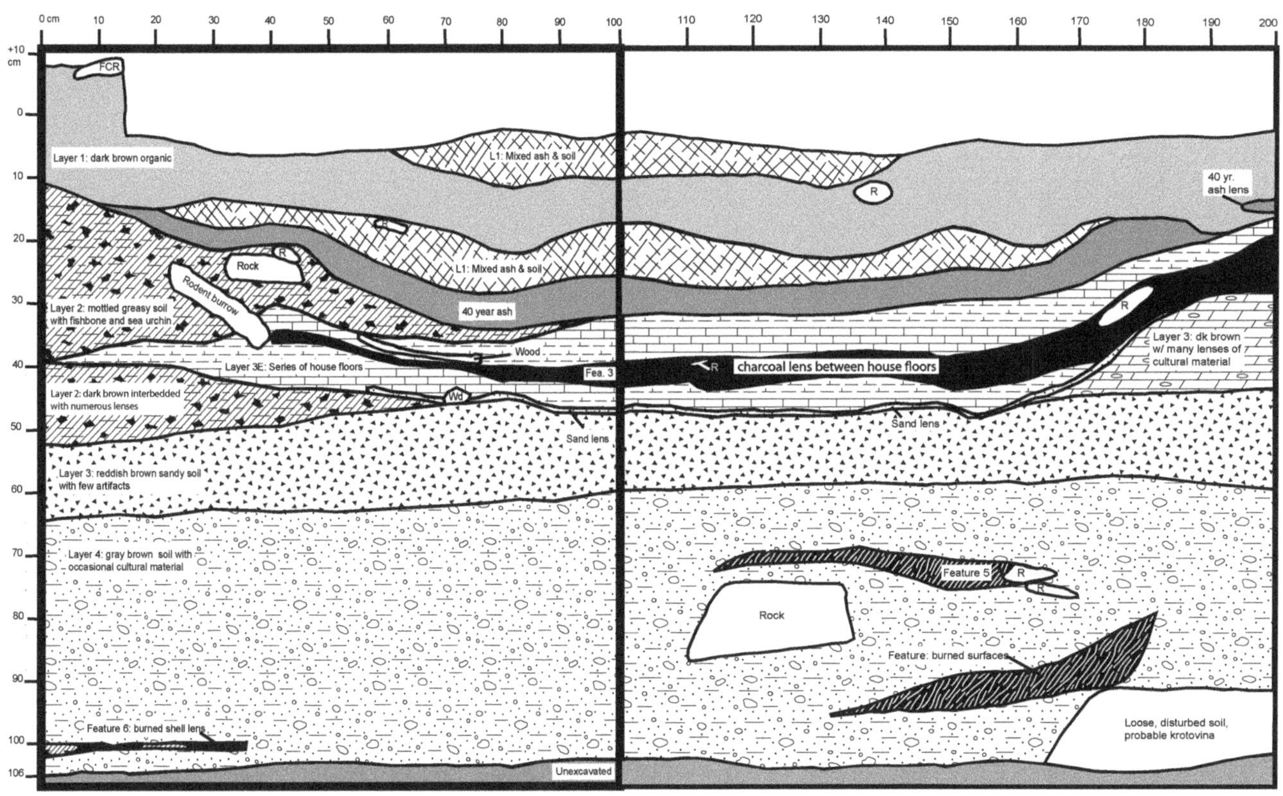

Figure 1.14. East profile of the ADK-011 house layers. (Drawn by Elizabeth Wilmerding)

1850±60 years BP (Beta- 132883) and 2900±40 years BP (Beta-133842). ADK-193 lies on a second terrace just 20m upslope from ADK-013. The site consists of several distinct pit features vegetated with wild rye grass. US Fish and Wildlife archaeologists discovered the site in 1999 while investigating ADK-013. During 2006 CAAPP explored both sites. Prior to the fieldwork, CAAPP hypothesized that the pits on the upper terrace should be much older than the pits located on the lower terrace closest to the beach. The assumption was that Unangan people first built houses on the upper terrace (ADK-193). When the low terrace was lifted above sea level, only then did ADK-013 become dry real estate for habitation.

Digging a series of tests in depressions on the lower terrace (ADK-013), CAAPP archaeologists found no clear stratigraphic layers or cultural debris. Although the depressions resemble prehistoric house pits, an absence of cultural debris and unclear stratigraphic layers suggest that natural formation processes, beach erosion, and possibly military construction created some, if not all, of the depressions on the lower terrace (see also Luttrell and Corbett 2000 for mention of military destruction in the vicinity).

ADK-193, located on a 20m high terrace above Sitkin Sound, is clearly cultural. CAAPP archaeologists excavated a single test pit in a depression approximately 10m from the bluff edge and recovered stone artifacts 40-70cm deep and below clear stratigraphic layers of 40 Year Ash. The USBIA dated charcoal from this site to 2900±40 BP (Luttrell and Corbett 2000). CAAPP recovered organic materials for dating from the test pit. Charcoal samples recovered 43cm and 57cm deep dated at 1770±20 BP (UCIAMS-61683) and 2485±20 BP (UCIAMS-61682) respectively. Elizabeth Wilmerding identified 13 stone artifacts from this site. These included: two pieces of pumice, four basalt thinning flakes, two andesite thinning flakes, a chert complex flake, a basalt simple flake, and three andesite tabular slabs.

ADK-084: Andrew Bay Site—test excavations

ADK-084 consists of 11 depressions on a narrow beach terrace overlooking a small, protected cove at Andrew Bay. Frolich and Kopjanski first reported this site in 1975. During 1999 archaeologists with the USFWS and USBIA investigated the site and noted the very large size of pit features (Luttrell and Corbett 2000). The 1999 USFWS team excavated 30cm X 30cm test pits to collect materials for radiocarbon dating. Charcoal from the site dated 500±40 years BP (Beta-135180). Organic sediment produced a date of 290± 40 years BP (Beta-133556).

During 2006 CAAPP archaeologists excavated five tests in separate pit features and collected charcoal and faunal bones. Charcoal from two test pits dated between 280±20 years BP (UCIAMS-61687) and 335±20 years BP (UCIAMS-61686). Elizabeth Wilmerding identified the ADK-084 stone artifacts. These included: six pieces of pumice, six basalt flakes, two chert flakes, one andesite flake, one manuport of unidentified material, and six andesite tabular slabs. Susan Crockford (Chapter 6 this volume) analyzed the animal bones from the site.

ADK-084 is intriguing because it is a comparatively recent site that chronologically corresponds to ADK-011, Component 2. Well preserved faunal remains at ADK-084 can provide useful information about the environment and central Aleutians economic patterns in relatively recent times. ADK-011, Component 2 and ADK-084 might suggest that late prehistoric populations, prior to Russian contact, were relatively large in the Andreanof Islands group or that social systems had changed over time with more people living in larger dwellings (ADK-084) or larger villages (ADK-011). Larger dwellings, like those found at ADK-084, could also suggest winter village living. ADK-084, among others on north Adak, is a site well worth revisiting.

In three field campaigns, between 2005 and 2007, the Central Aleutians Archaeological and Paleobiological Project conducted interdisciplinary research on north Adak and Kanaga Islands in the Andreanof group of the Aleutian Islands. The following chapters provide integrated geological, paleoenvironmental, and archaeological data about the North Pacific Rim and the peoples who inhabited the central part of that region from 6000 years ago to the recent period. Additionally, this research is a first attempt to link archeological and paleobiological information about Unangan populations who inhabited both ends of the archipelago.

Acknowledgements

This research was supported under the National Science Foundation's Arctic Social Sciences Program, Award # 0353092 to Lyn Gualtieri and Award # 0353065 to Dixie West. The Kansas University Biodiversity Research Center, Whitman College, and Vassar College provided financial support for undergraduate student participants. Excavation on north Adak was carried out with permits from the Aleut Corporation. The US Fish and Wildlife Service granted permission to excavate on the island in Clam Lagoon. The Aleutian Islands Maritime Refuge offered logistical support by shipping food and field supplies to Adak via the *RV Tiglax*. The Aleut Housing Authority and Adak citizens provided local logistical support. CAAPP returned all artifacts to the Aleut Corporation for permanent curation and storage.

References

Aigner, Jean S. 1966. Bone Tools and Decorative Motifs from Chaluka, Umnak Island. *Arctic Anthropology* 3(2), 57-83.

Aigner, Jean S. 1970. The Unifacial Core and Blade Site on Anangula Island, Aleutians. *Arctic Anthropology* 7(2), 59-88.

Aigner, Jean S. 1973. Studies in the Early Prehistory of

Nikolski Bay: 1937-1971. *Anthropological Papers of the University of Alaska* 16(1), 9-25.

Aigner, Jean S. 1976a. Dating the Early Holocene Maritime Village of Anangula. *Anthropological Papers of the University of Alaska* 18(1), 51-62.

Aigner, Jean S. 1976b. Early Holocene Evidence for the Aleut Marine Adaptation. *Arctic Anthropology* 13(1), 32-45.

Aigner, Jean S. 1977. Anangula: An 8500 BP Coastal Occupation in the Aleutian Islands. *Quartär* 27/28, 65-102.

Aigner, Jean S. 1978. Activity Zonation in a 4000 Year Old Aleut House, Chaluka Village, Umnak, Alaska. *Anthropological Papers of the University of Alaska.* 19(1), 11-25.

Aigner, Jean S. 1983. Idaliuk Bay Camp, Umnak Island, Circa 4200 BP. *North American Archaeologist* 4, 89-125.

AVO (Alaska Volcano Observatory) 2011. Kasatochi Eruptions.http://www.avo.alaska.edu/volcanoes/volcact.php?volcname=Kasatochi&eruptionid=605&page=basics.

Bacon, Glenn. 1977. A Preliminary Narrative Report on the 1977 Excavations at the Amaknak Bridge Site, Aleutian Islands, Alaska. Report to the Alaska Department of Transportation and Public Facilities.

Bamforth, Douglas. 1986. Technological Efficiency and Tool Curation. *American Antiquity* 51(1), 38-50.

Bank II, Theodore P. 1948-1951. Unpublished Field Maps. On file, Alaska State Office of History and Archaeology, Anchorage.

Bank II, Theodore P. 1956. *Birthplace of the Winds.* New York, Thomas Y. Crowell Company.

Black, Lydia T. 1981. Volcanism as a Factor in Human Ecology: the Aleutian Case. *Ethnohistory* 28(4), 313-333.

Black, Lydia T. 1984. *Atka, An Ethnohistory of the Western Aleutians.* Alaska History No. 24. Kingston, Ontario, Canada, The Limestone Press.

Black, Robert F. 1974a. Geology and Ancient Aleuts, Amchitka and Unimak Islands, Aleutians. *Arctic Anthropology* 11(2), 126-140.

Black, Robert F. 1974b. Late-Quaternary Sea Level Changes, Umnak Island, Aleutians—Their Effects on Ancient Aleuts and Their Causes. *Quaternary Research* 4, 264-281.

Black, Robert F. 1976a. Late-Quaternary Glacial Events, Aleutian Islands, Alaska. In D.J. Easterbrook and V. Sibrava (eds.), *Quaternary Glaciations in the Northern Hemisphere,* 285-301. Bellingham, Washington, International Union of Quaternary Research.

Black, Robert F. 1976b. Geology of Umnak Island, Eastern Aleutians, as Related to the Aleuts. *Arctic and Alpine Research* 8(1), 7-35.

Bouchet, Françoise, Christine Lefèvre, Dixie West, and Debra Corbett. 1999. First Paleoparasitological Analysis of a Midden in the Aleutian Islands (Alaska): Results and Limits. *Journal of Parasitology* 85 (2), 369-372.

Bouchet, Françoise, Christine Lefèvre, Dixie West, and Debra Corbett. 2001. Identification of Parasitoses in a Child Burial from Adak Island, Alaska, *Comptes Rendus de l'Académie des Sciences, Paris* 324, 123-127.

Bradley, Charles C. 1948. Geologic Notes on Adak Island and the Aleutian Chain, Alaska. *American Journal of Science* 246, 214-220.

Cameron, C. and D. Stone. 1970. *Outline Geology of the Aleutian Islands with Paleomagnetic Data from Shemya and Adak Islands.* University of Alaska Geophysical Institute, UAG-213,153 pp.

Causey, Douglas, Debra G. Corbett, Christine Lefèvre, Dixie West, Arkady B. Savinetsky, Nina K. Kiseleva, and Bulat F. Khassanov. 2005. The Paleoenvironment of Humans and Marine Birds of the Aleutian Islands: Three Millennia of Change. *Fisheries Oceanography* 14 (Supplement), 259-276.

Cook, John, E. James Dixon, and Charles E. Holmes. 1972. Archaeological Report Site 49 RAT 32 Amchitka Island, Alaska. Pullman, Washington, Arctic Research Section, Laboratory of Anthropology, Washington State University.

Corbett, Debra G. 1989. Shemya Island Prehistory. Manuscript on file, Anthropology Program 615, Fairbanks, University of Alaska.

Corbett, Debra G. 1990. Archaeological Survey and Testing on Shemya Island, Western Aleutians, Alaska in June 1990. Final Report to the Geist Fund, Fairbanks, University of Alaska Museum.

Corbett, Debra G. 1991. Aleut Settlement Patterns in the Western Aleutian Islands, Alaska. Unpublished MA Thesis, University of Alaska, Fairbanks.

Corbett, Debra G., Christine Lefèvre, and Douglas Siegel-Causey. 1997a. The Western Aleutians: Cultural Isolation and Environmental Change. *Human Ecology* 25(3), 459-479.

Corbett, Debra G., Christine Lefèvre, Thomas J. Corbett, Dixie West and Douglas Siegel Causey. 1997b. Excavations at KIS-008, Buldir Island: Evaluation and Potential. *Arctic Anthropology* 34 (2), 100-117.

Corbett, Debra G., Dixie L. West and Christine Lefèvre. 2001. Prehistoric Village Organization in the Western Aleutians. In Don Dumond (ed.), *Archaeology of the Aleut Zone of Alaska, Some Recent Research,* 251-266. University of Oregon Anthropological Papers No. 58. Eugene, Oregon, University of Oregon Press.

Corbett, Debra G., Douglas Causey, Mark Clementz, Paul Koch, Angela Doroff, Christine Lefèvre, Dixie West 2009. Aleut Hunters, Sea Otters, and Sea Cows: Three Thousand Years of Interactions in the Western Aleutian Islands, Alaska. In T.C. Rick and J.M. Erlandson (eds.) *Human Impacts on Ancient Marine Ecosystems: A Global Perspective,* 43-75. Berkeley, University of California Press.

Corbett, Debra G., Dixie West, and Christine Lefèvre. (eds.) 2010. *The People at the End of the World: The Western Aleutians Project and the Archaeology of Shemya Island.* AURORA, Alaska Anthropological Associations Monograph Series VIII. Anchorage, Alaska.

Crockford, Susan. 2011. Archaeozoology of Adak Island: 6000 Years of Subsistence History in the Central

Aleutians. Chapter 6 in D. West, V. Hatfield, E. Wilmerding, C. Lefèvre. L Gualtieri (eds), *The People Before: The Geology, Paleoecology and Archaeology of Adak Island, Alaska*. Oxford, British Archaeological Reports.

Crockford, S. and G. Frederick. 2000. Aleutian Fish Tell a Very Big Tale. Paper presented at the Eighth Annual Arctic Conference, University of Kansas, Nov. 3-4, 2000.

Dall, William H. 1877. On Succession in the Shell-Heaps of the Aleutian Islands. In Tribes of the Extreme Northwest, Vol. 1 *Contributions to North American Ethnology*, 41-91. Washington, DC, U.S. Government Printing Office.

Dall, William H. 1899. On the Remains of Later Pre-Historic Man, Obtained from Caves in the Catherina Archipelago, Alaska Territory and Especially from the Caves of the Aleutian Islands. *Smithsonian Contributions to Knowledge* 318.

Davis, Brian. 2001. Sea Mammal Hunting and the Neoglacial: Environmental Change and Subsistence Technology at Margaret Bay. In Don Dumond (ed.), *Archaeology in the Aleut Zone of Alaska, Some Recent Research*, 71-85. University of Oregon Anthropological Papers No. 58. Eugene, Oregon, University of Oregon Press.

Denniston, Glenda B. 1966. Cultural Change at Chaluka, Umnak Island: Stone Artifacts and Features. *Arctic Anthropology* 3 (2), 84-118.

Desautels, Roger J., Allen J. McCurdy, James D. Flynn and Robert R. Ellis. 1970. *Archaeological Report. Amchitka Island, Alaska* 1969-1970. United States Atomic Energy Commission, TID-25481, Los Angeles, Archaeological Research Inc.

Dinesman, Lev, Nina Kiseleva, Arkady B. Savinetsky, and Bulat Khassanov. 1999. *Secular Dynamics of Coastal Ecosystems of the Northeastern Chukchi Peninsula*. Tübingen, Germany, Mo Vince Verlag.

Dumond, Don. 1981. *Archaeology of the Alaska Peninsula: The Naknek Region*, 1960-1975. University of Oregon Anthropological Papers No.12. Eugene, Oregon, University of Oregon Press.

Dumond, Don. 1990. Notes on Some Early Blade Industries of the Alaska Peninsula Region, In *Chronostratigraphy of the Paleolithic in North, Central, East Asia and America,* 166-172. Novosibirsk: Institute of History, Philology and Philosophy, Siberian Branch of the Academy of Sciences of the U.S.S.R.

Dumond, Don. 1998. Maritime Adaptations on the Northern Alaska Peninsula. *Arctic Anthropology* 35(1), 187-203.

Dumond, Don E. and Richard A. Knecht. 2001. An Early Blade Site in the Eastern Aleutians. In Don Dumond (ed.), *Archaeology in the Aleut Zone of Alaska, Some Recent Research,* 9-34.University of Oregon Anthropological Papers No. 58. Eugene, Oregon, University of Oregon Press.

Fournelle, John H., Bruce D. Marsh, and James D. Myers. 1994. Age, Character, and Significance of Aleutian Arc Volcanism. In George Plafker and H.C. Berg, (eds.), The Geology of Alaska, Vol. G-I: *Geology of North America Series* V, 723-758. Geological Society of America Decade of North American Geology.

Frohlich, Bruno and David Kopjanski. 1975. Aleutian Site Survey, 1975. Preliminary Report to the Aleut Corporation, Anchorage, Alaska.

Funk, Caroline. 2006. Prehistory and Contact on Rat Island. Paper presented at the Alaska Anthropological Association Meeting. Kodiak, Alaska. March 2-4, 2006.

Garong, Ame, Chiozuru Takashima, Shin Nishida, Akihiro Kano, and Hiroko Koike. 2011. Carbon, Nitrogen, and Oxygen Isotopic Compositions of Bone Collagen and Tooth Enamel Carbonateof the Sea Otter *Enhydra lutris* from Archaeological Sites at Adak Island, Aleutian Archipelago. Chapter 8 in D. West, V. Hatfield, E. Wilmerding, C. Lefèvre, L. Gualtieri (eds), *The People Before: The Geology, Paleoecology and Archaeology of Adak Island, Alaska*. Oxford, British Archaeological Reports.

Gualtieri, Lyn, Brenn Sarata, Mitsuru Okuno, and Dixie West. 2011. Did Holocene Paleoenvironmental Factors Affect Ancient Aleut Occupation and Settlement in the Central Aleutian Islands? Chapter 3 in D. West, V. Hatfield, E. Wilmerding, C. Lefèvre, L. Gualtieri (eds), *The People Before: The Geology, Paleoecology and Archaeology of Adak Island, Alaska*. Oxford, British Archaeological Reports.

Hanson, Diane K., Debra G. Corbett, David P. Staley, and Kimberly Fleming. 2008. Upland Sites in the Aleutian Archipelago. Paper presented at the 73rd Annual Meeting of the Society for American Archaeology, Vancouver, British Columbia.

Hanson, Diane and Debra Corbett. 2010. Shifting Ground: Archaeological Surveys of Upland Adak Island, the Aleutian Islands, Alaska and Changing Assumptions of Unangan Land Use Patterns. *Polar Geography* 33, 165-178.

Heusser, Calvin J. 1978. Postglacial Vegetation on Adak Island, Aleutian Islands, Alaska. *Bulletin of the Torrey Botanical Club* 105(1), 18-23.

Heusser, Calvin J. 1990. Late Quaternary Vegetation of the Aleutian Islands, Southwest Alaska. *Canadian Journal of Botany 68*, 1320-1326.

Hoffman, Brian W. 1999. Agayadan Village: Household Archaeology on Unimak Island, Alaska. *Journal of Field Archaeology* 26, 147-161.

Holland, Kathryn. 2001. Regional Interaction as Seen from the Eastern Aleutians. In Don Dumond (ed.), *Archaeology in the Aleut Zone of Alaska, Some Recent Research*, 173-182. University of Oregon Anthropological Papers No. 58. Eugene, Oregon, University of Oregon Press.

Hrdlička, Aleš. 1945. *The Aleutian and Commander Islands and Their Inhabitants*. Philadelphia, Pennsylvania, The Wistar Institute of Anatomy and Biology.

Jeannotte, Richard, Kirsten Nicolaysen, Libin Yao, Floyd E. Dowell, Elizabeth Maghirang, Taylor Johnson, and Dixie West. 2011. Griddlestones: Their Provenance and Associated Organic Molecules, Adak Island, Alaska. Chapter 14 in D. West, V. Hatfield, E. Wilmerding, C. Lefèvre, L. Gualtieri (eds), *The People Before: The*

Geology, Paleoecology and Archaeology of Adak Island, Alaska. Oxford, British Archaeological Reports.

Jochelson, Waldemar. 1925 *Archaeological Investigations in the Aleutian Islands*. Washington, DC, Carnegie Institution of Washington Publication 367.

Jochelson, Waldemar. 1933. *History, Ethnology and Anthropology of the Aleut.* Washington, DC, Carnegie Institution of Washington, Publication 432 [Reprinted 1968 by Anthropological Publications, Oosterhout, N.B., The Netherlands]

Johnson, Lucy L. 1988. Archaeological Surveys of the Outer Shumagin Islands, Alaska, 1984 and 1986. *Arctic Anthropology* 25(2), 139-170.

Johnson, Lucy L. 1992a. Prehistoric Settlement Patterns and Population in the Shumagin Islands. *Anthropological Papers of the University of Alaska* 24(1-2), 73-88.

Johnson, Lucy L. 1992b. Pre-Contact and Contact-Period Aleut Settlements in the Shumagin Islands. In O.W. Frost (ed.), *Bering And Chirikov: The American Voyages and Their Impact.* Anchorage, Alaska. Alaska Historical Society.

Johnson, Lucy L. 2002. Natural Disasters and Cultural Change in the Shumagin Islands. In Robin Torrence and John Grattan (eds.), *Natural Disasters and Culture Change*, 193-203. London, Routledge.

Johnson, Lucy L. 2004. Prehistoric Fishing Technologies and Species Targeted in the Aleutian Islands: Archaeological and Ethnohistoric Evidence. *New Zealand Journal of Archaeology* 24, 45-59.

Johnson, Lucy L. and Margaret A. Winslow. 1991. Paleoshorelines and Prehistoric Settlements on Simeonof and Chernabura Islands, Outer Shumagin Islands, Alaska. In Lucy Johons and M. Stright (eds.) *Paleoshorelines and Prehistory: An Investigation of Method*, 171-186. Boca Raton, CRC Press.

Johnson, Lucy L. and Elizabeth G. Wilmerding. 2001. Bringing the House Down: Modeling Construction and Deconstruction of Aleut Semisubterranean Houses. In Don Dumond (ed.), *Archaeology in the Aleut Zone of Alaska, Some Recent Research*, 127-149. University of Oregon Anthropological Papers No. 58. Eugene, Oregon, University of Oregon Press.

Jordan, James and Herbert Maschner. 2000. Coastal Paleogeography and Human Occupation of the Western Alaska Peninsula. *Geoarchaeology* 15, 385-414.

Kay, Marvin. 2011. Clam Lagoon Archaeology and Technology. Chapter 13 in D. West, V. Hatfield, E. Wilmerding, C. Lefèvre, L. Gualtieri (eds), *The People Before: The Geology, Paleoecology and Archaeology of Adak Island, Alaska*. Oxford, British Archaeological Reports.

Kiriyanov, Vladimir Y. and T. Miller. 1997. Volcanic Ashes of Adak Island, Aleutian Islands, Alaska. *Volcanology and Seismology* 19, 65-78.

Knecht, Richard A. and Richard S. Davis. 2001a. A Prehistoric Sequence for the Eastern Aleutians. In Don Dumond (ed.), *Archaeology in the Aleut Zone of Alaska, Some Recent Research,* 269-288. University of Oregon Anthropological Papers No. 58. Eugene, Oregon, University of Oregon Press.

Knecht, Richard A., Richard S. Davis, and Gary A. Carver. 2001b. The Margaret Bay Site and Eastern Aleutian Prehistory. In Don Dumond (ed.), *Archaeology in the Aleut Zone of Alaska,* 35-85. University of Oregon Anthropological Papers No. 58. Eugene, Oregon, University of Oregon Press.

Koike, Hiroko, Chizuru Takashima and Akihiro Kano. 2011. The Aleut Cockle, *Clinocardium nuttallii* (Conrad, 1837) from ADK-011, Adak Island, Alaska. Chapter 9 in D. West, V. Hatfield, E. Wilmerding, C. Lefèvre, and L. Gualtieri (eds), *The People Before: The Geology, Paleoecology and Archaeology of Adak Island, Alaska.* Oxford, British Archaeological Reports.

Lantis, Margaret. 1970. The Aleut Social System, 1750 to 1810, from Early Historical Sources. In Margaret Lantis (ed.), *Ethnohistory in Southwestern Alaska and the Southern Yukon*, 139-295. Lexington, Kentucky, University Press of Kentucky.

Lantis, Margaret. 1984/ Aleut. In David Damas (ed.), *Handbook of North American Indians Vol.* 5. Arctic, 161-184. Washington, DC, Smithsonian Institution Press.

Laughlin, William S. 1951. A New View of the History of the Aleutians. *Arctic* 4, 75-88.

Laughlin, William S. 1958. Neo-Aleut and Paleo-Aleut Prehistory. In *Proceedings of the Thirty-second International Congress of Americanists*, 516-530. Copenhagen, Denmark.

Laughlin, William S. 1962. Archaeological Investigation at Umnak Island, Aleutians. *Arctic Anthropology* 1(1), 108-110.

Laughlin, William S. 1963. The Earliest Aleuts. *Anthropological Papers of the University of Alaska* 10 (2), 73-91.

Laughlin, William S. 1966. Aleutian Studies. *Arctic Anthropology* 3(2), 23-27.

Laughlin, William S. 1967. Human Migration and Permanent Occupation in the Bering Sea Chain. In D. M. Hopkins (ed.), *The Bering Land Bridge,* 409-450. Stanford, California, Stanford University Press.

Laughlin, William S. 1974/1975. Holocene History of Nikolski Bay Alaska and Aleut Evolution. *Folk* 16/17, 95-115.

Laughlin, William S. 1980. *Aleuts: Survivors of the Bering Land Bridge*. New York, Holt, Rinehart and Winston.

Lefèvre, Christine and Douglas Siegel-Causey. 1993. First Report on Bird Remains from Buldir Island, Aleutian Islands, Alaska. *Archaeofauna* 2, 83-96.

Lefèvre, Christine, Debra G. Corbett, Dixie West, and Douglas Siegel-Causey. 1997. A Zooarchaeological Study at Buldir Island, Western Aleutians, Alaska. *Arctic Anthropology* 34(2), 118-131.

Lefèvre, Christine, Dixie West and Debra G. Corbett. 2001. Archaeological Surveys in the Near Islands: Attu Island and Shemya Island. In Don Dumond (ed.), *Archaeology in the Aleut Zone of Alaska, Some Recent Research*, 235-250. University of Oregon Anthropological Papers No. 58. Eugene, Oregon, University of Oregon Press.

Lefèvre, Christine, Debra Corbett, Dixie West. In press. Zooarchaeological Analysis at ADK-011, Adak Island, Central Aleutian Islands, Alaska. In Caroline Funk and Debra Corbett, (eds.), Anthropology of the Western Aleutian Islands. *Arctic Anthropology*.

Liapunova, Roza G. 1996. *Essays on the Ethnography of the Aleuts (At the End of the Eighteenth and the First Half of the Nineteenth Century*. [J. Shelest, trans.]. Rasmuson Library Historical Translation Series, Vol. 9. Fairbanks, Alaska, University of Alaska Press.

Lippold, Lois, K. 1966. Chaluka: The Economic Base. *Arctic Anthropology* 3(2), 125-131.

Luttrell, Mark and Debra Corbett. 2000. Archaeological Investigations on Northeast Adak Island, Alaska, July 1999. Report to the US Fish and Wildlife Service, Anchorage, Alaska.

Maschner, Herbert, James Jordan, Brian Hoffman, and Tina M. Dochat. 1997. The Archaeology of the Lower Alaska Peninsula. University of Wisconsin Laboratory of Arctic and North Pacific Archaeology, Report 4. Madison, Wisconsin.

Maschner, Herbert and James Jordan. 2001. The Russell Creek Manifestation of the Arctic Small Tool Tradition on the Western Alaska Peninsula. In Don Dumond (ed.), *Archaeology in the Aleut Zone of Alaska, Some Recent Research*, 151-171. University of Oregon Anthropological Papers No. 58. Eugene, University of Oregon Press.

McCartney, Allen P. 1972. An Archaeological Site Survey and Inventory for the Aleutian Islands National Wildlife Refuge, Alaska, 1972. File Report 1-1-72, US Fish and Wildlife Service, Anchorage.

McCartney, Allen P. 1974. Prehistoric Cultural Integration Along the Alaska Peninsula. *Anthropological Papers of the University of Alaska 16* (1), 59-84.

McCartney, Allen P. 1977. Prehistoric Human Occupation of the Rat Islands. In Melvin L. Merritt and R. Glen Fuller (eds.), *The Environment of Amchitka Island, Alaska*, 59-113. Springfield, Virginia, Technical Information Center, Energy Research and Development Administration TID-26712.

McCartney, Allen P. 1984. Prehistory of the Aleutian Region. In D. Damas (ed.), *Handbook of North American Indians. Vol. 5*. Arctic, 119-135. Washington, DC, Smithsonian Institution Press.

Merritt, Melvin L. 1977. History, 1741-1967. In Melvin L. Merritt and R. Glenn Fuller (eds.), *The Environment of Amchitka Island, Alaska,* 115-140. Springfield, Virginia. Technical Information Center, Energy Research and Development Administration TID-26712

Nicolaysen, Kirsten, Taylor Johnson, Elizabeth Wilmerding, Virginia Hatfield, Dixie West, and Robert G. McGimsey. 2011. Provenance of Obsidian Fragments Recovered from Adak Island, Central Aleutian Islands: Evidence for Long-distance Transport of Raw Lithic Material. Chapter 11 in D. West, V. Hatfield, E. Wilmerding, C. Lefèvre, L. Gualtieri (eds), *The People Before: The Geology, Paleoecology and Archaeology of Adak Island, Alaska.* Oxford, British Archaeological Reports.

Nishida, Shin, Dixie West, Susan Crockford, and Hiroko Koike. 2011. Ancient DNA Analysis for the Sea Otter (*Enhydra lutris*) from Archaeological Sites on Adak, Aleutian Islands. Chapter 7 in D. West, V. Hatfield, E. Wilmerding, C. Lefèvre, L. Gualtieri (eds), *The People Before: The Geology, Paleoecology and Archaeology of Adak Island, Alaska.* Oxford, British Archaeological Reports.

Okada Hiroaki and Atsuko Okada. 1974. Preliminary Report of the 1972 Excavations at Port Moller, Alaska. *Arctic Anthropology* (Supplement) 11, 112-124.

Okada, Hiroaki, Atsuko Okada, Yoshinobu Kotani, and K. Hattori. 1976. *The Hot Springs Village Site* (2). Hachioji, Tokyo, Hiratsuka Printing Company.

Okuno, Mitsuru, Toshio Nakamura, Hiroki Kamata and Tetsuo Kobayashi. 2001. Radiocarbon Dating of Paleosol Intercalated with Tephra Layers in Japan. In E. Juvigné and J. P. Raynal(eds.), *TEPHRAS, Chronology/ chronologie, Archeology/archeologie,* 67-71. CDERAD éd, Clermont-Ferrand. (with French abstract)

Okuno, Mitsuru, and Toshio Nakamura. 2003. Radiocarbon Dating of Tephra Layers: Recent Progress in Japan. *Quaternary International* 105, 49-56.

Okuno, Mitsuru, Keiji Wada, Toshio Nakamura, Lyn Gualtieri, Brenn Sarata, Dixie West, Masayuki Torii. 2011. Holocene Tephra Layers on the Northern Half of Adak Island in the west-central Aleutians, Alaska. Chapter 3 in D. West, V. Hatfield, E. Wilmerding, C. Lefèvre, L. Gualtieri (eds), *The People Before: The Geology, Paleoecology and Archaeology of Adak Island, Alaska.* Oxford, British Archaeological Reports.

O'Leary, Matthew. 1998. New Site No. 2 ('Tutiakoff Site'). Unpublished report on file at US Department of the Interior, Bureau of Indian Affairs, ANCSA Projects Office, Anchorage, Alaska.

O'Leary, Matthew. 2001. Volcanic Ash Stratigraphy for Adak Island, Central Aleutian Archipelago. In Don Dumond (ed.) *Archaeology in the Aleut Zone of Alaska, Some Recent Research,* 215-234. University of Oregon Anthropological Papers No. 58. Eugene, University of Oregon Press.

Pinart, Alfonse. 1873. Voyage la côte nord-ouest d'Amérique d'Ounalashka à Kodiak (îles Aléoutiennes et péninsule d'Aliaska. *Bulletin de la Société de Géographie, Serie* VI, (6), 561-580.

Pinart, Alfonse. 1875. La caverne d'Aknanh, île d'Ounga (Archipel Shumagin, Alaska). Paris, E. Leroux..

Reedy-Maschner, Katherine. 2010. Where Did All the Aleut Men Go? Aleut Male Attrition and Related Patterns in Aleutian Historical Demography and Social Organization. In M. Crawford, D. West and D. O'Rourke (eds.), *Special Issue on the Origins of the Populations of the Aleutian Islands, Human Biology* 82 (5-6), 583-611.

Savinetsky, Arkady B., Dixie West, Zhanna A. Antipushina, Bulat F. Khassanov, Nina K. Kiseleva, Olga A. Krylovich, and Andrei M. Perelodov. 2011. The Reconstruction of Ecosystems History of Adak Island (Aleutian Islands) During the Holocene. Chapter 5 in D. West, V. Hatfield, E. Wilmerding, C. Lefèvre,

L. Gualtieri (eds), *The People Before: The Geology, Paleoecology and Archaeology of Adak Island, Alaska.* Oxford, British Archaeological Reports.

Savinetsky, Arkady B., Nina Kiseleva, and Bulat Khassanov. 2004. Dynamics of Sea Mammal and Bird Populations of the Bering Sea Region Over the Last Several Millennia. *Palaeogeography, Palaeoclimatology, Palaeoecology* 6, 335-352.

Savinetsky, Arkady B., Nina Kiseleva, and Bulat Khassanov. 2010. Paleoenvironment—Holocene Deposits from Shemya Island. In D. Corbett, D. West, and C. Lefèvre (eds.), *The People at the End of the World: The Western Aleutians Project and the Archaeology of Shemya Island*, 71-82. Anchorage, Aurora Monograph Series.

Schlanger, Nathan. 1990. Techniques as Human Action: Two Perspectives. *Archaeological Review from Cambridge* 9, 18–26.

Sekora, Palmer. 1973. Aleutian Islands National Wildlife Refuge Wilderness Study Report. Anchorage, Alaska, US Fish and Wildlife Service.

Sellet, Frederic. 1993. Chaîne Opératoire: The Concept and Its Applications. *Lithic Technology* 18, 106–112.

Siegel-Causey, Douglas, Christine Lefèvre and Arkady Savinetsky. 1991. Historical Diversity of Cormorants and Shags from Amchitka Island, Alaska. *The Condor* 93, 840-852.

Siegel-Causey, Douglas. Debra Corbett, and Christine Lefèvre. 1993. Report of the Buldir Island Expedition: Second Season in the Aleut Midden Site, June 1993. Report on file, National Geographic Society, Washington, D.C.

Siegel-Causey, Douglas, Debra Corbett, Christine Lefèvre, and Stephen Loring. 1994. Report of the Western Aleutians Archaeological Project: Shemya Island Excavations: August 1994. Report on file, National Geographic Society, Washington, D.C.

Spaulding, Albert. 1962. *Archeological Investigations on Agattu, Aleutian Islands.* Anthropological Papers, Museum of Anthropology, The University of Michigan 18, 1-79.

Thorson, Robert M. and Thomas D. Hamilton. 1986. Glacial Geology of the Aleutian Islands in Alaska. In T. D. Hamilton and R. M. Thorson (eds.), *Glaciation in Alaska: The Geological Record*, 171-191. Anchorage, Alaska Geological Society.

Turner, Christy G. II. 1972. Preliminary Report of Archaeological Survey and Test Excavations in the Eastern Aleutian Islands, Alaska. *Arctic Anthropology* 9(2), 32-35.

Turner, Christy G. II. 1976. The Aleuts of Akun Island. *Alaska Journal* 6(1), 25-31.

Turner, Christy G. II. 1981. Russian Influence on Aleut Ecology, Culture, and Physical Anthropology. *National Geographic Society Research Reports* 13, 623-633.

Turner, C. G. II, Jean Aigner and Linda Richards. 1974. Chaluka Stratigraphy, Umnak Island, Alaska. *Arctic Anthropology* 11 (Supplement), 125-142.

US BIA (US Bureau of Indian Affairs, A. O.) 1985. Report of Investigations for BLM AA-12080. The Aleut Corporation. By William L Sheppard and Debra Corbett. On File at the BIA ANCSA Office, Anchorage.

US BIA (US Bureau of Indian Affairs, A. O.) 1989. Report of Investigation for Shemya Island. The Aleut Corporation, BLM AA-11923, BLM AA-11924, BLM AA-11925, BLM AA-11926. The Aleut Corporation, BLM AA-11927 *et al.* Report on file, ANCSA Office, Anchorage.

US BIA (US Bureau of Indian Affairs, A. O.) 1991. Compilation of Investigation Reports, 1983 Adak Sites. By William L Sheppard *et al.* On File at BIA ANCSA Office, Anchorage.

US BIA (US Bureau of Indian Affairs, A. O.) 1992a. Report of Investigation for Rat Islands overview. The Aleut Corporation, BLM AA-11927. On File at the US Bureau of Indian Affairs, ANCSA Office, Anchorage.

US BIA (US Bureau of Indian Affairs, A. O.) 1992b. Report of Investigations for Adak Island, BLM AA-12088 *et al.*, The Aleut Corporation (a.k.a. Adak Overview). By Thomas J. Turck. On File at the US Bureau of Indian Affairs, ANCSA Office, Anchorage.

US BIA (US Bureau of Indian Affairs, A. O.) n.d. Radiocarbon dates for sites AA-11960, 11967, 12011, 12013. On File at the US Bureau of Indian Affairs, ANCSA Office, Anchorage.

Veltre, Douglas W. 1979. Korovinski: The Ethnohistoric Archaeology of an Aleut and Russian Settlement on Atka Island, Alaska. Unpublished Ph.D. Dissertation, Department of Anthropology, University of Connecticut, Storrs.

Veltre, Douglas W. 1990. Perspectives on Aleut Culture Change during the Russian Period. In Barbara Sweetland Smith and Redmond J. Barnett (eds.), *Russian America: The Forgotten Frontier,* 175-183. Tacoma, Washington, Washington State Historical Society.

Veltre, Douglas W. 1997. An Archaeological Survey of Selected Sites on Northern Adak Island, Aleutian Islands, Alaska. Report submitted to the Aleut Corporation, Anchorage, Alaska.

Veltre, Douglas W. 2001. Korovinski:Archaeological and Ethnohistorical Investigations of a Pre- and Post-Contact Aleut and Russian Settlement on Atka Island. In Don Dumond (ed.), *Archaeology in the Aleut Zone of Alaska: Some Recent Research,* 187-214. University of Oregon Anthropological Papers No. 58. Eugene, Oregon, University of Oregon Press.

Veltre, Douglas W. 2011. One Hundred Forty Years of Archaeology in the Central Aleutian Islands, Alaska. Chapter 2 in D. West, V. Hatfield, E. Wilmerding, C. Lefèvre, L. Gualtieri (eds), *The People Before: The Geology, Paleoecology and Archaeology of Adak Island, Alaska.* Oxford, British Archaeological Reports.

Veltre, Douglas, Allen McCartney and Jean Aigner. 1990. Margaret Bay Archaeological Site (UNL-048) Draft Mitigation Plan. Report to the Ounalashka Corporation for Mitigation of UNL-48.

Veltre, Douglas, Allen McCartney, Mary Veltre, and Jean Aigner. 1984. An Archaeological Site Survey of Amaknak and Unalaska Islands, Alaska. Anchorage,

Alaska Division of Parks and Outdoor Recreation.

Veniaminov, Ivan. 1984. *Notes on the Islands of the Unalashka District*. [L. Black and R. Geoghegan, trans.] Ontario, Limestone Press.

Waythomas, Christopher F. 1995. Surficial Geologic Map of Northern Adak Island, Alaska. USGS Open-File Report 95-128. Anchorage, Alaska.

West, Dixie, Debra Corbett, and Christine Lefèvre. 1997. Western Aleutians Archaeological and Paleobiological Project: Annual Report for 1997. Report on file, National Science Foundation, Washington, D.C. (Buldir Island)

West, Dixie, Debra Corbett, and Christine Lefèvre. 1998. The Western Aleutians Archaeological and Paleobiological Project: Annual Report for 1998. Report on file, National Science Foundation, Washington, D.C. (Attu Island)

West, Dixie, Debra Corbett, and Christine Lefèvre. 1999. The Western Aleutians Archaeological and Paleobiological Project: Annual Report for 1999. Report on file, National Science Foundation, Washington, D.C. (Shemya and Adak Islands)

West, Dixie, Debra Corbett, and Christine Lefèvre. 2000. The Western Aleutians Archaeological and Paleobiological Project: Annual Report for 2000. Report on file, National Science Foundation, Washington, D.C. (Attu Island)

West, Dixie, Debra Corbett, and Christine Lefèvre. 2001. The Western Aleutians Archaeological and Paleobiological Project: Annual Report for 2001. Report on file, National Science Foundation, Washington, D.C. (Buldir Island)

West, Dixie, Debra Corbett, and Christine Lefèvre. 2002. The Western Aleutians Archaeological and Paleobiological Project: Annual Report for 2002. Report on file, National Science Foundation, Washington, D.C. (Attu Island)

West, Dixie, Christine Lefèvre, Debra Corbett, and Arkady Savinetsky. 1999. Radiocarbon Dates for the Near Islands, Aleutian Islands, Alaska. *Current Research in the Pleistocene* 16, 83-85.

West, Dixie, Christine Lefèvre, Debra Corbett, and Susan Crockford. 2003. A Burial Cave in the Western Aleutian Islands, Alaska. *Arctic Anthropology* 40(1), 70-86.

West, Dixie, Joan Brenner Coltrain, M. Geoffrey Hayes, Dennis H. O'Rourke, Christine Lefèvre, Debra Corbett, Françoise Bouchet, and Arkady Savinetsky. 2011. A Human Burial from the Andreanof Islands, Alaska. Chapter 16 in D. West, V. Hatfield, E. Wilmerding, C. Lefèvre and L. Gualtieri (eds), *The People Before: The Geology, Paleoecology and Archaeology of Adak Island, Alaska*. Oxford, British Archaeological Reports.

West, Dixie, Arkady B. Savinetsky, Zhanna Antipushina, and Olga Krylovich. 2011. Invertebrate Remains: A Pan Aleutian Comparison. Chapter 10 in D. West, V. Hatfield, E. Wilmerding, C. Lefèvre and L. Gualtieri (eds), *The People Before: The Geology, Paleoecology and Archaeology of Adak Island, Alaska*. Oxford, British Archaeological Reports.

Wilmerding, Elizabeth G. 1993. A Brief Description of Buldir Island Lithic and Bone Tools and a Comparison with Material from the Rat and Near Islands. Paper presented at the 21st annual Alaska Anthropological Association, Juneau, Alaska.

Wilmerding, Elizabeth. 2005. The Culture Sequence at the Nunik Site, Chernabura Island, Alaska. Unpublished PhD dissertation, Department of Anthropology, Washington State University, Pullman, WA.

Wimerding, Elizabeth and Marvin Kay. In review Continuity and Change in Lithic Debitage and Artifacts Over Time at Three Localities on Clam Lagoon, Adak Island, Alaska. In Caroline Funk and Debra Corbett (eds.), Anthropology of the Western Aleutian Islands. *Arctic Anthropology*.

Wilmerding, Elizabeth and Virginia Hatfield. 2011. Six Thousand Years of Lithic Technology on Adak in a Broader Aleutian Context. Chapter 12 in D. West, V. Hatfield, E. Wilmerding, C. Lefèvre and L. Gualtieri (eds), *The People Before: The Geology, Paleoecology and Archaeology of Adak Island, Alaska*. Oxford, British Archaeological Reports.

Winslow, Margaret A. and L. Lewis Johnson. 1989. Prehistoric Human Settlement Patterns in a Tectonically Unstable Environment: Outer Shumagin Islands, Southwestern Alaska. *Geoarchaeology* 4(4), 297-318.

Workman, William B. 1966. Prehistory of Port Moller, Alaska Peninsula, in Light of Field Work in 1960. *Arctic Anthropology* 3(2), 132-153.

Yarborough, Michael. 1974. Analysis of Pottery from the Western Alaska Peninsula. *Anthropological Papers of the University of Alaska* 16(1), 85-89.

Yarborough, Michael. 1988. Archaeological Testing of UNL-048, The Margaret Bay Site, Unalaska. Report prepared for the Ounalashka Corporation.

Yesner, David. 1977. Prehistoric Subsistence and Settlements in the Aleutian Islands. Unpublished Ph.D. Dissertation, University of Connecticut.

Yesner, David. 1988. Significance Assessment for the Margaret Bay Site, Dutch Harbor, Unalaska. Report of Assessment to Ounalashka Corporation for the 'Margaret Bay Industrial Port' Project

Appendix 1A. Primary CAAPP researchers in alphabetical order.

Antipushina, Zhanna		Invertebrate analysis, excavation	Severtsov Laboratory, Russian Academy of Sciences, Moscow, Russia
Bouchet, Françoise		Paleoparisitology	Université de Reims, France
Coltrain, Joan		Human stable isotopes	Department of Anthropology, University of Utah, Salt Lake City, Utah
Corbett, Debra		Human burial archaeology	US Fish and Wildlife Service, Anchorage, AK
Crockford, Susan		Zooarcheology	Pacific Identifications, Inc. Victoria, British Columbia, Canada
Garong, Ame		Vertebrate stable isotopes	Laboratory of Biodiversity, Kyushu University, Fukuoka, Japan
Gualtieri, Lyn	Co-Principle Investigator (2005-2007)	Geomorphology	College of Science and Engineering, Seattle University, Seattle, WA
Hatfield, Virginia	Field supervisor (2006)	Lithic analysis	Dept. of Anthropology, Texas Tech University, Lubbock, TX
Hayes, M. Geoffery		Ancient human DNA	The Feinberg School of Medicine, Northwestern University, Evanston, Ill.
Jeannotte, Richard		Biochemical analysis	Lipidomics Research Center, Kansas State University, Manhattan, KS
Johnson, Taylor		Geology	Department of Geology, Whitman College, WA
Kano, Akihiro		Invertebrate analysis	Dept. of Biodiversity, Kyushu University, Japan
Kay, Marvin	Field supervisor (2007)	Lithic use wear analysis	Dept. and Anthropology, University of Arkansas, Fayetteville, ARK
Khassanov, Bulat		Historical ecology	Severtsov Laboratory, Russian Academy of Sciences, Moscow, Russia
Kiseleva, Nina		Historical ecology	Severtsov Laboratory, Russian Academy of Sciences, Moscow, Russia
Koike, Hiroko		Ancient vertebrate DNA, stable isotopes	Laboratory of Biodiversity, Kyushu University, Japan
Krylovich, Olga		Invertebrate identification, excavation	Severtsov Laboratory, Russian Academy of Sciences, Moscow, Russia
Lefèvre, Christine	Co-Principle Investigator (2005-2007)	Archives, editing	Muséum national d'Histoire Naturelle, Paris, France
Nakamura, Toshio		AMS dating	Center for Chronological Research, Nagoya University, Japan
Nicolaysen, Kirsten		Obsidian and raw material sourcing	Dept. of Geology, Whitman College, WA
Nishida, Shin		Ancient vertebrate DNA	Laboratory of Biodiversity, Kyushu University, Japan
Okuno, Mitsuru		Tephrachronology, AMS dating	Dept. of Earth System Science, Fukuoka University, Japan
O'Rourke, Dennis		Ancient human DNA	Dept. of Anthropology, University of Utah, Salt Lake City, Utah
Pereladov, Andrei		Diatom analysis	Severtsov Laboratory, Russian Academy of Sciences, Moscow, Russia
Sarata, Brenn		Geomorphology	Leidschendam, Netherlands
Savinetsky, Arkady		Paleoenvironmental analysis, AMS dating, zooarchaeology	Severtsov Laboratory, Russian Academy of Sciences, Moscow, Russia
Southon, John		AMS dating	Keck Laboratory, University of California, Irvine, CA
Takashima, Chizuru		Invertebrate analysis	Department of Biodiversity, Kyushu University, Japan
Torii, Masayuki		Tephrochronology	Kumamoto Gakuen University, Japan
Veltre, Douglas		History of Aleutian research	Department of Anthropology, University of Alaska, Anchorage, AK
Wada, Keiji		Terphrochronology	Dept. of Earth Science, Hokkaido University, Japan
West, Dixie	Principle Investigator (2005-2007)	Invertebrates	Biodiversity Research Center, University of Kansas, Lawrence, KS
Wilmerding, Elizabeth	Field supervisor (2005-2006)	Lithic analysis	Dept. of Anthropology, Vassar College, Poughkeepsie, NY

Appendix 1B. Radiocarbon dates from north Adak Island.

Lab number	*AHRS#*	*Site*	*Pit / Feat*	*Level*	*Conventional RCYBP*	*Material*
Beta-141261	ADK-009	Sweeper Cove Site	Unit 1	Midden 2 Level N	1180±70(1180+90*)	shell
Beta-135537	ADK-009	Sweeper Cove Site	Unit 1	Midden 1 Level 1	1720±70(1490+85*)	(possible) sea cow bone
Beta-141258	ADK-009	Sweeper Cove Site	Unit 1	Midden 2 Level A	1040±70(1040=90*)	shell
Beta-141259	ADK-009	Sweeper Cove Site	Unit 1	Midden 2 Level F	1180±60(1180+80*)	shell
Beta-141260	ADK-009	Sweeper Cove Site	Unit 1	Midden 2 Level K	1240±70(1240=90*)	shell
IEMAE-1242	ADK-009	Sweeper Cove Site	Unit 1	Midden 2 Level F	1360 ± 110*	fish
IEMAE-1265	ADK-009	Sweeper Cove Site	Unit 1	Midden 2 bottom of cultural layer N	2080 ± 55*	fish bone
IEMAE-1288	ADK-009	Sweeper Cove Site	Unit 1	Modden 2 Level K	1560 ± 80*	fish bone
IEMAE-1287	ADK-009	Sweeper Cove Site	Unit 1	Midden 1 Level 1	1700 ± 75*	fish bone
IEMAE-1264	ADK-009	Sweeper Cove Site	Unit 1	Midden 2Level A	970 ± 80*	fish bone
Beta-133557	ADK-010	Palisades Site	House Pit 26		150 ± 60	soil & charcoal
Beta-132880	ADK-011	Zeto Point	Feature 1	Unit 1 bottom	2160 ± 40	charred material
Beta-132879	ADK-011	Zeto Point	Feature 1	Unit 2 Level 5	2490 ± 50	charred material
Beta-132878	ADK-011	Zeto Point	Feature 1	Unit 1 Level 3	440 ± 40	charred material
Beta-133745	ADK-011	Zeto Point	Feature 1	Burial	840 ± 40	bone
UCIAMS-28896	ADK-011	Zeto Point	Feature 1	Unit 1 Level 1 Midden	1325 ± 14	shell
UCIAMS-28899	ADK-011	Zeto Point	Feature 1	Unit 1 Level 1 Midden	1235 ± 15	fish bone
UCIAMS-28900	ADK-011	Zeto Point	Feature 1	Unit 1 Level 1 Midden	335 ± 15	ptarmigan bone
UCIAMS-28902	ADK-011	Zeto Point	Feature 1	Unit 1 Level 1 Midden	320 ± 20	charcoal
Beta-132882	ADK-011	Zeto Point	Feature 2	Level 5	180 ± 60	charred material
Beta-132881	ADK-011	Zeto Point	Feature 2	Level 3- S corner	220 ± 50	charred material
UCIAMS-61695	ADK-011	Zeto Point	Feature 2	Unit F, Layer 2, Level 2	335 ± 20	wood
UCIAMS-61694	ADK-011	Zeto Point	Feature 3	Unit J, floor	370 ± 20	wood
UCIAMS-61691	ADK-011	Zeto Point	Feature 8	Unit J, Layer 2, Level 3	75 ± 20	wood
UCIAMS-61692	ADK-011	Zeto Point	Feature 8	Unit J, Layer 2, Level 4	265 ± 20	wood
UCIAMS-28905	ADK-011	Zeto Point	House	Unit J Feature 3	285 ± 15	charcoal
UCIAMS-28904	ADK-011	Zeto Point	House	Unit J Layer 2 Level 3	340 ± 15	charcoal
UCIAMS-42399	ADK-011	Zeto Point	House	Unit G Layer 1	325 ± 20	charcoal
UCIAMS-42400	ADK-011	Zeto Point	House	Unit G Layer , Level 2	340 ± 20	charcoal
UCIAMS-42401	ADK-011	Zeto Point	House	Unit H Layer 3 east Level 4	320 ± 20	charcoal
UCIAMS-42402	ADK-011	Zeto Point	House	Unit G Layer 4 Level 9 Feature 6	2440 ± 20	charcoal
UCIAMS-42403	ADK-011	Zeto Point	House	H170-190 below string, 1174	2590 ± 20	charcoal
UCIAMS-42411	ADK-011	Zeto Point	House	G2 Feat 1,	390 ± 20	ptarmigan bone
UCIAMS-42412	ADK-011	Zeto Point	House	G2 Level 3,	365 ± 20	bunting and finch bone
UCIAMS-28906	ADK-011	Zeto Point	House	Unit H, Layer 1	7760+20 **	charcoal

Lab number	*AHRS#*	*Site*	*Pit / Feat*	*Level*	*Conventional RCYBP*	*Material*
UCIAMS-28907	ADK-011	Zeto Point	House	Unit H Layer 2, Level 3	615 ± 15	charcoal
UCIAMS-28908	ADK-011	Zeto Point	House	Unit H, Layer 4 Level 7	2390 ± 15	charcoal
UCIAMS-28909	ADK-011	Zeto Point	House	Unit H, Layer 4, Level 8	2395 ± 15	charcoal
UCIAMS-28910	ADK-011	Zeto Point	House	Unit H, Layer 4, Level 9	2420 ± 15	charcoal
UCIAMS-42396	ADK-011	Zeto Point	Midden	F1 no level,	375 ± 20	charcoal
UCIAMS-42397	ADK-011	Zeto Point	Midden	F2 Level 2,	285 ± 20	charcoal
UCIAMS-42398	ADK-011	Zeto Point	Midden	F2 Level 7,	385 ± 25	charcoal
UCIAMS-42407	ADK-011	Zeto Point	Midden	F1 Level 2, 276a	170 ± 25	ptarmigan bone
UCIAMS-42408	ADK-011	Zeto Point	Midden	F2, Level 3,	415 ± 20	bunting bone
UCIAMS-42409	ADK-011	Zeto Point	Midden	D2 Level 4,	410 ± 20	ptarmigan bone
UCIAMS-42410	ADK-011	Zeto Point	Midden	F2 Level 5,	400 ± 20	ptarmigan bone
UCIAMS-61688	ADK-011	Zeto Point	House	Unit J, Layer 1, Level 2	435 ± 20	wood
UCIAMS-61689	ADK-011	Zeto Point	House	Unit I, Layer 1	315 ± 20	wood
UCIAMS-61690	ADK-011	Zeto Point	House	Unit H, Layer 2, Level 2	320 ± 25	wood
UCIAMS-61693	ADK-011	Zeto Point	House	Unit J, Layer 3 east, Level 4/5	160 ± 20	wood
UCIAMS-61775	ADK-011	Zeto Point	Midden	Unit F, Layer 2, Level 2	1180 ± 20	mammal bone
Beta-132884	ADK-012	Dozered Site	1		2510 ± 50	bone
UCIAMS-22159	ADK-012	Dozered Site	Unit 1	Layer 2 Mixed deposit	2085 ± 15	burned bone
UCIAMS-22161	ADK-012	Dozered Site	Unit 2		1865 ± 15	Thrush leg bone
UCIAMS-22162	ADK-012	Dozered Site	Unit 2		2455 ± 20	Rockfish bone
UCIAMS-22158	ADK-012	Dozered Site	Unit 2	Intact deposit	1515 ± 15	charcoal
UCIAMS-22273	ADK-012	Dozered Site	Unit 2		2470 ± 15	rockfish bone
UCIAMS-61685	ADK-012	Dozered Site		Unit 3/1	1480 ± 20	charcoal
Beta-132883	ADK-013	Ocean View Site	profile		1850 ± 60	organic sediment charred material
Beta-133842	ADK-013	Ocean View Site	profile		2900 ± 40	charred material
Beta-133554	ADK-014	Andrew Lake Site	Northern midden		340 ± 50	peat
Beta-133555	ADK-014	Andrew Lake Site	Southern midden		760 ± 50	wood
Beta-133556	ADK-084	Andrew Bay Site		16-30 cm. Below ash	290 ± 40	organic sediment
Beta-135180	ADK-084	Andrew Bay Site		16-30 cm below ash	500 ± 40	charred material
UCIAMS-61687	ADK-084	Andrew Bay Site	CH & EGW pit	- 13 - 25 cm below surface	280 ± 20	charcoal
UCIAMS-61686	ADK-084	Andrew Bay Site	Taylor pit	- 54 cm below surface	335 ± 20	charcoal
Beta-122574	ADK-171	Tutiakoff Site	Profile exposure	115-148 cm, upper shell layer	6410 ± 60	shell
UCIAMS-22153	ADK-171	Tutiakoff Site	Unit 3	Layer 2	5760 ± 20	charcoal
UCIAMS-22154	ADK-171	Tutiakoff Site	Unit 3	Layer 3	5735 ± 30	charcoal
UCIAMS-22274	ADK-171	Tutiakoff Site	Unit 55-1	N corner above ash (noncultural)	3210 ± 20	charcoal
UCIAMS-22151	ADK-171	Tutiakoff Site	Units 1 and 2	- 126 cm	6005 ± 20	charred material
UCIAMS-22275	ADK-171	Tutiakoff Site	Units 1 and 2	- 122 cm	5405 ± 1 15	charcoal
Beta-141262	ADK-171	Tutiakoff Site			5960±60(5960±80*)	shell

Lab number	***AHRS#***	***Site***	***Pit / Feat***	***Level***	***Conventional RCYBP***	***Material***
IEMAE-1281	ADK-171	Tutiakoff Site		(Upper) cultural layer Level 2	6141 ± 123	fish bone
IEMAE-1248	ADK-171	Tutiakoff Site		Level 1	6172 ± 192	fish bone
IEMAE-1296	ADK-171	Tutiakoff Site		level 1	6525 ± 94	fish bone
Beta-137934	ADK-177	Sweeper Cove 2 Site			840 ± 40	bone
Beta-137930	ADK-178	Sweeper Cove 3 Site	TP2		1360 ± 40	bone
Beta-137927	ADK-180	Peninsula Site	TP1	Northermost feature	2020 ± 40	charred material
Beta-137928	ADK-181	Grassy Knoll Site	TP3	Feature 16, upper locus	3300 ± 50	charred material
Beta-137929	ADK-181	Grassy Knoll Site	TP4	Feature E, lower locus	1950 ± 40	charred material
Beta-137933	ADK-182	Lake Site	TP1		1900 ± 40	charred material
Beta-137932	ADK-183	Overlook Site	TP1	Feature 4	450 ± 40	charred material
Beta-137931	ADK-185	Ow-ee Site	TP1	Feature 14	790 ± 40	charred material
Beta-137935	ADK-187	Dune Site		255-270 cm	3600 ± 50	charred material
UCIAMS-61682	ADK-193			- 57 cm below surface	2485 ± 20	charcoal
UCIAMS-61683	ADK-193			- 43 cm below surface	1770 ± 20	charcoal
UCIAMS-61684	ADK-283	Swetzof Site		Unit F, Level 3	1745 ± 25	charcoal
UCIAMS-42404	ADK-283	Swetzof Site	E	Level 7	1755 ± 20	charcoal
UCIAMS-42405	ADK-283	Swetzof Site	E	Level 3	1735 ± 20	charcoal
UCIAMS-22149	ADK-283	Swetzof Site	Test pit 1	- 50,5 cm	3835 ± 20	residue
UCIAMS-22150	ADK-283	Swetzof Site	Test pit 1	NE corner	890 ± 25	charcoal
IEMAE-1312		2/99 Clam Lagoon	Above ADK-171	98-100 cm	1848 ± 73	humus
IEMAE-1311		2/99 Clam Lagoon	Above ADK-171	45-52 cm	1417 ± 82	humus
IEMAE-1255		5/97 near Haven Lake		168-170 cm	7381 ± 149	humus
IEMAE-1309		6/97 Haven Lake		260-265 cm	8697 ± 201	plant remains
IEMAE-1308		6/97 Haven Lake		110-120 cm	4159 ± 136	plant remains
IEMAE-1307		6/97 Haven Lake		69-81 cm	3124 ± 91	plant remains
IEMAE-1306		6/97 Haven Lake		60-64 cm	2756 ± 91	plant remains
IEMAE-1305		6/97 Haven Lake		30-35 cm	modern	plant remains
IEMAE-1304		6/97 Haven Lake		96-98 cm	2887 ± 105	plant remains
IEMAE-1303		6/97 Haven Lake		135-140 cm	2114 ± 159	plant remains
IEMAE-1302		6/97 Haven Lake		245-250 cm	8658 ± 175	plant remains
IEMAE-1268		Peat bog Haven Lake		185-190 cm	8657 ± 86	humus

*Calibrated
** This date on charcoal from ADK-011 Layer 1 is anomalous and does not reflect the age of Layer 1.

Chapter 2
One Hundred Forty Years of Archaeology in the Central Aleutian Islands, Alaska

Douglas W. Veltre

Abstract: *The central Aleutian Islands stretch for some 900km and account for approximately half of the total precontact area of occupation of the Aleut (Unangax̂) people of Alaska. Although this region can be defined with reference to geographical boundaries, it also is one that was recognized as culturally and linguistically distinctive by Aleuts and others, at least since the recent late precontact and early contact periods. While archaeological research in Alaska had its beginnings in the central Aleutian Islands in the late 1800s, anthropological research of any kind in this important region over the last 140 years has nevertheless been very limited, restricted not only by substantial logistical difficulties but also by the vagaries of the research priorities of those investigating the Aleut past. A broad range of research questions, therefore, remains to be fully addressed; these include the nature of maritime adaptations over the several thousand year occupation of the islands; the definition, time-depth, and visibility of intra-regional political and linguistic Aleut groupings; the role and impacts of humans on the Aleutian Islands ecosystem; the mechanisms and effects of precontact trade and exchange between the eastern Aleut area, with its proximity to other Native cultures of mainland Alaska, and the largely isolated western Aleutian Islands; and the changes in Aleut culture which occurred as a result of the arrival of Russian fur hunters in the late eighteenth century.*

Keywords: *Aleut, Aleutian Islands, anthropology, archaeology, history*

Introduction

An interesting footnote to the history of anthropological and archaeological research in Alaska is that the first formal archaeological studies in the entire state began in the early 1870s with the pioneering work of William Healy Dall in the central Aleutian Islands. This fact is rendered somewhat remarkable considering that the central Aleutian Islands—along with the even farther removed western Aleutians—comprise the most geographically isolated part of Alaska to be occupied by people in precontact times. (Only the Pribilof Islands, in the Being Sea, and the Commander Islands, to the west of the Aleutians and now part of Russia, show no archaeological evidence of precontact human occupation. It is likely telling that these are the only Bering Sea islands which cannot be seen from another island or land mass.) Unlike the eastern Aleutian region, with its major fishing community of Unalaska (Dutch Harbor) and six other extant Aleut villages, few people live in the central archipelago. Only the approximately 90 Aleut (*Unangax̂*) residents of Atka village on Atka Island and about fifty in the newly created community at the former U.S. Navy base on Adak Island occupy the central Aleutian Islands today. Likewise, travel to the region by non-residents is rare, chiefly undertaken by U S Fish and Wildlife Service (FWS) employees, military personnel, and those with research or business interests.

This paper focuses on several tasks. First, the geographic and cultural features that render the central Aleutians a demonstrable sub-region of the Aleutian archipelago are considered. Second, a brief review is given of the archaeological work done over the last one hundred and forty years in the central Aleutian Islands. Previous summaries of the work done in this region include the broadly ranging discussion of Rat Islands archaeology by McCartney (1977), regional reviews of the entire archipelago (McCartney 1984; Veltre and Smith 2010), and a consideration of maritime adaptations in the central and western Aleutian Islands by Veltre (1998). To anticipate a conclusion that will be expanded upon later, the central region of the Aleutian archipelago—especially when compared to the eastern islands and the Alaska Peninsula—has witnessed only minimal archaeological research. Third, the precontact record of the region is placed in the context of our current understanding of the archaeology of the entire Aleut region. Finally, some general thoughts and concerns regarding the archaeology of the region are offered; these point both to certain shortcomings of past work as well as to some promising avenues for future research.

The Central Aleutian Islands Region

The central Aleutian Islands (Figure 2.1) stretch for some 900km and account for approximately half of the total area of precontact Aleut occupation in Alaska, itself including the western end of the Alaska Peninsula, the Shumagin Islands group, and the entire Aleutian Islands archipelago. The easternmost island in the central Aleutians region is Seguam, just east of Amlia Island. Westward from this are the remaining major islands of the Andreanof Islands group—Atka, Great Sitkin, Kagalaska, Adak, Kanaga, and Tanaga islands. Farther west, past the Andreanofs, lies the small Delarof Islands group, followed by the Rat

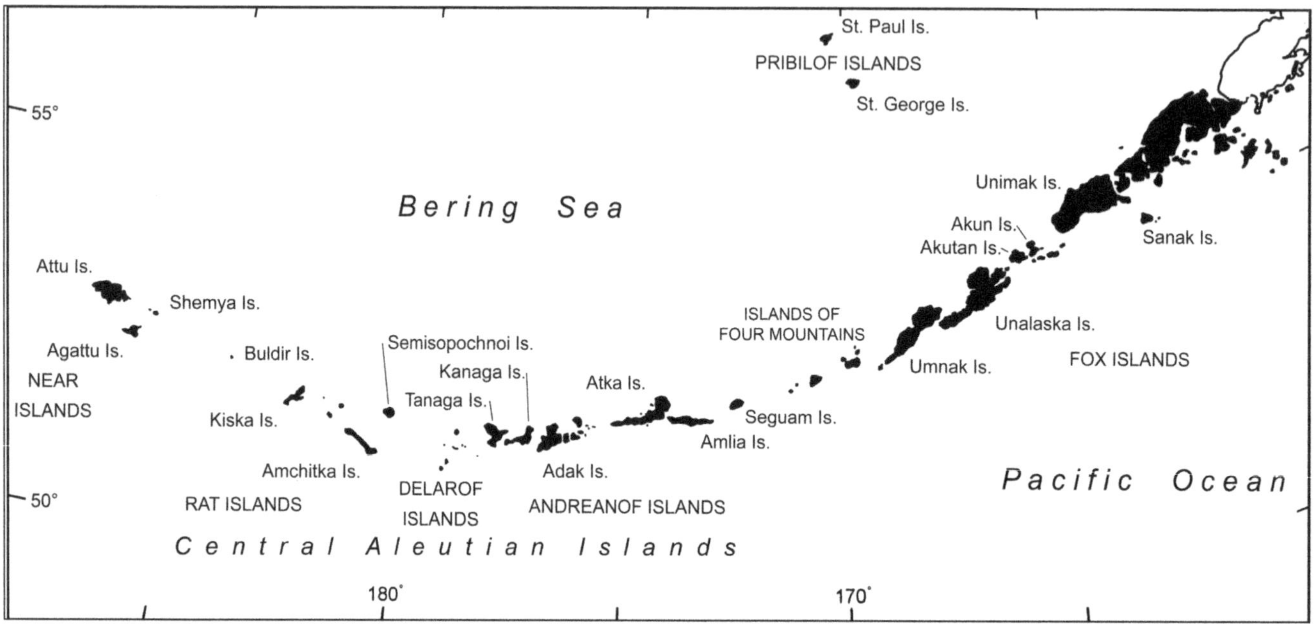

Figure 2.1. The Aleut cultural area (shaded) of the southwestern Alaska Peninsula, Pribilof Islands, and Aleutian Islands, Alaska. The central Aleutian Islands region covered in this paper (labeled) includes the Rat, Delarof, and Andreanof island groups.

Islands group, which includes Semisopochnoi, Amchitka, Rat, Little Sitkin, and Kiska islands. Between Kiska Island and the western Aleutians (also known as the Near Islands group) is tiny Buldir Island, often considered the westernmost of the central Aleutian Islands (Corbett *et al.* 1997a). Recent investigations (Corbett *et al.* 1997b, 475; Corbett *et al.* 1997a, 115) suggest that Buldir was occupied and utilized by Aleuts of both the Near Islands as well as the Rat Islands.

On several measures, anthropologists and others have described early contact period (i.e., late eighteenth century) Aleuts of the central Aleutian Islands as somewhat culturally distinct from their western and eastern Aleut neighbors. The apparent basis for this lies largely in the relatively great inter-island distances separating this region from islands both to the east and to the west. At the eastern end of the central Aleutians, Umnak and Amlia islands are separated—except for several intervening smaller and arguably less easily inhabitable islands—by some 100km; at the western end, Buldir Island and the islands of the Near Islands group are some 50km apart. These distances, although clearly not sufficient to have prevented precontact travel, are nevertheless greater than any others within the entire Aleutian archipelago and, therefore, must have restricted inter-regional contact by a meaningful, although currently unspecifiable, degree.

At the time of initial Russian contact in 1741, the geographical separation of the central Aleutians from regions to the east and west was paralleled, at least in general, linguistically. The single Aleut language was divided into at least three dialects, the central Aleut dialect corresponding to the region being described here (Bergsland and Dirks 1990, 5). However, it should be noted that this seemingly straightforward correspondence might, in fact, not be quite accurate for a number of reasons. For example, the original Russian publication of Russian Orthodox Priest Ivan Veniaminov's ethnographic account suggests that prior to 1825 the eastern boundary of the central dialect lay farther to the east, to include at least the southwestern part of Umnak Island (Bergsland 1959, 12; Black 1984, 44; Laughlin and Marsh 1951, 86-87). In addition, considerable uncertainty exists regarding our knowledge of the pre-Russian linguistic characteristics and relationships within the entire central Aleutians region generally (and the Rat Islands, specifically [Berge 2010. 564-572; Bergsland 1959, 12-13; Black 1984, 41-47]), and it is reasonable to suppose that prior to the devastating cultural and demographic disruptions of the Russian era (Veltre 1990), substantially more linguistic variation existed than was present by the early 1800s. Be that as it may, the central Aleutian Islands region may reasonably be defined along geographically-influenced linguistic dimensions, regardless of the precise numbers and locations of dialects and sub-dialects.

Another important line of evidence testifies to the integrity of the central Aleutians—the names applied by Aleuts themselves to residents of different parts of the entire Aleut region. Drawing upon Veniaminov (1984) and other sources, Bergsland and Dirks (1990, xviii, 2-5) have shown that the eastern and western boundaries of the central Aleutians, as used in this paper, coincide with early contact period boundaries between named Aleut regional groups.

A final contributor to the cultural distinctiveness of the central Aleutian region lies in certain characteristics of the natural environment that differentiate it from neighboring regions, in particular from the eastern Aleutians. Overall, the current mean climate of the entire Aleut region is remarkably uniform, although substantial local variability occurs (Rodionov, Overland, and Bond 2005), a feature largely attributable to the narrow range of latitude occupied by the islands (McCartney and Veltre 1999). However, while many of the marine mammal, fish, and avian species that were the focus of precontact Aleut subsistence are found throughout the archipelago, some notable exceptions exist, and these may have given substance to regional cultural differences. These include several species of large whales that concentrate their migrations between the Pacific Ocean and the Bering and Chukchi seas through the inter-island passes of the eastern Aleutians rather than farther west. Likewise, northern fur seals (*Callorhinus ursinus*), which migrate between the west coast of North America and the Pribilof Islands in the Bering Sea, swim predominantly through eastern Aleutian passes and have only rarely been sighted in appreciable numbers in the waters of the central Aleutians. Paralleling these observations, a recent study of temperature, salinity, and nutrient concentration of ocean waters of the eastern and central Aleutian Islands by Ladd *et al.* (2005) found support for the idea 'that Samalga Pass [immediately southwest of Umnak Island near the eastern boundary of the central Aleutian Islands region] forms a physical and biogeographic boundary between the eastern and central Aleutian maritime ecosystem' (2005, 22).

A final environmental feature of the central Aleutians that differentiates it from the eastern region is the complete absence of winter sea ice. Although essentially ice-free today, the easternmost Aleut region nonetheless is in close proximity to the southern limit of winter sea ice in the Bering Sea; even now, especially cold winters produce sea ice in some local areas. Because climate fluctuated over both short and long terms, the eastern Aleut region consequently witnessed variability in the presence of more northern, ice-related species, including walrus, ringed seal, and others. This has been shown from archaeological faunal analyses at the Margaret Bay site (Davis 2001; Knecht, Davis, and Carver 2001, 54) and the Amaknak Bridge site (Crockford *et al.* 2004, 76-77), both in Unalaska Bay. Therefore, the proximity of the eastern islands to the ice margin meant that Aleut subsistence strategies there were more highly sensitive to the effects of climate variability than were those of residents farther to the west, whose subsistence regimes were likely more stable.

A Chronological Overview of Archaeological Research

This section enumerates many of the archaeological field efforts undertaken in the central Aleutian Islands, beginning with that of Dall. For the most part, the methods and findings of these are not evaluated in detail, nor are all research enterprises (and their reports) listed, especially when initial results are subsumed in later, more comprehensive publications. In addition, subsequent analytical treatments of previously procured archaeological data are not included here.

From 1871 to 1874, William Healy Dall combined his hydrographic and geodetic reconnaissance for the U S Coast and Geodetic Survey in the Aleutians with excavations at a number of archaeological sites throughout the archipelago. In the central Aleutians, Dall undertook larger, albeit quickly done, excavations, on Amchitka, Adak, and Atka islands (1877, 44, 47) and observed 'shell-heaps,' likely testing some, on Kiska and Little Kiska islands (1877, 44). In addition, Dall excavated a burial cave on the eastern shore of Atka Island, from which he obtained remains representing at least eleven individuals (Dall 1877, 44, 67-68). Unlike even some later Aleutian archaeologists, Dall employed stratigraphic analysis in his examination of the deep midden deposits he trenched, methodology that led to his well-known three-stage evolutionary sequence for the development of precontact Aleut culture. This interpretation—which maintained that the earliest human occupation in the region was culturally impoverished, only later developing fishing and, ultimately, marine mammal hunting capabilities—was not out of line with unilineal cultural evolutionary schemes common to anthropological thinking of his time.

Dall also made a crude attempt at absolute dating of certain deposits he excavated on Amchitka Island by using the measured volume of a contemporary sea urchin, the computed volume of archaeological site deposits (made up largely of sea urchin remains), and estimates of the sea urchin consumption rates of ancient Aleuts and their population size to arrive at the time necessary to form the deposits he observed (1877, 51-52). Although methodologically flawed (see Jochelson 1925, 106), this extrapolation led Dall to an estimate of some 2200 years, coincidentally not particularly out of line with modern radiocarbon chronologies (Turner 1970, 126).

No other archaeological investigations occurred in the central Aleutians until 1909 and 1910, when the Imperial Russian Geographic Society sponsored the Aleut-Kamchatka Expedition, with Waldemar Jochelson leading its anthropological division. Accompanied by his wife, Dina Brodsky-Jochelson, herself a medical doctor, Jochelson undertook a broad spectrum of anthropological studies, including archaeological excavations, anthropometric studies of living Aleuts, and oral history recording on wax cylinders (Bergsland and Dirks 1990; Jochelson 1912, 1925, 1933). These efforts continued the overall goals of the turn of the century Jesup Expedition to northeast Asia and northwestern North America, of which Jochelson had been a part (Loring and Veltre 2003).

Confining his work in the central Aleutians to Atka Island in the summer of 1909, Jochelson extensively excavated at two sites, Atxalax̂ and Halaacha, and two burial caves, all within several kilometers of the modern village of Atka (1925, 26-29, 46-48; Bergsland 1994, 624-5). Jochelson obtained a large number of artifacts as well as some human remains from the two village sites and the remains of at least 28 individuals from the caves. The manner in which Jochelson reported his findings—brief descriptions and occasional depth measurements—makes interpreting his materials difficult, and it is particularly unfortunate that detailed site maps and plan drawings of the house features he dug through are lacking.

Nevertheless, based on his work on Atka and elsewhere in the Aleutians, Jochelson's primary contributions to the archaeology of the region lay in his recognition that, contrary to Dall's view, even the earliest Aleuts in the island chain had to have had a high degree of adaptational sophistication to have survived at all in such a harsh environment and that there was little archaeological evidence of substantial cultural change over the precontact period (Jochelson 1925, 101-110).

The summers of 1936 to 1938 saw the Smithsonian Institution's Aleš Hrdlička conduct archaeological research in the Aleut region, including on the central Aleutian islands of Atka, Kiska and Little Kiska in 1936 (1945, 219-237); Tanaga, Ilak, Adak, and Atka in 1937 (1945, 312-322); and Amlia, Kanaga, Ilak, and Amchitka (1945, 347-364) in 1938. For the most part, Hrdlička and his small crew made just brief visits (of hours) to sites in this area, with only his Kiska, Little Kiska, and Amchitka excavations occupying more than a day or two. As his interest was primarily in acquiring human remains, a substantial amount of his time was spent examining coastal cliffs for burial caves, of which several were found (Hrdlička 1941a, 1941b, 1945).

From World War II to the passage of the Alaska Native Claims Settlement Act in 1971, only a small number of projects—most devoted to archaeological survey, artifact collection, or limited excavation—took place at selected locales in the central Aleutians, but few were long-term or large-scale. During the war itself, Paul Guggenheim, one of Hrdlička's crew members in 1937, was stationed for ten months beginning in May, 1943, on Amchitka Island, where he served as a medical officer. Because of his earlier work and his medical training, Guggenheim understood the scientific importance of ancient remains on the island, although the recreational looting of sites by the military had, by his own measure, 'reached almost epidemic proportions' (Guggenheim 1945, 22). At Guggenheim's urging, the collection of artifacts was finally prohibited early in 1944, except for those encountered during normal military activities.

During his tenure on Amchitka, Guggenheim visited most of the forty known sites on the island, interviewed those who had found ancient specimens, and examined their collections. From some individuals, he was given both artifacts and human remains, and at his own expense (about $500) he also purchased certain items (but not human remains) he deemed particularly worthwhile (Guggenheim 1945, 30-31). Guggenheim sent most of these materials to the Smithsonian Institution, and in 2006 he returned to Alaska a few specimens and records he had kept.

From 1948-1949, Theodore Bank and his wife taught in the school in Atka. With a background in botany and sponsorship from the University of Michigan (and, later, the American Institute for Exploration), Bank began what would be several years (1948 to 1951 and 1954) of botanical and casual archaeological investigations on Atka, Adak, and several other islands in the central Aleutians. Although his botanical research has been published in some detail (Bank 1952, 1953a, 1953b), information concerning his archaeological efforts was disseminated largely in a variety of informal, unpublished forms (Bank n.d.; Bank *et al.* 1950). His popularized account of life in Atka (Bank 1956) reflects his lack of anthropological training and casts doubts on his other works (which in later years included a number of field seasons in the Unalaska area).

Other archaeological work in the central Aleutian Islands during this time included that by U S Geological Survey researchers Willis Nelson and Frank Barnett (1955) on the north coast of Kanaga Island. Inside a cave produced by coastal erosion, they found cultural materials, including a generally circular arrangement of boulders, wooden boat parts, an animal figure carved in wood, and the partial remains of at least seven individuals.

The first large-scale archaeological effort in the central Aleutians since that of Hrdlička had its roots in the early 1960s, when the U S Atomic Energy Commission (AEC) initiated preparations for underground nuclear testing on Amchitka Island (Kohlhoff 2002). The first detonation, Long Shot, occurred in 1965. Although diverse biological and other studies were undertaken before and after that test, archaeological concerns were not addressed until 1968. In that year, AEC hired Christy Turner and Richard Sense to conduct a survey of the archaeological resources on Amchitka (Sense and Turner 1970; Turner 1970). The work continued in 1969, when the agency contracted with Roger Desautels to undertake salvage excavations at five sites deemed threatened by the upcoming Milrow detonation, to take place later that year. Desautels and his crew identified various cultural features and collected over 9000 specimens from these midden sites as well as from several lithic scatters (Desautels *et al.* 1971). Altogether, these efforts on Amchitka Island meant that for the first time a single, large Aleutian island had been fully archaeologically surveyed, with nearly 80 sites identified, of which 21 produced over 1400 artifacts from surface collections and test excavations.

The final nuclear test on Amchitka, Cannikin, took place in late 1971. Because AEC had determined that one particular site, RAT032, was endangered by ground motion from the test, John Cook from the University of Alaska Fairbanks was contracted to conduct archaeological excavations at the site. Some twenty 2m x 2m units and four smaller units were excavated to the base of the cultural midden deposits, which in places were in excess of 2m deep (Cook, Dixon, and Holmes 1972, 10).

Of some five house pits that were exposed at RAT032, one, radiocarbon dated to approximately 415 BP, was excavated fully (Cook, Dixon, and Holmes 1972, 14-15). This house, whose outer dimensions were 7.5m x 6.0m, possessed a raised bedrock bench around its inside perimeter, a shallow trough closer to the center of the floor, and a rectangular, hard-packed flat area in the middle. Six large post holes and numerous smaller peg holes, for roof support posts and internal dividers, respectively, along with a variety of artifacts and a hearth area were apparent on the house floor (Cook, Dixon, and Holmes 1972, 91-101). To this date, it is still the only complete excavation of a house from the entire central Aleutians.

In the mid-1970s, I conducted several seasons of survey and excavation on eastern Atka Island, focusing on the archaeological manifestations of Russian and Aleut contact in the late 1700s and early 1800s (Veltre 1979, 2001). The primary focus of research was the site of Korovinski, 16km northwest of the contemporary village of Atka. The site was a locus of Russian activity during that time, serving eventually as the western office for the Russian American Company. Because testing in 1974 revealed Korovinski to have been occupied both before and during the Russian period, the initial expectation was that a continuous archaeological record from pre- to postcontact materials would allow for the entire contact period to be archaeologically examined.

This was not to be the case, however, as more extensive excavations in 1975 and 1976 later revealed. The earliest occupation of the site, represented by dense midden deposits over 2m deep, dated to 1900 radiocarbon years BP. These materials continued until the entire site was blanketed by a 25cm sequence of volcanic ashes some 500 radiocarbon years ago, an event which effectively brought occupation of the site to an end. Reoccupation did not occur until both Russians and Aleuts settled the site, possibly as late as the 1820s, some 50 years after contact on the island had occurred.

Based on the wide range of artifact types, the numerous house features, and the three-meter deep midden accumulation, Korovinski appears to have functioned as an Aleut village in the precontact period. After the arrival of the Russians and its new role as a colonial administrative center, the Korovinski settlement expanded to include a Russian Orthodox church and cemetery, extensive potato gardens, coastal boat slips, and some eighty Russian houses, warehouses, workshops, and other buildings. After its Aleut residents moved to the present village of Atka in the years immediately following the 1867 sale of Alaska to the United States, Korovinski continued to serve them as a hunting and trapping locale.

At the same time as the 1975 Korovinski excavations were in progress, the regional Aleut Corporation (TAC) supported a survey of archaeological sites on Attu, Adak, and Kagalaska islands as part of its preparation of documentation for culturally important places under Section 14(h)(1) of the Alaska Native Claims Settlement Act (ANCSA) of 1971 (Pratt 1992, 2009). Bruno Frohlich and David Kopjanski circumnavigated most of Adak and a part of Kagalaska, inspecting previously reported site locales and exploring for new ones (Frohlich and Kopjanski 1976; Frohlich 2002).

ANCSA archaeological work continued in 1983 and continued well into the 1990s, as Bureau of Indian Affairs (BIA) archaeologists undertook site surveys throughout a large part of the central Aleutian Islands as a prerequisite to transfer these properties to TAC. This work, although involving little subsurface testing, nevertheless provided a wealth of information—especially surface feature site maps—on over 200 sites in the region, many of which had never before even been seen by a trained archaeologist. Because sites applied for under 14(h)(1) had to be on land otherwise not selected by TAC or any of the region's village corporations, and because most contemporary villages are concentrated in the eastern Aleutians (the westernmost village in the Aleut region is Atka), a disproportionately large number of TAC's 14(h)(1) selections were in the central and western Aleutian Islands.

It is worth noting that although the formal ANCSA surveys are now complete, issues remain regarding the status of certain sites. Recently, Rita Miraglia's (2006) detailed account of the fate of one of TAC's site applications on Amlia Island is a lesson in frustration. Following an appeal by TAC, a site previously declared ineligible for conveyance was reinvestigated by BIA in 2001 and ultimately deemed eligible. However, due to nuances in post-ANCSA federal regulations, the site now appears unlikely to be conveyed.

Few other archaeological projects occurred in the central Aleutians until 1990, when a brief survey was done of a parcel on Amchitka Island prior to the construction of radar facility there (Lobdell 1990). In 1992, members of the Western Aleutian Archaeological Project moved their efforts eastward to Little Kiska Island. Their work, which provided comparative materials to their collections from Buldir Island, took place at the same site excavated in 1936 by Hrdlička (Corbett, Siegel-Causey, and Lefèvre 1995; Loring 1993). In 1995, Charles Mobley (1996) examined selected properties on Kiska, Little Kiska, and Semisopochnoi islands as part of a U S Army Corps of Engineers hazardous, toxic, and radioactive waste study. Because his work was directed largely at World War II materials, Mobley made few observations regarding more ancient cultural resources.

Additional small-scale projects in the region over the past several decades include a brief survey on Adak Island in 1997, done for the Aleut Corporation in part to inspect sites for looting (Veltre 1997), a 1999 joint investigation by the FWS and the BIA, under a contract with the U S Navy, to survey for archaeological sites on the military base on northern Adak Island (Luttrell and Corbett 2000), and several construction-related investigations on Atka

Island (DePew and Thompson 2009; Mobley 2003, 2006; Reynolds 1986; Veltre 1977).

In 1999, the Western Aleutians Archaeological and Paleobiological Project conducted one year of fieldwork on northern Adak, and more recently (2005-2007) its 'descendent' multidisciplinary program, the Central Aleutians Archaeological and Paleobiological Project, undertook archaeological, geological, and paleobiological investigations on that island at several locations accessible from the former Naval Station. These efforts (see West *et al.* Chapter 1 this volume) have shown that most sites on Adak date to no older than 3500 years, although one site (ADK-171) on the east side of Clam Lagoon has been radiocarbon dated to some 6000 BP, in accord with a date obtained at the same location in 1998 by BIA archaeologists (O'Leary 1998, 2001).

On Rat Island, two weeks of testing in 2003 by Debra Corbett and Caroline Funk at a site occupied both before and after Russian contact focused on the impacts that the introduction of rats to the island had made to the ecology of the island (Corbett and Funk 2010). In 2009, Funk (2010) and others returned to Rat Island for additional survey and limited testing; the team also undertook similar work on Kiska Island. Altogether, over 20 newly located sites of precontact and postcontact age were investigated on the two islands.

The most recent research initiative in the central Aleutians region is that by Diane Hanson, who in 2007 began several seasons of survey and testing on Adak Island. Her field work, largely focused on the discovery and elucidation of previously little-known upland and interior sites, will continue through 2011 and promises to add a new dimension to our understanding of Aleut settlement and subsistence patterns (Hanson *et al.* 2008a, 2008b; Hanson 2009; Hanson and Corbett 2010).

Observations on Previous Research

Looking back on this long history of archaeological investigations, a number of observations may be made regarding our view of Aleut history in the central Aleutian Islands. First, most investigators have noted that the broad features of Aleut maritime adaptation in this region show little marked change over either space or the last 3500 years, the period to which most investigated sites date (Veltre 1998). By comparison to the more complex picture of eastern Aleut precontact history emerging from the work of Maschner (1999), Maschner *et al.* (1997), Hoffman (1999), and others on the lower Alaska Peninsula and Unimak Island, and of Knecht, Davis, and Carver (2001) and McCartney and Veltre (2002) in the Unalaska area, this seems to be even more firmly established now than it was before. Even considering some of the limitations of research in the region, the artifact types, house forms, and general site characteristics seen throughout the central Aleutians are certainly more similar than dissimilar to one another.

Second, the earliest radiocarbon dates from the region support the long-held and non-controversial view that initial human colonization of the archipelago moved westward through the central Aleutian Islands, ultimately to the Near Islands (Dumond 2001, 302). Although there are few early dates upon which to establish a precise island-by-island chronology, the broad pattern is now quite clear: people had occupied the Andreanof Islands by about 6000 BP (O'Leary 1998, 2001, 2007) and the Rat Islands and Near Islands by about 3500 BP (Corbett, West, and Lefèvre 2001; West, Lefèvre, Corbett, and Savinetsky 1999). This scheme reasonably places the earliest occupation of the central and western Aleutians as occurring later than the Anangula core and blade tradition of the eastern archipelago, in accord with the absence of such technology west of the Umnak Island area.

Third, sites of the greatest antiquity (i.e., those in excess of about 2500 years) are, not surprisingly, few in number; hence, what little we know about the adaptations of the earliest migrants into the region is based more on logical supposition than on actual archaeological data. Because this entire region (unlike the eastern Aleutians) was insular throughout the Pleistocene and Holocene (Gard 1977, 1980; Thorson and Hamilton 1986) and, consequently, shows no evidence of substantial indigenous terrestrial fauna, it is unquestionable that from the earliest occupation of these islands humans had to have possessed an essentially coastal and maritime adaptation. What remains to be studied are the particulars of that adaptation—how it was expressed bay by bay, island by island, and so forth, as well as over time. These are the kinds of questions that are now being addressed by some of the latest work in the region.

One particularly intriguing possibility within the context of maritime exploitation exists, namely, whether the first people may have caused, or at least influenced, the local extinction of the Steller sea cow. This lumbering, shallow-water vegetarian is known from fossil evidence to have lived in the region over 100,000 years ago (Whitmore and Gard 1977), but by the time of Russian contact in the mid 1700s was likely confined to the uninhabited Commander Islands, where Russian hunters brought it to extinction in less than three decades (Stejneger 1887). Possible sea cow bones from recently investigated archaeological contexts on Buldir, Kiska, and Adak islands suggest the possibility that the earliest Aleut hunters in the region subsisted on sea cows and perhaps even contributed to their local extinction (Domning 1972; Domning, Thomason, and Corbett 2007; Corbett *et al.* 2008). Savinetsky, Kiseleva, and Khasanov (2004) make the strong argument that combined climatic change and human predation likely played a role in the extinction of the sea cow.

Finally, we know specific details concerning the artifact types and faunal remains from those sites where testing and larger excavations have been conducted, but these are quite few in number, especially in comparison to the number of sites investigated more fully in the eastern archipelago.

Further, particularly from the BIA surveys, we know a great deal about the numbers, sizes, and shapes of the most recent houses and, possibly, other structures at a much larger number of sites.

Beyond these things, however, we know relatively little, and many questions regarding the basic outlines of culture history remain to be addressed in future research. Likewise, there are critical methodological and analytical issues that need to be confronted.

First, we do not yet have full knowledge of the numbers of sites in the region. Not only have most islands never been fully surveyed, but even those which have still yield surprises. On Amchitka, for example, completely surveyed during the AEC work, BIA archaeologists nevertheless found an additional 26 midden sites and 19 blowout sites during their 14(h)(1) site investigations in 1985. More recently, Dixie West and her colleagues have located previously unreported sites in the Clam Lagoon area on Adak Island, virtually in the heart of an area of intense military utilization of that island. Finally, as noted above, Hanson's on-going research on inland sites on Adak Island is opening new possibilities for the archaeology of the whole archipelago. Extrapolating to the entire central region, the number of undocumented sites must be substantial.

Second, we know very little about either inter- or intra-site settlement patterns. Investigations to date have either been small-scale test excavations or have been larger, but site-specific efforts, and, at least to the extent undertaken so far, these do not provide the information necessary to address adequately questions of differential site function. Again, even from Amchitka Island, where we have the most complete site inventory, the relationships of sites to one another are essentially indefinite. It is most often simply unknown which sites were seasonal subsistence camps, which served as more sedentary base villages, whether, as suggested in some literature, there were differences in site types between the Pacific and Bering sea coasts, and whether sites are contemporaneous.

Third, we also know very little about intra-Aleutian political groupings, or polities. The map of regional groups by Bergsland and Dirks (1990, xviii) is often incorrectly interpreted as representing well-defined, named political entities. However, it actually portrays a mix of both actual Aleut regional autonyms as well as descriptive names given instead by those Aleuts outside of particular areas. In any case, at least three pertinent questions arise, the last two of which we really have no answers at all to: What was the nature of Aleut political groupings, what might their archaeological visibility be, and what time-depth can they be shown to have had?

The single most important methodological concern in assessing past archaeological research as well as in planning future efforts in the central Aleutian Islands is adequate sampling: that is, acquiring a collection of artifacts, fauna, and other site data that is large enough and sufficiently well controlled (whether randomly or otherwise) to enable us to say with some reliability what went on at a site, either at one time or over time. Because how large a sample must be in order to be adequate is based in part on the variability of analytical elements in the site itself, it seems clear that when one considers the obvious substantial variability in contents among excavation units at any single midden site, it is simply erroneous to suppose that any small number of test excavations in a large site will be at all representative of the site as a whole.

The question of adequate sample size for investigating middens is one that can be addressed through statistical methods using real or hypothetical data, although this has not been done in the Aleutians. Inadequate samples, I suggest, have contributed—and will continue to do so—to a certain kind of misinterpretation: an unwarranted exaggeration of the differences among archaeological assemblages (at whatever scale), especially with regard to the faunal component, which tends to be treated quantitatively to a greater degree than material cultural remains. Finally, while my observations on sampling are relevant to archaeological work throughout the Aleutians, they have particular bearing on the central (and western) regions, where far less work has been accomplished to date than in the eastern Aleut area and, hence, where interpretations can be more easily skewed.

I do not mean to suggest that no archaeology be done unless certain minimum sample sizes can be attained. Many projects are undertaken for resource management concerns, and these are often very small scale. However, I do suggest two things. First, archaeologists must be more aware of the limitations of their data, especially in the face of enormous complexity in the cultural and natural systems they are examining. In practical terms, this often will mean that we archaeologists need to reign in some of our enthusiasm when drawing (unwarranted) conclusions from unreliable data and, similarly, that we should critically appraise the work of others. Second, archaeologists must consciously contribute—whether through a succession of many smaller projects or through larger projects—to a more reliable picture of Aleut adaptations by using appropriate sampling procedures in their work. Because distortions can so easily arise from analyses of inadequate samples of artifacts and fauna from limited site testing, such testing in the Aleut region has—aside from the simple demonstration of cultural deposits—no real purpose and cannot be justified as an end in itself. Moreover, I will go so far as to suggest that if it is not followed up with more extensive excavations, it may actually be a disservice to our understanding of the Aleut past.

The issue of archaeological sample sizes is directly analogous to a problem in various ethnohistoric data as well. If, as I suggest, we must take small archaeological samples to be suspect (that is, they are not necessarily representative), then we must also utilize ethnohistoric data in the same critical manner. In the big picture of the Aleut region, there are precious few early Russian period house descriptions, hunting descriptions, kinship descriptions, and so on, and, of these, very few pertain to the central Aleutians. The temptation, of course, is to over-rely on those that exist: to accept them as truth and go on to interpret archaeological information as a somewhat inferior data set, what Lightfoot (1995, 205) has referred to as 'the privileging of written documents over archaeological remains.' However, simply put, ethnohistoric information is subject to the same fundamental limitations vis-à-vis small sample size as are other data, and we need to subject them all to critical evaluation and sufficient testing.

Perhaps the best large data set we now have for the central Aleutians is the collection of BIA site maps and associated data (including oral histories). We know the locations of sites, their layouts, and the sizes and configurations of many surface features. The work by Corbett, West, and Lefèvre (2001) comparing western Aleutian house forms to those in the Rat Islands is an excellent example of the use of this data. Additional work can and should be done with these materials, especially in light of information garnered over the last twenty years from the eastern Aleut region regarding variability in Aleut house size and complexity.

For example, it seems increasingly probable that the zone near the boundary between the eastern and central Aleutians was one where, in recent precontact times at least, Aleut social, political, and economic systems transitioned between an eastern Aleutian pattern—having a large and dense population, a high degree of social ranking or even stratification (reflected in part in larger and more complex house forms), and intensive economic procurement—to a central and western Aleutian pattern of lower population size and density, more egalitarian social systems (again, reflected in part in smaller, simpler house types), and less intensive food procurement. The BIA site data may prove helpful in elucidating this hypothesis.

Conclusions

It is not surprising that, while Alaskan archaeology had its beginnings there, very little anthropological or archaeological research has been conducted in the central Aleutian Islands. The size and isolation of the region contribute to substantial financial and logistical constraints for archaeological endeavors, particularly for the kinds of large-scale efforts necessary to address the issues discussed above. Except for the most general views regarding their initial occupation, fundamental maritime adaptations, and site locations, almost everything awaits future research. The central Aleutians will probably never shed light on the origins of maritime adaptations in the North, nor is the region likely to reveal the precontact cultural complexity that is now being seen in the eastern archipelago. However, while some broad cultural patterns are known, the area remains little understood archaeologically, and the opportunities for exploring how maritime adaptations were maintained and the ways in which people responded to (and were participants in) changing environmental regimes are wide open. Likewise, all other aspects of precontact and postcontact Aleut culture—social and political organization, religion, etc.—are worthy of archaeological study. We have a lot to learn.

References

Bank, Theodore P., II. n.d. [Various unpublished maps, in possession of the author.]

Bank, Theodore P., II. 1952, A Preliminary Account of the University of Michigan Aleutian Expeditions, 1950-1951. *The Asa Gray Bulletin* 1(3), 211-218. Ann Arbor.

Bank, Theodore P., II. 1953a. Cultural Succession in the Aleutians. *American Antiquity* 19(1), 40-49.

Bank, Theodore P., II. 1953b. Ecology of Prehistoric Aleutian Village Sites. *Ecology* 34(2), 246-264.

Bank, Theodore P., II. 1956. *Birthplace of the Winds*. New York, Thomas Y. Crowell Co.

Bank, Theodore P., II, Albert C. Spaulding, H. A. Miller, and Janet F. Bank. 1950. The University of Michigan Expedition to the Aleutian Islands, 1948-49. Preliminary Report to the Office of Naval Research, Department of the Navy. Botanical Gardens and Museum of Anthropology, University of Michigan, Ann Arbor.

Berge, Anna. 2010. Origins of Linguistic Diversity in the Aleutian Islands. *Human Biology* 82(5-6), 557-581.

Bergsland, Knut. 1959. Aleut Dialects of Atka and Attu. *Transactions of the American Philosophical Society* 49(3), Philadelphia.

Bergsland, Knut. 1994. *Aleut Dictionary*: Unangam Tunudgusii. Alaska Native Language Center, University of Alaska, Fairbanks.

Bergsland, Knut and Moses L. Dirks (editors). 1990. *Unangam Ungiikangin kayux Tunusangin/ Unangam Uniikangis ama Tunuzangis/Aleut Tales and Narratives, Collected 1909-1910 by Waldemar Jochelson*. University of Alaska, Fairbanks, Alaska Native Language Center.

Black, Lydia T. 1984. *Atka: An Ethnohistory of the Western Aleutians*. Kingston, Ontario, Limestone Press.

Cook, John P., E. James Dixon, Jr., and Charles E. Holmes. 1972. Archaeological Report, Site 49 Rat 32, Amchitka Island, Alaska. *USAEC Report HN-20-1045* Las Vegas, Holmes and Narver.

Corbett, Debra G., Douglas Causey, Mark Clementz, Paul L. Koch, Angela Doroff, Christine Lefèvre, and Dixie West. 2008. Aleut Hunters, Sea Otters, and Sea Cows: Three Thousand Years of Interactions in the Western Aleutian Islands, Alaska. In Torben C. Rick and Jon M. Erlandson (eds.), *Human Impacts on Ancient Marine Ecosystems: A Global Perspective*, 43-75. Berkeley and Los Angeles, University of California Press.

Corbett, Debra G., Christine Lefèvre, Thomas J. Corbett, Dixie West, and Douglas Siegel-Causey. 1997a. Excavations at KIS-008, Buldir Island: Evaluation and Potential. *Arctic Anthropology* 34(2), 100-117.

Corbett, Debra G., Christine Lefèvre, and Douglas Siegel-Causey. 1997b. The Western Aleutians: Cultural

Isolation and Environmental Change. *Human Ecology* 25(3), 459-479.

Corbett, Debra G., Douglas Siegel-Causey, and Christine Lefèvre. 1995. Update and Preliminary Results of the Western Aleutian Archaeological Project. Paper presented at the annual meetings of the Alaska Anthropological Association, Anchorage.

Corbett, Debra G., Dixie West, and Christine Lefèvre. 2001. Prehistoric Village Organization in the Western Aleutians. In Don Dumond (ed.), *Archaeology in the Aleut Zone of Alaska, Anthropological Papers of the University of Oregon* 58, 251-266. Eugene, University of Oregon Press.

Corbett, Debra G., and Caroline L. Funk. 2010. Uncovering the Past: Rat Island. Web page: http://alaska.fws.gov/nwr/akmar/whatwedo/uncover/ratisland.htm

Crockford, Susan, Gay Frederick, Rebecca Wigen, and Iain McKecknie. 2004. Analysis of the Vertebrate Fauna from Amaknak Bridge UNL-50, Unalaska, AK. Report prepared for the Museum of the Aleutians, Unalaska, AK and Alaska Department of Transportation], Anchorage, AK. Pacific Identifications, Inc., Victoria, B.C., Canada.

Dall, William H. 1877. On Succession in the Shell-heaps of the Aleutian Islands. In: Tribes of the Extreme Northwest. (*Contributions to North American Ethnology* 1), pp. 41-91. Washington, D.C., U.S. Government Printing Office.

Davis, Brian L. 2001. Sea Mammal Hunting and the Neoglacial: An Archaeofaunal Study of Environmental Change and Subsistence Technology at Margaret Bay, Unalaska. In Don Dumond (ed.), *Archaeology in the Aleut Zone of Alaska, University of Oregon Anthropological Papers* 58, 71-85. Eugene, University of Oregon Press.

DePew, Alan, and Daniel Thompson. 2009. *Cultural* Resource Monitoring of Water System Improvements for the Department of Environmental Conservation Village Safe Water Program in Atka, Alaska. Office of History and Archaeology Alaska Department of Natural Resources, Anchorage.

Desautels, Roger J., Albert J. McCurdy, James D. Flynn, and Robert R. Ellis. 1971. Archaeological Report, Amchitka Island, Alaska, 1969-1970. *US Atomic Energy Commission Report TID-25481*. Archaeological Research, Incorporated.

Domning, Daryl P. 1972. Steller's Sea Cow and the Origin of North Pacific Aboriginal Whaling. *Syesis* 5, 187-189.

Domning, Daryl P., James Thomason, and Debra G. Corbett. 2007. Steller's Sea Cow in the Aleutian Islands. *Marine Mammal Science* 23(4), 976-983.

Dumond, Don E. 2001. Toward a (Yet) Newer View of the (Pre)history of the Aleutians. In Don Dumond (ed.), *Archaeology in the Aleut Zone of Alaska, University of Oregon Anthropological Papers* 58, 289-309. Eugene, University of Oregon Press.

Frohlich, Bruno. 2002. Aleutian Settlement Distribution on Adak, Kagalaska, Buldir and Attu Islands. Aleutian Islands, Alaska. In B. Frohlich, A. B. Harper, and R. Gilberg (eds.), *To the Aleutians and Beyond: The Anthropology of William S. Laughlin*, 63-88. Ethnographical Series vol. 20, Copenhagen, The National Museum of Denmark.

Frohlich, Bruno and David Kopjanski. 1976. Aleut Site Survey, 1975: Preliminary Report. Submitted to the Aleut Corporation, Anchorage.

Funk, Caroline. 2010. Rats and Birds: Tracking Ecological Change with Evidence from Prehistoric to Historic Aleut Village Midden Test Excavations. Rat Island, Alaska. Report to US Fish and Wildlife Service, Anchorage.

Gard, Leonard M., Jr. 1977. Geologic History. In M. L. Merritt and R. G. Fuller (eds.), *The Environment of Amchitka Island, Alaska*, 13-34. National Technical Information Service, US Department of Commerce.

Gard, Leonard M., Jr. 1980. The Pleistocene Geology of Amchitka Island, Alaska. *Geological Survey Bulletin* 1478, Washington, DC, Government Printing Office.

Guggenheim, Paul. 1945. Anthropological Campaign on Amchitka. *The Scientific Monthly* 61(July), 21-32.

Hanson, Diane K. 2009. *Archaeological Survey of Adak Island in 2007 and 2008*. Report to the US Fish and Wildlife Service and the Aleut Corporation, Anchorage, Alaska.

Hanson, Diane K. and Debra G. Corbett. 2010. Shifting Ground: Archaeological Surveys of Upland Adak Island, the Aleutian Islands, Alaska and Changing Assumptions of Unangan Land Use Patterns. *Polar Geography* 33, 165-178.

Hanson, Diane K., Debra G. Corbett, David P. Staley, and Kimberly Fleming. 2008a. Upland Sites in the Aleutian Archipelago. Paper presented at the 73rd Annual Meeting of the Society for American Archaeology, Vancouver, British Columbia.

Hanson, Diane K., David P. Staley, Debra G. Corbett, and Kimberly Fleming. 2008b. Upshore Sites on Central Aleutian Islands. Paper Presented at the 35th Annual Meetings of the Alaska Anthropological Association, Anchorage, Alaska.

Hoffman, Brian W. 1999. Agayadan Village: Household Archaeology on Unimak Island, Alaska. *Journal of Field Archaeology* 26, 147-161.

Hrdlička, Aleš. 1941a. Exploration of Mummy Caves in the Aleutian Islands. Part I. Previous Knowledge of Such Caves. *The Scientific Monthly* 52, 5-23.

Hrdlička, Aleš. 1941b. Exploration of Mummy Caves in the Aleutian Islands. Part II. Further Exploration. *The Scientific Monthly* 113-130.

Hrdlička, Aleš. 1945. *The Aleutian and Commander Islands and Their Inhabitants*. Philadelphia, The Wistar Institute of Anatomy and Biology.

Jochelson, Waldemar. 1912. Scientific Results of the Ethnological Section of the Riabouschinsky Expedition of the Imperial Russian Geographical Society to the Aleutian Islands and Kamchatka. *Proceedings of the 18th International Congress of Americanists*, London, pp. 334-343.

Jochelson, Waldemar. 1925. *Archaeological Investigations in the Aleutian Islands*. Publication 367. Washington, Carnegie Institution of Washington.

Jochelson, Waldemar. 1933. *History, Ethnology and Anthropology of the Aleut*. Publication 432. Washington, Carnegie Institution of Washington.

Knecht, Richard A., Richard S. Davis, and Gary A. Carver. 2001. The Margaret Bay Site and Eastern Aleutian Prehistory. In Don Dumond (ed.), *Archaeology in the Aleut Zone of Alaska, University of Oregon Anthropological Papers* 58, 35-69. Eugene, University of Oregon Press.

Kohlhoff, Dean W. 2002. *Amchitka and the Bomb: Nuclear Testing in Alaska*. Seattle, University of Washington Press.

Ladd, Carol, George L. Hunt, Jr., Calvin W. Mordy, Sigrid A. Salo, and Phyllis J. Stabeno. 2005. Marine Environment of the Eastern and Central Aleutian Islands. *Fisheries Oceanography* 14 (Suppl.), 22-38.

Laughlin, William S., and Gordon H. Marsh. 1951. A New View of the History of the Aleutians. *Arctic* 4(2), 74-88.

Lightfoot, Kent G. 1995. Culture Contact Studies: Redefining the Relationship Between Prehistoric and Historical Archaeology. *American Antiquity* 60(2), 199-217.

Lobdell, John E. 1990. A Cultural Resources Assessment of the Proposed Relocatable Over-the-Horizon Radar (ROTHR) Installation, Amchitka Islands, Aleutian Archipelago, Alaska. Report produced for Dames and Moore, Ltd.

Loring, Stephen. 1993. Research in the Western Aleutian Islands. *Arctic Studies Center Newsletter* No.2, 4-5. Washington, D.C., Smithsonian Institution.

Loring, Stephen and Douglas W. Veltre. 2003. Aleut Archaeology and Cultural Heritage: The Legacy of the Jesup North Pacific Expedition. In L. Kendall and I. Krupnik (eds.), *Contributions to Circumpolar Anthropology* 4, 307-318. Washington, DC, Arctic Studies Center, National Museum of Natural History, Smithsonian Institution.

Luttrell, Mark, and Debra Corbett. 2000. Archaeological Investigations on Northeast Adak Island, Alaska. Report on file US Fish and Wildlife Service, Anchorage.

Maschner, Herbert D.G. 1999. Prologue to the Prehistory of the Lower Alaska Peninsula. *Arctic Anthropology* 36(1-2), 84-102.

Maschner, Herbert D.G., James W. Jordan, Brian W. Hoffman, and T. M. Dochat. 1997. The Archaeology of the Lower Alaska Peninsula. Laboratory of Arctic and North Pacific Archaeology, Report 4. Madison, University of Wisconsin.

McCartney, Allen P. 1977. Prehistoric Human Occupation of the Rat Islands. In M.L. Merritt and R.G. Fuller (eds.), *The Environment of Amchitka Island, Alaska*, 59-113. National Technical Information Service, US Department of Commerce.

McCartney, Allen P. 1984. Prehistory of the Aleutian Region. In D. Damas (ed.), *Arctic, Handbook of North American Indians*, vol. 5, 119-135. Washington, D. C., Smithsonian Institution.

McCartney, Allen P., and Douglas W. Veltre. 1999. Aleutian Island Prehistory: Living in Insular Extremes. *World Archaeology* 30(3), 503-515.

McCartney, Allen P., and Douglas W. Veltre. 2002. Longhouses of the Eastern Aleutian Islands, Alaska. In B. Frohlich, A. B. Harper, and R. Gilberg (eds.), *To the Aleutians and Beyond: The Anthropology of William S. Laughlin*, 249-265. Ethnographical Series vol. 20. Copenhagen, The National Museum of Denmark.

Miraglia, Rita A. 2006. Tukusngaluga: The Chief's Gathering Place. Paper presented at the annual meeting of the Alaska Anthropological Association, Kodiak.

Mobley, Charles M. 1996. Cultural Resource Investigations at Kiska, Little Kiska, and Semisopochnoi, Aleutian Islands, Alaska. Report to Dames and Moore, Anchorage, Alaska.

Mobley, Charles M. 2003. Atka Village Hydroelectric Project, Atka Island, Aleutian Islands, Alaska: Archaeological Inventory. Report prepared by Charles M. Mobley and Associates, Anchorage, under contract to CRW Engineering Group, Anchorage, for Alaska Energy Authority, Anchorage.

Mobley, Charles M. 2006. Atka Water and Sewer Improvements: Archaeological Survey, Atka Island, Alaska. Report prepared by Charles M. Mobley and Associates, Anchorage, Alaska, under contract to HDR, Alaska, Inc., Anchorage, for the City of Atka and the Alaska Village Safe Water Program, Anchorage.

Nelson, Willis H., and Frank Barnett. 1955. A Burial Cave on Kanaga Island, Aleutian Islands. *American Antiquity* 20(4), 387-392.

O'Leary, Matthew. 1998. New Site No. 2 ("Tutiakoff Site"). Unpublished report on file at US Department of the Interior, Bureau of Indian Affairs, Alaska Native Claims Settlement Act Projects Office, Anchorage.

O'Leary, Matthew. 2001. Volcanic Ash Stratigraphy for Adak Island, Central Aleutian Archipelago. In Don Dumond (ed.), *Archaeology in the Aleut Zone of Alaska*, Anthropological Papers of the University of Oregon 58, 215-233. Eugene, University of Oregon Press.

O'Leary, Matthew. 2007. A Roster of BIA ANCSA Radiocarbon Dates. *Alaska Journal of Anthropology* 5(1), 123-145.

Pratt, Kenneth L. 1992. Documenting Alaska Native Cultural History: ANCSA and the Role of the Bureau of Indian Affairs. *Arctic Research of the United States*, Vol. 6, 74-77. Arlington, Virginia.

Pratt, Kenneth L. 2009. A History of the ANCSA 14(h)(1) Program and Significant Reckoning Points, 1975-2008. In Kenneth L. Pratt (ed.), *Chasing the Dark: Perspectives on Place, History and Alaska Native Land Claims*, 3-43. Anchorage, Bureau of Indian Affairs, Alaska Region.

Reynolds, Georgeie L. 1986. Atka Small Boat Harbor Cultural Resources Reconnaissance. US Army Corps of Engineers, Alaska District report submitted to the Alaska Office of History and Archaeology, Anchorage, Alaska.

Rodionov, Sergie N., James E. Overland, and Nicholas A. Bond. 2005. Spatial and Temporal Variability of the

Aleutian Climate. *Fisheries Oceanography* 14 (Suppl. 1), 3-21.

Savinetsky, A.B., N. Kiseleva, and B. Khasanov. 2004. Dynamics of Sea Mammal and Bird Populations of the Bering Sea Region Over the Last Several Millenia. *Palaeogeograpy, Palaeoclima-tology, Palaeoecology* 209, 335-352.

Sense, Richard, and Christy G. Turner, II. 1970. Catalog of Archaeological Site Survey Records, Amchitka Island, Alaska. Holmes and Narver, Inc., On Continent Test Division. Report submitted to United States Atomic Energy Commission.

Stejneger, Leonhard. 1887. How the Great Northern Sea Cow (*Rytina*) Became Extinct. *American Naturalist* 21(12), 1047-1054.

Thorson, Robert M., and Thomas D. Hamilton. 1986. Glacial Geology of the Aleutian Islands (Based on the Contributions of Robert F. Black). In T. D. Hamilton, K. M. Reed, and R. M. Thorson (eds.), *Glaciation in Alaska: The Geologic Record*. 171-191. Anchorage, Alaska Geological Society.

Turner, Christy G., II. 1970. Archaeological Reconnaissance of Amchitka Island, Alaska. *Arctic Anthropology* 7(2), 118-128.

Veltre, Douglas W. 1977. A Report on the Potential Archaeological Impact of a Proposed US Public Health Service Water and Sewer Project in the Village of Atka, Alaska. Report for US Public Health Service, Anchorage.

Veltre, Douglas W. 1979. Korovinski: The Ethnohistorical Archaeology of an Aleut and Russian Settlement on Atka Island, Alaska. Ph.D. Dissertation, Department of Anthropology, University of Connecticut, Storrs.

Veltre, Douglas W. 1990. Perspectives on Aleut Culture Change During the Russian Period. In B. S. Smith and R. J. Barnett (eds.), *Russian America: The Forgotten Frontier*, 175-183. Tacoma, Washington State Historical Society.

Veltre, Douglas W. 1997. An Archaeological Survey of Selected Sites on Northern Adak Island, Aleutian Islands, Alaska. Submitted to the Aleut Corporation, Anchorage.

Veltre, Douglas W. 1998. Prehistoric Maritime Adaptations in the Western and Central Aleutian Islands, Alaska. *Arctic Anthropology* 35(1), 223-233.

Veltre, Douglas W. 2001. Korovinski: Archaeological and Ethnohistorical Investigations of a Pre- and Post-Contact Aleut and Russian Settlement on Atka Island. In Don Dumond (ed.), *Archaeology in the Aleut Zone of Alaska, Anthropological Papers of the University of Oregon* 58, 187-213. Eugene, University of Oregon.

Veltre, Douglas W., and Melvin A. Smith. 2010. Historical Overview of Archaeological Research in the Aleut Region of Alaska. *Human Biology* 82(5-6), 487-506.

Veniaminov, Ivan. 1984. *Notes on the Islands of the Unalashka District*. Translated by Lydia Black and R. H. Geoghegan. Kingston, Ontario, The Limestone Press.

West, Dixie, Christine Lefèvre, Debra Garland Corbett, and Arkady Savinetsky. 1999. Radiocarbon Dates from the Near Islands, Aleutian Islands, Alaska. *Current Research in the Pleistocene* 16, 83-85.

Whitmore, Frank C., Jr., and Leonard M. Gard, Jr. 1977. Steller's Sea Cow (*Hydrodamalis gigas*) of Late Pleistocene Age from Amchitka, Aleutian Islands, Alaska. *Geological Survey Professional Paper* 1036. Washington, D C, Government Printing Office.

Chapter 3
Did Holocene Paleoenvironmental Factors Affect Ancient Aleut Occupation and Settlement in the Central Aleutian Islands?

Lyn Gualtieri, Brenn Sarata, Mitsuru Okuno, and Dixie West

Abstract: *A geomorphological study of two central Aleutian Islands, Adak and Kanaga, was undertaken in order to investigate the possible environmental effects on ancient Aleut occupation during the Holocene. The environmental changes that may have affected the ancient Aleuts are general climate changes associated with warming and cooling events, tectonic uplift due to seismic activity, and volcanic eruptions. Multiple, thick layers of tephra cover most of the central Aleutian Islands. Tephra may have altered both local fresh and saltwater chemistry such that the Aleuts' nutritional resources were altered. Tectonic uplift most likely temporarily altered the availability and types of marine resources of food. Five terraces (2-3m; 5-6m; 7-10m; 13-15m; and 30-40m) are recognized across the central Aleutian Islands, though dating the terraces is difficult due to the lack of datable material. The absence of marine sediment on the terraces indicates that the terraces may have resulted from tectonic emergence. Radiocarbon and optically stimulated luminescence (OSL) dating techniques were used to establish a chronology of uplift on the Clam Lagoon spit and terraces around Adak Island. Emergence of the spit at Clam Lagoon on Adak Island was complete before 2045 ± 290 years and probably sooner. Our results confirm the interpretations of a complex general Holocene geologic history evidenced on the other Aleutian Islands and the Alaska Peninsula to the east. This study documents field evidence for paleoenvironmental change in the central Aleutian Islands throughout the Holocene, though none of the changes seem to have been marked enough to significantly affect the migration and/or demise of the ancient Aleuts.*

Keywords: *Aleutian Islands, geomorphology, Adak Island, Clam Lagoon, terraces*

Introduction

The question of how ancient peoples adapted to their environment is one that continues to intrigue researchers regardless of the cultures' geographic locations and time periods. Ancient peoples probably adapted to, even capitalized on, the changes in the climate, the waterways and food resources. Studies have indicated that past environmental changes must have been simply too abrupt, catastrophic or extreme for people to adapt, but more often than not the changes were gradual and steady, allowing for cultural adaptation.

Black (1976) first approached the question of the climatic effects on ancient Aleuts by looking at paleo temperature and wind proxies from the Aleutian Islands. He concluded there were minimal effects of glaciation on the Aleuts and no major effects on their food resources. Our study further tested the relationship between environmental factors and Aleut occupation and migration in the central Aleutian Islands by relating the geomorphology and chronology of paleoenvironmental change to patterns of humans on Adak Island.

The goal of this paper is to synthesize research that was conducted between 2005-2007 on Adak and Kanaga Islands (Figure 3.1) in conjunction with the Central Aleutians Archaeology and Paleobiology Project. This chapter summarizes the paleoenvironmental research, and in particular the chronology of uplift and emergence on northern Adak Island. In the past, geologic records have not been tied specifically to archaeological records in the central Aleutians Islands. The original impetus for this research was to investigate what caused the demise of the north Adak Aleut culture—was it environmental factors?

Knowing the geomorphology of the island is essential to place human history in context and to understand how changes in the physical environment affected the lives of humans in the central Aleutians. Paleoenvironmental data often provide a baseline for our understanding of the environment prior to Aleut occupation. Currently little is known about the Holocene paleoenvironmental history of the central Aleutian Islands. More is known about other areas of southern Alaska, where changes in sea level due to isostatic recovery and seismic activity most likely significantly affected the migration and settlement pattern of ancient populations.

Regional and Local Geology

Adak and Kanaga Islands are part of the active Aleutian volcanic arc, a series of volcanic islands that span 2500km of the North American plate margin where the Pacific Plate is being subducted. Paleozoic sedimentary and volcanic rocks are found on some Aleutian Islands, although most of the islands are composed of Mesozoic and early Tertiary volcanic sequences capped with Quaternary aged rocks

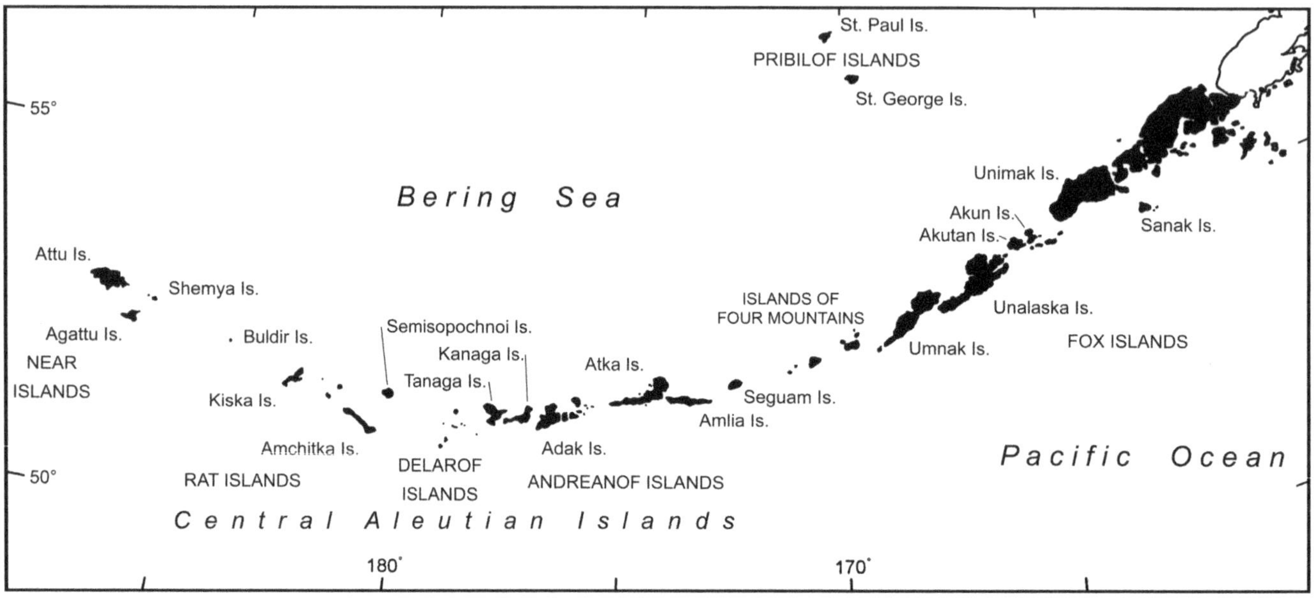

Figure 3.1. The location of Adak and Kanaga Islands within the central Aleutian Islands.

and sediments. Because of their location along an active plate margin, the central Aleutian Islands continue to be seismically active.

The geology and geomorphology of the central Aleutian Islands was described in early studies, but there was little absolute chronologic control. The first description of the geomorphology of Adak Island was published by Bradley (1948). His survey noted the joint and fault patterns characteristic of the south, wave cut terraces and benches, till, ash and soil sequences, dunes, glacial drift, beach ridges and local glaciation features of central and northern Adak Island. Bradley astutely recognized that the modern coastline had been submerged. He deduced the coast had recently emerged, probably due to elastic rebound after deglaciation/ice unloading or deeper crustal movements. Coats (1956) was the first to map the bedrock geology of Adak Island. His contributions include mapping the distribution of the pre-Quaternary aged Finger Bay volcanics, named after an outcrop on the north shore of Finger Bay on Adak Island. This unit, which consists of tuff-breccia with gabbro and rhyolite intrusions, is the dominant underlying bedrock of Adak Island. The rocks have been highly deformed and hydrothermally altered. Coats (1956) found evidence for glaciation in the deeply dissected southern highlands and near Mt. Moffett, a remnant volcanic cone that is Adak Island's highest topographic feature. That discovery led Coats to believe that volcanic summits were the source of the glaciers on Adak Island.

Our study focused on the abundant Holocene deposits on northern Adak Island, and north-central and northeastern areas of Kanaga Island because of known archaeological sites, and practical considerations (permitting and conservation zones). Waythomas (1995) mapped and described the unconsolidated glacial, pyroclastic, colluvial, eolian, beach lacustrine, alluvial and other deposits of Adak Island. Glacial deposits are rare on the islands, as most of the Pleistocene glacial sediment was deposited offshore and is now underwater. Suites of ash, pumice and soil mantle most of northern Adak and Kanaga Islands. O'Leary (2001) described the stratigraphy of the principle ashes and their relationships to the archaeology on Adak Island in detail. Kiriyanov and Miller (1997) published the chemical compositions and chronology of the ashes on Adak Island. Chapter 4 of this volume (Okuno *et al.* 2011) describes the origins and distribution of the ashes as they pertain to the archaeological sites around Clam Lagoon (CL) on northern Adak Island. The sand dunes fringing Clam Lagoon as described by Judson (1946) post-date all glacial deposits.

Previous Work

Marine deposits, terraces and wave-cut platforms have been identified on Adak and Kanaga Islands, but their ages are uncertain. On other Aleutian Islands many of the marine terraces have been faulted or displaced by different phenomena (eustatic sea level rise, isostatic rebound from glaciers, tectonic activity) causing the local sea level to rise and fall in the Aleutians (Jordan 2001). Just as ages of the marine deposits of the Aleutian Islands are not certain, the Holocene sea level history is, in many cases unknown or, where known, it varies greatly from island to island. There is evidence for both sea level rise (up to 5m) and fall during the mid Holocene (Black 1974). An additional problem with previous studies is that many radiocarbon dates from organic material interbedded with glacial and/or marine deposits were uncorrected for the reservoir effect, which is variable and could be quite large in the Aleutian Islands.

Thorson and Hamilton (1986) concluded that the lowlands of Adak Island might have been glaciated in the early Holocene, followed by valley glaciation in the mid-

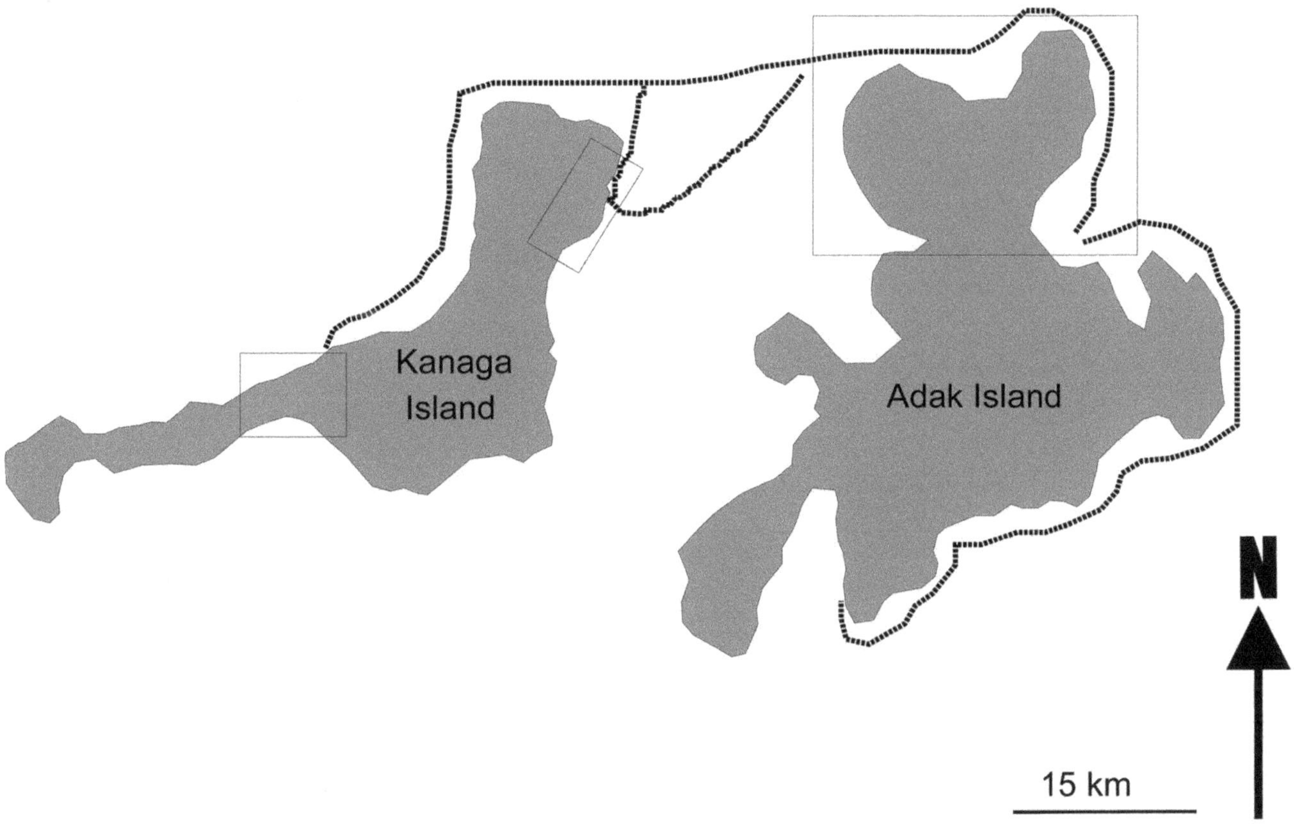

Figure 3.2. Adak and Kanaga Islands. Rectangles indicate focused areas of study. Dashed lines indicate boat surveys.

Holocene and cirque glacier advance in the late Holocene. More recently, Waythomas (1995) mapped the surficial geology of northern Adak Island and determined that that part of the island was deglaciated by 11ka; glaciers remained on Mt. Moffett during the Holocene. Although Waythomas (1995, 3) described and mapped glacial deposits, he states that 'the age of the glacial deposits on Adak is not precisely known.'

Methodology

The study area is shown in Figure 3.2. The northern part of Adak Island was explored by driving on publicly accessible roads and on foot. As shown in Figure 3.2, the geomorphology of the eastern and southern coastal areas of the island was explored in a one day boat survey. The research team travelled to Kanaga Island by private boat, set up camps and travelled over the island on foot. One camp was established on the north central part of the island (2006), and another camp was on the northeastern part of the island in 2007. Travel on foot was restricted both summers due to overgrown vegetation, swampy areas, and persistent fog, wind and rain.

Organic material for AMS radiocarbon dating was pretreated using the standard acid base: acid treatment at the University of Washington and analytical analyses were performed at the NSF Arizona AMS Facility. Nancy Bigelow, University of Alaska, identified seeds and plant material. Optically-stimulated luminescence samples were collected using 5-6cm diameter PVC pipe (15-20cm long) on quartz rich medium sands. Samples were analyzed at the University of Illinois Luminescence Dating Research Laboratory. Jeffrey Snyder, Bowling Green State University, analyzed diatoms. Terrace heights were measured using a Vaisala hand held altimeter-barometer using the local mean high tide as sea level (0m).

Study Area

Adak Island is a logistically appealing field area because Adak has an airport serviced by Alaska Airlines twice a week. However, during the first season of fieldwork in 2005 we learned that much of the island's surface had been altered by the military during and since WWII, when the Army and Navy first established operations there. In particular, much of landscape around Clam Lagoon has been altered. The first task for the geology team was to figure out what was disturbed and what was natural on the north side of Adak Island. We asked a local resident who also served on Adak in the 1970's to give us a tour. He pointed out areas that were disturbed and described the military activities at each location.

The second priority was to examine a nearby island not disturbed by military activity. In 2006 and 2007 we visited Kanaga Island, a nearby island west of Adak (Figure 3.1). The southwestern peninsula of Kanaga Island is a relatively low elevation, grassy, undulating platform with ephemeral streams. Shallow lakes dot the highest elevations. The lakes

Figure 3.3. Meltwater channels cut into bedrock. Finger Bay, Adak Island.

are underlain by perched clay layers. Tephra exposures can be found where streams or lake banks have eroded unconsolidated material. Bedrock was not found on the southwest peninsula, although angular volcanic ejecta were found in association with some tephra sections exposed along lake banks.

Paleoenvironmental Changes

Volcanic Events

It is apparent from the ubiquitous volcanic deposits that abundant Holocene volcanic activity affected the landscape, soil and water chemistry of Adak and Kanaga Islands. The most common volcanigenic sediment on Adak and Kanaga Islands is tephra. Tephra blankets the landscape and is usually seen in sequences with soil. There is usually a clear boundary of charred organic peat below the tephra. Three prominent tephra layers (the Main Ash, Intermediate Ash, and the Sandwich Ash), and two less distinctive ash layers (YBO Ash, 40-Year Ash) are present on the islands. Thin (1cm-thick) ash layers are also present in many sections on both Adak and Kanaga Islands. An ideal section for the complete tephra/soil stratigraphy was found at Palisades Lake on the northern part of Adak Island and is described below:

40-Year – 3cm thick, clayey, olive, no lithics.

YBO – 5.5cm thick, tan, clayey, no lithics.

Sandwich – Base is fine-grained. Red to orange. Upper part consists of blue-cheese like chunks (up to 1 cm in diameter) that are felsic with mafic inclusions. Often occurs as 4 distinct layers of clay+silt/lapilli/fine sand/lapilli.

Intermediate- ± 25cm thick, most variable ash, sometimes occurs as 8-9 layers and displays a flow-like structure, grayish, fine-grained with sub-angular clasts (7-8mm long axis.). This ash is possibly the result of a pyroclastic flow or lahar.

Main –± 25cm thick, but most uniform thickness of all ashes, orange color, clasts present. A plant stem underlying the Intermediate tephra at Palisades Lake yielded an age of 5670 ± 49 ^{14}C years (PL78, Table 3.1).

Volcanigenic sediment other than tephra was observed on the northern Adak Island coast west of Adagdak Point. The 2m thick volcanigenic sediment was most likely a lahar or debris flow from an active Adagdak volcano. The tephra section at the Loran Station is also an excellent type section for the northern Adak Island tephra/soil sequence.

Field number	*Laboratory number*	*Lat/Long*	*Material*	*^{14}C age radiocarbon years*
PL78	AA69158	51°54.743' N.,	plant stem	5670 ± 49
		176°36.050 W.		
CL94	AA69160	51°57.352' N.,	Carex or Polygonum seeds	846 ± 38
		176°33.804' W.		
CL46	AA69161	51°56.748' N.,	Graminoid stem bits	180 ± 130
		176°34.037' W.		
CL105	AA71231	51°56.748' N.,	Graminoid stem bits	213 ± 51
		176°34.037' W.		

Table 3.1. Radiocarbon ages for Adak Island material.

Here we observed the lahar/debris flow unit that we had not seen anywhere else on the island. The unit lies above the Intermediate and below the Sandwich Ash layers, is correlative with the lahar/debris flow unit at Adagdak Point, and can be traced across the Loran Station terrace stratigraphy site.

The thickness and distribution of tephra was recorded because the tephra is sure to have been important in Aleut life. Tephra raining down from the sky might not have been enough to drive away the Aleuts, but it likely affected the water chemistry and the Aleuts' aquatic food resources. In an informal experiment at Candlestick Bridge, Adak Island we found that some of the larger tephra clasts float on the water for a few minutes while some sink immediately. How did this affect the marine and freshwater shellfish? How long did this effect last and would it have been enough to dramatically alter the aquatic populations? Black (1981) proposed that volcanic activity led to periods of hardship amongst the Aleuts. Black cited examples of habitat destruction due to volcanic and tectonic processes on eastern Aleutian Islands and non-Aleutian islands when writing about the effects of volcanism on Aleutian ecology. Specifically she cites habitat disruptions such as saltwater intrusion into freshwater sources, alteration of the geography of the coastline, change in seawater salinity and temperature, overproduction of plankton, fish kills, tsunamis, slides and rock falls damming lakes and rivers, gaseous emissions and chemical pollution as potential consequences of volcanic and/or tectonic activity. Although quite possible, we did not find direct field evidence in the geological record for any of these events significantly affecting the Aleuts on Adak Island during the Holocene.

Climatic Change Events

The most likely model for glaciation in the central Aleutian Islands is that of local ice caps coalescing within island groups and possibly a floating ice shelf (Mann and Hamilton 1995). After a major climate transition between 11-9ka, most islands were ice free by 9ka (Thorson and Hamilton 1986) and experienced a mid-Holocene warm period from 8-6ka in some areas and from 6-4ka in others (Drewes *et al.* 1961). After 4k, the climate was generally cooler and wetter. A Neoglacial advance occurred on some islands and

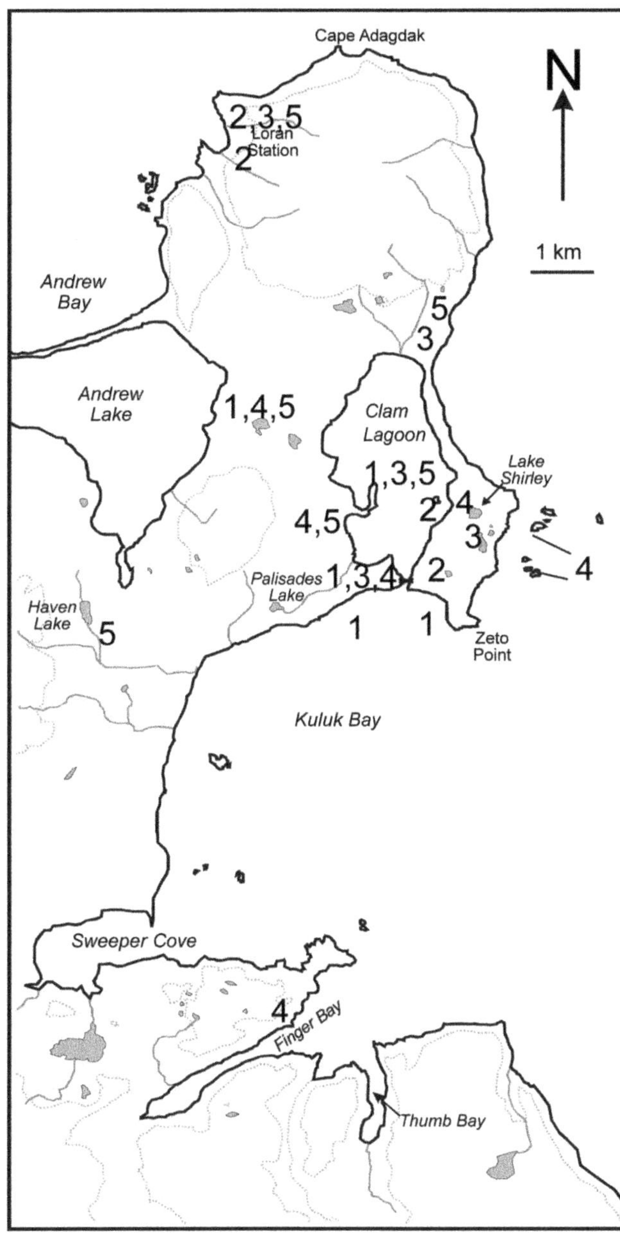

Figure 3.4. Summary of terrace locations and elevations for northern Adak Island. 1= 1st terrace, 2= 2nd terrace, 3 = 3rd terrace, 4 = 4th terrace, 5 = 5th terrace.

Figure 3.5. Terraces at the Loran Station on Adak Island.

on the Alaska Peninsula around 3ka (Black 1976). Holocene sea level rose to present by 7.5ka on Adak Island (Thorson and Hamilton 1986), pre-dating known Aleut occupation.

In the Finger Bay area on Adak Island we observed bedrock joints, striated bedrock, meltwater channels, uplifted terraces, small U-shaped valleys and an ice-scoured landscape (Figure 3.3). At some locations sub-angular pebbles and cobbles in a clay/silt matrix underlie a 2m-thick soil-tephra sequence. The underlying sediment is most likely a till. At Palisades Lake, southwest of CL we also identified till-like sediment underlying the Main Ash. The sediment was composed of sub-angular cobbles of varying lithologies and diameters up to 15cm.

Seismic/Emergence/Terrace Events

We identified five sets of terraces on Adak and Kanaga Islands (Figure 3.4). The elevations of the terraces are given as ranges, and ages of the terraces are difficult to assess because there was no dateable material on the terraces themselves. Examples of three of the five terraces are found at the Loran Station on Adak Island and can be seen in Figure 3.5.

Investigation Areas

Clam Lagoon Spit, Adak Island

One of the original project goals was to constrain spit emergence and the formation of CL, thus changing the food resources and coastal environment for the Aleuts (Figure 3.6). Although the spit has been altered by the military to create a landing strip, disturbance is restricted to only the uppermost material and to the central interior part of the spit. Sections were described and sampled on the inner and outer, undisturbed parts of the spit. The rocks that comprise the spit are well-rounded and of varying igneous lithologies typical of beach deposits. We did not observe angular or moraine-like beach deposits as suggested by Waythomas *et al.* 1994.

Inner spit, 51° 56.748'N, 176° 34.037'W

(Site 1 on Fig. 3.6)

Figure 3.7 shows the detailed upper and lower spit stratigraphy. Based on two distinct peat horizons, which overlie two distinct cobble layers, we believe there were two emergence events. Optically stimulated luminescence

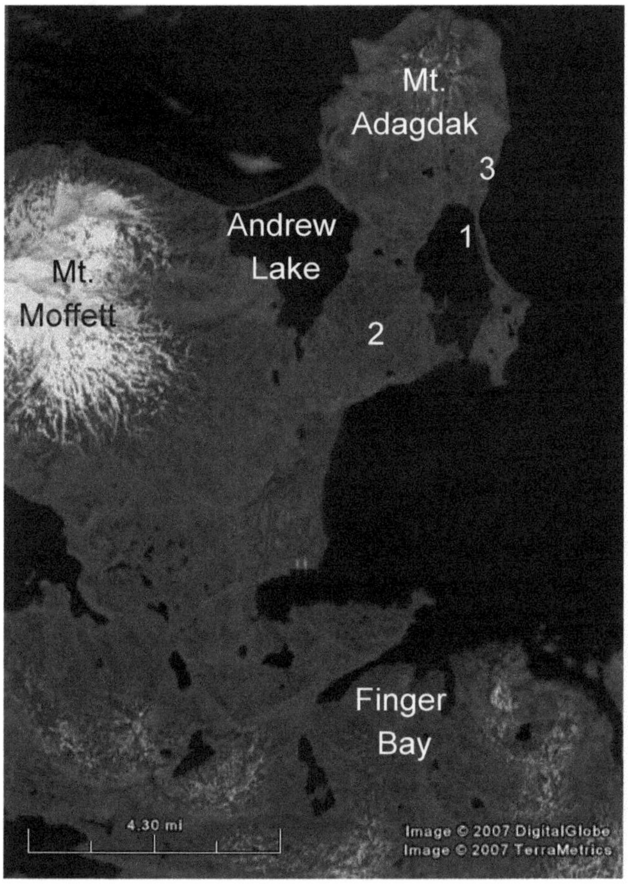

Figure 3.6. Adak Island site locations described in text. 1 = Spit site, 2 = Clam Lagoon marsh site, 3 = Blue Clay site.

dating of the sandy sediments was not successful. Nonetheless, a plant stem was collected in 2005 from the lowermost peat horizon and it yielded an age of 180 ± 130 ^{14}C years (CL46, Table 3.1). A radiocarbon date on grass on the same peat horizon was collected in 2006 and yielded an age of 213 ± 51 ^{14}C years (CL105, Table 3.1). We believe the ash in the upper part of the section to be either the 40 Year or YBO. The sand dunes on the southern interior margin of the spit were investigated for tephra and organic material, but neither was found.

Outer spit, 51° 57.352'N, 176° 33.804'W

(Site 3 on Fig. 3.6)

Blue clay was observed at the base of the 7.8m terrace on the seaward side of the spit's northern end (between archaeological sites ADK 012 and ADK 192). See Figure 3.8 for the outer spit stratigraphy.

Our initial interpretation in the field was that the blue clay was emergent marine sediment. However, the diatom species in the clay indicate freshwater and possibly slightly brackish conditions (see list below analyzed by Jeffrey Snyder). A seed from within the clay was dated and yielded an age of 846 ± 38.^{14}C years (CL94, Table 3.1). Clay from the base of the section yielded an OSL age of 1345 ± 220 years (CL113, Table 3.2). We believe that the OSL dates are not (greatly) affected by the phenomenon of anomalous fading in samples that contain volcanigenic feldspars that cause low or 'too young' OSL ages (Wintle 1973), because corresponding radiocarbon ages are also 'young'.

Diatoms in Blue Clay from CL spit

Caloneis bacillum (Grunow) Cleve

Gomphonema acuminatum Ehrenberg

Fragilaria exigua Grunow

Pinularia viridis (Nitzsch) Ehrenberg

Eunotia praerupta Ehrenber

Nitzschia cf. vermicularis (Kutzing) Hantzsch

Hantzschia amphioxys (Ehrenberg) Grunow

Pinnularia borealis Ehrenberg

Marsh pit, 51° 55.065'N, 176° 35.076'W

(Site 2 on Figure 3.6)

The Marsh pit, which lies 3.2m asl, represents the time the area landward of the CL spit emerged. An OSL sample from the basal medium to coarse sand in the Marsh pit yielded an age of 2045 ± 290 years (CL111, Table 3.2). This date gives a minimum date of emergence.

Terraces on Adak and Kanaga Islands

Multiple terraces were observed on many Aleutian Islands. Although only Adak and Kanaga Islands were studied in detail, observations of other islands were made via boat in 2002. Due to the lack of marine sediment on the terraces we interpret them to be the result of seismic uplift. Some terrace surfaces are composed of bedrock capped by tephra and soil. In some places (Finger Bay and Palisades Lake) till underlies the tephra/soil sequence. Local variations in terrace height are due to dune formation and vegetation patterns. The main question regarding the terraces that we were trying to answer was: did the surfaces uplift and then accumulate the tephra/soil sequence, or did they uplift after the tephra/soil sequence was in place?

There are a series of terraces around the northern part of Adak Island. They are present at:

1st terrace: 2-3m
2nd terrace: 5-6m
3rd terrace: 7-10m
4th terrace: 13-15m
5th terrace: 30-40m

Field number	Laboratory number	Lat/Long	Fraction	Equivalent	U	Th	K2O	Cosmic dose	Dose rate	OSL age
			(μm)	dose (Gray)a	(ppm)b	(ppm)b	(%)b	(mGray/yr)c	(mGray/yr)	(yr)d
AD99	UIC1840	51°58.864' N.,	150-250	1.18 ± 0.36	1.9 ± 0.1	2.5 ± 0.1	0.79 ± 0.02	0.20 ± 0.02	1.31 ± 0.05	805 ± 270
		176°37.187' W.								
CL111	UIC1835	51°55.065' N.,	150-251	3.27 ± 0.45	1.9 ± 0.1	2.5 ± 0.1	1.00 ± 0.02	0.19 ± 0.02	1.47 ± 0.06	2045 ± 290
		176°35.076' W.								
CL113	UIC1838	51°57.352' N.,	150-252	2.99 ± 0.42	2.0 ± 0.1	4.7 ± 0.1	1.5 ± 0.02	0.18 ± 0.018	2.19 ± 0.09	1345 ± 220
		176°33.804' W.								

aQuartz grains analyzed under blue-light excitation (470 ± 20 nm).
bU, Th and K20 content analyzed by inductively coupled plasma-mass spectrometry at Activation Laboratory LTD, Ontario, Canada.
cFrom Prescott and Hutton 1994.
dAges are calculated from reference year 2000. All errors are at 1 sigma.

Table 3.2. OSL ages for Adak Island sediments.

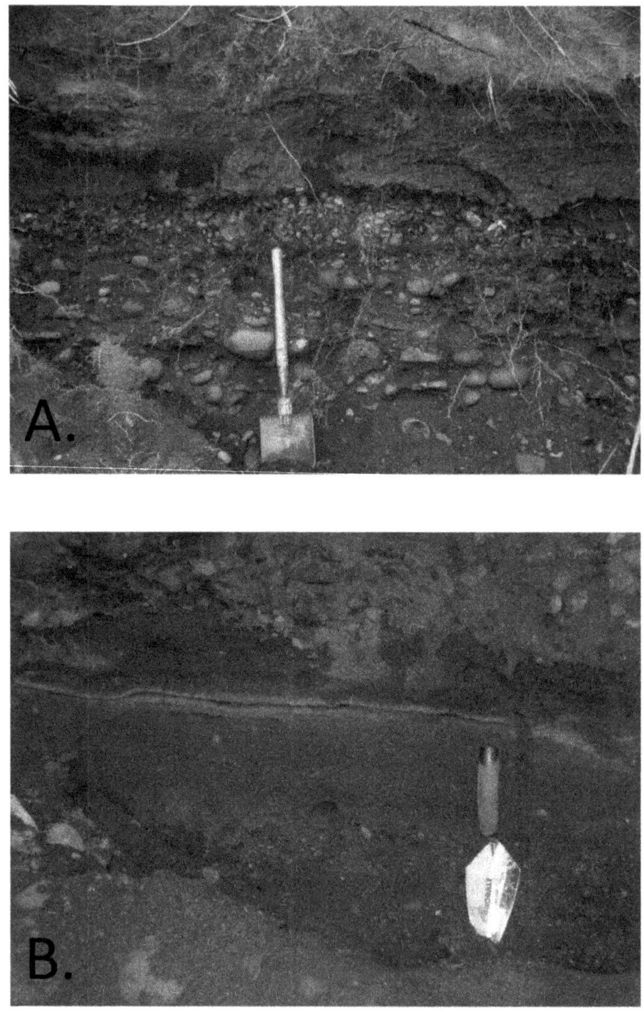

Figure 3.7. Clam Lagoon Spit stratigraphy.
A = upper, B = lower.

For the most part, terraces are not well preserved on the eastern and southern parts of Adak Island. The eastern coastline of Adak Island along Kagalaska Strait and south to the Bay of Waterfalls is mostly rocky. Exceptions include a 10m terrace at Adak Bight and Boot Bay as well as a 13m terrace at Chapel Cove/Bay of Waterfalls. Possible terraces (15-30m) on Kagalaska Island were observed by boat. Terrace data on northern Adak and Kanaga Islands are summarized in Table 3.3 and also displayed on Figure 3.4.

An OSL age of 805 ± 270 years (AD 99, Table 3.2) was obtained from a sand unit from the base of the 6.0m terrace at the Loran Station. The terrace is composed of well-rounded beach cobbles and sand at the base, which is covered by a lithic layer containing subangular fragments of volcanigenic sediment. An ash (possibly YBO) is present in the upper part of the section. In 2005 an attempt was made to date organic material by ^{14}C however, no organic material remained after the samples were washed during preparation. Emergence of the 6.0m terrace (2nd terrace) was dated to 805 years (OSL).

Figure 3.8. Blue Clay site (3m asl) at north end of Clam Lagoon.

Jordan (2001) described a series of high sea level stands on the western Alaska Peninsula following the Last Glacial Maximum (LGM). The youngest high sea level stands are at 2.1 and 0.6-1.8ka ago, with sea levels at 6m and 2-3m asl, respectively. The difference between the western Alaska Peninsula and the central Aleutian Islands during the LGM is the extent of ice. There is no evidence for crustal depression in the central Aleutian Islands simply because there was a lack of large ice sheets, whereas, the western Alaska Peninsula experienced significant crustal loading at the height of the local LGM. Although Jordan describes high sea level stands and terraces associated with them, he also cites 'sudden tectonic displacements' playing a role in the reorganization of waterways in the late Holocene. Jordan (2001) attributes a gap in the archaeological record at approximately 2.1ka to a transgression caused by tectonic influence. Although such an/this abrupt tectonic event may have affected those living in the western Alaska Peninsula, we found no field evidence that similar events affected the Aleuts in the same way on Adak.

The question as to whether tsunamis generated by earthquakes along the seismically active Aleutian margin could have led to the demise of Aleut populations has been raised by some researchers. Given that tsunami deposits have been identified in the Aleutian Islands, we hoped to address this question. A tsunami that would significantly and adversely affect an Aleut community, would most likely be widespread over a large area and not seen at just one location. Most often tsunamis deposits feature a sheet-like layer of sand in strata that fine away from the coastline. Although they vary based on location and tsunami origin, tsunami deposits can include sand, sediment eroded from adjacent beaches, rip-up clasts, fish remains, benthic foraminifera, and poorly sorted boulders (Pinegina and Bourgeois 2001; Dawson and Stewart 2007). A problem inherent in identifying tsunami deposits is that they are often indistinguishable from storm deposits. There may be tsunami deposits on Adak Island, but we did not see any deposits which could be definitively attributed to tsunamis rather than any other possible source. We do not think tsunamis had a major impact on Aleut life in the central Aleutian Islands.

Conclusions

Our research documents geologic evidence of volcanic activity, local climate change events and land emergence/shoreline displacement in the central Aleutian Islands in the mid to late Holocene. These environmental changes may have been both slowly and episodically they surely did not go unnoticed to the Aleuts. Instead, it is likely the changes invited the Aleuts to adapt and innovate, just as modern cultures may choose in respond to current environmental changes. Additionally, this research contributed numerical dates constraining the emergence of the Clam Lagoon spit on Adak Island and the 'second highest'/5-6m terrace.

Our initial hypothesis was that when the spit closed and isolated CL from the ocean, the food resources of the ancient Aleuts would have changed significantly, which could have led to a change in settlement patterns, or even to migration. Our research suggests that this was not the case. It seems that the spit emerged and closed off Clam Lagoon sometime before 2045 ± 290 years ago, and that the change had little effect on the Aleuts living on Adak. This study suggests that major environmental factors can be 'ruled out' as the causes of the migration or demise of the Aleuts in the central Aleutian Islands in the Holocene.

Acknowledgements

This research was supported under the National Science Foundation's Arctic Social Sciences Program, Award # 0353092 to Lyn Gualtieri and Award # 0353065 to Dixie West. Permission to work on Adak and Kanaga Islands was granted by the Aleut Corporation and the Alaska Maritime National Wildlife Refuge. We thank the US Fish and Wildlife Service as well as VECO Polar Resources for logistical field support. Radiocarbon samples were measured at the University of Arizona AMS Laboratory and OSL samples were measured at University of Illinois. Al Giddings, Rod Whitehead and Dan Nystrom provided boat support for island surveys. We also thank the City of Adak, Alaska and its citizens for welcoming us into their community during the summers of 2005-2007.

References

Black, Lydia T. 1981. Volcanism as a Factor in Human Ecology: The Aleutian Case. *Ethnohistory* 28 (4), 313-340.

Black, Robert F. 1976a. Influence of Holocene Climatic Changes on Aleut Expansion into the Aleutian Islands, Alaska. *Anthropological Papers of the University of Alaska* 18-1 pp. 31-42.

Black, Robert F. 1976b. Pre-Quaternary Glacial Events, Aleutian Islands, Alaska. In D. J. Easterbrook and V. Sibrava (eds.), *Quaternary Glaciations in the Northern Hemisphere*, I.U.G.S.-U.N.E.S.C.O. International Geological Correlation Programme Project No. 73/1/24, Report No. 3, pp. 285-301. [publ. Geological Survey of Czechoslovakia, Prague].

Bradley, Charles C. 1948. Geologic Notes on Adak Island and the Aleutian Chain, Alaska. *American Journal of Science* 246, 214-220.

Coats, Robert R. 1956. *US Geologic Survey Bulletin* 1028-C p. 45-67. 1 sheet, 1:50000.

Dawson, Alastair G. and Iain Stewart. 2007. Tsunami Deposits in the Geological Record. *Sedimentary Geology* 200, 166-183.

Drewes, Harald, G. D. Fraser, G. L. Snyder, and H. F.Barnett. 1996. Geology of Unalaska Island and Adjacent Insular Shelf, Aleutian Islands, Alaska. *US Geological Survey Bulletin* 1028-S, 583-676.

Jordan, James W. 2001. Late Quaternary Sea-Level Change in Southern Beringia: Postglacial Emergence of the Western Alaska Peninsula. *Quaternary Science Review* 20(1), 509-523.

Judson, Sheldon. 1946. Late-Glacial and Postglacial Chronology on Adak. *Journal of Geology* 54(6), 376-385.

Kiriyanov, Vladimir Y. and Thomas P. Miller. 1997. Volcanic Ashes of Adak Island, Aleutian Islands, Alaska. *Volcanology and Seismology* 19, 65-78.

Mann, Daniel H. and T. D. Hamilton. 1986. Late Pleistocene and Holocene Paleoenvironments of the North Pacific Coast. *Quaternary Science Reviews* 14(5), 449-471.

Okuno, Mitsuru, Keiji Wada, Toshio Nakamura, Lyn Gualtieri, Brenn Sarata., Dixie West and Masayuki Torii. 2011. Holocene Tephra Layers on Northern Adak Island, West-central Aleutian Islands, Alaska. Chapter 4 in D. West, V. Hatfield, E. Wilmerding, C. Lefèvre, and L. Gualtieri (eds.). *The People Before: The Geology, Paleoecology and Archaeology of Adak Island, Alaska.* Oxford, British Archaeological Reports.

O'Leary, Matthew. 2001. Volcanic Ash Stratigraphy for Adak Island, Central Aleutian Archipelago. In D. Dumond (ed.), *Archaeology in the Aleut Zone of Alaska, Some Recent Research*, 215-233. University of Oregon Anthropological Papers No. 58. Eugene, Oregon. University of Oregon Press.

Pinegina, Tatiana K. and Jody Bourgeois. 2001. Historical

and Paleo-tsunami Deposits on Kamchatka, Russia: Long-term Chronologies and Long-distance Correlations. *Natural Hazards and Earth System Sciences* 1, 177-185.

Prescott, John R. and J. T. Hutton. 1994. Cosmic Ray Contributions to Dose Rates for Luminescence and ESR Dating: Large Depths and Long-term Time Variations. *Radiation Measurements* 23, 497-500.

Thorson, Robert M. and Thomas D. Hamilton. 1986. Glacial Geology of the Aleutian Islands. In K. M. R. T. D. Hamilton and R. M. Thorson (eds.), *Glaciation in Alaska: The Geological Record*, 171-191. Anchorage, Alaska. Alaska Geological Society.

Waythomas, Christopher F. 1995. Surficial Geologic Map of Northern Adak Island, Alaska. US Geological Survey Open File Report 95-128.

Wintle, Ann G. 1973. Anomalous Fading of Thermoluminescence in Mineral Samples. *Nature* 244, 143-144.

Chapter 4
Holocene Tephra Layers on the Northern Half of Adak Island in The West-Central Aleutian Islands, Alaska

Mitsuru Okuno, Keiji Wada, Toshio Nakamura, Lyn Gualtieri, Brenn Sarata, Dixie West, and Masayuki Torii

Abstract: *Adak Island, Aleutian Islands, Alaska is covered with Holocene sequences of soil-tephra complexes. When identified and placed within a chronostratigraphic framework, these deposits are a useful, independent dating technique for archaeological sites on the island. Black (1976) thought that the three conspicuous tephra deposits (the Main, Intermediate and Sandwich) erupted from the Kanaga Island volcano. Waythomas et al. (2001), however, inferred that a possible source volcano of the tephras was Mount Moffett, Adak Island. We conducted a tephrostratigraphic study on the island to re-evaluate the chronological framework and to determine the sources of the tephras. The tephra deposits become thicker from south to north but not towards Mounts Moffet and Adagdak volcanoes. Intermediate tephra thickness and maximum diameter of lithic fragments (ML) on the northeastern part of the island suggests that the source for this tephra is a now submerged volcano nearby. Chemical compositions of volcanic glass in the tephras were determined by electron probe micro-analysis (EPMA). Radiocarbon dates were determined using accelerator mass spectrometry (AMS). Eruptions of the Intermediate, Sandwich, YBO and Forty Year tephra deposits were dated to approximately 7.2, 4.7, 3.6 and 0.4 cal ka BP, respectively.*

Keywords: *Adak Island, tephrochronology, dispersal maps, AMS radiocarbon dating, EPMA*

Introduction

Adak Island is a member of the Andreanof Islands in the Aleutian Islands, Alaska (Figure 4.1). Three volcanoes on northern Adak, Mount Moffett (1200m above sea level), Andrew Bay (350m above sea level) and Mount Adagdak (650m above sea level), formed during the Pleistocene (Alaska Volcano Observatory 1998; Coats 1956; Marsh 1990; Myers and Frost 1994). The oldest volcano, Andrew, is largely eroded and is filled with waters of Andrew Bay and Lake Andrew (Coats 1956; Myers and Frost 1994). The next youngest volcano, Mount Adagdak, formed through three major eruptive stages. Mount Moffet, the largest volcanic edifice, consists of both a main and a parasitic cone (Coats 1956; Myers and Frost 1994).

Throughout the Holocene, Adak has been covered by a sequence of soil-tephra complexes (Coats 1956). These complexes formed when tephra buried existing soil and vegetation, and subsequent soil formation processes began in the new tephra layer. Black (1976) described three conspicuous tephra deposits: the Main, Intermediate and Sandwich, and established the tephrostratigraphy and radiocarbon (^{14}C) chronology for the northern part of the island (Figure 4.1). He also suggested that these three tephra deposits did not originate from local Adak volcanoes, but were most likely derived from Kanaga volcano 30km to the west or Great Sitkin volcano 35km to the east of Adak Island (Figure 4.1: Black 1976, 1980; Kiriyanov and Miller 1997; Romick *et al.* 1992). Black (1976) also described two fine-grained tephra deposits: the YBO (Yellow-Brown-Olive) and Forty Year in the upper horizons of the Adak soil-tephra profiles. Both have compositions similar to analyzed tephras from historical eruptions of Great Sitkin volcano (Romick *et al.* 1992). Kiriyanov and Miller (1997) reported mineral and chemical compositions along with grain-size distribution for Adak deposits and identified more than 60 ash layers in the Holocene sequence. Using tephrostratigraphy with ^{14}C dates, Waythomas *et al.* (2001) hypothesized that the YBO tephra originated from Kanaga volcano and that Mount Moffett is a likely possible source of the Main, Intermediate and Sandwich tephras.

We began a tephrostratigraphic survey in 2005 with two goals in mind: 1) to provide a chronological framework for archaeological work on the island and, 2) to determine the distribution and sources of the major Adak tephras. Tephrochronology is a useful, independent dating technique for archaeological sites in the Aleutian Islands (O'Leary 2001). Geochemical analysis combined with stratigraphy and distribution patterns of tephra deposits can reveal source volcanoes.

Chemical compositions were analyzed using an Electron Probe Micro Analyzer (EPMA), and ^{14}C dates were obtained with Accelerator Mass Spectrometry (AMS). In 2007, we conducted a field survey at the northeastern coast of Kanaga Island to evaluate Black's (1976) interpretations of tephra sources. This paper presents the distribution maps, chemical compositions and ^{14}C ages for the major Adak tephra deposits and discusses their ages and potential sources in geological and archaeological contexts.

Tephrostratigraphic Survey and Sampling

Tephra occurrences and distributions were surveyed across

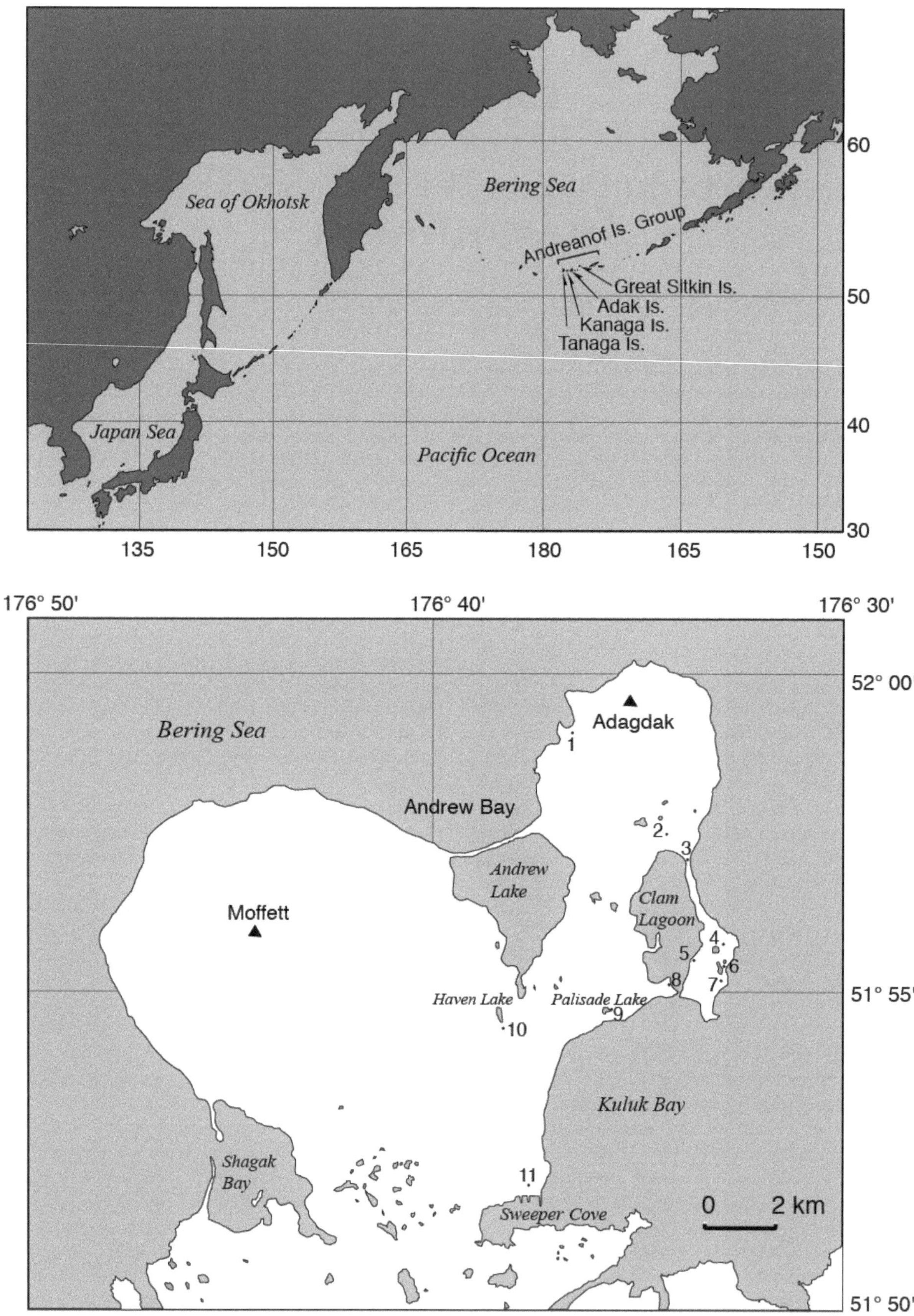

Figure 4.1. Index maps. (Top) Location of Adak and adjacent islands, (Bottom) Map showing northern part of Adak Island. Closed circles with numbers 1 to 11 indicate site localities described in this paper.

northern Adak. Samples were collected at numerous sites including archaeological sites (Figures 4.1-4.4) and Black's type section (Figure 4.5, Site 10) near Haven Lake (Black 1976; Heusser 1978). We adopted this section as the type section for chemical analysis by EPMA. The field characteristics of conspicuous tephra deposits allowed us to correlate our samples with previous descriptions (e.g., Black 1976; Kiriyanov and Miller 1997). This paper also includes a previously published section at Site 4 (Figure 4.2; Okuno *et al.* 2007, 2009). Figures 4.3 and 4.4 show the stratigraphic position of samples collected for EPMA analysis and AMS ^{14}C dating.

Typical Holocene soil-tephra sequences cover glacial till deposits at Site 4 (Figure 4.2; Okuno *et al.* 2007, 2009). This section is easily accessible and is located near the archaeological Sites 5 (ADK-171) and 6 (ADK-011) (Figure 4.4). The major tephra deposits at Site 4 are briefly described from oldest to youngest as follows (Okuno *et al.* 2009): The Main tephra-fall deposit covers multiple tephra-soil complexes (thickness = 14cm) of multiple, fine, thin tephra and soil (Waythomas *et al.* 2001) and glacial till deposits. The Main tephra-fall deposit is 30cm thick and consists of coarse pumice (maximum pumice size, MP = 1.5cm) and a consolidated ash layer as thick as 5cm in the upper part. A sequence of tephra-soil complexes between the Main and Intermediate tephras measures 48cm thick. The Intermediate tephra-fall deposit consists of pumice and lithic fragments with a cross-laminated ash bed. It is 43cm thick and contains a large amount of dense lithics with a maximum size (ML) of 3.8cm. The Sandwich tephra-fall deposit covers a 74cm thick sequence of tephra-soil complexes, is 20cm thick, and comprises alternating lenses of ash and pumice (MP is 0.9cm). As Kiriyanov and Miller (1997) noted, the thickness and lithology of the Sandwich tephra vary from section to section, and humus loams are intercalated in the tephra in some places. The ash and pumice beds of the Sandwich tephra infill the erosion gully in the humic soil layer, which contains the Sandwich pumice lapilli at Site 2 on the southern slope of Adagdak volcano (Figure 4.6). We attribute this humus layer to a contemporaneous reworking associated with the eruption of the Sandwich tephra and thus consider this humic layer as a pseudo-soil. Two fine grained tephra layers: the YBO (5cm thick) and Forty Year (6cm thick) are located as tephra-soil complexes between the Sandwich tephra and the ground surface (Figures 4.2 and 4.5).

A sequence of pumice and debris-rich layers covers the Intermediate tephra at the western foot of Adagdak volcano (Figure 4.1 Site 1). This stratigraphic sequence was not observed elsewhere on the island. Pumice clasts measure up to 30cm in diameter, and display surface textures resembling water-chilled pumice (Kano *et al.* 1996, Stewart and McPhile 2004). The ballistic dense clast makes an impact sag (Figure 4.7).

Although distinguishing primary tephra from reworked lenses is important in Adak tephra sequences (Thorson

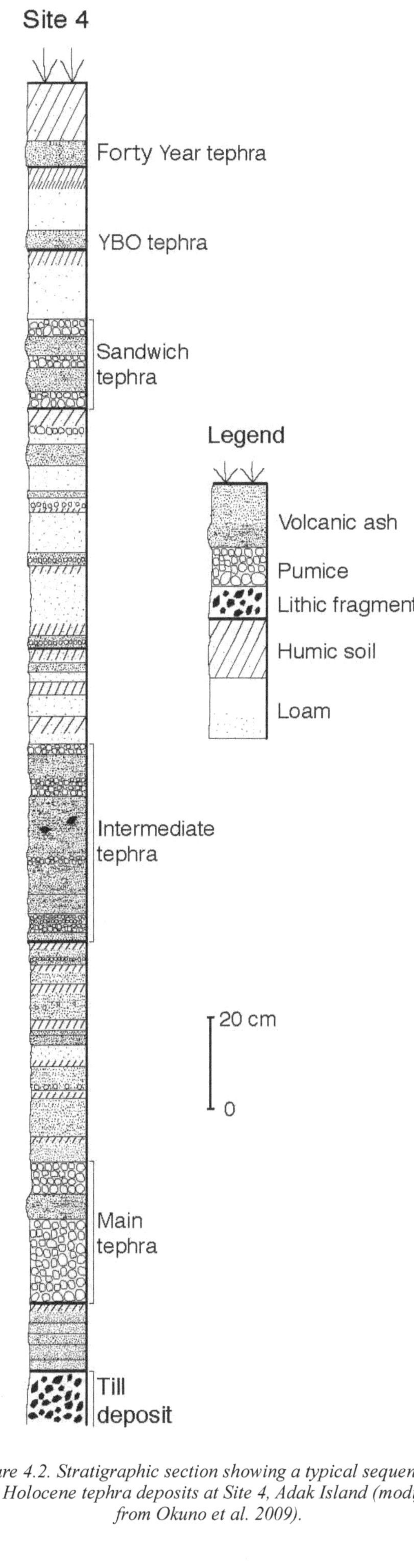

Figure 4.2. Stratigraphic section showing a typical sequence of the Holocene tephra deposits at Site 4, Adak Island (modified from Okuno et al. 2009).

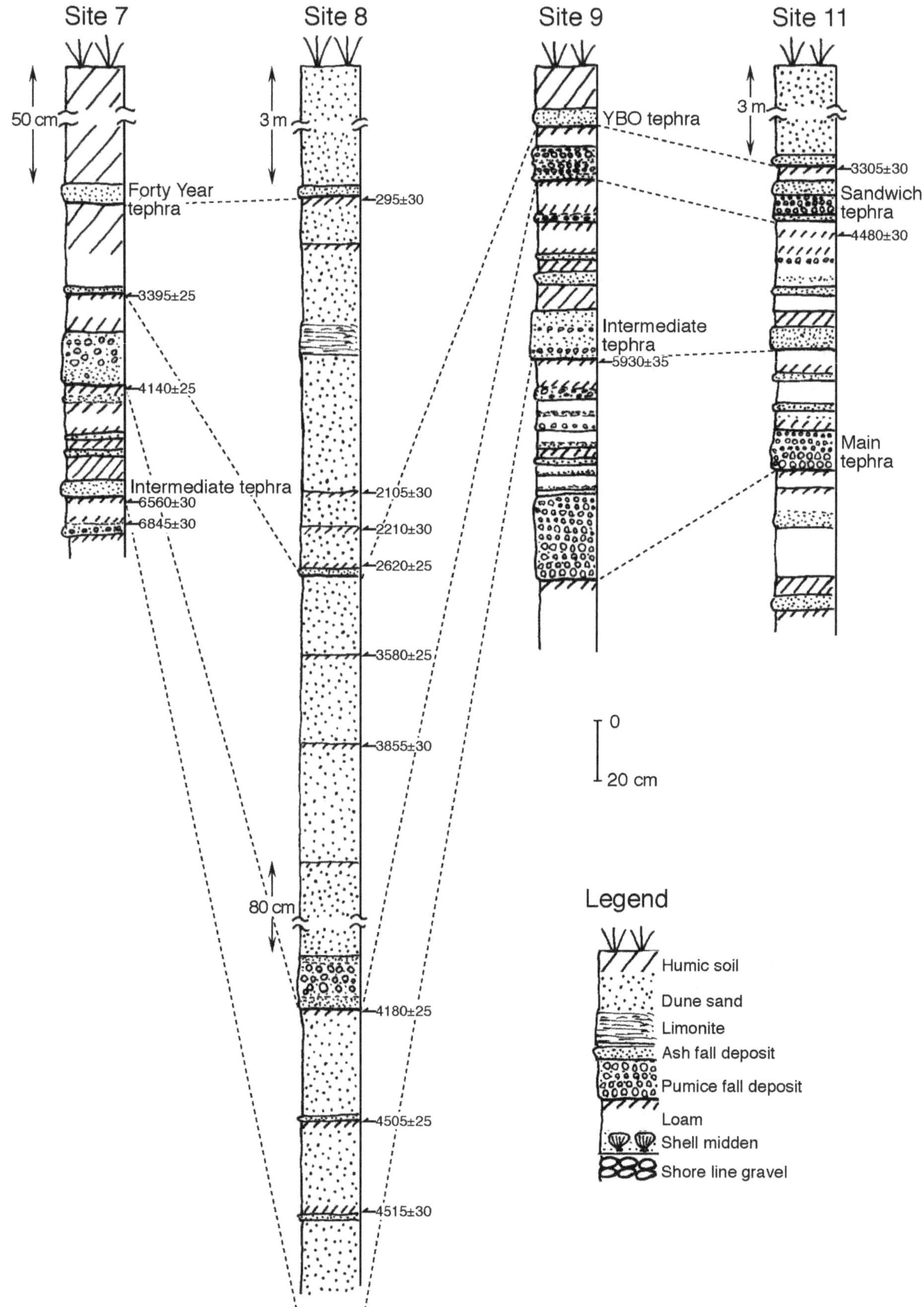

Figure 4.3. Stratigraphic sections of the primary Holocene sequences at Sites 7, 8. 9 and 11 on Adak Island. Arrows indicate sampling horizons for radiocarbon dating and geochemical analysis.

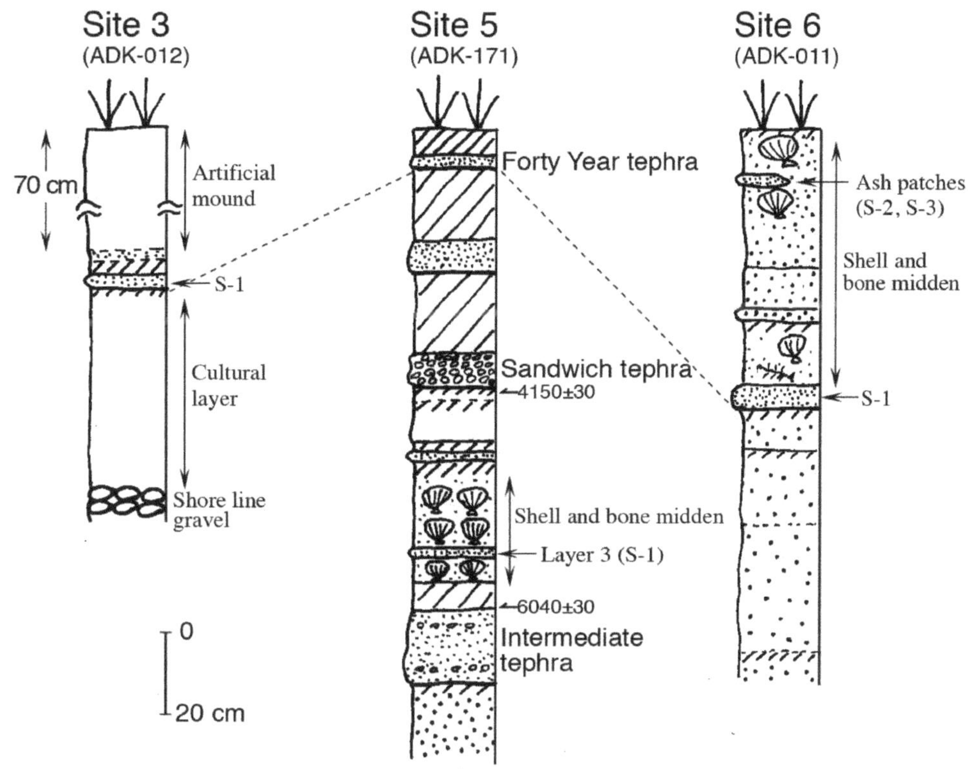

Figure 4.4. Stratigraphic sections at sampled archaeological sites 3, 5 and 6. Arrows indicate sampling horizons for radiocarbon dating and geochemical analysis.

and Hamilton 1986), it is difficult to do so in the field. This problem will be discussed in reference to the unnamed ash (Figure 4.4, Site 5, Layer 3, S-1) that is intercalated with a shell and bone midden (Okuno *et al.* 2007) at the archaeological Site 5 (ADK-171).

Radiocarbon dates of organic materials immediately below a tephra bed indicate its eruption age because these organics represent *in situ* vegetation at the time the ash fell (Okuno *et al.* 1997, Okuno and Nakamura 2003). We collected charcoal fragments from strata immediately below and when possible, above and between, tephra deposits (Figure 4.3) at Sites 1, 5, 7, 8, 9 and 11 (Figure 4.1).

The tephra sequence at Site 8, the southern entrance to Clam Lagoon, was particularly suited for accurate ^{14}C dating (Figures 4.1 and 4.3). The tephra-sand sequence is in a dune (Judson 1946), and the thick sand provides a barrier against penetration of roots from surface vegetation. Thus, living roots did not contaminate these samples (*e.g.*, Black 1976; Thorson and Hamilton 1986).

Experimental Procedures

Chemical analyses of plagioclases and volcanic glasses

The chemical compositions of the major elements in plagioclases and volcanic glasses of the representative tephra samples were determined by EPMA using a JEOL 8600 microprobe at Hokkaido University of Education at Asahikawa. Plagioclase phases were analyzed for major elements using a 15 KeV, 15 nA, focused beam and standard oxide ZAF (atomic number (Z), absorption (A) and fluorescence (F)) corrections. Considering the Na counting loss by beam irradiation, we chose the low probe currents (15 KeV, 8 nA) and a beam scanning an area of 5 to 10 micron-wide when analyzing the volcanic glasses. The accuracy and precision of these analyses were simultaneously checked by chemical standards of both obsidian glass (Astimex Scientific Ltd.) and dacite glass (Spfa-1 pumice erupted from Shikotsu caldera, Hokkaido, Japan). Both yielded values that were in good agreement with the recommended values (Wada *et al.* 2003).

Plagioclase, a group of silicate minerals within the feldspar family, is used to identify the composition and origin of volcanic rocks. In the mineral's crystal structure, sodium (Na) and calcium (Ca) atoms can substitute for one another. This substitution is used to define the type of plagioclase mineral. Plagioclase is typically defined by the general percentage of the Ca-rich anorthite (An) or the more Na-rich albite (Ab). The plagioclase compositions are presented by the An content = 100Ca/(Ca+Na).

AMS radiocarbon dating

Charcoal fragments and the humin fraction of a soil sample were purified by acid-alkali-acid (AAA) treatments

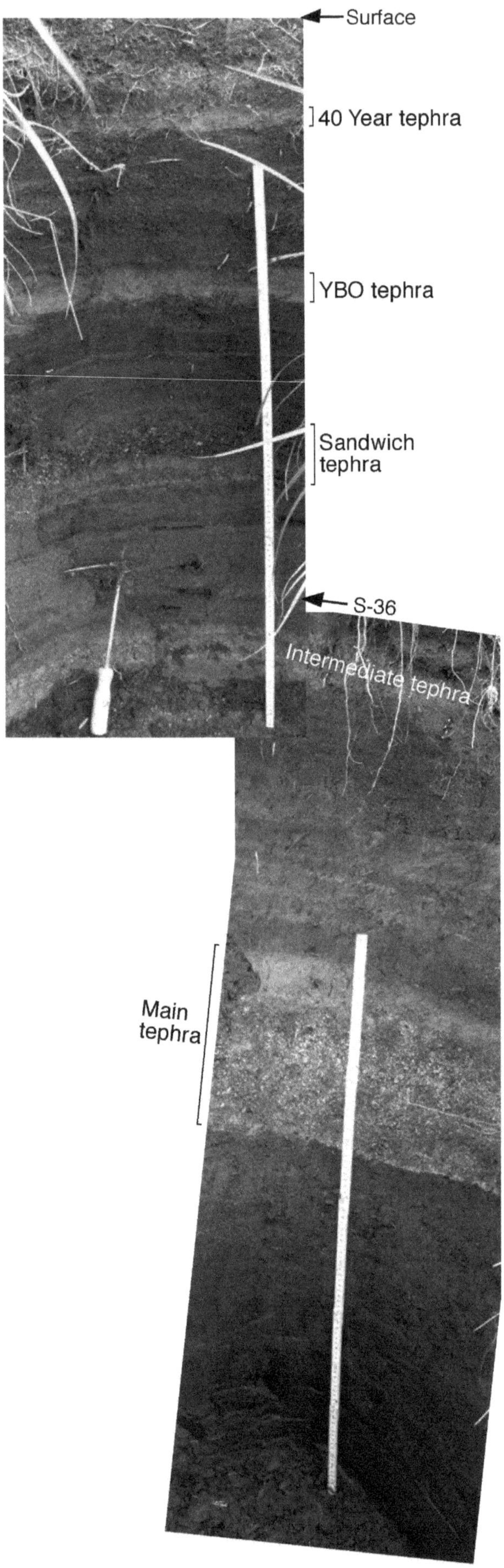

Figure 4.5. Mosaic of photographs showing the Holocene tephra sequence at Site 10, the type section from Black (1976). Each scale bar is 1m.

following the procedure by Okuno *et al.* (1997, 2001). The pretreated material was oxidized by heating with CuO at 850° C for 6 hours in a sealed Vycor® tube. The CO_2 that was produced was catalytically reduced to graphite on Fe-powder with hydrogen gas in a sealed Vycor® tube (Kitagawa *et al.* 1993). The amount of ^{14}C on the graphite targets was measured using an HVEE Tandetron AMS system at Nagoya University with NIST oxalic acid (HoxII) as standards (Nakamura *et al.* 2000). We corrected for carbon isotopic fractionation using the ^{13}C / ^{12}C ratio ($\delta^{13}C_{PDB}$). To estimate the ^{14}C background level, the ^{14}C concentration of commercial graphite powder (dead carbon) was also measured in the same analytic sequence of sample measurements.

Results and Discussion

1. Distribution of the Main, Intermediate and Sandwich tephra deposits on Adak Island

The Main, Intermediate and Sandwich tephra deposits tend to increase in thickness to the north but not toward Mounts Moffett and Adagdak, as is evident on the dispersal maps of Figures 4.8 – 4.10. Dispersal axes are not apparent on the dispersal maps. The thickest deposit with maximum ML area of the Intermediate tephra is located on the northeastern part of the island (Figure 4.9). We confirmed Waythomas *et al.*'s (2001) description of the tephra stratigraphy at the eastern foot of Kanaga volcano. Thus, it seems that the source vent for the Main, Intermediate, and Sandwich tephras was not the Kanaga volcano.

2. Chemical characteristics of tephra deposits and correlation

Table 4.1 shows the chemical compositions of the volcanic glasses in pumices and glass shards in fine-grained tephra deposits from the Adak sections (Figures 4.2 and 4.3). The data were normalized to a water-free 100% to compare with these chemical analyses.

Four samples from Black's (1976) type section at Site 10 (Figure 4.5) were analyzed by EPMA. The Main, Intermediate, and Sandwich tephras chemically resemble each other in that all display rhyolitic glass (SiO_2 = 77 – 78 weight percent). However, the glasses from the Sandwich tephra are slightly lower in weight percent SiO_2 than the glasses of the Main and Intermediate tephras (Figure 4.11 and Table 4.1). The Intermediate tephra glasses are slightly higher in weight percent TiO_2, FeO and K_2O and lower in weight percent CaO than those of the Main tephra. By contrast, the Forty Year tephra shows dacitic glass compositions (69 weight percent SiO_2).

Locality	**10**			**10**			**10**			**10**			**10**		
Tephra	**Forty Year**			**Sandwich**			**S-36**			**Intermediate**			**Main**		
	analyzed value	**100% - standarized**		**analyzed value**	**100% - standarized**		**analyzed value**	**100% - standarized**		**analyzed value**	**100% - standarized**		**analyzed value**	**100% - standarized**	
	average (N=19)	**average (N=19)**	**std. dev. (1s)**	**average (N=69)**	**average (N=69)**	**std. dev. (1s)**	**average (N=11)**	**average (N=11)**	**std. dev. (1s)**	**average (N=25)**	**average (N=25)**	**std. dev. (1s)**	**average (N=61)**	**average (N=61)**	**std. dev. (1s)**
SiO2	69,67	69,49	0,86	75,45	77,04	0,29	58,23	58,96	0,79	75,50	78,18	0,38	75,26	78,13	0,30
TiO2	0,49	0,49	0,06	0,22	0,23	0,04	0,96	0,97	0,10	0,22	0,23	0,04	0,18	0,18	0,03
Al2O3	15,36	15,32	0,36	12,54	12,80	0,13	17,36	17,58	1,25	11,78	12,20	0,20	11,95	12,41	0,18
FeO	3,56	3,55	0,42	1,33	1,36	0,14	6,81	6,89	0,98	1,15	1,19	0,13	1,09	1,13	0,13
MnO	0,12	0,12	0,07	0,10	0,10	0,06	0,17	0,18	0,06	0,09	0,09	0,05	0,09	0,10	0,06
MgO	0,86	0,86	0,13	0,29	0,30	0,03	1,77	1,79	0,35	0,22	0,23	0,04	0,22	0,22	0,02
CaO	2,92	2,91	0,35	1,55	1,59	0,07	7,10	7,19	0,93	1,24	1,29	0,08	1,36	1,41	0,08
Na2O	4,40	4,39	0,27	3,55	3,63	0,15	3,95	4,00	0,45	3,40	3,52	0,19	3,38	3,51	0,19
K2O	2,58	2,57	0,20	2,70	2,76	0,15	2,19	2,22	0,31	2,76	2,86	0,19	2,59	2,68	0,14
Cl	0,29	0,29	0,03	0,19	0,20	0,03	0,23	0,23	0,04	0,19	0,20	0,03	0,21	0,22	0,03
Total	**100,25**	**100,00**		**97,93**	**100,00**		**98,76**	**100,00**		**96,56**	**100,00**		**96,33**	**100,00**	

Locality	**5**			**6**			**6**			**6**		
Sample	**S-1**			**S-1**			**S-2**			**S-3**		
	analyzed value	**100%-standarized**		**analyzed value**	**100%-standarized**		**analyzed value**	**100%-standarized**		**analyzed value**	**100%-standarized**	
	average (N=9)	**average (N=9)**	**std. dev. (1s)**	**average (N=23)**	**average (N=23)**	**std. dev. (1s)**	**average (N=16)**	**average (N=16)**	**std. dev. (1s)**	**average (N=19)**	**average (N=19)**	**std. dev. (1s)**
SiO2	77,40	78,12	0,75	69,37	69,58	0,97	69,63	69,53	0,40	68,87	68,69	0,89
TiO2	0,25	0,25	0,05	0,50	0,50	0,06	0,49	0,49	0,05	0,54	0,54	0,06
Al2O3	12,02	12,13	0,46	15,32	15,37	0,74	15,24	15,22	0,28	15,66	15,62	0,51
FeO	1,23	1,24	0,21	3,61	3,63	0,37	3,70	3,69	0,22	3,87	3,86	0,34
MnO	0,07	0,08	0,06	0,12	0,12	0,06	0,12	0,12	0,06	0,12	0,12	0,06
MgO	0,23	0,23	0,09	0,85	0,86	0,19	0,90	0,90	0,10	0,96	0,95	0,14
CaO	1,52	1,53	0,27	3,05	3,06	0,37	2,98	2,97	0,22	3,15	3,14	0,30
Na2O	3,30	3,33	0,32	4,06	4,07	0,46	4,21	4,20	0,18	4,29	4,28	0,16
K2O	2,96	2,99	0,57	2,54	2,55	0,25	2,58	2,58	0,16	2,55	2,54	0,23
Cl	0,11	0,11	0,10	0,27	0,27	0,04	0,31	0,31	0,04	0,26	0,26	0,03
Total	**99,08**	**100,00**		**99,70**	**100,00**		**100,14**	**100,00**		**100,26**	**100,00**	

Table 4.1. Results of EPMA major-element analyses for Adak tephra deposits.

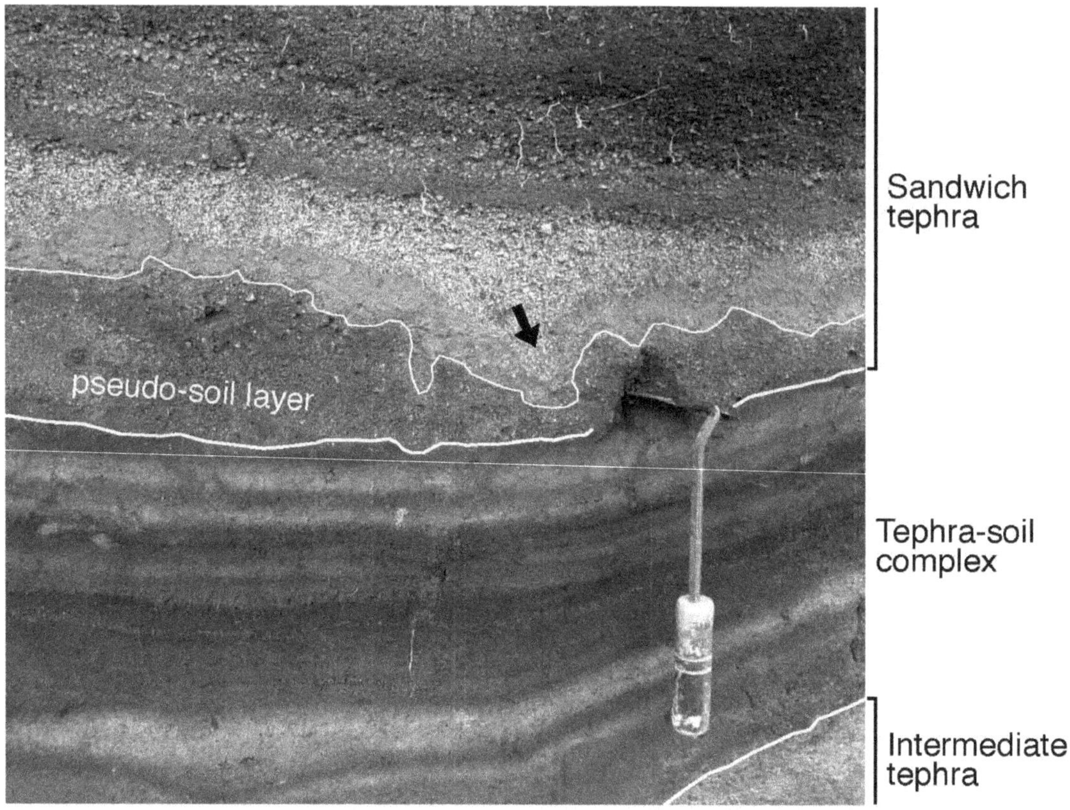

Figure 4.6. Photograph showing occurrence of the Sandwich tephra deposit at Site 2. Solid arrow indicates the erosion gully in the pseudo-soil layer. Length of tool is 27cm.

At archaeological Site 5, ADK-171-*Clinocardium* midden, (West *et al.* Chapter 1 and Savinetsky *et al.* Chapter 5 in this volume), a loosely packed cockle layer containing animal bones and human artifacts is separated by a thin lens of unidentified volcanic ash (Layer 3, S-1) from a more densely packed and highly deteriorated shell layer below (Figure 4.4). The chemical composition of this ash lens (78 weight percent SiO_2) is similar to that of the Intermediate tephra. On the other hand, glass shards from an unnamed ash layer (S-36), covering the peat and Intermediate tephra deposits at Site 10 show andesitic glass compositions (59 weight percent SiO_2) (Figure 4.11). The chemical compositions of the tephras at the archaeological Site 5 and those at Site 10 do not match. Glass composition obtained using microscopic particle analysis of the unknown ash from Site 5 midden confirms that the unknown ash lens is redeposited Intermediate tephra. This suggests that some unknown geological/climatic event redistributed the Intermediate tephra while Site 5 was occupied. This interpretation also implies that the Site 5 shell midden was created soon after the eruption of Intermediate tephra. The tephra-fall event could have destroyed most of surface vegetation on north Adak. A scarcity of vegetation and associated susceptibility to erosion may have enhanced the reworking of the Intermediate tephra while humans occupied the site. This result also implies that the thin ash layers in the tephra-soil complex may represent a reworked deposit.

Glass shards from ash patches (S-2 and S-3) at Site 6 (Figure 4.4) compare favorably in terms of composition to those of the Forty Year tephra in primary deposits, although the S-3 glasses differ slightly in SiO_2 content (Figure 4.11). We can identify these ashes as the Forty Year tephra.

The Sandwich, Intermediate, and Main tephras contain plagioclase crystals that consist mostly of An42-48, An42-46 and An40-48, respectively (Figure 4.12). These three tephras show similar An-poor (<50) plagioclase compositions, but the Sandwich tephra also contains several plagioclase crystals with slightly higher An-rich (An54-68), which is consistent with the lower weight percent SiO_2 glass compositions of the Sandwich tephra.

Although the glass composition of the ash in Site 3 (Figure 4.4) was not obtained, the ash may be correlated with the Forty Year tephra on the basis of field characteristics and microscopic inspection.

3. Eruption ages inferred from AMS radiocarbon ages

The AMS ^{14}C dates for the Intermediate, Sandwich, YBO and Forty Year tephra deposits are shown in Table 4.2. The conventional ^{14}C dates are calculated using Libby's half-life of 5568 years. Our results appear reliable because the ^{14}C dates are nearly consistent with the stratigraphy (Figures 4.2-4.6). These conventional ^{14}C dates are also calibrated

Site No.	Stratigraphic Position Relative to Tephra Deposit	Radiocarbon Material	d13C (PDB)	14C date (BP)	Lab. No. (NUTA2-)	Calibrated years range (2s, probability %)	Remarks
8	Just below 40 Year	Charcoal	-27,9	295±25	13942	292-334 (31.26)	
						349-439 (66.48)	
						443-455 (2.27)	
8	26 cm above YBO	Charcoal	-29,0	2105±30	13941	1999-2147 (100)	
8	14 cm above YBO	Charcoal	-27,7	2210±30	13939	2149-2325 (100)	
8	Just above YBO	Charcoal	-28,3	2620±25	13938	2724-2772 (100)	
7	Just below YBO	Charcoal	-28,4	3395±25	13931	3574-3698 (100)	
11	3 cm above Sandwich	Charcoal	-25,9	3305±30	10970	3454-3590 (97.72)	Okuno *et al.* (2009)
						3800-3612 (2.28)	
8	26 cm belowYBO	Charcoal	-26,9	3580±25	13937	3782-3786 (0.65)	
						3829-3933 (89.89)	
						3939-3971 (9.46)	
8	57 cm below YBO	Charcoal	-31,1	3855±30	13936	4157-4205 (16.88)	
						4223-4408 (83.12)	
5	Just below Sandwich	Soil	-27,9	4150±30	10966	4579-4770 (80.45)	Okuno *et al.* (2007)
						4780-4824 (19.55)	
7	Just below Sandwich	Charcoal	-26,5	4140±25	13930	4573-4744 (70.65)	
						4747-4821 (29.35)	
8	Just below Sandwich	Charcoal	-27,3	4180±25	13935	4618-4765 (77.77)	
						4785-5296 (34.37)	
11	2 cm below Sandwich	Charcoal	-25,9	4480±30	10972	4878-5008 (4.74)	Okuno *et al.* (2009)
						5036-5145 (36.96)	
						5153-5288 (58.30)	
8	38 cm below Sandwich	Charcoal	-27,9	4505±25	13934	5048-5199 (65.63)	
						5211-5296 (34.37)	
8	71 cm below Sandwich	Charcoal	-29,6	4515±30	13933	5050-5193 (66.76)	
						5213-5303 (33.24)	
5	Just above Intermediate	Charcoal	-27,6	6040±30	10967	6794-6966 (100)	Okuno *et al.* (2007)
1	In Intermediate	Charcoal	-28,7	6245±30	13943	7025-7060 (7.43)	
						7065-7116 (11.06)	
						7154-7257 (81.52)	
9	Just below Intermediate	Charcoal	-35,2	5935±35	10971	6669-6803 (90.24)	Okuno *et al.* (2009)
						6813-6849 (9.76)	
7	2 cm below Intermediate	Charcoal	-28,7	6560±30	13929	7426-7509 (95.88)	
						7544-7555 (4.12)	
7	8 cm below Intermediate	Charcoal	-22,9	6845±30	13928	7612-7730 (100)	

Table 4.2. Results of AMS 14C dating for tephra-fall deposits located on Adak, Alaska.

to ranges of calendar years using the calibration data set IntCal04 (Reimer *et al.* 2004) and the *CALIB 5.0* program (Stuiver and Reimer 1993).

Okuno *et al.* (2007) reported two AMS ^{14}C dates from Site 5 (Figure 4.4), which indicate eruption ages of the Intermediate and Sandwich tephra deposits as 6.8 cal ka BP and 4.6 cal ka BP (within the mode of calibrated probability distribution), respectively. In this paper, we describe five dates for the Intermediate, seven dates for the Sandwich, six dates for the YBO, and one date for the Forty Year tephra deposits (Table 4.2), including the five previous dates (Okuno *et al.* 2007, 2009). For the Intermediate tephra, a raw AMS ^{14}C date of 6245±30 BP from the tephra deposit at Site 1 is the most reliable (Figures 4.1 and 4.7). Thus, we adopted an age of 7.2 cal ka BP for the Intermediate tephra. AMS ^{14}C dates, 4180±25 BP, 4140±25 BP, 4150±30 BP, on charcoal or soil immediately below the Sandwich tephra agree with each other within one sigma range. Therefore, we calculated an age of 4.7 cal ka BP for the Sandwich tephra. Ages of both deposits generally agree with Black (1976). In contrast, only one ^{14}C date, 3395±25 BP, was obtained from charcoal immediately below the YBO tephra. Five AMS ^{14}C dates support this age, and we propose an age of 3.6 cal ka BP for the YBO tephra. This age is inconsistent with any previous studies; however, it agrees with the T7 tephra at Kanaga Island (Waythomas *et al.* 2001). The eruption age of the Forty Year tephra is calculated at 0.4 cal ka BP (=16th century). This is based on one AMS ^{14}C date, 295±25 BP, from charcoal immediately below the tephra at the sand dune section (Figure 4.1 Site 8).

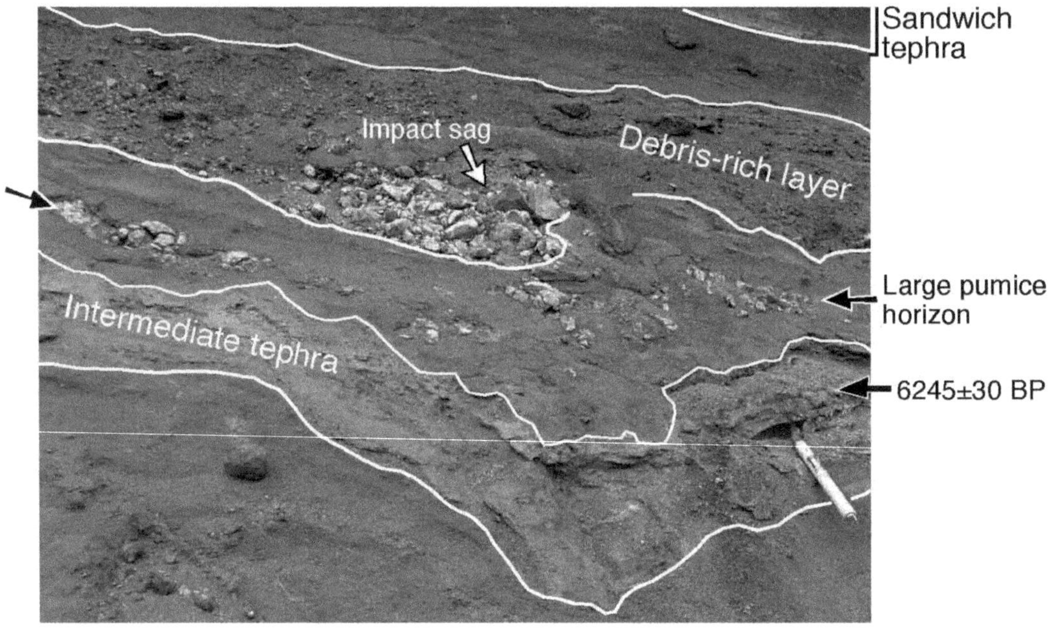

Figure 4.7. Photograph showing an occurrence of large pumice horizon and debris-rich layer between the Intermediate and Sandwach tephra deposits at Site 1. An impact sag is shown by an open arrow. An AMS radiocarbon age of 6245±30 BP was obtained from a charcoal fragment incorporated in the Intermediate tephra. Length of tool is 35 cm.

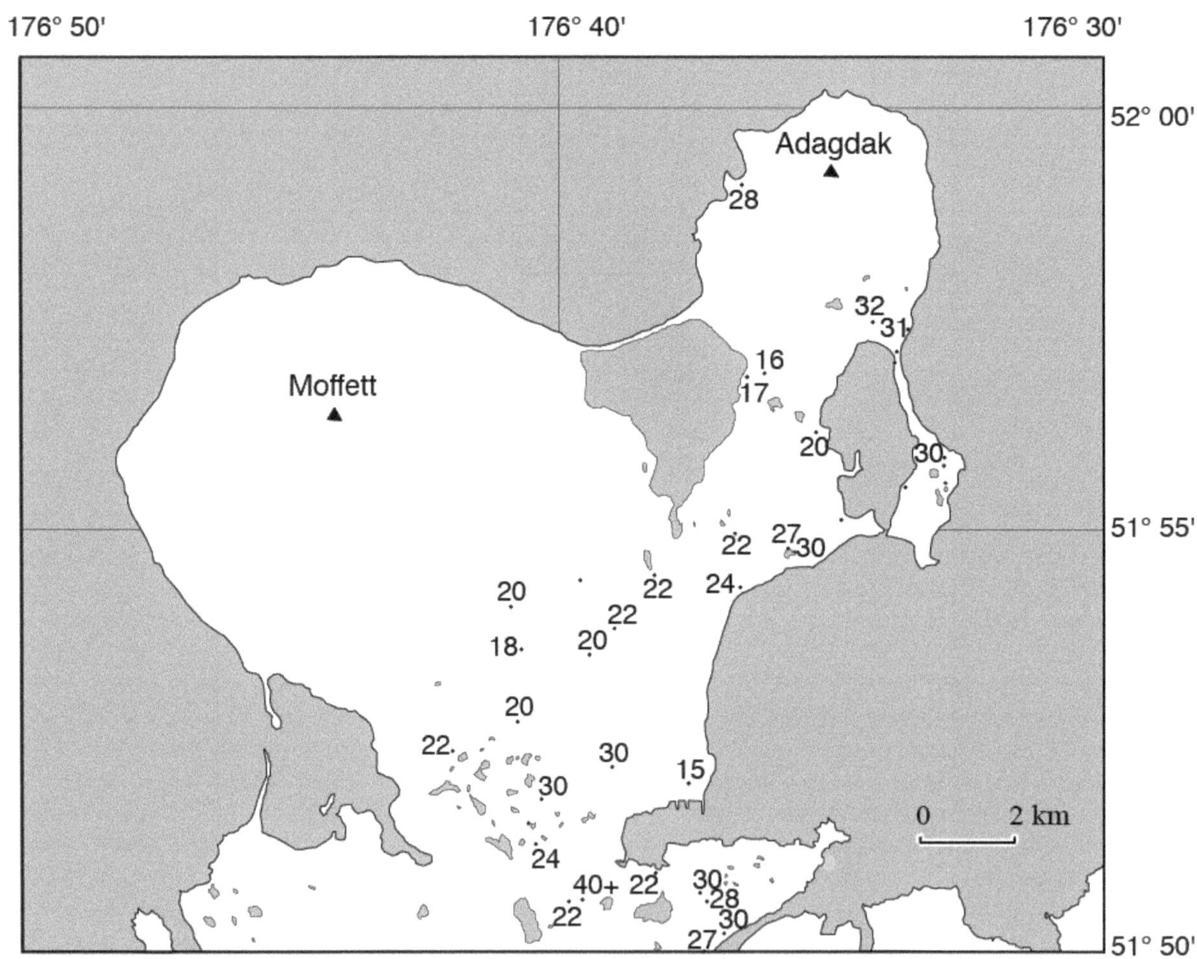

Figure 4.8. Thickness distribution of the Main tephra in cm (after Okuno et al. 2009).

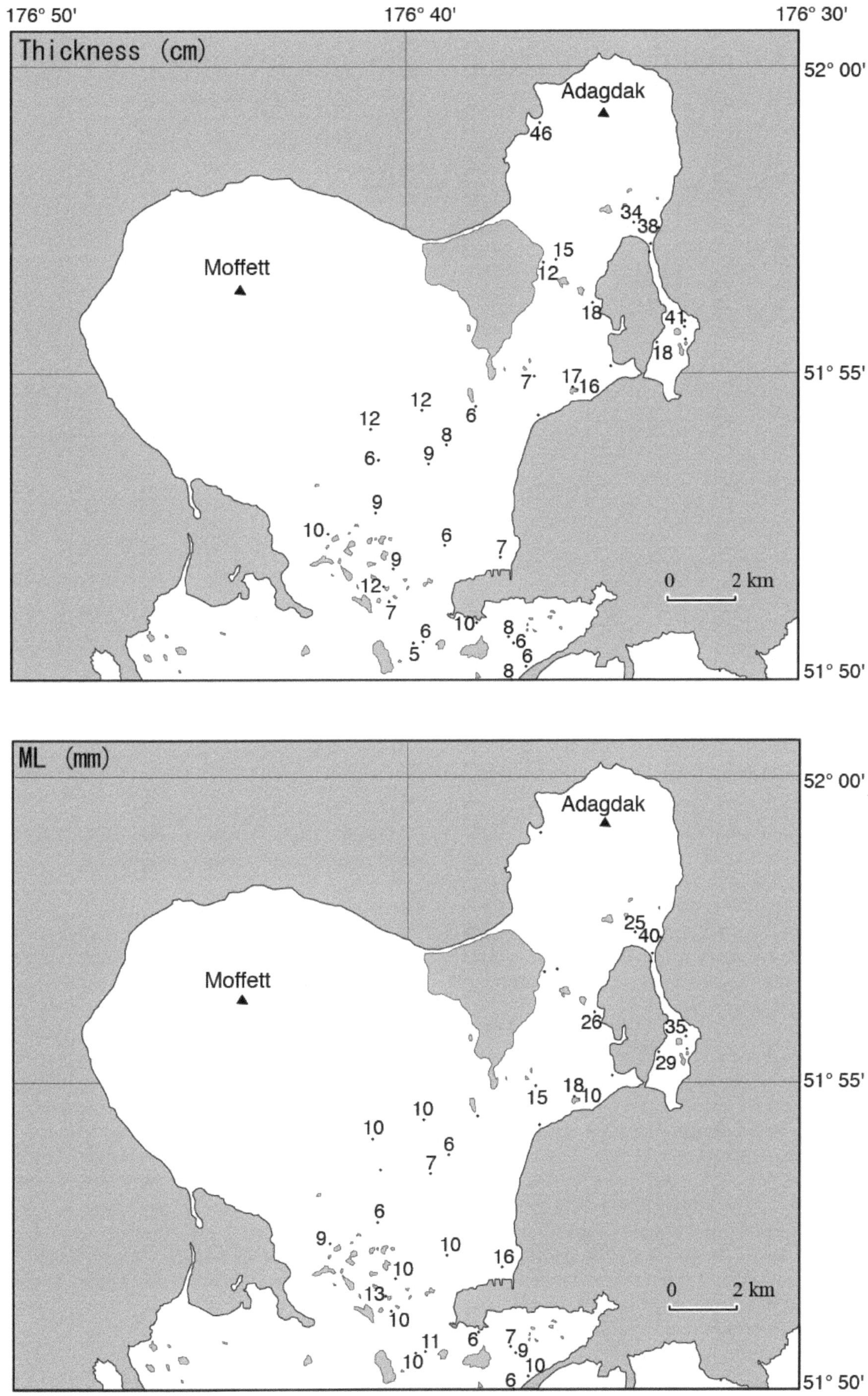

Figure 4.9. Dispersal maps for the Intermediate tephra (after Okuno et al. 2009). (Top) Thickness in cm, (Bottom) Maximum lithic size (ML) in mm.

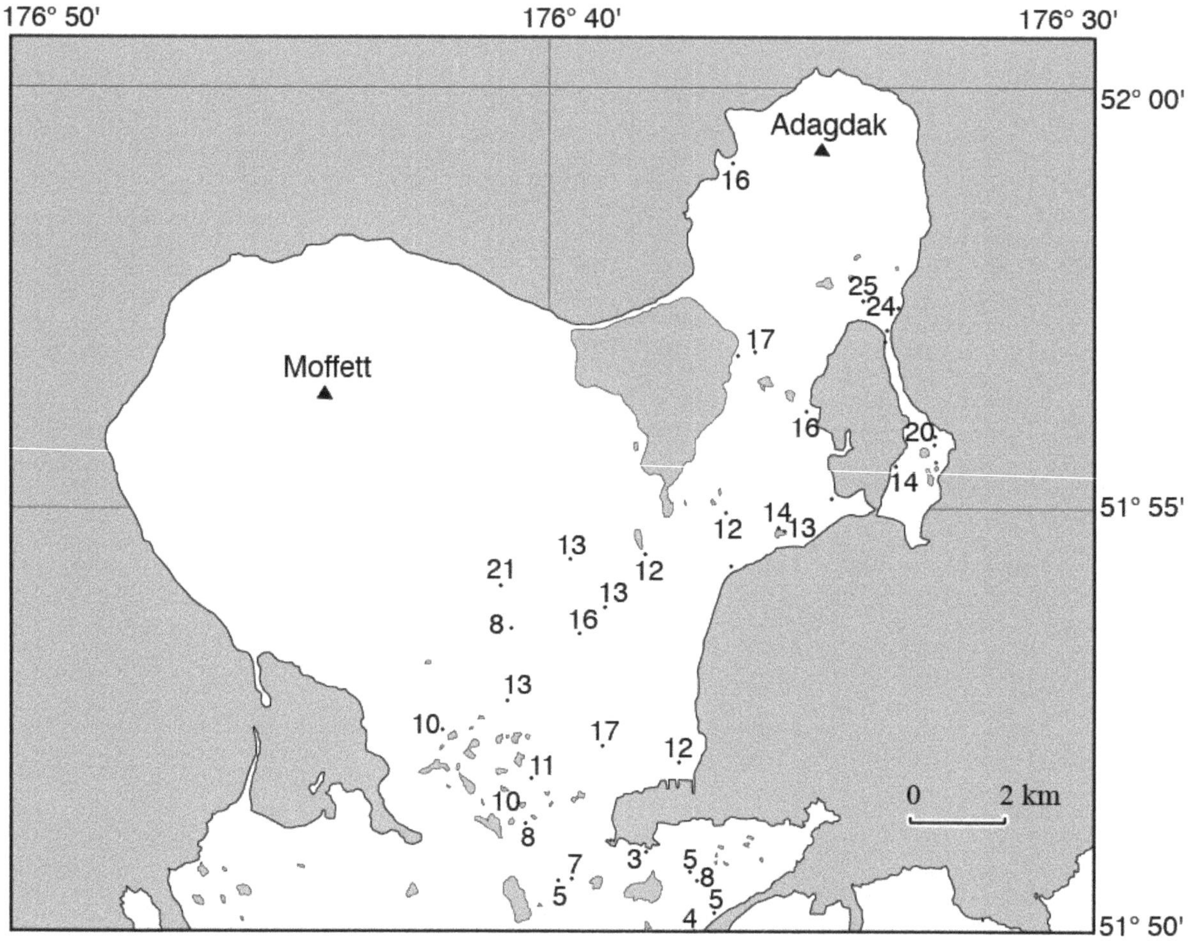

Figure 4.10. Thickness distribution of the Sandwich tephra in cm (after Okuno et al. 2009).

4. Source volcanoes of conspicuous tephra deposits

Dispersal maps for the Main, Intermediate and Sandwich tephras indicate that the source volcano is not Mount Moffett. Furthermore, the stratigraphy that we have recently established for Kanaga volcano and that of Waythomas *et al.* (2001) suggest that these deposits are also not attributed to the eruptions of Kanaga volcano.

Based on distribution and field characteristics, we propose that the possible source for these tephras is located close to the northern part of Adak Island. The Main tephra contains a well-sorted pumice-fall deposit and, in the upper portion, a consolidated ash unit (Figure 4.2). This may be due to contact with water (*e.g.*, sea water, snow/ice) during the final stage of eruption. The Intermediate tephra consists of laminated ash and pumice and contains a large amount of dense lithic fragments. This tephra was probably ejected in a phreatomagmatic eruption where magma and water interact. The large pumice clasts (Figure 4.7) may have been generated by a submarine eruption. For any deposit the range of ballistic fragments is restricted to a few kilometers from the vent (e.g., Cas and Wright 1988). The Sandwich tephra has a complex lithology at various sites. This tephra was most likely water-flushed (Walker 1981) by seawater or snow and/or alpine ice during its eruption. Therefore, it appears that deposition and reworking occurred simultaneously at the time of eruption.

The similarity of glass and plagioclase major element compositions indicates that the Main, Intermediate, and Sandwich tephras originated from the same source volcano. Based on similarity of high-silica content with Japanese widespread tephras (Machida and Arai 1983; Machida 1999), the high silica content implies that the source is most likely a caldera volcano. These arguments coupled with geological field evidence support the hypothesis that the source vent was located in a submerged area near the northern part of Adak Island. The SiO_2 content in the Sandwich tephra is lower than other tephras. There is a slight difference in Ti, Fe, Ca, K contents of the Intermediate and Main tephras. These slight chemical variations indicate magma evolution during periods of repose. Although more detailed research is needed, the current evidence suggests that these deposits originated from the same source volcano.

Conclusions

The following conclusions were reached from this study:

1. Among Holocene tephra deposits on Adak Island, the Intermediate, Sandwich, YBO and Forty Year tephra

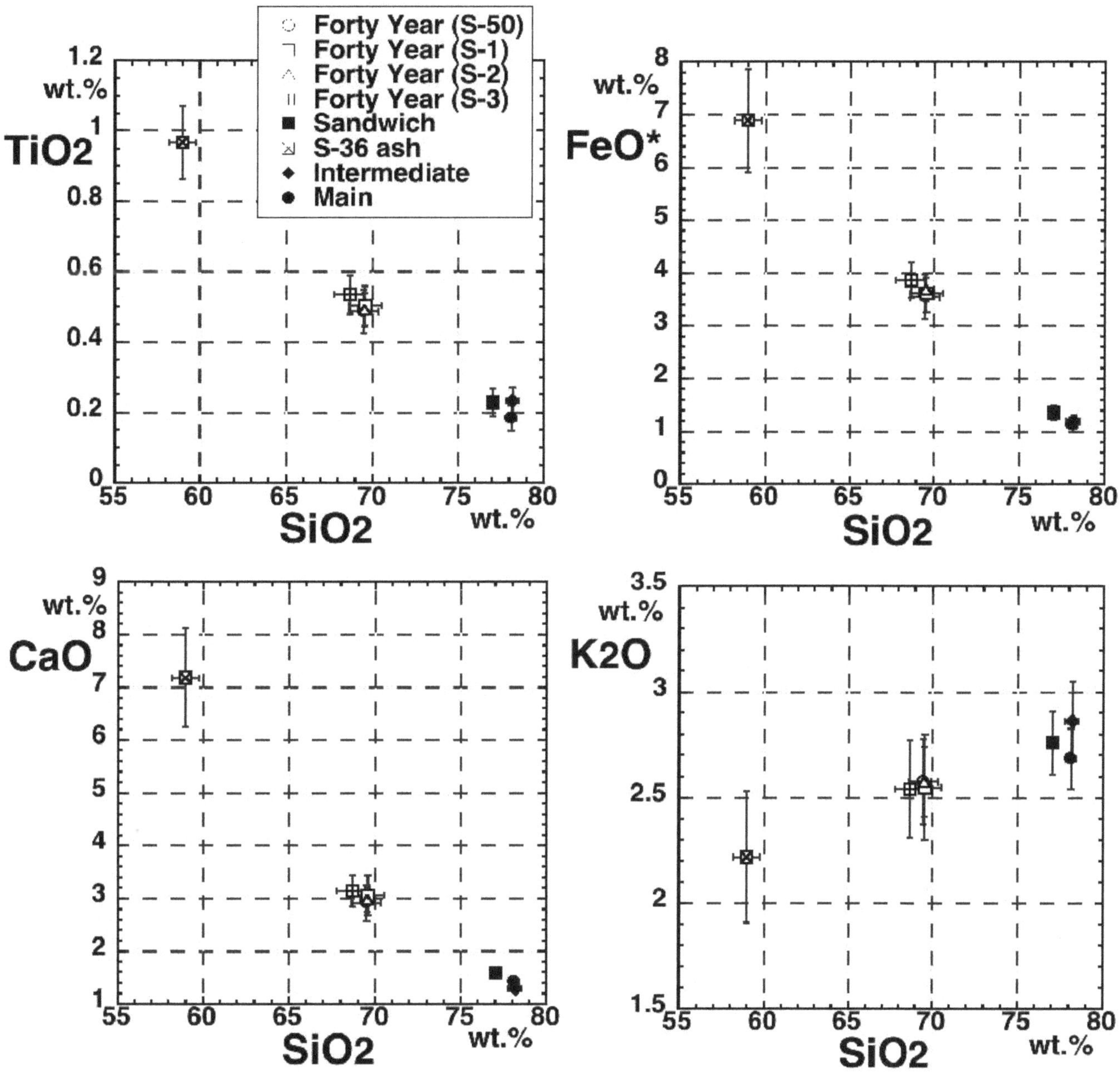

Figure 4.11. Plots showing EPMA major-oxide geochemistry of the volcanic glasses from conspicuous tephra deposits on Adak Island. Data are average values with error bars representing 1 sigma standard deviation (see also Table 4.1).

deposits are intercalated with Aleut archaeological sites. AMS ^{14}C dating of mostly charcoal fragments immediately below tephra deposits reveal eruptive ages of the Intermediate, Sandwich, YBO and Forty Year tephra deposits to be approximately 7.2, 4.7, 3.6 and 0.4 cal ka BP, respectively. The tephra-sand sequence in a dune provided suitable organics for precise ^{14}C dating using the AMS method, because the thick sand layer protected against the penetration and potential contamination of roots from surface vegetation.

2. Based on the stratigraphic relation with tephra deposits in Site 5 (ADK-171) of the *Clinocardium* midden (Figure 4.4), we conclude that the archaeological site was settled immediately after the Intermediate eruption. The Forty Year tephra provides a 16th-century time-marker at archaeological Sites 6 (ADK-011) and 3 (ADK-012) (Figure 4.4).
3. On the basis of distribution and lithology, we propose that the source of the Main, Intermediate, and Sandwich tephra deposits is a submerged volcano near the northern part of Adak Island. More detailed studies are needed to determine the exact location of the source volcano.

Acknowledgements

We thank Dr. Ma. Hannah T. Mirabueno of the Philippine Institute of Volcanology and Seismology (PHIVOLCS) and Kristi Wallace of the Alaska Volcano Observatory, US Geological Survey for constructive comments that benefited improvement of the manuscript. This research

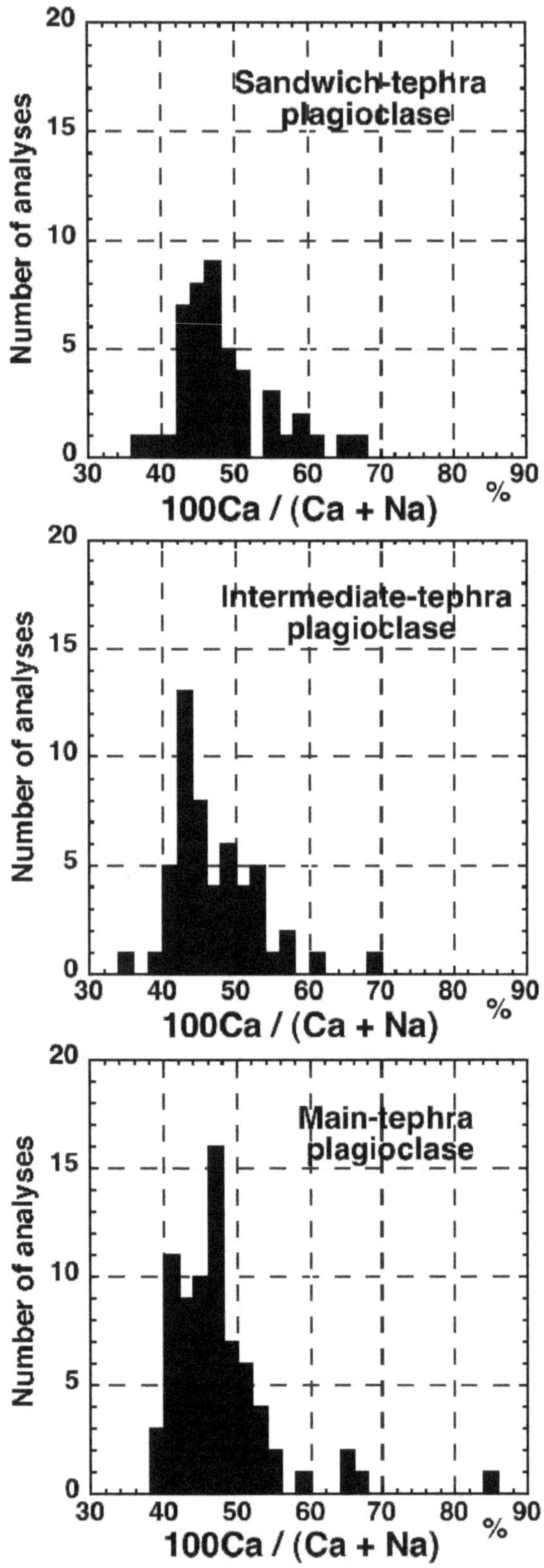

Figure 4.12. The An contents (100Ca/(Ca+Na)) of the plagioclase crystals from the Main, Intermediate and Sandwich tephras.

was supported with a collaborative grant awarded to the University of Kansas and the University of Washington by the Office of Polar Programs, Arctic Social Sciences Division, National Science Foundation (Award # 0353065 to Dixie West, Award # 0353092 to Lyn Gualtieri). This study was also supported by a Grant-in-Aid for Scientific Research from the Japan Society for the Promotion of Science (Award # 17300295 to Koji Okumura) and the Fukutake Foundation of Japan (to Mitsuru Okuno). Permission to work on Adak and Kanaga Islands was granted by the Aleut Corporation and the Alaska Maritime National Wildlife Refuge. We thank the US Fish and Wildlife Service and VECO Polar Resources for logistical field support. We also thank the City of Adak, Alaska and its citizens for welcoming us into their community during the summers of 2005-2007.

References

Alaska Volcano Observatory. 1998. *Volcanoes of Alaska: a Cooperative Program of the United States Geological Survey, University of Geophysical Institute, and Division of Geological and Geophysical Surveys.* Information Circular 38. Alaska Division of Geological and Geophysical Surveys, Fairbanks.

Black, Robert F. 1976. Late Quaternary Glacial Events, Aleutian Islands, Alaska. In D. J. Easterbrook and V. Sibrava (eds.), *Quaternary Glaciations in the Northern Hemisphere: IUGS-UNESCO International Geological Correlation Program, Project* 73-1-24, 285-301. Bellingham, International Union of Quaternary Research.

Black, Robert F. 1980. Isostatic, Tectonic, and Eustatic Movements of Sea Level in the Aleutian Islands, Alaska. In N. A. Morner (ed.), *Earth Rheology, Isostasy and Eustasy,* 231-248, New York, John Wily and Sons.

Cas, Ray A. F. and John V. Wright. 1988. *Volcanic Successions: Modern and Ancient: A Geological Approach to Processes, Products and Successions.* London, Chapman and Hall.

Coats, Robert R. 1956. Geology of Northern Adak Island, Alaska. *US Geological Survey Bulletin* 1028-C, 45-67. Washington, DC, US Government Printing Office.

Heusser, Calvin J. 1978. Postglacial Vegetation on Adak Island, Aleutian Islands, Alaska. *Bulletin of the Torrey Botanical Club* 105 (1), 18-23

Judson, Sheldon. 1946. Late-glacial and Postglacial Chronology on Adak. *Journal of Geology* 54, 376-385.

Kano, Kazuhiko, Takahiro Yamamoto and Koji Ono. 1996. Subaqueous Eruption and Emplacement of the Shinjima Pumice, Shinjima (Moeshima) Island, Kagoshima Bay, SW Japan. *Journal of Volcanology and Geothermal Research* 71 (1-2), 187-206.

Kiriyanov, V. Yu and Thomas P. Miller. 1997. Volcanic Ashes of Adak Island, Aleutian Islands, Alaska. *Volcanology and Seismology* 19, 65-78.

Kitagawa, Hiroyuki, Toshiyuki Masuzawa, Toshio Nakamura and Eiji Matsumoto. 1993. A Batch Preparation Method for Graphite Targets with Low

Background for AMS ^{14}C Measurements. *Radiocarbon* 35 (2), 295-300.

Machida, Hiroshi. 1999. The Stratigraphy, Chronology and Distribution of Distal Maker-tephras in and around Japan. *Global and Planetary Change* 21 (1-3), 71-94.

Machida, Hiroshi and Fusao Arai. 1983. Extensive Ash Falls in and around the Sea of Japan from Large Late Quaternary Eruptions. *Journal of Volcanology and Geothermal Research* 17 (1-4), 151-64.

Marsh, Bruce D. 1990. Moffett and Adagdak. In C. A. Wood and J. Kienle (eds.), *Volcanoes of North America: United States and Canada*, 25-26. Cambridge, Cambridge University Press.

Myers, James D. and Carol D. Frost. 1994. A Petrologic Re-investigation of the Adak Volcanic Center, Central Aleutian Arc, Alaska. *Journal of Volcanology and Geothermal Research* 60 (2), 109-146.

Nakamura, Toshio, Etsuko Niu, Hirotaka Oda, Akiko Ikeda, Masayu Minami, Hiroshi Takahashi, Mamoru Adachi, Ludi Pals, Andreas Gottdang, and Noriyoshi Suya. 2000. The HVEE Tandetron AMS System at Nagoya University. Nuclear Instruments and Methods. *Physics Research Section B: Beam Interactions with Materials and Atoms* 172 (1-4), 52-57.

Okuno, Mitsuru and Toshio Nakamura. 2003. Radiocarbon Dating of Tephra Layers: Recent Progress in Japan. *Quaternary International* 105 (1), 49-56.

Okuno, Mitsuru, Toshio Nakamura, Horishi Moriwaki and Tetsuo Kobayashi. 1997. AMS Radiocarbon Dating of the Sakurajima Tephra Group, southern Kyushu, Japan. *Nuclear Instruments and Methods in Physics Research B: Beam Interactions with Materials and Atoms* 123 (1-4), 470-474.

Okuno, Mitsuru, Toshio Nakamura, Hiroki Kamata and Tetsuo Kobayashi. 2001. Radiocarbon Dating of Paleosol Intercalated with Tephra Layers in Japan. In E. Juvigné and J. P. Raynal (eds.), *TEPHRAS, Chronology/ chronologie, Archeology/archeologie,* 67-71. CDERAD éd, Clermont-Ferrand. (with French abstract)

Okuno, Mitsuru, Lyn Gualtieri, Dixie West, Elizabeth G. Wilmerding and Toshio Nakamura. 2007. Aleut Shell Mounds Intercalated with Tephra Layers in Adak Island and Adjacent Volcanoes in the West-central Aleutians, Alaska, USA. *The Journal of the Geological Society of Japan* 113 (8), XI-XII.

Okuno, Mitsuru, Lyn Gualtieri, Brenn Sarata, Dixie West, Toshio Nakamura, Keiji Wada and Masayuki Torii. 2009. Stratigraphy of the Holocene Tephra Layers in Adak Island, Central Aleutians, USA. In *Nature, Society, and Human–Studying Geography* , edited by Department of Geography, Kanazawa University pp. 47-53, Tokyo, Kokon-shoin Publishing. (in Japanese)

O'Leary, Matthew. 2001. Volcanic Ash Stratigraphy for Adak Island, Central Aleutian Archipelago. In Don Dumond (ed.), *Archaeology in the Aleut Zone of Alaska: Some Recent Research.* 215-233. University of Oregon Anthropological Papers 58. Eugene, Oregon. University of Oregon Press.

Reimer, Paula J., Mike G. L. Baillie, Edouard Bard, Alex Bayliss, J. Warren Beck, Chanda J. H. Bertrand, Paul G. Blackwell, Caitlin E. Buck, George S. Burr, Kirsten B. Cutler, Paul E. Damon, R. Lawrence Edwards, Richard G. Fairbanks, Michael Friedrich, Thomas P. Guilderson, Alan G. Hogg, Konrad A. Hughen, Bernd Kromer, Gerry McCormac, Stuart Manning, Christopher Bronk Ramsey, Ron W. Reimer, Sabine Remmele, John R. Southon, Minze Stuiver, Sahra Talamo, F. W. Taylor, Johannes van der Plicht and Constanze E. Weyhenmeyer. 2004. IntCal04 Terrestrial Radiocarbon Age Calibration, 0-26 Cal Kyr BP. *Radiocarbon* 46 (3): 1029-1058.

Romick, Jay D., Suzanne Mahlburg Kay and Robert W. Kay. 1992. The Influence of Amphibole Fractionation on the Evolution of Calc-alkaline Andesite and Dacite Tephra from the Central Aleutians, Alaska. *Contribution to Mineralogy and Petrology* 112, 101-118.

Savinetsky, Arkady B., Dixie West, Zhanna A. Antipushina, Bulat F. Khassanov, Nina K. Kiseleva, Olga A Krylovich, and Andrei M. Perelodov. 2011. The Reconstruction of Ecosystems History of Adak Island (Aleutian Islands) During the Holocene. Chapter 5 in D. West, V. Hatfield, E. Wilmerding, L. C. Lefèvre, L. Gualtieri (eds), *The People Before: The Geology, Paleoecology and Archaeology of Adak Island, Alaska*. Oxford, British Archaeological Reports.

Stewart, Andrew L. and Jocelyn McPhile. 2004. An Upper Pliocene Coarse Pumice Breccia Generated by a Shallow Submarine Explosive Eruption, Milos, Greece. *Bulletin of Volcanology* 66 (1), 15-28.

Stuiver, Minze and Paula J. Reimer. 1993. Extended ^{14}C Data Base and Revised CALIB 3.0 ^{14}C Age Calibration Program. *Radiocarbon* 35 (1): 215-230.

Thorson, Robert M. and Thomas D. Hamilton. 1986. Glacial Geology of the Aleutian Islands: Based on the Contribution of Robert F. Black. In T. D. Hamilton, K. M. Reed and R. M. Thorson (eds.), *Glaciation in Alaska - The Geologic Record,* 171-192. Anchorage, Alaska Geological Society.

Wada, Keiji, Masayuki Mukai and Osamu Takeda, 2003. Chemical Compositions of Obsidian Glasses Determined by EPMA: The Origin of Obsidians from the Remains at the Mouth of Tokoro River, Eastern Hokkaido, Japan. *Reports of the Taisetsuzan Institute of Science* 37, 59-70. (In Japanese with English abstract)

Walker, George P. L. 1981. Characteristics of Two Phreatoplinian Ashes, and Their Water Flushed Origin. *Journal of Volcanology and Geothermal Research* 9 (4), 395-407.

Waythomas, Christopher F., Thomas P. Miller, Christopher J. Nye. 2001. Geology and Late Quaternary Eruptive History of Kanaga Volcano, A Calc-alkaline Stratovolcano in the Western Aleutian Islands, Alaska. *US Geological Survey Professional Paper* 1678, 181-197.

West, Dixie, Christine Lefèvre, Elizabeth Wilmerding, Virginia Hatfield, and Lyn Gualtieri. 2011. The Central Aleutians Archaeological and Paleobiological Project. Chapter 1, in D. West, V. Hatfield, E. Wilmerding, C. Lefèvre, L. Gualtieri (eds), *The People Before: The*

Geology, Paleoecology and Archaeology of Adak Island, Alaska. Oxford, British Archaeological Reports.

Chapter 5
The Reconstruction of Ecosystems History of Adak Island (Aleutian Islands) During the Holocene

Arkady B. Savinetsky, Dixie L. West, Zhanna A. Antipushina, Bulat F. Khassanov, Nina K. Kiseleva, Olga A. Krylovich, and Andrei M. Pereladov

Abstract: *To reconstruct the ancient environment of Adak and the surrounding Bering Sea, the Russian Group of Historical Ecology studied natural and archaeological deposits on the north side of Adak Island, Aleutian Islands, Alaska. The stratigraphy and composition of a sequence in the Haven Lake peat bog was compared with those from two archaeological sites, ADK-171 and ADK-009. This approach reveals new information about the Holocene history of ecosystems of Adak and the surrounding area. Radiocarbon dates indicate that the cultural midden at ADK-171 was formed about 7000 cal BP. Diatom analysis and saffron cod remains identification suggest that the ADK-171 site was occupied during the coldest period in the Central Aleutians in the last 10,000 years. The ADK-009 midden accumulated between 2,080 and 970 years ago. Stable isotope analysis of barnacle plates indicates both cool and warm intervals during the time ADK-009 was occupied. Cooler periods coincide with high biodiversity and productivity.*

Keywords: *Adak Island, peat bog analysis, diatoms, stable isotopes, saffron cod*

Introduction

To fully understand the prehistoric populations of Adak Island, Alaska, and the Bering Sea, it is necessary to reconstruct the natural world in which they settled, adapted and flourished. The Central Aleutians are a rich resource for the study of the ancient maritime environment of the North Pacific Rim. Biology and geology of the region have been extensively studied (e.g., Black 1976; Cameron and Stone 1970; Heusser 1978, 1990; Panuska 1980; Rubenstone 1984; Thorson and Hamilton 1986; Waythomas 1995); however, this research is not tied to the archaeological record of human settlement and adaption. To reconstruct the early Bering Sea environment within a cultural context, our scientific group at the Laboratory of Historical Ecology, Severtsov Institute of Ecology and Evolution, Russian Academy of Sciences, studied natural and archaeological deposits on the north side of Adak Island.

The stratigraphy and contents of a peat bog deposit at Haven Lake on Adak Island provided a framework for Holocene environmental change. Radiocarbon dating, morphological structure, and botanical remains of the Haven Lake profile defined the biological and geologic processes during the formation of this deposit. These data from the Haven Lake deposit were compared to similar data obtained from ADK-171 and ADK-009, two archaeological sites on the north side of Adak Island (Figure 5.1).

The Haven Lake sequence provides a chronological and environmental framework for the last 10,000 years on Adak Island. Our archaeological stratigraphy and faunal analyses were interpreted within the context of that framework. Comparison of these three sites, one natural and two cultural, ultimately provides some insight into climate and landscape change on Adak during the Holocene. In addition to scientific interpretation, the methods employed were: diatom analysis, pedology, geomorphology, paleozoology, paleobotany, stable carbon and oxygen isotope analysis, and radiocarbon dating.

The Haven Lake profile (6/97) is situated on the banks of a stream flowing from the lake (51°54'8"N, 176°38'15"W). Beginning in the 1940's, military construction modified the surrounding landscape. This resulted in increased runoff that eroded the original stream and exposed a 2.7m thick bank of peaty material interbedded with distinctive volcanic ashes. Previous investigations (Heusser 1978; Black 1980; Kiriyanov and Miller 1997; Okuno *et al.* 2007, O'Leary 2001) suggested that the highly visible soil and ash deposits on Adak Island were formed during the Holocene. Volcanic activity on surrounding islands form the most visible ash layers on Adak; volcanoes on the northern half of the island—Moffet and Adagdak—have been dormant during the Holocene (Fraser and Snyder 1959; Okuno *et al.* Chapter 4 this volume; Waythomas 1995). The thickness and clarity of the soil and ash layers on Adak provided information about the: 1) location of active volcanoes that produced the ash layers on Adak and, 2) macro- and mesorelief on Adak when the ash was deposited. Black (1976) initially studied the Haven Lake deposit; subsequently, Heusser (1978) and Black (1980) radiocarbon dated and analyzed the palynomorphs of the sediments. In 1997, using Black's

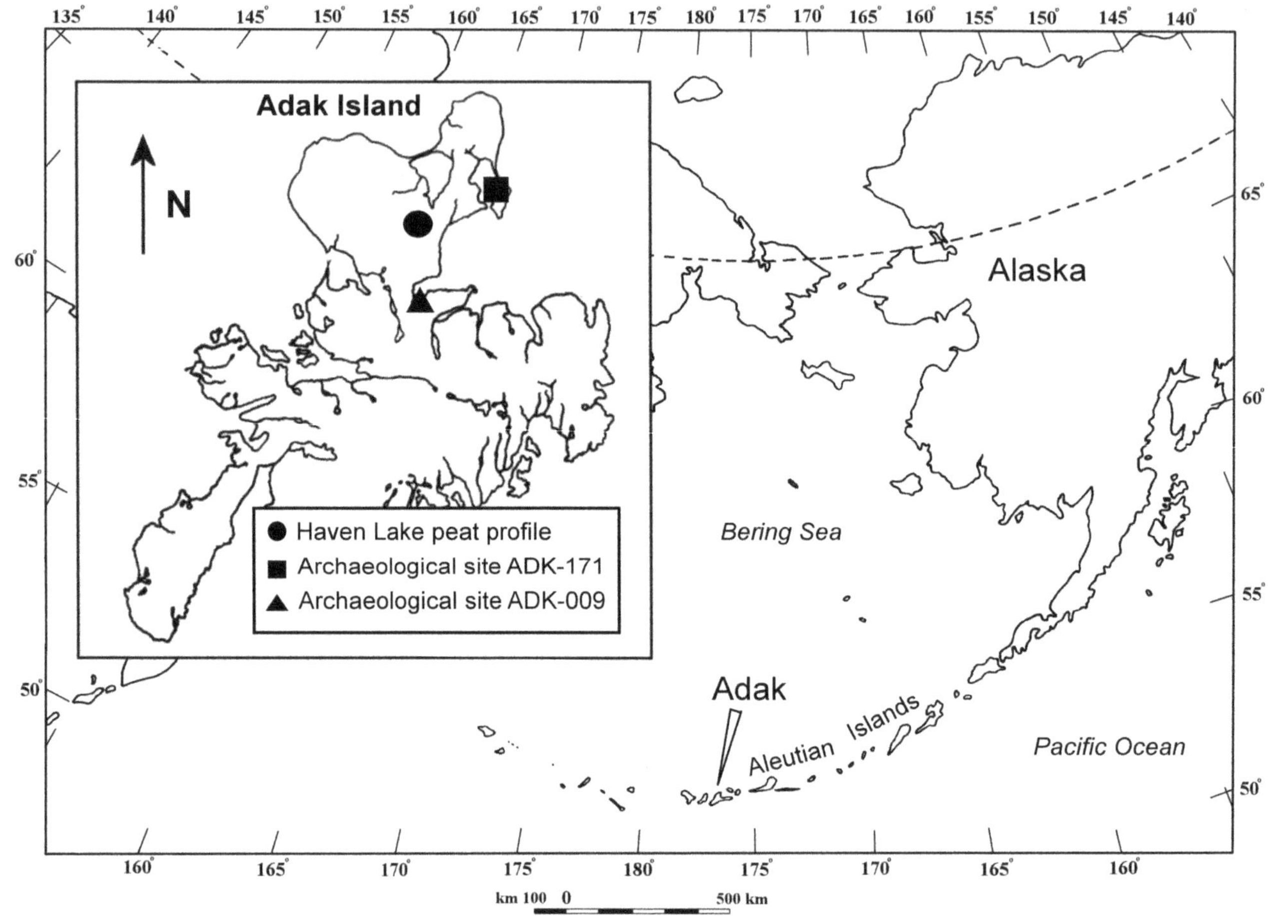

Figure 5.1. Map of Adak Island, Alaska, showing the general location of Haven Lake, ADK-171, and ADK-009.

description, we relocated, excavated, and collected samples near his original (1976) profile at Haven Lake.

The second profile (ADK-171/99), is on the east side of Clam Lagoon, and contains natural and cultural layers affiliated with a prehistoric Aleut village, ADK-171. In 1997 Vincent Tutiakoff of the Aleut Corporation noticed shells eroding from a steep bluff cut by road construction during World War II. In 1998 US BIA archaeologists recognized the shell exposure as an archaeological kitchen midden (Pratt and O'Leary 1999). Matt O'Leary, with the US BIA, described two distinct cultural layers—a loosely packed cockle layer containing animal bones and human artifacts separated by a thin lens of volcanic ash from a more densely packed and highly deteriorated shell layer below (Luttrell and Corbett 2000; O'Leary 1998, 2001). Shell from the deposit was radiocarbon dated at approximately 6000 BP (Luttrell and Corbett 2000). During 1999 US Fish and Wildlife archaeologists mapped 55 features in 'three general clusters' at ADK-171 (Luttrell and Corbett 2000, 48). The same year, due to a ship accident, which did not allow the researchers to reach the planned destination of Attu Island, the Russian team (authors of this report) worked on the more easily accessible Adak. They excavated a profile on the eroding road cut and collected bone, shell, and sediment samples. The purpose of the ADK-171 excavation was twofold: to retrieve and analyze faunal materials for paleoenvironmental analysis and to correlate the stratigraphy of ADK-171 with that at the Haven Lake profile. The 1999 profile revealed what then appeared to be two cultural layers filled with shells and animal bones 100-141cm below the present surface. During 2005 we cleared another profile (ADK-171/05/unit3) adjacent to the 1999 excavation to increase the size of the faunal sample.

A third profile, archaeological site ADK-009, is located near the head of Sweeper Cove on a bedrock shoreline with a gravel beach (Luttrell and Corbett 2000). This site was initially reported by McCartney (1972) and mapped in 1976 by Frolich and Kopjanski. Luttrell and Corbett (2000) described ADK-009 as a small site with seventeen features. Fourteen are on an 'elevated mound with abundant midden' overlooking Sweeper Cove and three are on a rocky islet, accessible intertidally, 20m from the shoreline (Luttrell and Corbett 2000, 16). In 1999, a 1m X 1m test pit, 1.92m deep, was excavated in Feature 12 at ADK-009. Artifacts (88) and fragments of 44 griddlestones were recovered (Luttrell and Corbett 2000), along with numerous shells and animal bones from a cultural deposit 166cm thick. Unlike ADK-171, the ADK-009 profile contained five distinctive, cultural layers that accumulated from the beginning of the 6th to the beginning of 17th century.

Methods and Materials

In the field, detailed profiles were drawn and organic samples collected for radiocarbon dating at the noncultural sequence, Haven Lake, and the two cultural deposits—ADK-171 and ADK-009. Samples from each stratigraphic layer of the peat bog at Haven Lake were collected to determine the loss on ignition as well as granulometric, mineralogical, diatom, and radiocarbon analyses. Vertebrate and invertebrate remains from cultural layers at the two archaeological sites were retrieved by wet sieving. Laboratory studies were subsequently conducted at the Laboratory of Historical Ecology of the Severtsov Institute of Ecology and Evolution, Russian Academy of Sciences (SIEE RAS), Moscow.

To identify the transition between organic and mineral layers and the mineral content of each peat layer, each sample collected from the sequence at Haven Lake was heated to a specific temperature (550°C; Arinushkina 1970). This heating removed all volatile substances, including water, organic matter, and plant residues. This heating was continued until the sample mass was constant. This procedure is referred to as LOI (Loss on ignition). The amount and kind of minerals obtained by LOI, coupled with redeposition of sediments in peat layers is used to infer the rate and amount of rainfall. Temperature change over time is inferred based on the rate of decomposition of samples.

Granulometric and mineralogical analyzes (Kachinsky 1965; Parfenova and Yarilova 1962) were performed to determine sedimentologic, (i.e. particle sizes and ratios) and mineral composition of samples from each layer. Particle sizes and ratios indicate the fractions of different mineral particles in the inorganic part of the layers and record the intensity of surface runoff, a process dependent upon the rate of precipitation. The composition and inferred origin of the minerals comes from the identification of the minerals and their physical condition in each layer sampled.

Diatoms, a common member of the phytoplankton, exist either as single cells or in colonies, and different species have different environmental requirements. Each diatom species is encased in a morphologically distinct siliceous skeleton. Upon death this skeleton becomes part of the accumulating sediment and the geologic record. By identifying the diatom species and radiocarbon dating sediments containing them a record of climatic change through time can be inferred.

Samples for diatom analysis were prepared using the standard hydrogen peroxide method (Battarbee 1986). Diatom skeletons were studied using a light microscope and a scanning electron microscope (SEM). Using the Shannon Index of Diversity (Magurran 1988) an estimate of the biodiversity of diatom species in the Bering Sea region was obtained. Diatoms were divided into ecological types based on: (1) preferred habitats (e.g., planktonic, tychoplanktonic, epithytic and benthic groups); (2) salinity; (3) pH tolerance (e.g., acidophilous: $pH<7$; alkaliphilous: $pH>7$; and indifferent; (4) saprobic conditions (e.g., xenosaprobic, oligosaprobic, betamesosaprobic and alphamesosaprobic); and (5) geographic (climatic) distribution based on temperature (Zhuze 1939).

The diversity of a community should account for both species richness and the evenness with which individuals are distributed among species. This can be done using a proportional abundance index. A particular diversity index is chosen with respect to the goals of the study (e.g., abundant vs. rare species) and to what extent sampling can be assumed to be random. Using specifically developed mathematical formulas, we use two diversity indices in our study: the Shannon Index and the Brillouin Index. The Shannon Index assumes that all species are represented in a sample and that the sample was obtained randomly. The most important source of error in the Shannon Index is failing to include all species from the community in the sample. Consequently, a species area curve assessment is created before the Shannon Index formula is used. Values of the Shannon diversity index for faunal communities usually fall between 1.5 and 3.5. When the randomness of a sample could not be guaranteed, we used the Brillouin Index for diversity.

Different diatom assemblages characterize three geographical regions: north-alpine, boreal and cosmopolitan. North-alpine species live above the Arctic Circle and in alpine mountain zones; boreal diatoms are typical in temperate climates; cosmopolitan species inhabit a variety of climatic zones. Because of Aleutian temperature constraints, the ratio between boreal and north-alpine species is the most useful comparison for diatom analysis on Adak.

All radiocarbon dates were obtained from two scintillation counters with dynamic stabilization of measurement of sample activity (Sulerzhitsky *et al.* 1984). The benzene standard of the Geological Institute, Russian Academy of Sciences, equivalent to the International Standard (Arslanov *et al.* 1987), was used. The radioactivity of all samples was corrected for isotope fractionation. The value of $\delta^{13}C$ equal to –25‰ was accepted for all terrestrial samples. Several measurements of marine fish bones from ADK-171 gave the average value of $\delta^{13}C$ –12,8‰, while the same figure for marine mammals bones turned out to be –13,9‰ (Krylovich, unpublished data). These values were used for corrections of marine fish and mammal bones respectively. Radiocarbon dates of marine bone samples from ADK-009 and ADK-171 were calibrated using the OxCal 3.10 calibration program – marine98.14c calibration curve. To calibrate marine measurements and take into the account of the marine reservoir effect, we set the ΔR value at 220±50 (Robinson and Thompson 1981). Bone from archaeological sites was dated using collagen (Longin 1971). Using these dates, deposit growth graphs for ADK-009 were created by plotting in coordinates of time versus thickness, and the chronological limit of each layer of each deposit was determined from these graphs.

Data on the oxygen and carbon stable isotopes in each stratigraphic layer of archaeological site ADK-009 were obtained. The isotopic composition of skeletal carbonate of different calcifying organisms (including benthic foraminifers, corals, brachiopods, mollusks, ostracods, barnacles and sea urchins) is a function of the organism's environment and metabolism. Oxygen isotopes undergo temperature-dependent fractionation during calcification; increasing temperature results in a decreasing fractionation between ^{16}O and ^{18}O, which produces a lower ^{18}O value (Epstein *et al.* 1951). The carbon isotope profiles are more complex than those of oxygen. Two main factors affecting ^{13}C values in shells include changes in invertebrate metabolism (due to changes in temperature, productivity, or gametogenesis) and the occurrence of phytoplankton blooms, which preferentially utilize $^{12}CO_2$ during photosynthesis (Kirby 1998). The plates of the barnacle *Semibalanus cariosus* were analyzed for oxygen and carbon isotopes. This particular barnacle was chosen because its plates contain only calcite (Killingley and Newrnan 1982; Smith *et al.* 1988) and because this barnacle is restricted to the littoral zone and does not inhabit the deeper, and subsequently colder, sublittoral zone like other barnacles (Pilsbry 1916; Tarasov and Zevina 1957).

Oxygen and carbon isotope analyzes were performed using standard techniques (Hu *et al.* 2001; Jones and Quitmyer 1996). Several unbroken barnacle plates were selected from each layer of archaeological site ADK-009. The plates were cleaned with distilled water, treated with a 5% solution of HCl for 3 minutes, rinsed again, dried, and powdered. Sample weights averaged between 150 and 200µg. For oxygen and carbon isotope analyzes, samples were reacted at 50°C with concentrated H_3PO_4. The isotopic composition of the evolved CO_2 was analyzed with a Thermo-Finnigan DELTA-V Plus mass spectrometer (SIEE RAS, Moscow). The total number of analyzed samples was 42 and the total number of laboratory standard replicates was 11 with a standard deviation of 0.045‰ for $\delta^{18}O$ and 0.035‰ for $\delta^{13}C$. All isotope data are expressed in the conventional delta (δ) notation δ ‰= ((Ratio sample – Ratio reference)/ (Ratio reference)) X 1000) (Epstein *et al.* 1951), where the isotope ratios of $^{18}O/^{16}O$ are reported relative to the SMOW standard and the isotope ratios of $^{13}C/^{12}C$ are reported to the Vienna PDB standard. SMOW or Standard Mean Ocean Water is the accepted zero point standard for expression of hydrogen and oxygen isotopes of water samples in delta units. PDB or Vienna-Pee Dee Belemnit is a belemnite from the Cretaceous Pee Dee formation of South Carolina, which is used as the accepted zero point standard for expression of carbon and oxygen isotope abundance in delta units.

All vertebrate and invertebrate remains from the cultural layers at ADK-171 and ADK-009 were wet screened through a 0.3cm mesh sieve to recover the maximum number of remains. The recovered remains were compared with known specimens in our laboratory at the SIEERAS. Radiocarbon dates indicate that the Adak kitchen middens formed over a long time with slow accumulation of remains, and the probability of finding separate bones or shells representing one individual is very low. Thus, the number of identified specimens (NISP) was used to quantify the vertebrates (Dinesman *et al.* 1999; Savinetsky *et al.* 2004), as well as to calculate the number of chitons (Polyplacophora). Bivalves were quantified using the total number of left and right umbos, barnacles (Cirripedia) using the number of plates, and brachiopods (Brachiopoda) using the number of valves (Antipushina 2008a). The number of sea urchins (Echinoderm) was determined by counting the total number of Aristotle's lantern, genital plates, and madreporits. Gastropods were quantified by counting shells of which more than half was present (Bird *et al.* 2002). Bryozoans, polychaetes and decapods were not counted; only their presence was noted.

An estimate of the catch of each species in each stratigraphic layer was determined by calculating the depositional rate of bones and shells. A relative index of the number of species collected or captured by the occupants of each cultural layer was obtained using the ratio NISP/t. This ratio of the number of remains of a species in the sample (NISP) to the temporal length of the interval of accumulation (t) is the accumulation rate (NISP/t) of that species. Temporal boundaries of the periods of accumulation can be determined from a graph of the growth of the cultural layer. Such a graph was constructed using the field descriptions of the profiles and the radiocarbon dates of the samples of bones from each cultural layer. The total number of remains, and their changing percentages among different layers, provided useful information about temporal shifts in ancient hunting gathering practices.

Estimates of prehistoric, natural populations of prey species based on numbers of bone or shell remains recovered from an archaeological site is, at best, tentative. Still, different methods have been used to calculate natural populations. Modifications of DeLury's (1947) method have been used to estimate numbers of modern cetaceans and pinnipeds when calculating harvest quotas (Allen 1966; Allen and Kirkwood 1988; Bockstoce and Botkin 1983). Chapman (1974) proposed that the average number of animals (N) during period t, the quantity of the hunted prey (C), and the hunting effort (f) spent on obtaining prey can be incorporated in the equation $N_t = C_t/qf_t$, where q is the hunting coefficient (the catch), the magnitude of which is determined by the methods of hunting. The ratio C_t/f_t calculates the number of prey per unit of hunting effort, or, in other words, the effectiveness of the hunt, or catch per unit effort (CPUE). With stability of the hunting coefficient q, the value CPUE changes in proportion to the abundance of the population being hunted, independent of the replenishment of the latter and its overall death rate (Ricker 1975), and in this case it can serve as a relative index of the number of animals.

Depth of horizon (cm)	*Material dated*	*Laboratory number IEMAE-*	*Radiocarbon age (years BP)*
30-35cm	plant remains	1305	modern
60-64cm	plant remains	1306	2755±90
69-81cm	plant remains	1307	3125±90
96-98cm	plant remains	1304	2890±105*
110-120cm	plant remains	1308	4160±135
135-140cm	plant remains	1303	5115±160*
185-190cm	humus	1268	8660±85
245-250cm	plant remains	1302	8660±175
260-265cm	plant remains	1309	8700±200

*Possibly this radiocarbon date is 'younger' due to penetration of younger roots.

Table 5.1. Radiocarbon dates for the Haven Lake deposit, Adak Island, Alaska.

The number of animal bones (NISP) accumulating over a particular period of time (t) at an archaeological site can be used to create a relative index of the number of animals obtained by early hunters. This calculation can be obtained with a sufficiently representative sample of radiocarbon dated bones. More problematic is the quantification of changes in hunting effort, estimated on the basis of changes in the paleopopulations of humans. With consideration of the very general assumptions of paleodemography, the average number of paleopopulations during time t is directly proportional to the quantity of burials (Acsadi and Nemeskeri 1970; Ubelaker 1978) and, in some cases to the accumulation rate of the cultural layer (see Dinesman *et al.* 1999; Savinetsky *et al.* 2004).

By comparing the number of remains of different species collected from different stratigraphic layers, changes in the biodiversity were estimated using two ecological indices: Shannon and Brillouin. Diatom diversity was determined using the Shannon Index, the most widely employed diversity index (Magurran 1988). The Brillouin index was used to estimate the past diversity of shellfish (invertebrates) because randomness of selection cannot be guaranteed.

Results

Profile 6/97 (Haven Lake)

One of the chief puzzles in the study of the formation of Holocene deposits on Adak Island is the timing of principal ash falls. Black (1976) first recognized and dated these volcanic deposits (i.e. tephra) during a study of a peaty-ash deposit near Haven Lake on the north side of the island. Black recognized five distinct tephra layers based on morphologic characteristics of the ash, and he timed their formation based on 17 radiocarbon dates. Haven Lake has become a key profile for studying Adak ash falls and climate. Subsequent researchers (Kiriyanov and Miller 1997; Okuno *et al.* 2007; O'Leary 2001) collected and dated samples from Black's layers; their results differ somewhat from those determined by Black. The inconsistencies in the radiocarbon dates might be due to penetration of young roots into ancient layers, redeposition of organic sediments, or other unknown reasons. The range of all these dates for the Adak ash beds is, however, very close to the original estimates by Black, as follows: 1) 8000 years BP – Main Ash; 2) 6000 years BP – Intermediate Ash; 3) 4000 years BP – Sandwich Ash; 4) 2000 years BP – Yellow-brown-olive (YBO) Ash; and 5) the first half of the twentieth century – the Forty-Year Old Ash. We have followed the chronology of the ash beds established by Black; however, based on the results of our radiocarbon analyses the time frame has been refined. (For additional information on tephra dates, see Okuno *et al.* Chapter 4 this volume).

The cleaned Haven Lake profile (Table 5.1, Appendix 5A) exposed a complex deposit of peat-tephra layers above sandy-loam glacial sediments. From the bottom to the top of the profile three buried peat bog soils (I - 182-265cm, II - 135-167cm, III - 96-124cm) are separated by three well-defined, largely coarse grained (>5mm) tephra layers (167-182cm, 124-135cm, and 85-96cm). At the top of the sequence, the 0-85cm stratum represents a package of largely fine-grained, peaty-ash layers with turf layers less then 10cm thick.

The most well-defined tephra layer, the Main Ash (Black 1976), is 167-182cm below the present day ground level. Radiocarbon dates of layers below the Main Ash are: 185-190cm (8660±85 yr BP, IEMAE-1268), 245-250cm (8660±175 yr BP, IEMAE-1302) and 260-265cm (8700±200 yr BP, IEMAE-1309) (Appendix 5A). These dates indicate that the Main Ash was deposited approximately 8000 years ago, as stated by Black (1976). The Intermediate Ash (124-135cm) consists of heterogeneous ash dominated by fine-grained particles. According to Black (1976) this ash was deposited approximately 6000 yr BP. Radiocarbon dates obtained for deposits immediately under this ash layer (135-140cm, 5115±160 yr BP, IEMAE-1303) are somewhat 'younger' than the date proposed by Black. It is possible that any roots in our sample are slightly younger than the layer of ash. The ash layer 85-94cm below the surface is the Sandwich Ash. Black (1976) estimated that this ash was deposited approximately 4000 yr BP. The Sandwich Ash, a heterogeneous layer, consists of fine-grained sediment

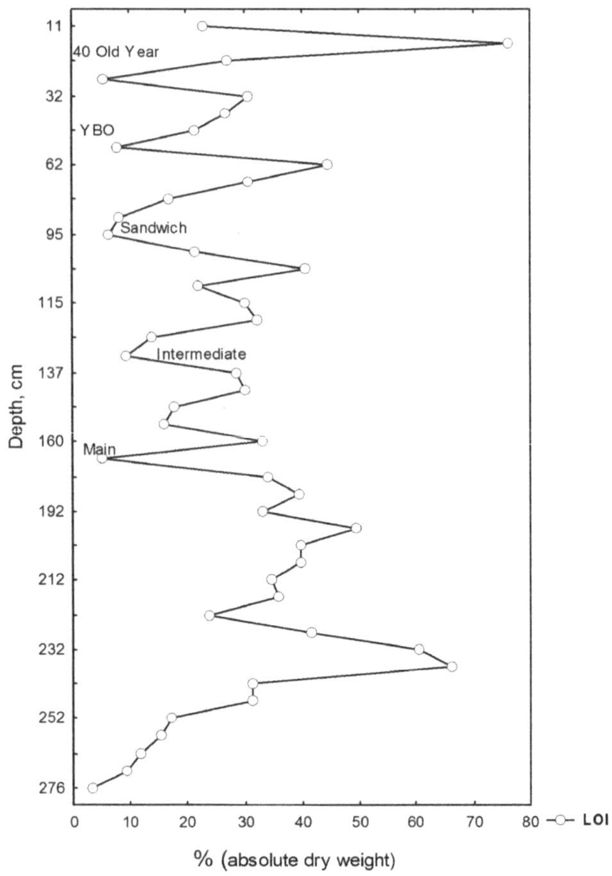

Figure 5.2 Loss on ignition (LOI) and ash content of the Haven Lake profile, Adak Island, Alaska.

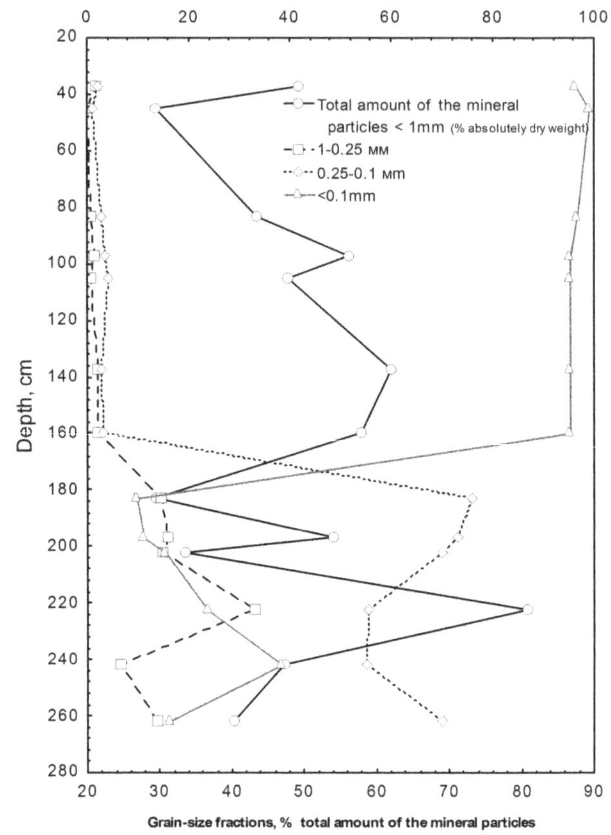

Figure 5.3. Grain size composition of peat and tephra in the Haven Lake profile, Adak Island, Alaska.

squeezed between two layers of coarse-grained tephra. Our radiocarbon date (4160±135 yr BP, IEMAE-1308) of the peat layer (110-120cm) below the Sandwich Ash supports the original date for this ash bed. However, the date (2890+105, IEMAE-1304) of the layer at 96-98cm, lying immediately below the Sandwich Ash, is probably too young, because younger roots penetrated the layer. Two radiocarbon dates of samples from layers 69-81cm (3125± 90 yr BP, IEMAE-1307) and 60-64cm (2755±90 yr BP, IEMAE-1306) indicate the YBO tephra layer is located 53-60cm below ground surface. Based on sediment structure, the 40 Year Old Ash is located 25-30cm deep. The loss on ignition of the samples from the Haven Lake profile verifies the location of these ash layers (Figure 5.2).

The ages of principal ash falls on Adak expressed in radiocarbon years have to be transformed into calendar dates for comparison with other data. This task, however, is far from complete due to wide range of radiocarbon dates obtained for each event so far. Therefore, the following results are only approximate estimations. The deposition of the Main Ash took place in the first half of the 7th millennium BC. The more precise date cannot be established due to the form of calibration curve, which is characterized by noticeable plateau around 8000 radiocarbon years ago. The Intermediate Ash fell around 5000 years BC, while the Sandwich Ash was deposited around 2700 years BC. Okuno and coauthors (2007), confirm this timing using AMS to date charcoals found immediately under the ash beds. These dates are 6040 ± 30 (5020 – 4840 years BC) and 4150 ± 30 (2880 – 2620 years BC) respectively. The YBO Ash was deposited at the very beginning of the Common Era.

Based on granulometric analysis (Figure 5.3) the mineral fractions from the Haven profile changed from 16% to 80.9%. Based on changes in the ratios and on the different sizes of mineral particles, two sections are recognized. The fine-grain fraction (<0.1mm) dominates in the upper section of the deposit (0-162cm), and the coarse-grain fraction in the lower. The mineralogical composition in this profile also changes. Volcanic ash, both unchanged and weathered, dominates the upper part (0-162cm), and plagioclase and pyroxenes are the most abundant minerals. Volcanic ash in fine-grained layers consisted largely of weathered and disintegrated pyroclastics. Approximately 50% of the minerals are a weakly coherent, friable formation of plagioclase and pyroxene phenocrystals altered by pyroclastics. Approximately 30% of the minerals are free grains of pyroxene with additions of ash; approximately 20% is tuff, sand, and tuff dust. In the lower part of the sequence, below 162cm, pyroxenes increase and there is almost no ash. Instead of loose ash, there was dense debris

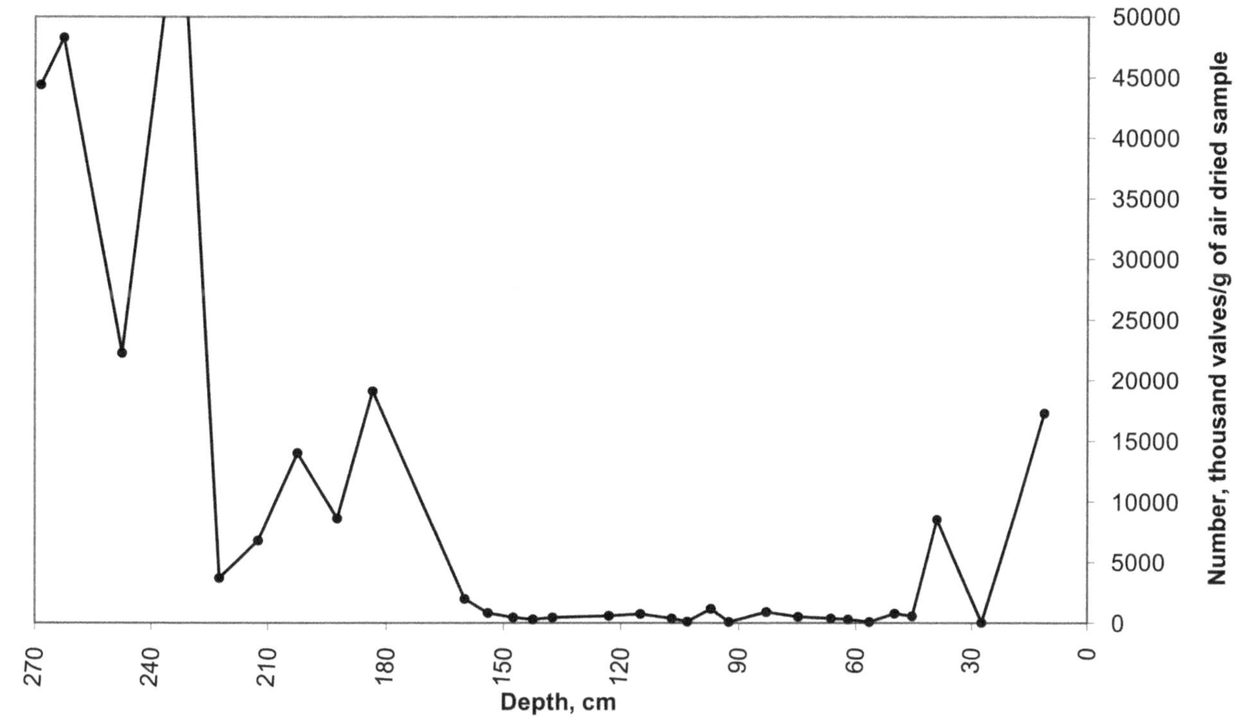

Figure 5.4. Changes in diatom numbers in the Haven Lake profile, Adak Island, Alaska.

of volcanic material, including tuff, with clear signs of water impact.

Of the 103 identified diatom taxa in the Haven Lake profile the most abundant genera were: *Pinnularia* (21), *Navicula* (12), *Fragilaria* (10), *Eunotia* (10) and *Cymbella* (8) (In brackets the number of species is reported). Diatom abundance changed from nearly zero to 61,000 valves/g dry wt over time (Figure 5.4). Diatom biodiversity increased from six taxa (140-145cm) to 47 taxa (265-272cm); the Shannon Index increased from 0.29 to 1.37 (Figure 5.5), with the lowest value near the middle of the profile.

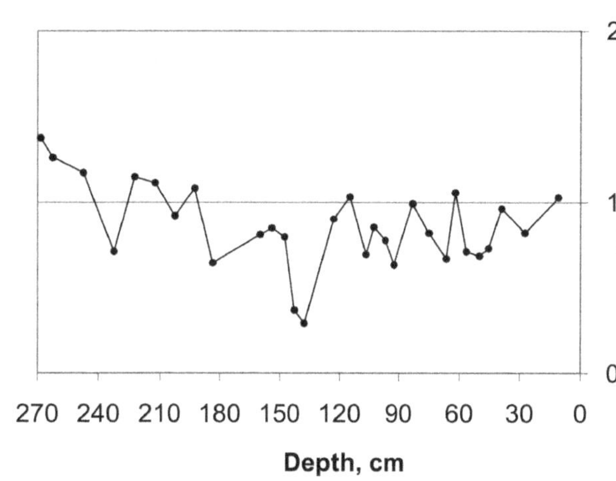

Figure 5.5. Changes in the Shannon Index (H') of diatoms in the Haven Lake profile, Adak Island, Alaska.

Based on species composition and ecological parameters, eight diatom assemblage zones were recognized. The oldest zone (260-272cm) is dominated by *Melosira italica* et.var. (21.4%), *Fragilaria construens et.var.* (16.4%), *F. pinnata* (11%), *F. brevistriata* (7%), *Eunotia praerupta* (5.8%), and *Pinnularia borealis* (4%). In the oldest zone, the abundance (44.4-48.3 thousand valves/g dry wt) and biodiversity (H': 1.26-1.37) of diatoms are very high and boreal diatoms dominate (Figure 5.6). Planktonic, tychoplanctonic, betamesosaprobe, and alkaliphilic diatoms characterize this zone. Based on these data, the initial depositional environment at Haven Lake was a shallow, freshwater alkalescent pool in a boreal climate.

The maximum abundance of diatoms, up to 61,000 valves/g dry wt, occurs in the second zone (230- 260cm) dominated by *Navicula pelliculosa* (24.8 %), *Eunotia exigua* (15.8 %), *Pinnularia microstauron var. brebissonii* (13.3 %), *P. interrupta f. minor* (10.5 %), and *P. viridis* (5 %). However, the biodiversity is much lower than in the oldest zone (H' - 0.71-1.17). This decrease in biodiversity suggests colder conditions, and the habitat was more oligotrophic (nutrient poor) and more neutral in acidity and salinity than during the oldest period.

The third zone (182-230cm) was formed just prior to the Main Ash, approximately 8000 years ago. Diatom abundance is still very high (3.7-19.1 thousand valves/g dry wt), and the biodiversity (H') increased from 0.64 to 1.15.

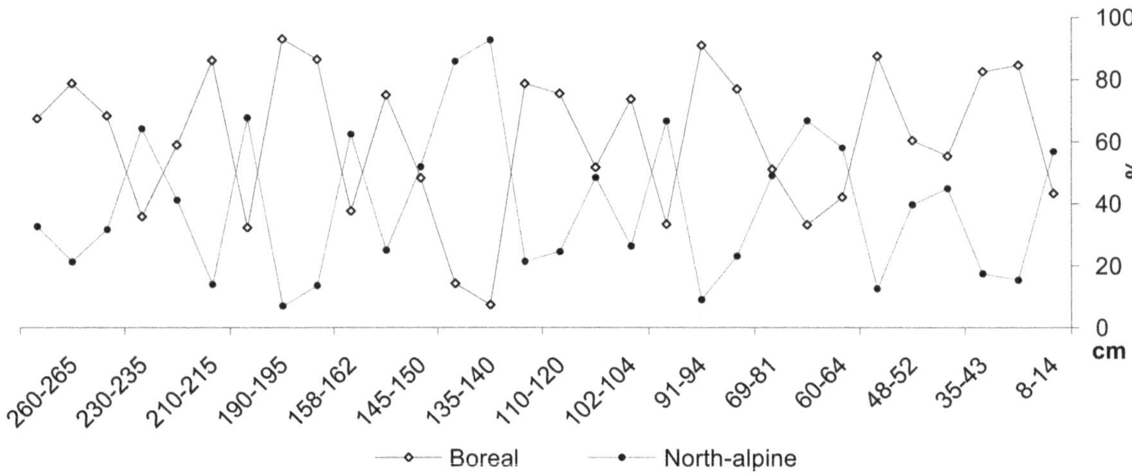

Figure 5.6. Ratio of boreal and northern alpine diatoms at the Haven Lake profile, Adak Island, Alaska.

Pinnularia viridis et.var. (24.3 %), *P. microstauron et.var.* (7.2 %), *P. subcapitata* (5 %,) *Navicula placenta* (15.2 %) and *Eunotia exigua* (9%) dominated this assemblage. These data suggest that climatic conditions changed twice during the deposition of this interval (182cm and 230cm), from warmer boreal to colder north alpine and back to boreal.

In the fourth zone (145-185cm) diatom abundance decreases from 2.0 to 0.85 thousand valves/g dry wt.; the dominant taxon is *Pinnularia*, and includes: *P. borealis* (31.7 %), *P. viridis et.var.* (18.2 %), *P. subcapitata* (11 %), *P. intermedia* (6 %) along with *Hantzschia amphioxys* (18 %) and *Eunotia praerupta* (6.3 %). These diatoms indicate that the habitat was oligotrophic and neutral, and climatic conditions were transitional, ranging from north-alpine to boreal.

Formed approximately 6000 years ago, the fifth zone (135-145cm) is characterized by a very low diatom abundance and biodiversity. *Pinnularia borealis* overwhelmingly dominates (82%) the assemblage, and a maximum number of north-alpine species are represented, testifying to extremely cold climatic conditions.

Diatom abundance remained low in the sixth zone (96-124cm), although climatic conditions were warmer and more stable. *Pinnularia* species dominated including: *P. viridis et.var.* (28.8 %), *P. borealis et.var.* (27.8 %), *P. subcapitata* (7.6 %), *P. intermedia* (6.3 %) along with *Hantzschia amphioxys* (9.7 %). The seventh zone (43-94cm) is similar to the sixth zone. However, difference can be seen in larger number of the alfamesosaprobe species and greater climatic variations. The amplitude of climate change was higher than previous periods. The uppermost, eighth zone is characterized by an increasing number of diatoms (8.5-17.3 thousand valves/g dry wt) living in comparatively warm conditions.

Profile ADK-171

Three pits at archaeological site ADK-171 were excavated in 1999 and 2005. Profiles ADK-171/99 and ADK-171/05 unit 3 (Appendices 5B and 5C), located on an eroding road cut at Clam Lagoon, contained archaeological layers representing a kitchen midden with animal bones, shells and some stone tools (see Wilmerding and Hatfield Chapter 12 this volume). A 1m X 2m pit located away from the slope at ADK-171/05 unit 1 was excavated. Only a thin charcoal lens and several culturally modified stones and debitage (see Wilmerding and Hatfield, Chapter 12 this volume) were found in this 1.8m deep pit; no bones or shell were found at this third site.

Three radiocarbon dates of marine fish and mammals bone collagen were run for the ADK-171 settlement: 6620±115 yr BP, (IEMAE-1281); 6620±140 yr BP, (IEMAE-1284); 6770±100 yr BP, (IEMAE-1296). The calibration of these dates shows that the cultural midden at ADK-171 was formed approximately 7000 years ago (5200-4830 BC). These new dates confirm that ADK-171 is the oldest known prehistoric settlement in the Central Aleutian Islands and equally important, the oldest site in the Aleutian archipelago containing significant deposits of faunal remains.

The 4207 invertebrate remains that were identified are listed in Appendix 5D. Bivalves were the most abundant, 66.5% (N=2799) and 99.8% (95.4kg) by weight of all invertebrate remains. Sea urchins constituted 16.3% (N=685) of total invertebrate remains and barnacles represent 14.0% (N=588). Both gastropods and chitons were rare, 1.1% (N=48) and 2.1% (N=87), respectively.

The eight bivalve taxa are: foolish mussel (*Mytilus trossulus)*, arctic hiatella (*Hiatella arctica)*, Nuttall's cockle (*Clinocardium nuttallii)*, butter clam (*Saxidomus giganteus)*, Pacific littleneck clam (*Protothaca staminea)*, arctic surf clam (*Mactromeris polynima)*, softshell-clam (*Mya truncata)*, and Cardiidae (genus and species undetermined) (Appendix 5D). *Clinocardium nuttallii*, is the dominant species [34.5% (N=1452)] of all the invertebrate remains. Foolish mussels and butter clams represent 24.3% (N=1022) and 7.0% (N=295), respectively. Remains of four species (*Hiatella arctica* (N=15), *Mactromeris polynima* (N=11), *Mya truncata* (N=2) and *Protothaca staminea* (N=1) represent less then 0.7% of the invertebrate assemblage.

Three gastropod species were identified: limpet *Lottia pelta* (0.5%; N=23), periwinkle *Littorina sitkana* (0.4%; N=19) and *Nucella lima* (0.1%; N=6), and two chiton species: the very common *Katharina tunicata* (1.5%; N=65) and the more rare *Tonicella submaemorea* (0.5%; N=22). Four barnacle species were also recognized: *Balanus crenatus, Semibalanus balanoides*, *S. cariosus* and *Hesperibalanus hesperius*; and *Balanus crenatus* was dominant. Most fragments of sea urchins (total N=685) were from *Strongylocentrotus polyacanthus* (16.1%; N=677). A few fragments of the sand dollar (*Echinarachnius parma*) were recognized.

Percentages of barnacle species differed between the lower and upper layers at ADK-171. *Balanus crenatus* dominated in the lower layer, but this species declined significantly in the upper layer, replaced by *Semibalanus cariosus*. Currently, *Balanus crenatus* is rare in the littoral zone of the Bering Sea; it prefers colder waters (Pilsbry 1916; Tarasov and Zevina 1957). This change in barnacle species suggests that at approximately 7000 cal BP the climate was initially cooler than present, but the climate became somewhat warmer by the time the ADK-171 midden was abandoned.

There was also a significant change in the mollusks over time at ADK-171 (Figure 5.7). Foolish mussels and other epifaunal species (chitons, gastropods and barnacles) decreased by half from the lower to upper layer. Conversely, Nuttall's cockle, other bivalves and sea urchins increased.

A total of 749 sea mammal bones were collected, but only 138 were identified to family or species (an identification rate of 18.4%) (Appendix 5E). Of the mammal bones, 34.1% (N =47) were the northern fur seal (*Callorhinus ursinus*). Using well preserved fur seal bone fragments, we roughly determined the approximate age of 38 individual fur seals. Of the 38 bone fragments, 20 were interpreted as newborns and/or juveniles and 18 as subadults and/or adults. Nearly 30% (N=33) of all sea mammal remains were from the sea otter (*Enhydra lutris*). Eight otter bone fragments were from juveniles, and 25 from adults and/or subadults. Poor preservation prevented the determination of the sex of the sea otters. The northern sea lion (*Eumetopias jubatus*) was represented by 18.8% (N=26) of the sea mammal bones. Four bones belonged to juveniles, and 22 were subadults and/or adults. Twenty-two bones of the true seal (*Phoca* sp.) accounted for 15.9% of the assemblage, and seven fragments were from the harbor seal (*Phoca vitulina*). The species or age of most true seal bones could not be determined either because of poor preservation or the bone fragments were not species specific. However, two true seal bones were interpreted as juvenile/subadult animals. Two dolphin vertebrae were also recovered.

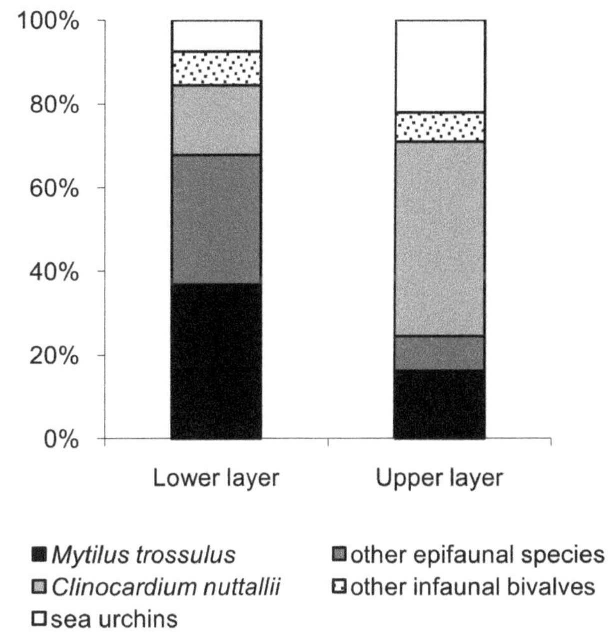

Figure 5.7. Percentages of invertebrates in upper and lower layers at ADK-171, Adak Island, Alaska.

At ADK-171 a thin layer/lens of ash separates the upper and lower midden. This ephemeral ash layer is identified as redeposited Intermediate Ash (Okuno *et al.* Chapter 4 this volume). This suggests that the two stratigraphic layers in the midden are a single deposition event. However, this ash conveniently divides early deposition from later deposition. Temporal changes in animal abundance could be inferred by comparing the animal bones from these two midden layers. Although our conclusions are based on a very small sample size, some general trends through time can be inferred. Over time sea otters, true seals and fur seals decreased slightly, and sea lions increased (Figure 5.8).

Percentages of animal bones at an archaeological site suggest the availability of desired species, preferences by humans for certain species, or both, but it does not necessarily reflect the density of the natural population. Consequently, it is impossible to determine if the Aleuts on Adak simply began to hunt more sea lions or if sea lion populations increased over time. Aleuts living at ADK-171 primarily hunted fur seals and sea otters and relied less on sea lions and seals, and hunted all age groups of all four species. Bones of newborn fur seals indicate the existence of a fur seal rookery in the vicinity of Clam Lagoon 7000 cal BP. Today, most fur seals give birth in the Pribilof Islands, although there is a modest breeding colony in the Commander Islands, Russia, and a small, recently established satellite colony on San Miguel Island, California. Even in the early historic period, fur seal rookeries have not been reported in the Aleutian Islands (Crockford *et al.* 2002).

At ADK-171, 30,332 fish bones and otoliths (ear stones)

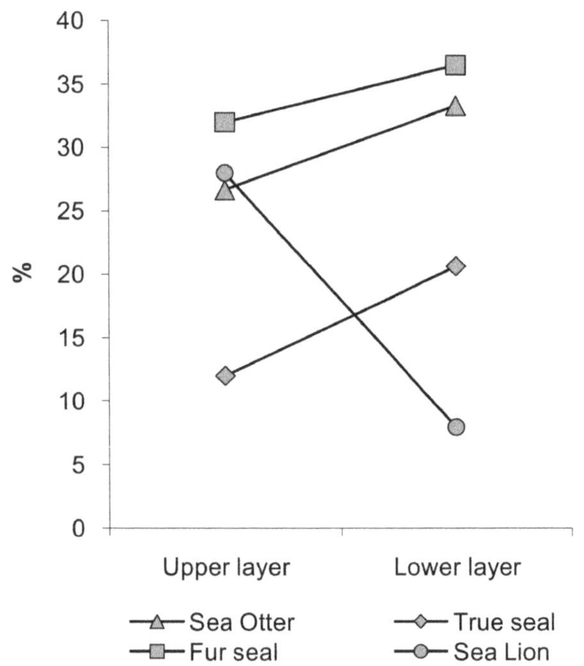

Figure 5.8. Frequencies of sea otter, seal, sea lion and fur seals at ADK-171, Adak Island, Alaska.

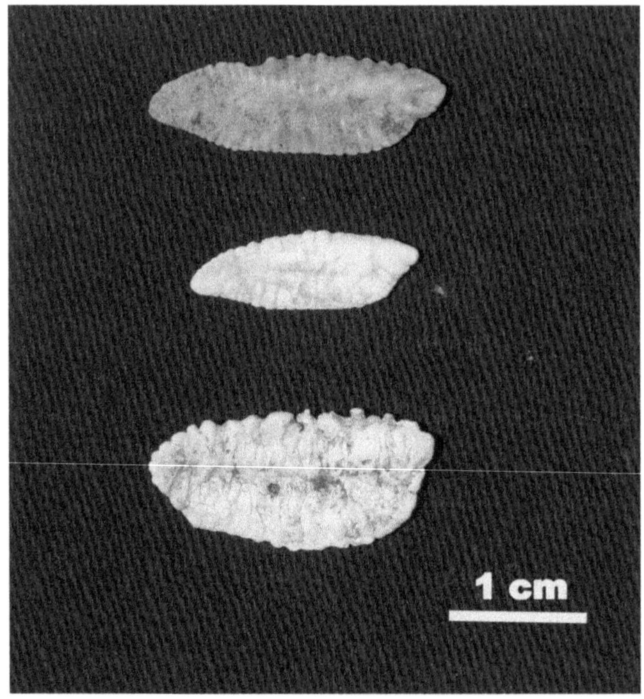

Figure 5.9: Top: Saffron cod otolith recovered from ADK-171, Adak Island, Alaska. Middle: Modern saffron cod otolith. Bottom: Modern Pacific cod otolith.

were recovered, and the bones were identified to family, genus, and species when possible. To quantify this large number of fish bones and otoliths the number of identified specimens (NISP) was used to calculate the percentage contributed by each species. Approximately 13% of the fish remains were identified to the family level or better (Appendix 5F). Eight fish taxa were identified in the assemblage.

The overwhelming majority of the identifiable fish bone represents greenlings (*Hexagrammos*) and cods (family Gadidae), both for which species could not often be determined. Approximately half of all the identified fish were greenlings, over 4114 bones. Currently, four species of *Hexagrammos* are abundant along the Aleutian archipelago (Mecklenburg *et al.* 2002). Bones of this genus are difficult to identify, and are generally lumped as 'greenling,' but it does not include the common hexagrammid species of the Bering Sea, the Atka mackerel (*Pleurogrammus monopterygius*). Of the identified NISP (N=3430) 40.7% were cod. All 26 fragments identified as Pacific cod (*Gadus macrocephalus*) indicate very large specimens. Generally, the fish bones from ADK-171 were undistinguished cod bones of small individuals of one or more species, either Pacific cod, saffron cod, or a mixture of both. Saffron cod (*Eleginus gracilis*; NISP=454; 13.4%) is represented by head bones, a large number of otoliths (Figure 5.9), and some vertebrae (Figure 5.10). Interestingly, all of the otoliths at ADK-171 belong to this species.

Flatfish, all small individuals, comprise 9.2% of the identified fish remains (N=775). Two species of flatfish could be readily identified: halibut (23.7%; N=184), due to its large size and distinctive shape, and starry flounder (11.0%; N=85), due to the dermal skin scutes that are unique to this species. Complete halibut bones were rare; most were fragmented. However, one halibut vertebra from the upper layer was 35mm in diameter and probably represents a 20-year-old fish that would have been more than 1.5m long.

Rockfishes (Sebastidae) and sculpins (Cottidae) contribute 0.5% and 0.8% respectively to the total fish assemblage. Salmonids are rare at ADK-171, and only five *Oncorhynchus* sp. bones were identified.

Approximately 7000 cal yrs BP the Aleuts living at Clam Lagoon caught large quantities of greenlings and small gadoid fishes but caught far fewer flatfish, sculpins and rockfish. Residents rarely caught large Pacific cod and halibut. Greenlings decreased over time, but some species of flatfish increased. The total percentage of cod changed little over time, but Pacific cod decreased, as did halibut and starry flounder.

The most surprising discovery at ADK-171 was saffron cod (*Eleginus gracilis*). Saffron cod have not been identified in later archaeological sites on Adak Island (Crockford 2006; Crockford Chapter 6 this volume; Orchard 2003) or from any archaeozoological assemblages in the Aleutian Islands (Aigner 1976; Crockford 2006; Crockford *et al.* 2004; Orchard 2003). Conversely, Pacific cod are a large percentage of the fish bones from Aleutian archaeological sites on different islands at different times (Aigner 1976;

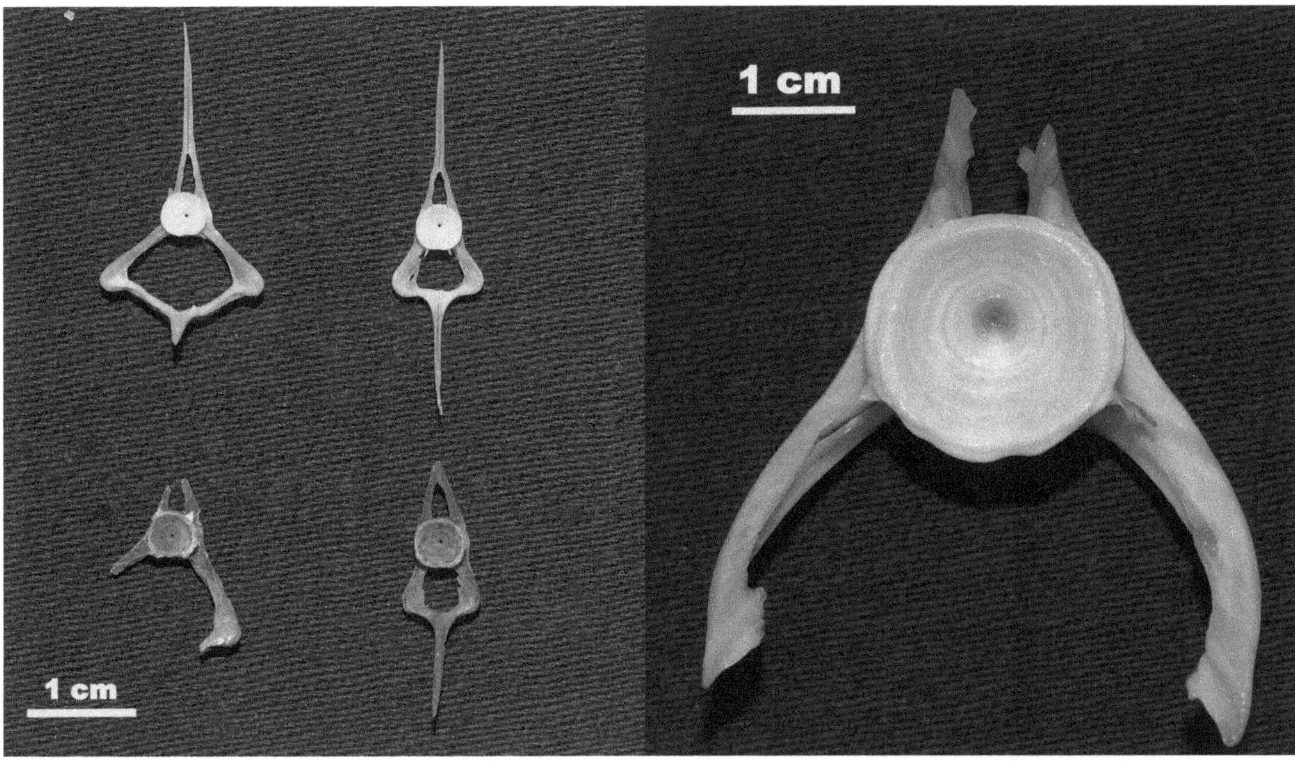

Figure 5.10. Top left: Modern saffron cod vertebrae. Bottom left: Saffron cod vertebrae recovered from ADK-171, Adak Island, Alaska. Right: Modern Pacific cod vertebra.

Crockford *et al.* 2004; Crockford 2006; Lefèvre *et al.* 1997; Orchard 2003).

Saffron cod feed on the deep ocean bottom, but enter brackish waters and rivers up to the limit of tidal influence. Adults move inshore during the winter to spawn and offshore during the summer to feed. Saffron cod spawn where tidal currents are strong and temperatures are cold (down to –1.8°C). Today the preferred habitat of saffron cod is the shelf areas in the Bering Sea where winter shore ice forms. Today saffron cod are absent or very rare in the western and central Aleutian archipelago; this cold loving species is more common in the eastern Aleutians and coastal waters of the Alaska Peninsula (Howes 1991; Mecklenburg *et al.* 2002). Interestingly, bones of saffron cod have not been identified in prehistoric faunal assemblages in the eastern Aleutian Islands (Crockford *et al.* 2004).

Like saffron cod, Pacific cod is also a demersal and pelagic dweller. Pacific cod, a temperate climate species, spawns in deep waters during winter and migrates to shallow waters to fatten during spring and summer (Cohen *et al.* 1990). Today, Pacific cod occur along the entire Aleutian chain. Because saffron cod prefer a colder climate than Pacific cod it is reasonable to suggest that when Aleuts inhabited ADK-171 the climate was much colder. Perhaps Clam Lagoon, with its soft sandy bottom, was a desirable spawning area for saffron cod.

Fish at archaeological sites represent a part of the natural population; they subsequently display characteristics that can be compared to that of past, wild populations. The size and age structure of fish recovered from an archaeological site can reflect hunting activities, competition pressure, or environmental changes (Begon *et al.* 1996). With the relatively large sample size and excellent preservation of bones and otoliths, it was possible to estimate the size and weight of greenlings and saffron cod.

The size (total length - TL) (Figure 5.11) and weight (Figure 5.12) of greenlings was estimated using the methodology developed by Orchard (2003). Measurements of the skull bones of the fish were obtained from the vomer (Mea.#1), dentale (Mea.#1 and #3), articulare (Mea.#1), praemaxillare (Mea.#1 and #2), quadratum (Mea.#4), epihyale (Mea.#1 and #3), and hyomandibulare (Mea.#1 and #2). Only one measurement from each bone was used. Based on preservation, only measurements with the lowest variability were used. Size was estimated using 278 bones; 267 bones were used to estimate weight. Medium-sized fish were dominant at ADK-171 (mean total length 31±0.3cm and mean weight 580±22g). The minimum length was 18cm and weight was 36g; the maximum length was 47cm and weight was 2400g. There was no significant difference in size (length) or weight of greenlings in the two layers at ADK-171 (Mann-Whitney U Test p-level for length - 0.64, for weight - 0.22). Apparently, ancient Aleut fishing pressure did not impact greenling populations during the ADK-171 occupation.

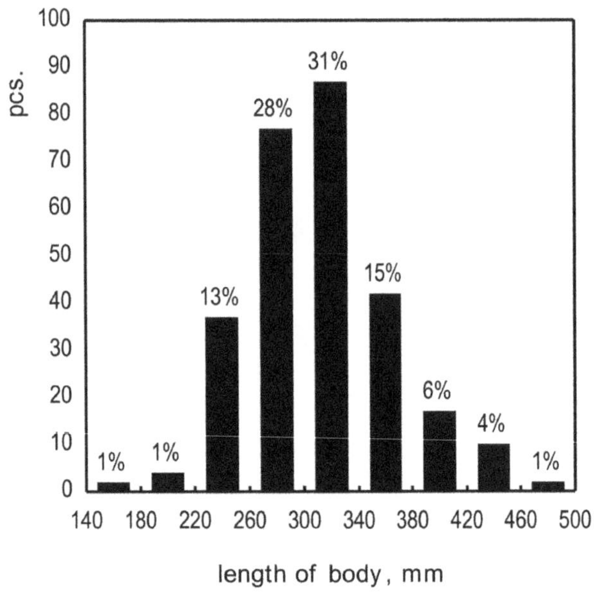

Figure 5.11. Percentages of greenling (Hexagrammos sp.) lengths at ADK-171, Adak Island, Alaska.

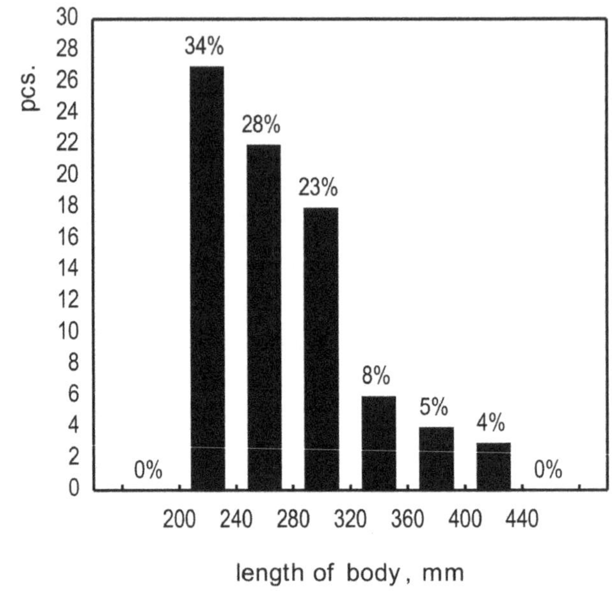

Figure 5.13 Percentages of saffron cod lengths from ADK-171, Adak Island, Alaska.

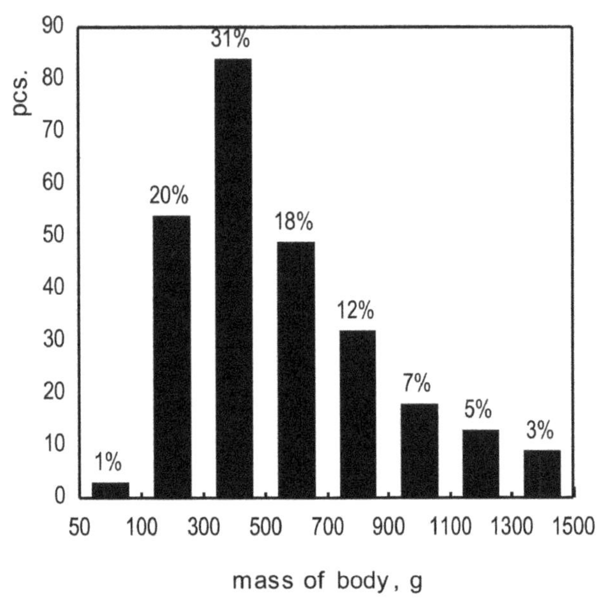

Figure 5.12. Percentages of greenling (Hexagrammos sp.) weights at ADK-171, Adak Island, Alaska.

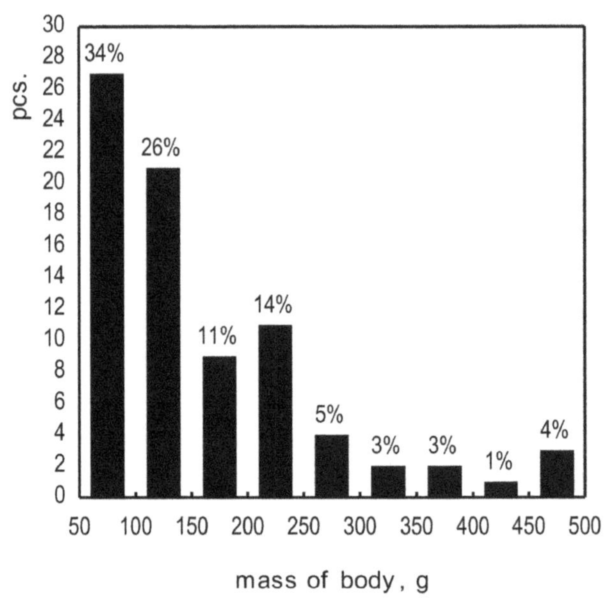

Figure 5.14. Percentages of saffron cod weights from ADK-171, Adak Island, Alaska.

Adult greenlings caught by the earliest Aleuts at Clam Lagoon were relatively large. Greenlings, the most abundant fish species of Aleutian inshore waters (Isakson *et al.* 1971; Simenstad and Nakatani 1977; Simenstad *et al.* 1978), prefer shallow, rocky areas or kelp beds (Mecklenburg *et al.* 2002). During their first year, juveniles can reach 15-20cm in length; after their pelagic stage, they live in inshore communities where they can grow up to 61cm long. Data on the sexual maturity and age structure of modern populations of greenlings in the Bering Sea is scarce. But, in general, greenlings are sexually mature in 2-4 years when they reach a length of 20-35cm, depending on the species (Antonenko and Gnuybkina 2001). Today, sports fishermen often angle for greenlings, but these fish are not commercially important (Eschmeyer *et al.* 1983).

Of the 101 saffron cod otoliths from ADK-171, 80 of the best preserved and most complete were measured to estimate the length (Figure 5.13) and weight (Figure 5.14) of these fish. The height of the otolith and appropriate

Depth, cm	*Laboratory Code, IEMAE-*	*Conventional age*	*Calibrated dates, years AD, (2σ)*
23-38	1264	970±80	1400-1760
45-79	1242	1360±110	1020-1450
115-127	1288	1560±80	860-1270
150-160	1265	2080±55	370-690

Table 5.2. Radiocarbon dates at ADK-009, Adak Island, Alaska.

regression equations were used to calculate the length and weight of this species (Gudkov *et al.* 2005). Saffron cod at ADK-171 averaged 23-31cm (mean 27±5cm) in length and 90-210g (mean 165±12g) in weight. Some were smaller, 20.7cm long and weighing 63g; others larger, 40.8cm long and weighing 485g. Overall, the 7000-year-old saffron cod at Adak Island are slightly larger than similar fish harvested today from the Okhotsk Sea, but smaller than those from Amur Bay (Gudkov *et al.* 2005). All saffron cod in our sample represent mature individuals; we found no 'yearlings' 20cm or less in length.

Adult saffron cod move seasonally, swimming inshore during winter to spawn and moving offshore during summer to feed. Saffron cod, younger than one year, spend winters apart from adults in water 100-200m deep. Because the saffron cod in our sample are adults, it is reasonable to infer that ancient Aleuts fished for saffron cod during the winter when this species moved inshore to spawn.

Profile ADK-009

During 1999, we excavated a 1m X 1m square pit to a depth of 1.92m in Feature 12 at ADK-009 (Appendix 5G). Four radiocarbon dates (Table 5.2), from fish bone collagen, indicate the cultural layers were formed between the 6th and 17th centuries AD. These radiocarbon dates indicate that the formation of the ADK-009 shell midden was uneven. Between the 6th and the middle of the 10th centuries (layers IV and V) the midden accumulated slowly, but formed more rapidly between the 12th and the 14th centuries (layer II).

Remains of 34,971 invertebrate, representing 61 taxa, were identified in the ADK-009 shell midden. Nine taxonomic groups (bryozoans, polychaetes, chitons, gastropods, bivalves, brachiopods, barnacles, decapods and sea urchins) were common (Appendix 5H); the other 52 taxa were rare. Ethnographic records (Zubkova 1948; McCartney 1975; Veniaminov 1984) indicate that 18 species of those identified in the Sweeper Cove sample were important Aleut food resources. These include: chitons (*Katharina tunicata* and *Cryptochiton stelleri)*, gastropods (*Lottia pelta*, *Littorina* spp., *Nucella lima*, *Buccinum percrassum* and *Volutopsius castaneus)*, bivalves (*Modiolus modiolus*, *Mytilus trossulus*, *Monia macrochisma*, *Clinocardium nuttallii*, *Serripes groenlandicus*, *Macoma* sp., *Mactromeris polynima* and *Mya truncata)*, barnacles (*Semibalanus cariosus)* and sea urchins *(Strongylocentrotus polyacanthus)*. Other species were occasionally used.

Layers	*Age, cal AD*	*Brillouin index*
I	1320-1600	1.43
II	1100-1320	1.47
III	940-1100	1.13
IV	760-940	0.65
V	500-760	1.37

Table 5.3. Biodiversity index data for the layers of shell midden ADK-009.

In the samples we analyzed, chitons (*Katharina tunicata)*, gastropods (*Littorina sitkana* and *Lottia pelta)*, foolish mussel (*Mytilus trossulus)*, barnacles (*Semibalanus cariosus)* and sea urchins (*Strongylocentrotus polyacanthus)* were the dominant species. Sea urchins represent 50-70% of the total number of invertebrates, with the foolish mussel representing 9-25% of the total invertebrate assemblage and nearly 99% of bivalves. The barnacle (*Semibalanus cariosus)* comprised 12-18% of invertebrates and 60-89% of the barnacles. Gastropods were mainly *Littorina sitkana* (3.7%) and *Lottia pelta* (2.5%), and chitons were mainly *Katharina tunicata* (1%).

Invertebrate biodiversity (Table 5.3) was relatively high during the 6th century (layer V), decreased by half between the mid-8th and the mid-10th centuries (layer IV) and again increased, reaching a peak between the 12th and the mid-14th centuries (layer II). The proportion (95%) of the six main species did not change between the 6th to the 12th centuries (layers III, IV, and V). Between the 12th and the mid-14th centuries (layer II), this percentage decreased to 88% and then increased to 92% (layer I). The greatest number of invertebrates at ADK-009 occurs between the 12th and the 14th centuries (layer II), and indicates an increase in the productivity of the littoral zone during this time. Productivity in the Bering Sea usually increases during relatively cold periods (Savinetsky *et al.* 2004).

Today, the Brillouin index of biodiversity for the littoral zone of Sweeper Cove is 1.44; this is comparable to values obtained for the period from the 12th to the 17th centuries (layers I and II). The faunal composition of the upper layers of the ADK-009 deposit does not significantly differ from the modern fauna living in the littoral zone at Sweeper Cove (Antipushina 2006a; 2008b). Today, *Semibalanus cariosus* and *Mytilus trossulus* are dominant species in the tidal zone with an average density of 2300 ind/m^2 and 600 ind/m^2,

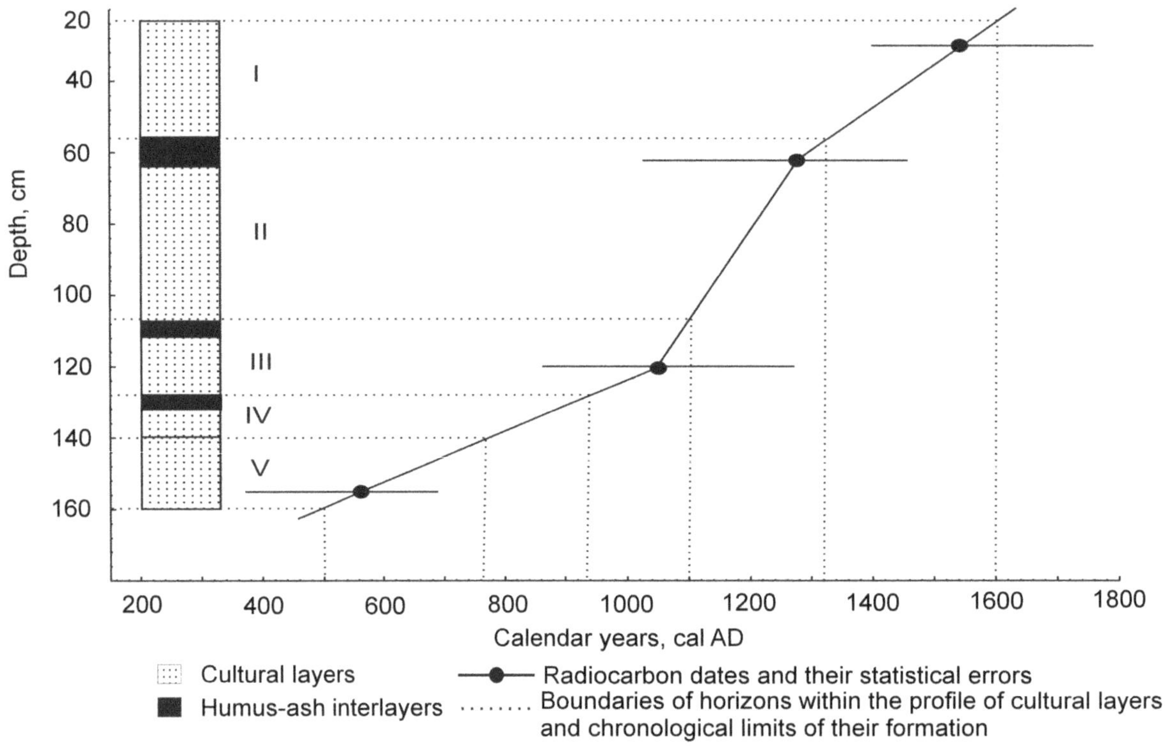

Figure 5.15. Temporal growth and duration of the five cultural layers at ADK-009, Adak Island, Alaska.

Number of layer, age of formation	***V 500-760 AD***	***IV 760-940 AD***	***III 940-1100 AD***	***II 1100-1320 AD***	***I 1320-1600 AD***
Accumulation rate of layer (cm/100yrs)	8.0	7.0	13.0	23.2	12.0
Taxonomic group					
Katharina tunicata	4.6* 60	13.3 200	33.7 257	59.5 257	31.4 267
Lottia pelta	3.0 40	16.7 250	43.7 333	116.4 502	76.4 648
Littorina sitkana	19.6 255	29.5 442	16.9 128	278.6 1202	57.5 488
Mytilus trossulus	171.1 2225	108.3 1625	317.5 2419	872.7 3765	633.6 5376
Semibalanus cariosus	117.7 1530	172.2 2583	668.1 5090	1106.4 4772	363.6 3085
Strongylocentrotus polyacanthus	333.8 4340	913.3 13700	2427.5 18495	4284.0 18480	1645.3 13960

* accumulation rate (sp/100yrs*m2)
concentration of remains (sp/m3)

Table 5.4. Accumulation rate and concentration of most abundant invertebrate species collected at ADK-009, Adak Island, Alaska.

respectively. Currently, there are 560 to 2480 individuals/m^2 of *Littorina sitkana*, and sea urchins are rare in the modern littoral zone; this could be due to sea otter predation (Estes and Duggins 1995).

The accumulation rate of invertebrate remains (NISP per 100 years) reflects the number of animals hunted or gathered, the 'catch value.' Thus, a change in this rate reflects changes in the 'catch value' per 100 years. It is accepted that the concentration (number) of remains of animal species reflects their relative number in the animal population at a specific time, and changes in the relative number of individuals in the population from period to period is an approximation of the changes in the number of animals in the population. Changes in the remains of invertebrates in the ADK-009 shell midden indicate changes in 'catch value' of ancient Aleuts, and changes in the concentration of invertebrate remains illustrates the temporal changes in the invertebrate populations. According to results of analysis (Figure 5.15 and Table 5.4)

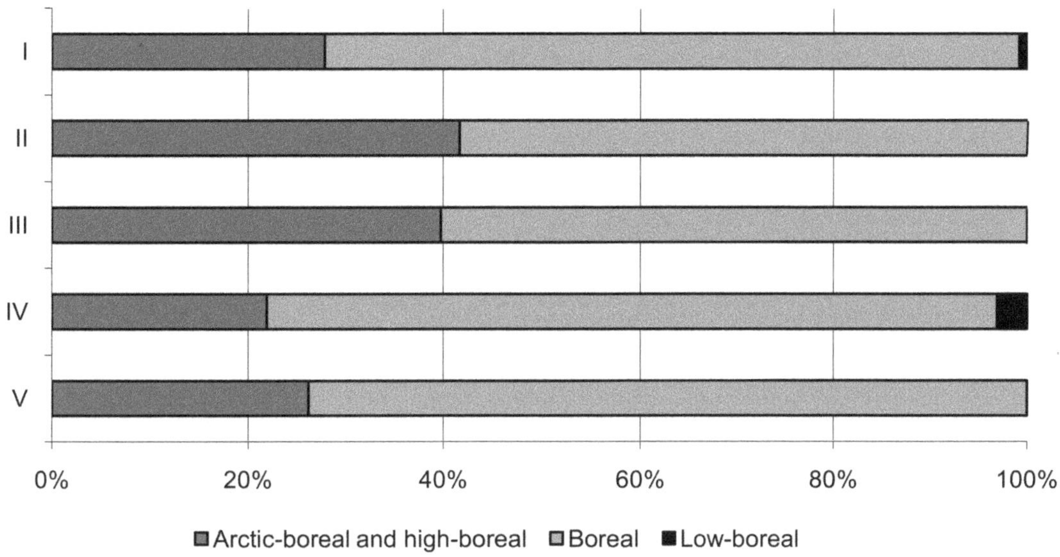

Figure 5.16. Zoogeographic spectra of the mollusks in the five layers of the shell midden at ADK-009, Adak Island, Alaska.

collection increased from the 6th until the 12th centuries (layers III, IV, and V), reached a maximum from the 12th until the 14th centuries (layer II) and then decreased. The number of the main prey species of invertebrates did not significantly change despite the intensification of collecting during the 12th to the 14th centuries (layer II). The decrease in biodiversity between the 6th and the mid-10th centuries (layers IV and V) cannot be attributed to the human impact on invertebrate populations, because the intensity of shellfish collecting was relatively low in this period. The traditional use of the littoral resources by the ancient Aleuts did not cause the decrease in the biodiversity of invertebrate species (Antipushina 2008b).

The faunal composition of the ADK-009 shell midden does not significantly differ from that of the modern fauna living in the littoral zone at Sweeper Cove. To demonstrate this, we divided all shellfish species in the shell midden into two groups: 1) the epifaunal types that prefer rocky substrates and, 2) infaunal types that bury themselves in loose sediments. Remains of epifaunal mollusks dominated in all layers of the ADK-009 shell midden, representing more than 90% of all mollusk remains. This percentage does not significantly change from layer to layer. This suggests that the rocky intertidal zone at Sweeper Cove has not dramatically changed from the 6th century to the present (Antipushina 2006b).

The proportion of mollusks in three zoogeographic groups (arctic-boreal, high-boreal and low-boreal) was not constant over time (Figure 5.16). The percentages of arctic-boreal and high-boreal species were lower in archaeological layers IV and I. Moreover, remains of the low-boreal gastropod *Nucella heyseana*, atypical in the Bering Sea, were found in the same layers. On the other hand, the percentage of arctic-boreal and high-boreal species increased in layers III and II. This change is probably due to changing temperature over time. By analyzing the taxonomic composition of the molluscs, it is possible to infer changes in temperature during the development of this shell midden. The interval from the 6th to the middle of the 8th centuries (layer V) was relatively cold. The warmer period from the middle of the 8th to the middle of the 10th centuries (layer IV) encouraged the migration of the low-boreal gastropod *Nucella heyseana* into the higher intertidal zone resulting in a decrease in psychrophilic (low temperature tolerant) species. An increase of arctic-boreal and high-boreal species from the middle of the 10th to the middle of the 14th centuries (layers II and III) suggests a colder temperature at this time (Cook *et al.* 2004; Savinetsky *et al.* 2004).

To estimate whether changes in zoogeographic structure corresponded to climatic changes we analyzed changes in $\delta^{18}O$ and $\delta^{13}C$ values in the plates of the littoral barnacle species *Semibalanus cariosus*. The results of these analyses show synchronous changes in $\delta^{18}O$ and $\delta^{13}C$ (Figure 5.17). The value of $\delta^{18}O$ is higher in layers V, III and II of the midden, and decreases in layers IV and I. As stated above the changes in the $\delta^{18}O$ values indicate temperature changes; changes in $\delta^{13}C$ show a similar trend. The $\delta^{13}C$ minimum in layer IV may be the result of higher metabolic rates due to increasing temperature, and the increase of $\delta^{13}C$ in layers V, II, and II may indicate a temperature decrease. The timing of phytoplankton blooms can also influence $\delta^{13}C$ values. These data suggest that conditions from the 6th to the mid-8th (layer V) and from the mid-10th to the mid-14th centuries (layer II and III) were colder, and from the mid-8th to the mid-10th (layer IV) and from the mid-14th to the 17th centuries (layer I) were warmer. In many respects these results correspond with the isotopic data from ostracode valves from lake sediments in Alaska (Hu *et al.* 2001) and with the dendrochronological data from Russian Arctic (Hantemirov and Shiyatov 2002).

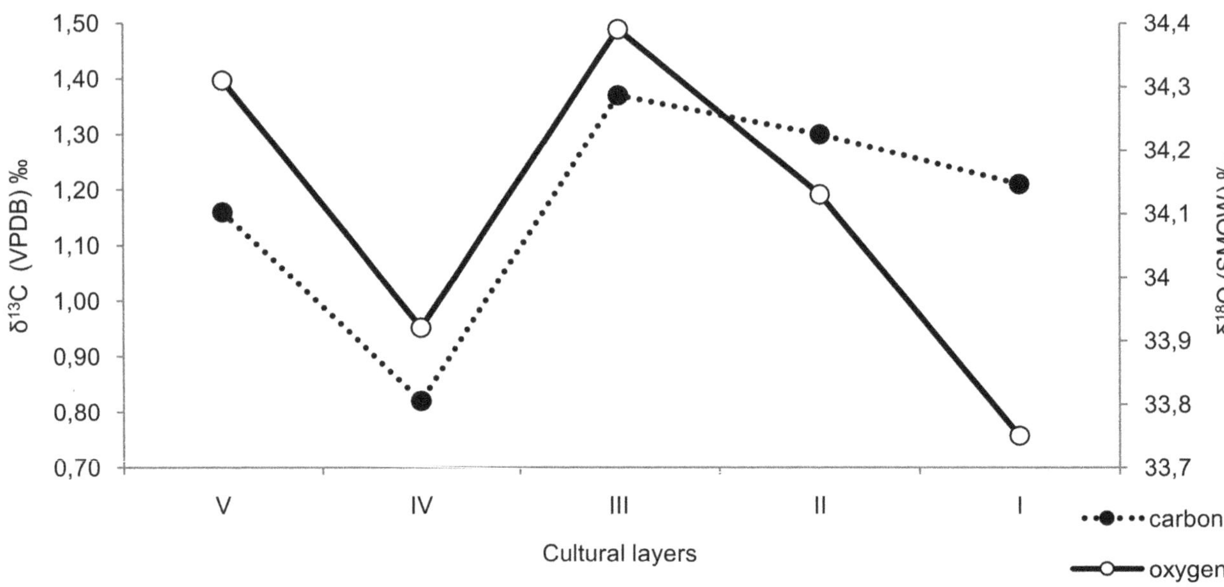

Figure 5.17. Changes in the $\delta^{18}O$ and $\delta^{13}C$ in the plates of barnacle Semibalanus cariosus at ADK-009, Adak Island, Alaska.

Ethnographic data indicate the ancient Aleuts collected littoral zone invertebrates at low tide and did not use any gear for harvesting them in shallow waters (Veniaminov 1984). However, the shell midden contained some sublittoral species such as *Hesperibalanus hesperius, Balanus balanus, Cyclocardia ventricosa, Aulacofusus periscelidus, Volutopsius castaneus, Boreotrophon cepula,* and *Puncturella longifissa*. Their proportion in layers V (500-760 AD), IV (760–940 AD), and III (940–1100 AD) was only 1.5–2%; it increased to 6% in layer II (1100–1320 AD) and then decreased by half, to 3% in layer I (1320-1600 AD) (Figure 5.18). The presence of these species in the shell midden may be explained by their migration from the sublittoral to the littoral zone, which could be a response to a cooler climate. Changes in the proportion of sublittoral species in the layers of the midden do not correspond exactly with changes in the zoogeographic and isotopic data. However, the maximum peak in the number and proportion of sublittoral species (Figure 5.18) occurred during a period of cooling between the 12th and the mid-14th centuries (layer II), not unlike that indicated by zoogeographic and isotopic data.

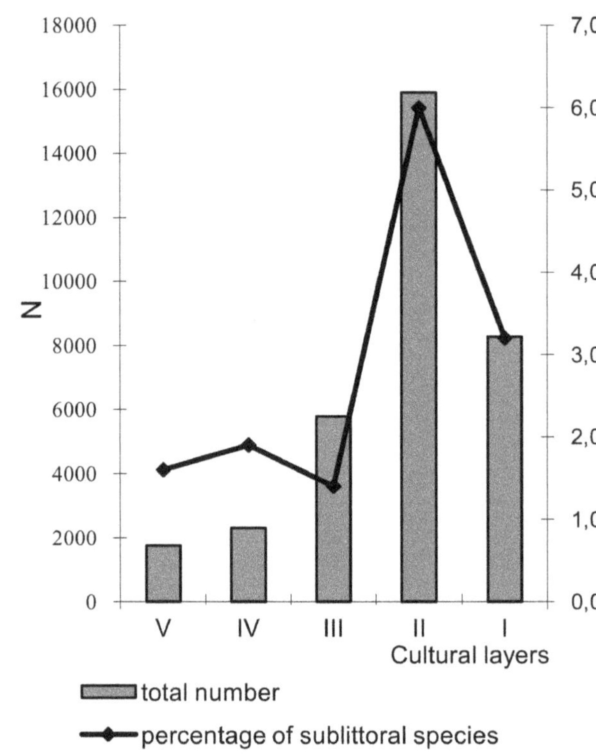

Figure 5.18. Changes in the sublittoral species in the five layers in the shell midden at ADK-009, Adak Island, Alaska.

Thus, the stable isotope analyses suggest that some long-term temperature changes may have taken place during the formation of the midden. The interval from the 6th to the mid-8th centuries (layer V) appears to have been relatively cold. Climatic warming took place between the mid-8th and the mid-10th centuries (layer IV) and between the mid-14th and the 17th centuries (layer I), with an intervening cooling period (from the mid-10th to the 14th centuries (layers II and III). This cooling was most pronounced between the 12th and the 14th centuries (layer II). Periods of cooling coincide with periods of high biodiversity and productivity.

Study of the invertebrate remains from the cultural layer of the ancient Aleut settlement on Adak Island showed that the essential diversity of invertebrates at ADK-009 did not significantly change during the Late Holocene. Ancient Aleuts actively harvested sea urchins *S. polyacanthus*, mussels *M. trossulus*, barnacles *S. cariosus*, and gastropods *Littorina sitkana* and *Lottia pelta*. However, their traditional mode of natural resource use had no impact on the abundance of valuable species and on biodiversity in general.

The littoral zone of Sweeper Cove was chiefly stony, with a certain proportion of sandy ground; this has not significantly changed since the 6th century. Zooarchaeological analysis of the remains of vertebrate animals from ADK-009 material represents mammals only. A total of 40 sea mammal bones were identified to family or better, out of a combined site total of 119 bones (identification rate about 30%). Most bones were fragmented and complete bones are very rare. Because there are so few mammal bones, we combined data from all layers (Appendix 5I).

Most of the mammal remains represent sea otter (*Enhydra lutris*). Two fragments belong to juvenile animals; all others represent adult or subadult animals. Poor preservation of the remains precludes a determination of the sex of any sea otter elements. The next most common species, by NISP, is true seal (*Phoca* sp.). Both poor preservation and the fragmentary nature of the remains precluded species determination and age estimation of the phocid seal sample. Five bone fragments represent northern sea lion (*Eumetopias jubatus*). One bone belongs to juvenile animal; all others represent subadult and adult animals. Northern fur seal (*Callorhinus ursinus*) is represented by only two bones, both from newborns.

Summary and Conclusions

To reconstruct the ancient environment of Adak and the surrounding Bering Sea natural and archaeological deposits on the north side of Adak Island, Aleutian Islands, Alaska were studied. The stratigraphy and composition of a sequence in a peat bog was compared with that from two archaeological sites. This approach revealed new information about the Holocene history of the ecosystems of Adak and the surrounding area.

Previous investigations (Black 1980; Heusser 1978; Kiriyanov and Miller 1997; Okuno *et al.* 2007; O'Leary 2001) suggested that the most highly visible soil-ash deposits on Adak Island were formed during the last 10,000 years. Ash layers on Adak were the result of volcanic activity on the surrounding islands; volcanoes on the northern half of the island—Moffet and Adagdak—were dormant during this time (Fraser and Snyder 1959; Okuno *et al.* Chapter 4 this volume; Waythomas 1995). The thickness and composition of the soil-ash layers provide information about the location of the active volcanoes that produced the ash on Adak and the geomorphology of the island when the ash was deposited (see also Okuno *et al.* Chapter 4 this volume)

Haven Lake, a key soil profile for understanding the environment and climate of Adak, contains the most complete number of ash layers and has been the subject of much previous research and analyses (Black 1976, 1980; Heusser 1978). A complex deposit of peat-tephra layers was revealed in the carefully cleaned profile at Haven Lake. Three buried peat bog soils separated by four well defined tephras were exposed. The soil and peat formation in the bog was interrupted approximately 9000-8500 years ago by a thick ash layer, the Main Ash, located 167-182cm below the surface. Kiriyanov and Miller (1997) stated that the physical characteristics of this ash indicated that the responsible eruption was about 100km from Adak Island. Above this, at 124-135cm, is the Intermediate Ash, which was deposited approximately 7000 years ago. The uppermost major ash layer is the Sandwich Ash (85-94cm) deposited approximately 4700 years ago.

Eight diatom zones were recognized, and it can be inferred from the diversity of diatom species in the early (lower) Haven Lake samples that initial deposition in the immediate area was in a shallow, freshwater pool in a relatively warm boreal climate. Diatom diversity above this level suggests that cold, alpine-like conditions prevailed in a habitat with few nutrients. The third diatom zone was formed just prior to deposition of the Main Ash, approximately 8500 years ago, where the abundance and diversity of diatom species suggest a climate shift back to a warm boreal climate. Just above the Main Ash layer, the low diatom abundance and diversity suggest that climatic conditions were transitional between cold alpine and boreal. The fifth diatom zone, which formed approximately 7000 years ago, is characterized by a very low diatom abundance and biodiversity. The maximum numbers of cold alpine species represented in this zone indicate extremely severe climatic conditions at this time. Above this level, in the sixth and seventh zones, diatom abundance indicates a return to a warmer, more stable, climate. The uppermost, eighth zone is characterized by an abundance and variety of diatoms indicative of warm conditions comparable to the present climate.

Significantly, the diatoms indicate that the ADK-171 site was occupied, just after the Intermediate Ash fall, during the coldest period in the Central Aleutians in the last 10,000 years. Based on three radiocarbon dates from fish bone collagen, the cultural midden at ADK-171 was formed approximately 7000 years ago. These new dates confirm that ADK-171 is the oldest known prehistoric settlement in the Central Aleutians. Equally important, it is also the oldest site in the entire Aleutian archipelago that contains animal remains.

Analysis of soil formation processes near Clam Lagoon suggest that, prior to the ADK-171 human occupation, conditions were favorable for peat and soil formation; these conditions did not significantly differ from those occurring at the present time. However, the soil formation at the coastal ecosystems, particularly near Clam Lagoon, was interrupted by the formation of dune sand. Dune sand could only accumulate when a large area became sea free due to low sea level for an extensive time during this period. The current tidal regime cannot form such deposits. This dune sand accumulation most likely, suggests a lake was present instead of a lagoon prior to human habitation in the area. Ultimately, with rising sea level, the ancient lake became a lagoon, and Intermediate Ash covered the sand dunes associated with the former lake. Soon after the deposition of Intermediate Ash and the increase in sea level the site at ADK-171 was occupied.

A thin line of ash, identified as ephemeral redeposited Intermediate Ash (Okuno *et al.* Chapter 4 this volume) conveniently separates the midden at ADK-171 into upper and lower layers. The unknown geological event that caused redeposition of the Intermediate Ash apparently did not impact local animal communities. Using this stratigraphic separation, animal remains from the two different levels (lower/early vs. upper/late) were compared to detect any possible changes in species abundance over time. Despite gaps in the archaeological record, our analysis of the fish, mammal and invertebrate remains suggests that the animal community around Clam Lagoon has not changed appreciably over the last 7000 years.

Bivalves dominate the invertebrate sample, and Nuttall's cockle, *Clinocardium nuttallii*, was the most abundant bivalve species. Barnacles and sea urchins were less abundant than cockles, and gastropods and chitons were rare. There was a noticeable change in the relative abundance of invertebrate species at ADK-171 over time. For example, the percentage of different barnacle species changed from the lower to the upper layer: *Balanus crenatus,* a cold-tolerant species, dominated the lower layer, but in the upper layer, *Semibalanus cariosus* is the most common. This change in dominant barnacle species suggests that at approximately 5000 BC, when the site was first occupied, the climate was markedly cooler than today but became warmer over time.

The chief mammalian prey species at ADK-171 were sea lions, sea otters, and fur and harbor seals. Remains of unweaned fur seal pups, which must have been hunted from late summer through fall, indicate a fur seal rookery nearby. An increase in the abundance of clams and sea urchins over time suggests that these food items were either harvested more heavily over time (out of necessity or preference) or that clams and urchins were actually more abundant due to better living conditions or decreased predation by sea otters. However, there is no hard evidence that the number of sea otters markedly declined during the time that the site was occupied. Although our conclusions are based on a very small sample, some temporal trends were documented. Relative proportions of sea otter, harbor seal and fur seal remains decreased slightly over time but those from sea lions increased.

Fish remains from ADK-171 indicate that ancient Aleuts caught their prey close to the shore. The overwhelming majority of the identifiable fish bones represent cods and greenlings (which could include remains of any or all four species in the genus *Hexagrammos*) in almost equal amounts. In contrast to most other prehistoric sites in the Aleutians, most of the cods at ADK-171 were relatively small individuals. Flatfish species, including halibut and starry flounder, were fairly common, but rockfishes (Sebastidae) and sculpins (Cottidae) were poorly represented. Salmonid bones were rare.

Remains of saffron cod (*Eleginus gracilis*), a small gadoid fish with very distinctive bones, were surprisingly common given the rarity of their occurrence in the Aleutians today. Because saffron cod prefer cold inshore waters for winter spawning, the modest abundance of their remains not only indicates winter fishing, but also infers that a cooler environment existed at the time ADK-171 was initially occupied. However, saffron cod are known to enter brackish waters and rivers up to the limit of tidal influence. It is possible that Clam Lagoon, with its soft sandy bottom and cold water, was especially suitable as a winter spawning ground at the time the human settlement was established, and when the diatom data indicate the climate was particularly cold. The fact that the relative proportion of saffron cod remains increased at ADK-171 over time suggests saffron cod continued to use the lagoon for winter spawning even though other indicators suggest the overall climate was warming. In contrast, the abundance of greenlings and Pacific cod decreased slightly over time, while all flatfishes increased.

On northern Adak Island there was apparently an archaeological hiatus between 7000 and 4000 years ago. Little is known about this period. However, compared with the previous period, it can be characterized as a time of unstable, warmer conditions, perhaps with an increase in precipitation. Possibly people simply did not live on the northern part of Adak between 7000 and 4000 years ago, and chose to settle on nearby Kanaga, Kagalaska, Tanaga, or the southern shores of Adak. Or perhaps the most ancient sites on Adak are covered with meters of overburden and we simply have not found them. At any rate, circa 4700 years ago, any village site on northern Adak would have been covered with Sandwich Ash.

Current evidence suggests the area at Clam Lagoon was reoccupied beginning 3590 years ago (Table 5.5; Luttrell and Corbett 2000). Villages sprang up on the Clam Lagoon shores and north half of Adak Island and persisted until approximately 2000 BP; this 2000-year time period is characterized by comparatively stable precipitation.

Archaeological site ADK-009 is located near the head of Sweeper Cove on a bedrock shoreline fronted by a gravel beach (Luttrell and Corbett 2000). Four radiocarbon dates suggest the cultural layers were formed between approximately 2080 and 970 years ago (starting about the 6th century AD). Radiocarbon dates suggest that the midden slowly accumulated between the 6th and middle of the 10th centuries and accumulated much more quickly between the 12th and the 14th centuries.

Sea urchins were the dominant invertebrates, between 50 and 70% of the total counted. Foolish mussel was relatively common and was the most common bivalve. The barnacle (*Semibalanus cariosus*) was also relatively common;

Site number	Location	Dates
ADK-171	Eastern Clam Lagoon	7000 BP
	2000 Year Cultural Hiatus?	
ADK-187	Eastern Clam Lagoon	3590±50 BP
ADK-181*	Southwest Clam Lagoon	3300±50 BP
ADK-194	West Clam Lagoon	3050±70 BP
ADK-193	Northeast Clam Lagoon	2900±40 BP
ADK-011*	Zeto Point	2500 BP
ADK-SW	Clam Lagoon island	2500 BP
ADK-012**	West Clam Lagoon	2470±15 to 1515±15 BP
ADK-180	Clam Lagoon peninsula	2060±40 BP
ADK-181	Southwest Clam lagoon	1950±40 BP
ADK-182	Lake north of Zeto Point	1920±40 BP
ADK-009*	Sweeper Cove	1888±50 BP
ADK-009	Sweeper Cove	1360±74 to 1126±103 BP
ADK-185	North of Clam Lagoon	810±40 BP
ADK-009	Sweeper Cove	755±72 BP
ADK-183	North of Clam Lagoon	460±40 BP
ADK-011	Zeto Point	250 to170 BP

*Site was occupied during more than one time period
** Most dates for ADK-012 are 2450 to 2300 BP

Table 5.5. Radiocarbon dates for settlements (from oldest to youngest) located on the north side of Adak Island (from Luttrell and Corbett 2000 and the Central Aleutians Project).

gastropods were relatively rare. The composition of the invertebrate fauna does not differ significantly from the modern fauna living in the littoral zone at Sweeper Cove (Antipushina 2006a; 2008b).

All shellfish species in the shell midden were divided into two groups: 1) epifaunal types and, 2) infaunal types. Epifaunal mollusks dominated in all layers of the midden and did not change significantly from layer to layer. This suggests that the rocky intertidal zone of Sweeper Cove has not changed much since the 6th century AD (Antipushina 2006b).

Changes in the relative proportion of $\delta^{18}O$ and $\delta^{13}C$ in plates of the littoral barnacle *Semibalanus cariosus* show that $\delta^{18}O$ and $\delta^{13}C$ in the carbonate of the plates changed over time. It suggests that the intervals from 6th to the mid-8th and from the mid-10th to the 14th centuries were colder. The intervals from the mid-8th to the mid-10th and from the mid-14th to the 17th centuries were warmer. The cooler periods appear to coincide with high biodiversity and productivity.

Most of the 40 identifiable sea mammal bones recovered were highly fragmented, and some specimens were not well preserved. Although the poor condition of the bone precludes conclusive species determination, most of the remains appear to be sea otters (*Enhydra lutris*) and possible harbor seals (*Phoca* cf. *vitulina*). A few fragments of the northern sea lion (*Eumetopias jubatus*) were present and only two bones of the northern fur seal (*Callorhinus ursinus*). However, both of the fur seal bones are remains of newborn (unweaned) individuals.

Acknowledgements

This research was made possible with a grant from the National Science Foundation, Office of Polar Programs, Arctic Social Sciences Program (Award # 0353065), Grant RFBR No. 09-04-00196 and the Russian Programs "Origin and Evolution of Biosphere", "Fundamental Basics of management of Biological Resources", "Biological Diversity". The Aleut Corporation granted permission to work in the Clam Lagoon vicinity on north Adak Island. Much thanks goes to Ken Pratt, USBIA, Anchorage, Alaska, and Ron West, professor emeritus, Kansas State University for reviewing and revising the manuscript.

References

Acsadi, György, and János Nemeskeri. 1970. *History of Human Lifespan and Mortality*. Budapest, Academical Kiedo.

Aigner, Jean S. 1976. Early Holocene Evidence for the Aleut Maritime Adaptation. *Arctic Anthropology* 13 (2), 32-45.

Allen, Kenneth R. 1966. Some Methods for Estimating Exploited Populations. *Journal of Fisheries Research Board of Canada* 23, 1553-1574.

Allen, Kenneth R., and Geoffrey P. Kirkwood. 1988. Marine Mammals. In J. A. Gulland (ed.), *Fish Population Dynamics: The Implications for Management*, 151–269. Chinchester, Wiley.

Antipushina, Zhanna A. 2006a. *Recent and Holocene Invertebrates on Adak, the Aleutian Islands*. Paper presented at the Conference of Young Scientists and

Postgraduates of the Severtsov Institute of Ecology and Evolution, Russian Academy of Sciences. [in Russian].

Antipushina, Zhanna A. 2006b. Analysis of Holocene Invertebrates from the ADK-009 Site on Adak, the Aleutian Islands. In *The Holocene Dynamics of Recent Ecosystems*, 24–29, KMK Scientific Press Ltd, Moscow. [in Russian].

Antipushina, Zhanna A. 2008a. *Quantitative Account of Invertebrate Remains in Archaeological Sites.* Paper presented at the Conference of Young Scientists and Postgraduates of the Severtsov Institute of Ecology and Evolution, Russian Academy of Sciences. [in Russian].

Antipushina, Zhanna A. 2008b. *Dynamics of Invertebrate Biodiversity in the Littoral Zone of the Commander–Aleutian Range in the Holocene.* Paper presented at the VIII International Conference 'Biodiversity Conservation in Kamchatka and Surrounding Seas' dedicated to the 275th Anniversary of the Second Kamchatka Expedition, 1732–1733, Petropavlovsk-Kamchatskii. [in Russian].

Antonenko Dmitry V., and Valentina P. Gnuybkina. 2001. Some Peculiarities of the Early Ontogeny of the Alaska Greenling *Hexagrammos octogrammus* and Fat Greenling *H. stelleri* from the Peter the Great Bay (Sea of Japan). *Journal of Ichthyology* 41(6), 799-803 [in Russian].

Arinushkina, Evdokiya V. 1970. *Manual to Chemical Analysis of Soil*. Moscow, Moscow State University Press.[in Russian].

Arslanov, Hikmatulla A., Sergej B. Chernov, and Tatjana V. Tertychnaya. 1987. Calibration of the All-Union Radiocarbon Standard by the NBS Standard and Pine Annual Rings of 1846-1855. In *Methods of Radioisotope Geology*, part 2, 254-255, Moscow [in Russian].

Battarbee, Richard W. 1986. Diatom Analysis. In B. E. Berglund (ed.), *Handbook of Holocene Palaeoecology and Palaeohydrology*, 527-570. Chinchester, Wiley.

Begon, Michael., John L. Harper, and Colin R. Townsend. 1996. *Ecology: Individuals, Populations and Communities.* 3rd Edition. New York, Blackwell Science.

Bird, Douglas W., Jennifer L. Richardson, Peter M. Veth, and Anthony J. Barham. 2002. Explaining Shellfish Variability in Middens on the Meriam Islands, Torres Strait, Australia. *Journal of Archaeological Science* 29, 457-469.

Black, Robert F. 1976. Late-Quaternary Glacial Events, Aleutian Islands, Alaska. *Quaternary Glaciations in the Northern Hemisphere*. International Union of Geological Sciences United Nations Educational, Scientific, Educational, Scientific, and Cultural Report No. 3, 285-301.

Black, Robert F. 1980. Isostatic, Tectonic, and Eustatic Movements of Sea Level in the Aleutian Islands, Alaska. In Nils- Axel Morner (ed.), *Earth Rheology, Isostasy and Eustasy,* 231-248. New York, John Wiley and Sons.

Bockstoce, John R., and Daniel B. Botkin. 1983. The Historical Status and Reduction of the Western Arctic Bowhead Whale Population by the Pelagic Whaling Industry. *Reports of the International Whaling Commission Special Issue* 5, 107-142.

Cameron, Christopher P., and David B. Stone. 1970. Outline Geology of the Aleutian Islands with Paleomagnetic Data from Shemya and Adak Islands. *University of Alaska Fairbanks Geophysical Institute Report UAG-R* 213, 1-153.

Chapman, Douglas G. 1974. Estimation of Population Size and Sustainable Yield of Sei Whales in the Antarctic. *Reports of the International Whaling Commission* 24, 82-90.

Cohen, Daniel M., Tadashi Inada, Tomio Iwamoto, and Nadia Scialabba. 1990. FAO Species Catalogue. Gadiform Fishes of the World (Order Gadiformes). An Annotated and Illustrated Catalogue of Cods, Hakes, Grenadiers and other Gadiform Fishes Known to Date. *Food and Agriculture Organization of the United Nations Fisheries Synopsis* 10, 1-125.

Cook, Edward R., Jan Esper, and Rosanne D.D. Arrigo. 2004. Extra-Tropical Northern Hemisphere Land Temperature Variability over the Past 1000 Years. *Quaternary Science Reviews* 23(20-22), 2063-2074.

Crockford, Susan. 2006. *Faunal Remains from Adak Island Test Excavations 2005*. Report prepared for D. West, University of Kansas, Principal Investigator. Unpublished Manuscript.

Crockford, Susan. 2011. Archaeozoology of Adak Island: 6000 Years of Subsistence History in the Central Aleutians. Chapter 6, in D. West, V. Hatfield, E. Wilmerding, C. Lefèvre and L. Gualtieri (eds), *The People Before: The Geology, Paleoecology and Archaeology of Adak Island, Alaska*. Oxford, British Archaeological Reports.

Crockford, Susan, Frederick Gay, Wigen Rebecca, and McKechnie Iain. 2004. Final Report on the Analysis of the Vertebrate Fauna from Amaknak Bridge, Unalaska, AK, UNL050. Supplement to Knecht, R. and Davis, R. 2004: Amaknak Bridge Site Data Recovery Project Final Report. Report on file, Museum of the Aleutians, Unalaska, Alaska.

DeLury, Daniel B. 1947. On the Estimation of Biological Populations. *Biometrics* 3, 145-167.

Dinesman, Lev G., Nina K. Kiseleva, Arkady B. Savinetsky, and Bulat F. Khassanov. 1999. *Secular Dynamics of Coastal Zone of Northeast Chukotka.*
Tubingen, Mo, Vince Verlag,

Epstein, Samuel, Ralph Buchsbaum, Heinz A. Lowenstam, and Harold C. Urey. 1951. Carbonate-water Isotopic Temperature Scale. *Bulletin of the Geological Society of America* 62, 417–426.

Eschmeyer, William N., Earl S. Herald, and Howard E. Hammann. 1983. *A Field Guide to Pacific Coast Fishes of North America*. Boston, Houghton Mifflin Company.

Estes, James A., and David O. Duggins. 1995. Sea Otters and Kelp Forests in Alaska: Generality and Variation in a Community Ecological Paradigm. *Ecological Monographs* 65(1), 75-100.

Fraser, George D., and George L. Snyder. 1959. Geology of Southern Adak Island and Kagalaska Island, Alaska.

Investigations of Alaskan Volcanoes. *US Geological Survey Bulletin* 1028-M, 371-408.

Frohlich, Bruno and David Kopjanski. 1975. Aleutian Site Survey, 1975. Preliminary Report to the Aleut Corporation, Anchorage, Alaska.

Gudkov, Pavel K., Michail V. Nazarkin, and Yuriy E. Vostretsov. 2005. Reconstruction by Fossil Otoliths of the Structure of Populations of Pacific *Navaga Eleginus gracilis* (Gadidae) that Inhabited Amur Bay 2400–2450 Years Ago. *Journal of Ichthyology* 45(4), 295-300.

Hantemirov, R. M. and S. G. Shiyatov. 2002. A Continuous Multimillennial Ring-width Chronology in Yamal, Northwestern Siberia. *The Holocene* 12 (6), 717-726.

Heusser, Calvin J. 1978. Postglacial Vegetation on Adak Island, Aleutian Islands, Alaska. *Bulletin of the Torrey Botanical Club* 105(1), 18-23.

Howes, Gordon J. 1991. Biogeography of Gadoid Fishes. *Journal of Biogeography* 18(6), 595-622.

Hu, Feng Sheng, Emi Ito, Thomas A. Brown, B. Brandon Curry, and Daniel R. Engstrom. 2001. Pronounced Climatic Variations in Alaska During the Last Two Millennia. *The Proceedings of the National Academy of Sciences of the United States of America* 98(19), 10552-10556.

Isakson, J. S., C. A. Simenstad and R. L. Burgner. 1971. Fish Communities and Food Chains in the Amchitka Area. *Bioscience* 21 (12), 666-670.

Jones, Douglas S., and Irvy R. Quitmyer. 1996. Marking Time with Bivalve Shells: Oxygen Isotopes and Season of Annual Increment Formation. *PALAIOS* 11, 340–346.

Kachinsky Nikodim A. 1965. *Texture of Soil and Its Microaggregate Composition, Methods of Exploration.* Academy of Sciences Moscow, USSR Press. [in Russian]

Killingley, John S., and William A. Newnan. 1982. O-18 Fractionation in Barnacle Calcite: A Barnacle Paleotemperature Equation. *Journal of Marine Research 40*, 893-902.

Kirby, Michael X., Thomas M. Soniat, and Howard J. Spero. 1998. Stable Isotope Sclerochronology of Pleistocene and Recent Oyster Shells (*Crassostrea virginica*). *PALAIOS* 13, 560–569.

Kiriyanov, Vladimir Yu., and Thomas P. Miller. 1997. Volcanic Ashes of Adak Island, Aleutian Islands, Alaska. *Volcanology and Seismology* 19, 65-78. [in Russian]

Lefèvre, Christine, Debra G. Corbett, Dixie West, and Douglas Siegel-Causey. 1997. A Zooarchaeological Study at Buldir Island, Western Aleutians, Alaska. *Arctic Anthropology* 34(2), 118-131.

Longin, Robert. 1971. New Method of Collagen Extraction for Radiocarbon Dating. *Nature* 230, 241-242.

Luttrell Mark, and Debra Corbett. 2000. Archaeological Investigations on Northeast Adak Island, Alaska July, 1999. Unpublished Manuscript, on file Archives, US Fish and Wildlife Service, Anchorage.

Magurran Anne E. 1988. *Ecological Diversity and Its Measure*. Princeton, New Jersey, Princeton University Press.

McCartney, Allen P. 1972. An Archaeological Site Survey and Inventory for the Aleutian Islands National Wildlife Refuge, Alaska, 1972. File Report 1-1-72, US Fish and Wildlife Service, Anchorage.

McCartney, Allen P. 1975. Maritime Adaptations in Cold Archipelagos: An Analysis of Environment and Culture in the Aleutian and Other Island Chains. In William Fitzhugh (ed.), *Prehistoric Maritime Adaptations of the Circumpolar Zone*, 281-338. Mouton, Chicago, The Hague.

Mecklenburg, Catherine W., Anthony A. Mecklenburg, and Lyman K. Thorsteinson. 2002. *Fishes of Alaska*. Bethesda, Maryland, American Fisheries Society.

Okuno, Mitsuru, Lyn Gualtieri, Dixie West, Elizabeth G. Wilmerding, and Toshio Nakamura. 2007. Aleut Shell Mounds Intercalated with Tephra Layers in Adak Island and Adjacent Volcanoes in the West-central Aleutians, Alaska, USA. *Journal of the Geological Society of Japan* 113(8), XI-XII.

Okuno, Mitsuru, Keiji Wada, Toshio Nakamura, Lyn Gualtieri, Brenn Sarata, Dixie West, Masayuki Torii. 2011. Holocene Tephra Layers on the Northern Half of Adak Island in the west-central Aleutians, Alaska. Chapter 3 in D. West, V. Hatfield, E. Wilmerding, L. C. Lefèvre, L. Gualtieri (eds), *The People Before: The Geology, Paleoecology and Archaeology of Adak Island, Alaska.* Oxford, British Archaeological Reports.

O'Leary, Matthew. 1998. New Site No. 2 (Tutiakoff Site). Unpublished report on file at US Department of the Interior, Bureau of Indian Affairs, ANCSA Projects Office, Anchorage, Alaska.

O'Leary, Matthew. 2001. Volcanic Ash Stratigraphy for Adak Island, Central Aleutian Archipelago. In Don Dumond (ed.), *Archaeology in the Aleut Zone of Alaska. University of Oregon Anthropological Papers* 58, 215–234. Eugene, University of Oregon Press.

Orchard, Trevor J. 2003. An Application of the Linear Regression Technique for Determining Length and Weight of Six Fish Taxa. The Role of Selected Fish Species in Aleut Paleodiet. *British Archaeological Reports International Series* 1772, 1-123.

Panuska, Bruce C. 1980. *Stratigraphy and Sedimentary Petrology of the Kiska Harbor Formation and Its Relationship to the Near Island Amchitka Lineament, Aleutian Islands.* Unpublished M.S. thesis, University of Alaska Fairbanks.

Parfenova Elena. I. and Ekaterina Yarilova A. 1962. *Mineralogical Study in Soil Science*. Moscow, Academy of Sciences USSR Press. [in Russian]

Pilsbry, Henry A. 1916. The Sessile Barnacles (Cirripedia) Contained in the Collections of the US National Museum; Including a Monograph of the American Species. *Bulletin of the United States National Museum* 93, 1-366.

Ricker, William E. 1975. *Computation and Interpretation of Biological Statistics of Fish Populations.* Ottawa, Canadian Department of Environment, Fisheries and Marine Service.

Robinson, Stephen W., and Gail Thompson. 1981. Radiocarbon Corrections for Marine Shell Dates with

Application to Southern Pacific Northwest Coast Prehistory. *Syesis* 14, 45–57.

Rubenstone, James L. 1984. *Geology and Geochemistry of the Early Tertiary Submarine Volcanic Rocks of the Aleutian Islands and Their Bearing on the Development of the Aleutian Island Arc*. Unpublished Ph.D. Dissertation, Cornell University, Ithaca.

Savinetsky, Arkady B., Nina K. Kiseleva, and Bulat F. Khassanov. 2004. Dynamics of Sea Mammal and Bird Populations of the Bering Sea Region Over the Last Several Millennia. *Palaeogeography, Palaeoclimatology, Palaeoecology* 209(1-4), 335-352.

Simenstad, Charles A., and Roy E. Nakatani. 1977. Nearshore Fish Communities of Attu Island, Alaska. Annual Report, June 1976 - December 1976, to US Fish and Wildlife Service, Unpublished Manuscript, on file Archives, Fisheries Research Institute, Seattle, Washington.

Simenstad, Charles A., Robert M. Mayer, and Roy E. Nakatani. 1978. Nearshore Fish and Macroinvertebrate Communities of Attu Island, Alaska. Annual Report, January 1977 - December 1977, to US Fish and Wildlife Service. Unpublished Manuscript, on file Archives, Fisheries Research Institute, Seattle, Washington.

Smith, Stuart H., Monique Delafontaine, and Burghard W. Flemming. 1988. Intertidal Barnacles, Assessment of Their Use as Paleo-environment Indicators Using Mg, Sr, $^{18}O/^{16}O$ and $^{13}C/^{12}C$ Variations. *Chemical Geology* 73(3), 211-220

Sulerzhitsky, Leopold D., Alexander L. Ryabinin, and Lev G. Dinesman. 1984. Dynamic Method of Measuring Activity of Samples as a Way to the Increasing of the Reliability and Accuracy of Radiocarbon Dates. In V.E. Sokolov and L. G. Dinesman (ed.), *Problems of Study of History of Recent Biogeocenoses*, 102-117, Moscow, Institute of Evolutionary Morphology and Animal Ecology. [in Russian]

Tarasov, Nikolay I., and Galina B. Zevina. 1957. Barnacles (Cirripedia, Thoracica) of the Seas of the USSR. *Fauna of the USSR. Crustacea*, Vol. 6, issue 1, Academy of Sciences of the USSR, Moscow [in Russian].

Thorson, Robert M., and Thomas D. Hamilton. 1986. Glacial Geology of the Aleutian Islands. In T. Hamilton and R. Thorson (eds.), *Glaciation in Alaska: the Geological Record,* 171-191. Anchorage, Alaska Geological Society.

Ubelaker, Douglas.H. 1978. *Human Skeletal Remains*. Chicago, Aldine Publishing Company.

Veniaminov, Ivan. 1984. *Zapiski ob ostrovakh Unalashkinskago otdela [Notes on the Islands of the Unalaska District*]. [Lydia T. Black and R. H. Geoghegan, trans.] Richard A. Pierce (ed.). The Elmer E. Rasmuson Library Translation Program, University of Alaska Fairbanks and Kingston, Ontario, Canada. Limestone Press.

Waythomas, Christopher F. 1995. Surficial Geologic Map of Northern Adak Island, Alaska. US Geological Survey Open-File Report 95-128

Wilmerding, Elizabeth and Virginia Hatfield. 2011. Six Thousand Years of Lithic Technology on Adak in a Broader Aleutian Context. Chapter 12 in D. West, V. Hatfield, E. Wilmerding, C. Lefèvre and L. Gualtieri (eds), *The People Before: The Geology, Paleoecology and Archaeology of Adak Island, Alaska*. Oxford, British Archaeological Reports.

Zhuze, Anastasiya P. 1939. The Paleogeography of Reservoirs on the Basis of Diatom Analysis. *Trudy Verhnevolzhskoy ekspeditsii* 4, 1-86 [in Russian].

Zubkova, Zinaida. N. 1948. *The Aleutians*. Moscow, Association of State Publisher. [in Russian]

Appendix 5A. Haven Lake Profile: Adak Island, Alaska

Deposit depth	***Deposit description***
0-25cm	Present day sod.
0-8cm	Litter, densely pierced by roots.
8-15cm	Peaty, from gray to light chestnut, friable, crumb, densely pierced by roots, plant remains are well decomposed. The boundary is distinct; transition to the underlying horizon is even.
15-20cm	Peaty, light brown, poorly decomposed. The boundary is even, transition is abrupt.
20-25cm	Peaty, grayish brown to black, well decomposed, rich in mineral particles, pierced by roots. The boundary is abrupt and even.
25-30cm	**40-Year Ash** Olive-brown (whitish upon drying) mineral layer, fine-grained sand to loam in texture with numerous aggregations of bright ochreous iron pellet-like nodules and dark gray amorphous new formations (Manganese?); pierced by very fine roots. The transition is clear and even.
30-85cm	Stratified peat-mineral stratum. **YBO Volcanic Ash.**
30-35, 43-52, 60-64, 60-69cm	Peat, light gray to light chestnut, well decomposed, compacted, plastic, pierced by fine and very fine roots. The transition is gradual.
35-43, 53-60, 69-81cm	Loamy, grayish olive to brown, impregnated with well decomposed organic matter, sticky, many fine roots. The transition is clear and even (tephra layers?).
81-85cm	Fine- grained sandy (to loamy-sandy), pale gray, compacted, unstructured with dispersed small dark peaty mottles; a thin (2-3 mm) whitish layer at the contact with the overlying layer. The transition is abrupt and even.
85-96cm	**Sandwich Ash.** Light pale to whitish, compacted coarse (to 0.5mm) and fine tephras. At 91-94 cm they are replaced by bluish gray-olive fine-grained tephra and at (94-96 cm) replaced by mixed clastic and fine-grained ash; intensively ferruginous at the boundary with the underlying horizon.
96-124cm	**III buried peat bog soil**
96-104cm	Peat, dark gray to dark brown, weakly decomposed, ferruginous, friable, sticky, a whitish fine-grained sandy layer at the contact with the underlying layer.
104-124cm	Loamy, dark brown, sticky, many fine roots; impregnated with poorly decomposed organic sediment; at a depth of 120-122cm, a well expressed ferruginous layer; a brown peat streak at the contact with underlying horizon.
124-135cm	**Intermediate Ash.** Bluish white, heterogeneous in texture: fine-grained tephra saturated with coarse clastic ash, many roots and ferruginous concretions. The boundary is abrupt and even.
135-167cm	**II buried peat bog soil**
135-162cm	Peat, dark brown, well decomposed. The transition is distinct.
162-167cm	Heavy loam, blue (wet) to olive and grayish-pale (dry); fine sorted material without ferruginous new formations, ductile, saturated with weakly decomposed plant remains; many mottles composed of peat. The transition is clear and even.
167-182cm	**Main Ash.** Coarse clastic tephra, whitish-pale, compacted with many iron concretions; abrupt, smooth boundary.
182-265cm	**I buried peat bog soil**
182-245cm	Peat, dark brown, well decomposed, compacted, plastic; lenses of mineral deposit concealed due to impregnation with organic material.
245-265cm	Peat with mineral layers unclearly visible due to impregnation with organic sediments. The transition to the underlying horizon is abrupt
265-270cm and below	Glacial deposits? Shallow sandy layer replaced by till.

Appendix 5B. ADK-171, Unit 1 Profile, Adak Island, Alaska.

Deposit Depth	*Deposit Description*
0-5cm	Present-day sod. Peaty sod, grayish chestnut, friable, unstructured, pierced by roots; transition to the underlying horizon is distinct and even.
6–18cm	**40-Year Ash** Between tussocks – light brown to grayish olive-brown, loamy, ductile, adhesive, unstructured; pierced by roots; transition to the underlying horizon is gradual, the boundary uneven. Under tussocks – dark gray to chestnut becoming lighter with depth, loamy, ductile, adhesive, unstructured; penetrated by roots; transition to the underlying horizon is gradual.
18–50cm	Peaty soil Peaty, dark chestnut, homogeneous, compacted, ductile, adhesive; densely pierced by roots; plant remains are well decomposed; transition to the underlying horizon is abrupt and even.
50-61cm	**Sandwich Ash** Heterogeneous tephra layers. Coarse-grained clastic ash (50-54cm) is replaced by inequigranular sandy volcanic ash with coarse clastic concretions (54-57cm). At 57-60cm the layer contains fine- grained, olive colored volcanic ash; the boundary is abrupt and even.
61-100cm	Stratified mineral stratum Heterogeneous; interbedded with tephra. Mixed volcanic ash layers, from brown chestnut to light olive tinted with black; ash layers are fine- grained, ductile, adhesive; many fine roots; lower boundary is abrupt and even.
100-140cm	**Cultural layer**
100-125cm	Layer of densely packed large Clinocardia shells and bones, little fine-grained mineral material.
125-129cm	Layer of volcanic ash, grayish-pink, fine-grained to loamy, adhesive.
129-141cm	Layer of small mussel shells. Shells are mixed with tiny charcoal fragments. In the NW part of the pit side and along slopes the layer flattens and consists only of broken shells. Shell fragments dominant at the contact with lower horizon, sediment is sooty.
140-150cm	Sandy loam, brown-pink to grayish brown, compacted, ductile, adhesive, friable upon drying, possible volcanic ash; impregnated with well decomposed organic matter; the boundary is abrupt and even.
150-166cm	**Intermediate Ash** Heterogeneous stratified tephra stratum; the layers alternate with the lenses of volcanic ash; mineral sediment is inequigranular to well sorted, saturated with coarse clastic ash; variegated- light olive to light brown; ductile, adhesive; the boundary is abrupt and even.
166-230cm	Dune sand. Homogenous sand, unstructured, medium sized grained, whitish gray; indistinct ferruginous streaks and spots at 166-170cm and 222-230cm; streaks of black-minerals; the boundary is abrupt and even.
230-245cm	Loam. Loam, black blue (wet) to brown and grayish-pale (dry), compacted, ductile, unstructured; the layer is saturated with numerous aggregations of bright ochreous iron pellet-like nodules; transition to the underlying horizon is distinct and wavy.
245-260cm	**Main Ash** Coarse clastic tephra, medium sized grained, light brown to bright ochreous; the boundary is abrupt and even.
260-270cm	Glacial deposits? Brown adhesive loam, is deposited on rock base; thin peaty streak marks the contact (260cm) with the overlying ash layer.

Appendix 5C. ADK-171, Unit 3 Profile, Adak, Alaska.

Deposit Depth	***Deposit Description***
0-5cm	Present-day sod. Peaty sod, grayish chestnut, friable, unstructured, penetrated by roots; transition to the underlying horizon is distinct and even.
6-115cm	Peaty soil. Peaty, dark chestnut, homogeneous, compacted, ductile, adhesive; densely pierced by roots; plant remains are well decomposed; transition to the underlying horizon is abrupt and even.
71-80cm	**Sandwich Ash**
116-147cm	**Cultural layer**
116-137cm	Layer of densely packed large *Clinocardium* shells and bones, fine-grained mineral sediment nearly absent.
138-142cm	Layer of volcanic ash, grayish-pink, fine-grained to loamy, adhesive. In lower level mixed with small bone fragments and mussel shells.
143-147cm	Layer of small mussel shells. Shells are mixed with tiny charcoal fragments. Tiny shell fragments dominate at the contact with lower horizon, material becomes sooty.
148-150cm	Light brown sand.
151-165cm	**Intermediate Ash** Heterogeneous stratified tephra stratum; the layers alternate with the lenses of volcanic ash; inequigranular to well sorted, saturated with coarse clastic ash; variegated- light olive to light brown; ductile, adhesive; the boundary is abrupt and even.
166cm and lower	Dune sand. Homogenous sand, unstructured, medium sized grained, whitish grey. Excavation ends at 180cm.

Appendix 5D. Invertebrate NISP and Percentages from ADK-171, Adak Island, Alaska.

Taxonomic group	*Scientific name*	*Lower layer*	*Upper layer*	*Total*
Chitons		32* / 1.9	55 / 2.2	87 / 2.1
	Tonicella submaemorea	3 / 9.4	19 / 34.5	22 / 25.3
	Katharina tunicata	29 / 90.6	36 / 65.5	65 / 74.7
Gastropods		18 / 1.0	30 / 1.2	48 / 1.1
	Littorina sitkana	3 / 16.7	16 / 53.3	19 / 39.6
	Lottia pelta	9 / 50.0	14 / 46.7	23 / 47.9
	Nucella lima	6 / 33.3	0	6 / 12.5
Bivalves		1030 / 62.1	1769 / 69.4	2799 / 66.5
	Mytilus trossulus	612 / 59.4	410 / 23.2	1022 / 36.5
	Hiatella arctica	5 / 0.5	10 / 0.6	15 / 0.5
	Clinocardium nuttallii	279 / 27.1	1173 / 66.3	1452 / 51.9
	Saxidomus giganteus	128 / 12.4	167 / 9.4	295 / 10.5
	Protothaca staminea	0	1 / <0.1	1 / <0.1
	Cardiidae gen. et sp. indet	0	1 / <0.1	1 / <0.1
	Mactromeris polynima	5 / 0.5	6 / 0.3	11 / 0.4
	Mya truncata	1 / <0.1	1 / <0.1	2 / 0.1
Barnacles		453 / 27.3	135 / 5.3	588 / 14
	Balanus crenatus	419 / 92.6	54 / 40.0	473 / 80.4
	Semibalanus balanoides	10 / 2.2	7 / 5.2	17 / 2.9
	S. cariosus	12 / 2.6	49 / 36.3	61 / 10.4
	Hesperibalanus hesperius	12 / 2.6	25 / 18.5	37 / 6.3
Sea urchins		125 / 7.5	560 / 22.0	685 / 16.3
	Strongylocentrotus polyacanthus	119 / 95.2	558 / 99.6	677 / 98.8
	Echinarachnius parma	6 / 4.8	2 / 0.3	8 / 1.2
Total		***1658***	***2549***	***4207***

* N/%

Appendix 5D. Invertebrate NISP and Percentages from ADK-171, Adak Island, Alaska.

Taxonomic group	*Scientific name*	*Lower layer*	*Upper layer*	*Total*
Chitons		**32** **1.9**	**55** **2.2**	**87** **2.1**
	Tonicella submaemorea	3 9.4	19 34.5	22 25.3
	Katharina tunicata	29 90.6	36 65.5	65 74.7
Gastropods		**18** **1.0**	**30** **1.2**	**48** **1.1**
	Littorina sitkana	3 16.7	16 53.3	19 39.6
	Lottia pelta	9 50.0	14 46.7	23 47.9
	Nucella lima	6 33.3	0	6 12.5
Bivalves		**1030** **62.1**	**1769** **69.4**	**2799** **66.5**
	Mytilus trossulus	612 59.4	410 23.2	1022 36.5
	Hiatella arctica	5 0.5	10 0.6	15 0.5
	Clinocardium nuttallii	279 27.1	1173 66.3	1452 51.9
	Saxidomus giganteus	128 12.4	167 9.4	295 10.5
	Protothaca staminea	0	1 <0.1	1 <0.1
	Cardiidae gen. et sp. indet	0	1 <0.1	1 <0.1
	Mactromeris polynima	5 0.5	6 0.3	11 0.4
	Mya truncata	1 <0.1	1 <0.1	2 0.1
Barnacles		**453** **27.3**	**135** **5.3**	**588** **14**
	Balanus crenatus	419 92.6	54 40.0	473 80.4
	Semibalanus balanoides	10 2.2	7 5.2	17 2.9
	S. cariosus	12 2.6	49 36.3	61 10.4
	Hesperibalanus hesperius	12 2.6	25 18.5	37 6.3
Sea urchins		**125** **7.5**	**560** **22.0**	**685** **16.3**
	Strongylocentrotus polyacanthus	119 95.2	558 99.6	677 98.8
	Echinarachnius parma	6 4.8	2 0.3	8 1.2
Total		**1658**	**2549**	**4207**

* N/%

Appendix 5E. Mammal NISP and Percentages from ADK-171, Adak Island, Alaska.

Common name	*Scientific"name*	*Lower layer*		*Upper layer*		*Total*	
		NISP	*%*	*NISP*	*%*	*NISP*	*%*
Sea Otter	*Enhydra lutris*	21	33.3	20	26.7	41	29.7
True seal	*Phoca* sp.	13	20.6	9	12.0	22	15.9
Sea Lion	*Eumetopias jubatus*	5	7.9	21	28.0	26	18.8
Fur seal	*Callorhinus ursinus*	23	36.5	24	32.0	47	34.1
Dolphin	*Delphinidae*	1	1.6	1	1.3	2	1.4
Total identified (NISP)		63	100	75	100	138	100
Unidentified mammal		350		261		611	
Total examined (NSP)		413		336		749	

Appendix 5F. Fish NISP and Percentages from ADK-171, Adak Island, Alaska.

Family name	*Genus and species*	*Lower layer*		*Upper layer*		*Total identified*	
		NISP	*%*	*NISP*	*%*	*NISP*	*%*
Hexagrammidae (Greenlings)		**1053**	**55.4**	**3061**	**47.0**	**4114**	**48.9**
	Hexagrammos sp. (Greenling)	1053	100	3061	100	4114	100
Gadidae (Codfishes)		**736**	**38.7**	**2694**	**41.3**	**3430**	**40.8**
	Eleginus gracilis (Saffron Cod)	93	12.6	361	13.4	454	13.2
	Gadus macrocephalus (Pacific Cod)	17	2.3	9	0.3	26	0,8
	Codfishes unspecified	626	85.1	2324	86.3	2950	86.0
Pleuronectidae (Flatfishes)		**100**	**5.3**	**675**	**10.4**	**775**	**9.2**
	Hippoglossus stenolepis (Pacific Halibut)	33	33.0	151	22.4	184	23.7
	Platichthys stellatus (Starry flounder)	15	15.0	70	10.4	85	11.0
	Flatfishes unspecified	52	52.0	454	67.2	506	65.3
Sebastidae (Rockfishes)		**3**	**0.2**	**34**	**0.5**	**37**	**0.4**
Cottidae (Sculpins)		**3**	**0.2**	**50**	**0.8**	**53**	**0.6**
Salmonidae (Salmonids)		**3**	**0.2**	**2**	**<0.1**	**5**	**0.1**
Total identified (NISP)		1898		6516		8414	
Total examined (NISP)		7528		22804		30332	

Appendix 5G. ADK-009 Profile, Adak Island, Alaska.

Deposit Depth	*Deposit Description*
1-13cm	**Redeposited culture layer:** Dark-gray to black, culture layer covered by turf and pierced by roots,saturated with crushed shells and bones.
13-18cm	**Buried sod** Dark-brown to brown buried mottled sod with inclusions of light-brown (presumably ash) spots.
18-20cm	**40-year-old ash** Light-brown finely-sorted compressed ash layer of uneven thickness
20-22cm	**Humus streak** Light-brown to brown smeared humus streak.
22-166cm	**Cultural layer**
22-56cm	**Layer I**
22-35cm	Dark-gray to black fine-grained mineral sediment filled with crushed shells, pierced by roots, containing few bones. Lens of sea urchin plates in the bottom NE corner.
35-37cm	Streak of fine-grained sediment: dark-brown to brown with black tint, compressed, unstructured, adhesive, few bones and shells.
37-56cm	Heterogeneous horizon of crushed shells and bones interspersed with lenses of fine-grained sediment.
56-107cm	**Layer II**
56-64cm	Fine-grained sediment: compressed, ductile, adhesive, filled with crushed shells. Layer thickness decreases toward SW corner, the stratigraphy is becoming disrupted.
64-107cm	Heterogeneous layer: sea urchin plates prevail in the upper part, charcoal streaks are found 71-73cm and 93-97cm, gastropod shells appear below 99cm.
107-123cm	**Layer III**
107-112cm	Fine-grained sediment: brown, wet, ductile, adhesive, filled with bones.
112-123cm	Layer of crushed shells and sea urchin plates, lacking fine-grained sediment.
123-140cm	**Layer IV**
123-128cm	Brown fine-grained sediment filled with crushed shells and bones.
128-140cm	Heterogeneous midden layer: crushed shells and fine-grained sediment comprise the upper part, the lens of sea urchin plates is found in the SW corner.
140-166cm	**Layer V**
140-150cm	Brown fine-grained sediment filled with crushed shells and bones.
150-166cm	Light-brown fine-grained sediment, unstructured, adhesive, with sparse grains of clastic ash. The upper 5cm are darker, filled with bones only.
166-173cm	**Underlying bedrock**
166-168cm	Pale-yellow finely-sorted adhesive unstructured silt with sparse grains of clastic ash.
168-173cm	Brown to light-chestnut finely-sorted ferruginous silt deposited on the bedrock.

Appendix 5H. Invertebrate Number and Percentages from ADK-009, Adak Island, Alaska.

Number of layer, age of formation, yr AD	***V*** ***500-760 AD***	***IV*** ***760-940 AD***	***III*** ***940-1100 AD***	***II*** ***1100-1320 AD***	***I*** ***1320-1600 AD***
Taxonomic group					
Bryozoans					
Tubulipora sp.	-	+*	-	-	-
Ctenostomata fam., gen. et sp. indet.	+	-	-	-	-
Membranipora sp.	-	-	+	-	+
Callopora sp.	-	-	+	+	-
Tegella sp.	-	+	-	+	+
Porella sp.	-	-	-	+	+
Escharopsis sp.	-	-	-	+	-
Smittina sp.	-	-	-	-	+
Schizoporella sp.	-	-	+	-	+
Stylopoma sp.	-	+	-	-	+
Ellisina sp.	-	-	+	-	-
Polychaetes					
Polidora sp.	+	+	+	+	+
Serpulidae gen. et sp. indet.	+	+	+	+	+
Chitons	12** 0.7	24 1.0	58 1.0	142 0.9	104 1.2
Lepidozona sp.	-	-	-	2 1.4	5 4.8
Boreochiton beringensis beringensis	-	-	3 5.2	6 4.2	10 9.6
Tonicella submarmorea	-	-	-	1 0.7	-
Mopalia wosnessenskii	-	-	-	2 1.4	1 1.0
Katharina tunicata	12 100	24 100	54 93.1	131 92.3	88 84.6
Cryptochiton stelleri	-	-	1 1.7	-	-
Gastropods	82 4.6	98 4.2	183 3.1	1024 6.2	476 5.7
Problacmaea sybaritica	-	-	1 0.5	-	-
Lottia digitalis	5 6.1	-	4 2.2	6 0.6	-
Lottia pelta	8 9.8	30 30.6	70 38.3	256 25.0	214 45.0
Testudinalia scutum	3 3.7	-	9 4.9	12 1.2	10 2.1
Puncturella longifissa	-	-	2 1.1	5 0.5	-
Crepidula sp.	1 1.2	-	-	-	-
Ariadnaria insignis	-	-	1 0.5	-	1 0.2
Trichotropis bicarinata bicarinata	-	2 2.0	1 0.5	-	-
Littorina aleutica	6 7.3	5 5.1	21 11.5	73 7.1	19 4.0
Littorina sitkana	51 62.2	53 54.1	27 14.8	613 59.9	161 33.8
Cryptonatica sp.	-	-	1 0.5	-	1 0.2
Onoba aurivillii	-	-	13 7.1	22 2.1	6 1.3

Number of layer, age of formation, yr AD	V 500-760 AD	IV 760-940 AD	III 940-1100 AD	II 1100-1320 AD	I 1320-1600 AD
Onoba sp.	-	-	2 1.1	-	-
Fusitriton oregonensis	-	-	-	-	1 0.2
Boreotrophon cepula	-	-	2 1.1	9 0.9	5 1.1
Nucella heyseana	-	1 1.0	-	-	1 0.2
Nucella lima	7 8.5	2 2.0	13 7.1	22 2.1	44 9.2
Buccinum percrassum	-	-	1 0.5	-	-
Aulacofusus periscelidus	-	2 2.0	7 3.8	-	5 1.1
Colus sp.	-	-	-	-	1 0.2
Neptunea sp.	-	-	1 0.5	-	-
Volutopsius castaneus	-	-	-	-	1 0.2
Astyris sp.	1 1.2	3 3.1	7 3.8	6 0.6	6 1.3
Bivalves	449 25.4	200 8.5	520 8.9	1941 11.7	1803 21.4
Modiolus modiolus	-	-	1 0.2	2 0.1	-
Mytilus trossulus	445 99.1	195 97.5	508 97.7	1920 98.9	1774 98.4
Monia macrochisma	1 0.2	2 1.0	3 0.6	-	-
Hiatella arctica	3 0.7	-	4 0.8	16 0.8	10 0.6
Kellia laperousii	-	1 0.5	1 0.2	1 <0.1	-
Cyclocardia ventricosa	-	-	1 0.2	-	-
Clinocardium nuttallii	-	-	1 0.2	1 <0.1	9 0,5
Serripes groenlandicus	-	1 0.5	-	-	-
Macoma sp.	-	-	1 0.2	-	-
Mactromeris polynima	-	-	-	1 <0.1	10 0.6
Mya truncata	-	1 0.5	-	-	-
Barnacles	356 20.1	382 16.3	1205 20.6	4055 24.4	1419 16.9
Balanus balanus	-	7 1.8	8 0.7	9 0.2	-
Balanus Crenatus	3 0.8	19 5.0	1 0.1	167 4.1	68 4.8
Balanus rostratus	-	-	7 0.6	-	7 0.5
Semibalanus cariosus	306 86.0	310 81.2	1069 88.7	2434 60.0	1018 71.7
Semibalanus balanoides	19 5.3	10 2.6	58 4.8	513 12.7	75 5.3
Hesperibalanus hesperius	28 7.9	36 9.4	62 5.1	932 23.0	251 17.7

Number of layer, age of formation, yr AD	*V 500-760 AD*	*IV 760-940 AD*	*III 940-1100 AD*	*II 1100-1320 AD*	*I 1320-1600 AD*
Decapods					
Decapoda gen. et. sp. indet.	-	-	-	+	-
Crustacea gen. et. sp. indet.	-	-	-	+	-
Brachiopods					
Hemithyris psittacea	1 0.1	1 <0.1	-	8 <0.1	2 <0.1
Sea urchins					
Strongylocentrotus polyacanthus	868 49.1	1644 70.0	3884 66.4	9425 56.8	4607 54.8
Total N	***1768***	***2349***	***5850***	***16595***	***8411***

* presence in faunal spectrum
** N/%

Appendix 5I. Mammal NISP and Percentages from ADK-009, Adak Island, Alaska.

Common name	*Scientific name*	*Summary*	
		NISP/NSP	*%NISP*
Sea Otter	*Enhydra lutris*	18	45.0
True seal	*Phoca* sp.	15	37.5
Sea Lion	*Eumetopias jubatus*	5	12.5
Fur seal	*Callorhinus ursinus*	2	5.0
Total identified (NISP)		40	100
Unidentified mammal		79	
Total examined (NSP)		119	

Chapter 6
Archaeozoology of Adak Island: 6000 Years of Subsistence History in the Central Aleutians

Susan J. Crockford

Abstract: *A comprehensive archaeozoological analysis is presented of all vertebrate skeletal remains recovered from two Adak Island archaeological sites: The "Dozered" site at Clam Lagoon (ADK-012), dated c.1865 to 1515 BP and Zeto Point (ADK-011), dated c.415 to 170 BP. In addition, preliminary faunal data from test excavations at a late settlement site at Andrew Bay (ADK-084), dated c.480-250 BP, and fish remains from a single unit at Sweeper Cove (ADK-009), dated c.2080-970 BP, are included. Where applicable, these fish, bird, and/or mammal figures are compared to faunal data from other prehistoric sites in the Aleutians reported by other researchers, including two Adak Island sites (Tutiakoff, ADK-171, dated c.7000-6000 BP and Sweeper Cove, ADK-009), four eastern Aleutian sites (Unalaska, UNL-048, UNL-050, and UNL-055; Umnak, Chaluka site 66000155 NPS Registry), and five western Aleutian sites (Shemya, ATU-003, ATU-022, and ATU-061; Buldir, KIS-008; Amchitka, RAT-031), to generate a regional picture of subsistence use over time for Adak relative to the rest of the Aleutians.*

Keywords: *faunal analysis, sea otter, sea lion, fur seal, Pacific cod, rock greenling, kelp, ancient murrelet, Aleutian cackling goose, emperor goose, common eider*

Introduction

As ancient Unangan people hunted for food and fat to eat, oil to burn, skins and sinew to make clothing, and bone to make tools, they selected particular animals out of the surrounding populations of fish, birds and mammals. Therefore, vertebrate faunal remains discarded and preserved in the archaeological record reflect Unangan preference (cultural choice) as well as relative abundance of animals available within the landscape (prehistoric animal population status). Both of these factors were subject to change over time as cultural circumstances and environmental conditions varied across the Aleutians (Crockford and Frederick 2007; Savinetsky, Kiseleva and Khassanov 2004; Veltre Chapter 2 this volume; West *et al.* Chapter 1 this volume).

Although preservation of vertebrate faunal remains excavated from ancient Unangan shell midden sites is generally very good, until recently so few complete archaeozoological analyses have been available that it was impossible to construct even a preliminary description of ancient harvests of mammals, birds, and fish across the Aleutian chain. Reporting for most faunal assemblages is unfortunately uneven amongst taxonomic orders, with many site reports missing fish data, while others lack figures for birds or mammals. For example, the oldest known site in the central Aleutians (and the oldest Aleutian site with fauna), the Tutiakoff site on Adak Island (ADK-171), has so far had only mammal and fish remains reported (Savinetsky *et al.* Chapter 5 this volume), while the second-oldest Aleutian site with fauna, the Margaret Bay site on Unalaska Island (UNL-048), has had only mammalian remains analyzed (Davis 2001). Only fish dentary bones from Amaknak Spit (UNL-055) were analyzed and these fish (as well as bird remains) were identified only to family (Yesner, in Knecht and Davis 2003), making these data difficult to use in comparative context. Over the past few years, however, researchers working at either end of the archipelago have published comprehensive and detailed faunal analyses of all vertebrate taxa from several sites: two on Shemya Island in the west (Corbett, West and Lefèvre 2010) and the impressive Amaknak Bridge site on Unalaska in the east (Crockford *et al.* 2004; Knecht and Davis 2008).

Here I add a detailed report on the vertebrate faunal analyses of several central Aleutian sites, the result of my collaboration in the Central Aleutians Archaeological and Paleobiological Project. I present data on the fish, bird and mammal remains recovered from two sites on the northeast corner of Adak Island (Figure 6.1): remnant deposits from the largely disturbed Dozered site (ADK-012, dated *c.*1865-1515 years BP) that lies on the barrier beach that separates Clam Lagoon from the open ocean and Zeto Point (ADK-011, dated *c.* 400-170 years BP), a large village site across from Clam Lagoon that lies in a protected cove on Sitkin Sound. In addition, data from test excavations at a late prehistoric village site on Andrew Bay to the northwest (ADK-084, dated *c.* 480-250 years BP) and fish remains from a site to the south at Sweeper Cove (ADK-009, dated *c.* 2080-970 years BP) are provided. Sweeper Cove had almost two meters of deposit, representing an occupation

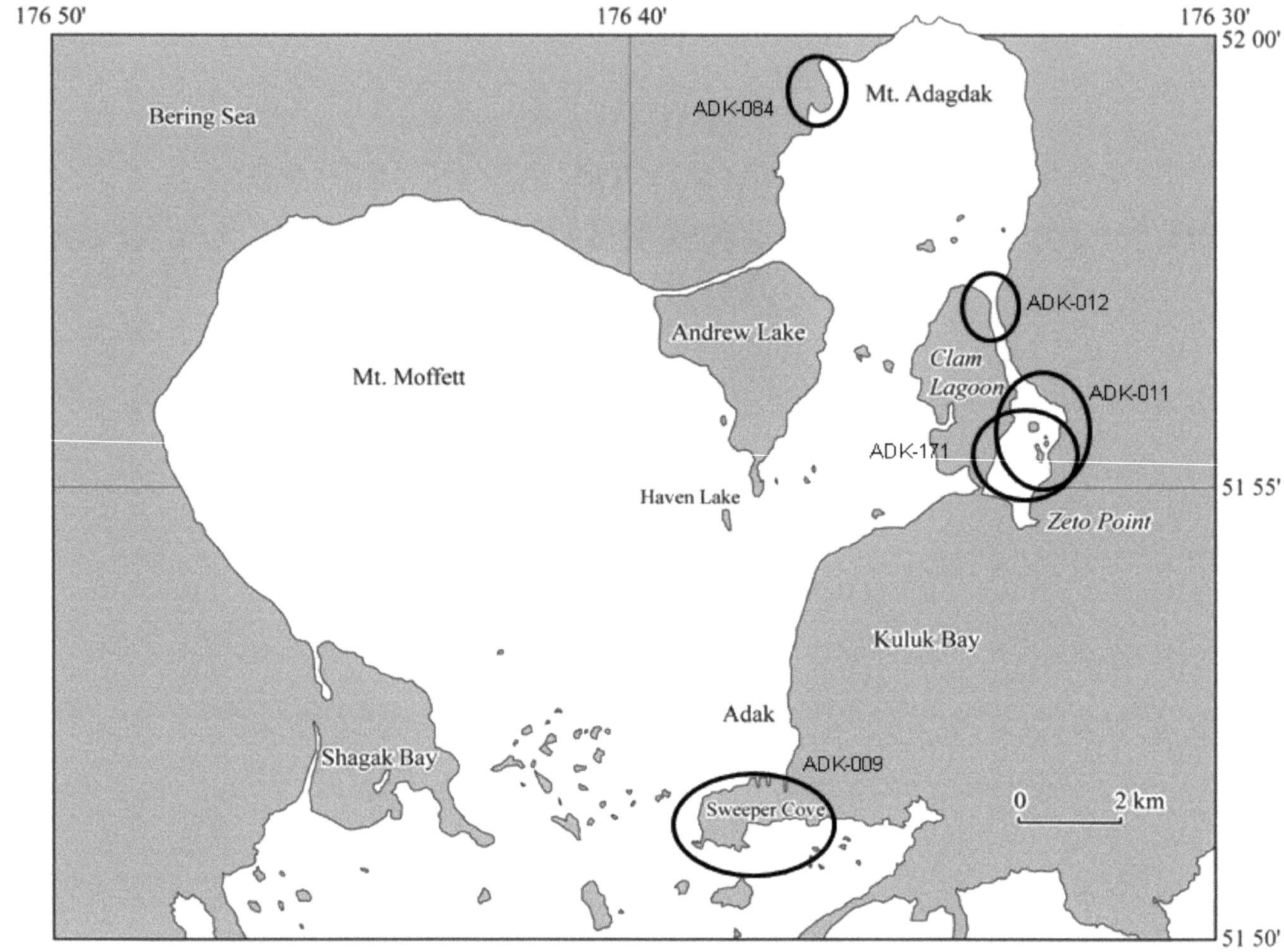

Figure 6.1. Map of North Adak, showing locations of landmarks mentioned in the text and habitat areas adjacent to the sites reported. (Map courtesy Mitsuru Okuno).

that spanned about 1100 years (Savinetsky *et al.* Chapter 5 this volume), overlapping chronologically with the occupation of the Dozered site at Clam Lagoon. To these fish, bird, and/or mammal figures are added the mammal and fish data from the 6000-7000 year old Tutiakoff site on northeast Adak (ADK-171) and the mammal data from Sweeper Cover (ADK-009) reported by Savinetsky *et al.* (Chapter 5 this volume) (Figures 6.1-6.5).

These Adak data are compared to faunal data from a number of Aleutian sites that have been reported by other researchers, including assemblages from four eastern Aleutian sites (Unalaska Island UNL-048, UNL-050, UNL-055; Umnak (Chaluka) site 66000155 NPS Registry), and five western Aleutian sites (Shemya Island ATU-003, ATU-022, and ATU-061; Buldir KIS-008; Amchitka RAT-031), to generate the most complete picture yet of subsistence use over time in the central Aleutians relative to the entire Aleutian archipelago. Essentially, the addition of comprehensive faunal data from a number of well-dated Adak Island sites means that there is now enough information to make a preliminary comparison of vertebrate harvests across the Aleutians over time, from the Near and Rat Islands in the west (including Shemya, Buldir and Amchitka) through the Andreanof Islands in the central region (including Adak) and the Fox Islands in the east (including Unalaska and Umnak). This comparison reveals some interesting patterns and one or two startling conclusions. Dates referred to in this report are predominantly uncalibrated 14C yrs BP, unless stated otherwise. Detailed lists of official radiocarbon dates for many of the Aleutian sites mentioned here are available elsewhere (Corbett *et al.* 2008; West *et al.* Chapter 1 Appendix 1.B this volume).

Methods

New Analyses of Adak Vertebrates

Vertebrate fauna reported in this section were identified to the least inclusive taxon possible using the comparative faunal collection at the Zooarchaeology Laboratory in the Department of Anthropology at the University of Victoria (British Columbia, Canada). Additional bird collections were consulted for identification of ADK-011 material, including the Burke Museum, University of Washington (Seattle, USA) and the Connor Museum, Washington State University (Pullman, WA, USA). A list of taxa identified is presented in Table 6.1. Tables 6.2-6.13 present the fish, birds, and mammals discussed throughout this chapter.

For a few ADK-011 specimens, confirmation of species identity was obtained from mitochondrial DNA (mtDNA) analysis made possible through collaboration with geneticists specializing in ancient DNA. Dr. Shin Nishida (Kyushu University, Fukuoka, Japan) undertook genetic analysis of several whale specimens that could not be confidently identified on morphological criteria, in conjunction with isotope analysis of Adak sea otters (Garong *et al.* Chapter 8 this volume). In addition, Dr. Brian Kemp and Bryan Wilson (Washington State University, Pullman) undertook ancient DNA analysis of a subsample of adult and nestling goose remains to determine species identity. Summary results of the goose DNA analyses are incorporated into this report but are reported in detail elsewhere (Wilson *et al.* 2011). Quantification is by Number of Identified Specimens (NISP). Data files have been archived and are available, along with several supplemental data tables (designated S1-S6), on request from the author and senior PI of this project, Dixie West, University of Kansas. Details on the nature of the sample for each site are given below.

ADK-012—Only three units were excavated (1-3) but all of the material recovered (fish, bird and mammal) was analyzed in detail. NISP totals reported here are for the entire site assemblage. Materials were water-screened in the field through 1/8 inch mesh.

ADK-011—Both midden deposits and house floor material were recovered from this site. Only NISP totals for the site are provided here (unit and layer totals are available as supplemental data, Tables S1-5). As only a handful of remains (virtually all unidentified mammal) are attributable to the lower component, all ADK-011 material reported here is from the most recent upper component. Mammal remains were analyzed from ten units (B through K) and bird remains from four units (D through G). Tables comparing bird and mammal remains from midden and house deposits are reported as supplemental data (Tables S1-S3). Although extensive deposits of fish bone were present throughout the site, excavated remains were not counted. Fish bones frequencies per taxon are reported here for a single midden unit only (EU E), which were examined thoroughly (identified to species where possible) but reported as an estimated relative abundance that could be compared to other sites. Material was water-screened in the field, units B through F, and I through K, with 1/8 inch mesh and units G and H with 1/4 inch mesh.

ADK-009—Fish remains were analyzed from a single 1m x 2m pit. A combination of sampling techniques was employed for fish bone recovery: both level bag samples (not screened) and matrix samples (30% subsample of the level total by volume, screened through 1/8" mesh) were collected and analyzed. Results of the matrix samples were volume-adjusted to 100% and added to the level sample total for each level to give a level plus matrix NISP for each taxon.

ADK-084—Vertebrate fauna (primarily fish) were analyzed from a single 30cm x 30cm test pit. Samples were hand collected at this site.

Adak Data from Other Analysts and Previously Reported Aleutian Fauna

Material previously analyzed by myself from other sites or fauna analyzed by other researchers that are included in comparative tables or figures come from the following sources:

Adak sites: ADK-171 mammals and fish (Savinetsky *et al.* Chapter 5 this volume); ADK-009 mammals (Savinetsky *et al.* Chapter 5 this volume); Aleutian sites UNL-048 mammals (Davis 2001); UNL-050 vertebrates (Crockford *et al.* 2004); UNL-055 mammals (Yesner, in Knecht and Davis 2003); Chaluka birds and mammals (Laughlin and Reeder 1962; Lippold 1966; Yesner 1976, 1988); RAT-031 birds and mammals (Causey *et al.* 2005; Corbett *et al.* 2008 citing mammal data from Desautels *et al.* 1971, 315; bird data supplemented by original figures from Warter (1970), cited in Funk 2010, 50); ATU-003 and ATU-061 mammals (Christine Lefèvre, pers. comm.; Corbett, West and Lefèvre 2010); ATU-003 and ATU-061 birds (Causey *et al.* 2005; Corbett, West and Lefèvre 2010); ATU-003 and ATU-061 fish (Crockford 1997b, 1999; Corbett, West and Lefèvre 2010); KIS-008 vertebrates (Causey *et al.* 2005; Lefèvre *et al.* 1997; Orchard 2003 - fish data exclude a small sample from *c.* 1160 BP and mammal data are from upper levels only). Where detailed fish and/or bird analyses have not been done (or were not available), general summaries for Figures 6.8-6.11 were obtained from the following sources: RAT-031 fish (Simenstad, Estes and Kenyon 1978; Isakson, Simenstad and Burgner 1971), RAT-036 birds (from MNI calculations provided by Harrington (1987), cited in Funk 2010, 56), UNL-055 fish and birds (from family level identifications provided by Yesner, in Knecht and Davis 2003).

Non-Aleutian sites, sea mammals only: (Table 6.13) Cape Addington, AK site 49-CRG-188 data from Moss (2004), Moresby Island, BC site 699T data from Orchard (2007), Hesquiat Harbor, BC site DiSo1-IV data from Calvert (1980), Ozette Village, WA site 45CA24 data from Huelsbeck (1983), and Seal Rock, OR site SRI data is from Lyman (1989).

Results of New Analyses

Tables 6.2-6.13 present the fish, bird and mammal data for the Aleutians. The bone counts for the two main village site assemblages were tallied separately and discussed below as "Adak Early" (ADK-012) and "Adak Late" (ADK-011). The extensive fish remains from Sweeper Cove (ADK-009) are designated "Adak Early2," as this deep deposit

Figure 6.2. Aerial photo of Clam Lagoon, Barrier Beach and Zeto Point. (Photo courtesy Mitsuru Okuno).

spans the period of occupation of ADK-012. The small fish assemblage from Andrew Bay is designated "Adak Late2," as the occupation of this site overlaps that of ADK-011. Site totals are listed below for each of the major vertebrate groups (fish, birds, mammals).

Adak Early (Dozered Site, ADK-012, dated *c.*1865-1515 BP)--Total sample size for all analyzed vertebrates was 2892, of which 87% (2562) were fish. The sample sizes for bird and mammal were 185 and 145, respectively.

Adak Early2 (Sweeper Cove, ADK-009, dated *c.*2080-970 BP)--The level plus matrix NISP for fish remains from this site was 24,825.

Adak Late (Zeto Point, ADK-011, dated *c.*400-170 BP)--The bird and mammal remains generated a total sample of 2507 (mammal 1234, bird 1273). Fish remains were not counted.

Adak Late2 (Andrew Bay ADK-084, dated *c.*480-250 BP)--Virtually the entire sample was fish bone (94), with only one piece of identifiable bird bone recovered (total sample 95).

Fish (ADK-009, ADK-011, ADK-012, ADK-084)

Adak fish remains are reported in Tables 6.2 and 6.9.

Adak (Early)--Pacific cod remains were the most common (63%), while greenling were the next most common (20%). Virtually all of the greenling that could be identified to species represents rock greenling. Relatively uncommon taxa (found at frequencies of 4% or less) included salmon, rockfish, a few flounder species (mostly starry flounder, some rock sole), Irish lord and large great sculpins, and herring. A number of species were very rare (< 1%), including Atka mackerel, a few additional species of small sculpin, longnose lancetfish, skate, and salmon shark. Halibut and sablefish were not identified.

Adak (Early2)--The most common species was rock greenling (48%), while Pacific cod had a frequency of only 31%. Large sculpins as a group (e.g., Irish lords plus great sculpin) were the next-most common category (9%), while flatfish other than halibut were uncommon (5%). Herring was also relatively uncommon (4%). All other taxa were represented at frequencies of 1% or less, including: Pacific salmon (species undetermined), Dolly varden, kokanee

Table 6.1. Species Identified for ADK-011 and ADK-012 and ADK-084 and ADK-009 fish.

COMMON NAME	**SPECIES NAME**	***FAMILY NAME***
Birds		
Common loon	*Gavia immer*	Gaviidae
Red-throated loon	*Gavia stellata*	Gaviidae
Shorttailed Albatross	*Phoebastria albatrus*	Diomedeidae
Northern fulmar	*Fulmarus glacialis*	Procellariidae
Short-tailed Shearwater	*Puffinus tenuirostris*	Procellariidae
Leach's storm petrel	*Oceanodroma leucorhoa*	Hydrobatidae
Double-crested cormorant	*Phalacrocorax auritus*	Phalacrocoracidae
Pelagic cormorant	*Phalacrocorax pelagicus*	Phalacrocoracidae
Red-faced cormorant?	*Phalacrocorax cf. urile*	Phalacrocoracidae
Emperor goose	*Chen canagica*	Anatidae
Aleutian cackling goose	*Branta hutchinsii leucopareia*	Anatidae
Goose sp. (lg.)	*Branta cf. canadensis*	Anatidae
Surf scoter	*Melanitta perspicillata*	Anatidae
Black scoter	*Melanitta nigra*	Anatidae
White-winged scoter	*Melanitta fusca*	Anatidae
Goldeneye	*Bucephala clangula/islandica*	Anatidae
Bufflehead	*Bucephala albeola*	Anatidae
Greater scaup?	*cf. Aythya marila*	Anatidae
Oldsquaw	*Clangula hyemalis*	Anatidae
Harlequin duck	*Histrionicus histrionicus*	Anatidae
Common eider	*Somateria mollissima*	Anatidae
King eider?	*cf. Somateria spectabilis*	Anatidae
Common/King eider	*S. mollissima/spectabilis*	Anatidae
Bald eagle	*Haliaeetus leucocephalus*	Accipitridae
Rock ptarmigan?	*Lagopus cf. mutus*	Phasianidae
Sandpiper sp.	*Charadriiform* sp.	Charadriiformes
Gull (lg)	*Larus* sp. *(lg)*	Laridae
Common murre	*Uria aalge*	Alcidae
Pigeon guillemot	*Cepphus columba*	Alcidae
Ancient murrelet	*Synthliboramphus antiquus*	Alcidae
Marbled murrelet	*Brachyrhamphus marmoratus*	Alcidae
Whiskered auklet	*probably Aethia pygmaea*	Alcidae
Crested-type auklet	*Aethia cristatella and/or Cyclorrhynchus* sp.	Alcidae
Least auklet	*Aethia pusilla*	Alcidae
Cassin's auklet	*Ptychoramphus aleuticus*	Alcidae
Rhinoceros auklet	*Cerorhinca monocerata*	Alcidae
Puffin sp.	*Fratercula* sp.	Alcidae
Robin?	*cf. Turdus migratorius*	Passeriformes
Snow bunting?	*cf. Plectrophenax nivalis*	Passeriformes
Winter wren	*Troglodytest troglodytes.*	Passeriformes
Gray-crowned rosy finch?	*cf. Leucosticted tephrocotis*	Passeriformes
Mammmals		
Norway rat?	*Rattus cf. norvegicus*	Muridae
Sea otter	*Enhydra lutris*	Mustelidae
Northern fur seal	*Callorhinus ursinus*	Otariidae
Steller's sea lion	*Eumatopias jubata*	Otariidae
Ringed seal	*Phoca hispida*	Phocidae
Harbor seal/Spotted seal	*could be Phoca largha &/or vitulina*	Phocidae
Ribbon seal	*Phoca fasciata*	Phocidae
Harbor seal	*Phoca vitulina*	Phocidae
Harbor porpoise	*Phocoena phocoena*	Phocoenidae
Dall's porpoise	*Phocoena dalli*	Phocoenidae
Stejneger's beaked whale?	*probably Mesoplodon stejnegeri*	Ziphiidae

COMMON NAME	SPECIES NAME	*FAMILY NAME*
Minke whale Humpback whale	*Balaenoptera acutorostrata* *Megaptera novaeangliae*	Balaenopteridae Balaenopteridae
Fish		
Skate sp.	*Raja* sp.	Rajidae
Salmon shark?	*cf. Lamna ditropis*	Lamnidae
Pacific herring	*Clupea pallasii*	Clupeidae
Pacific salmon	*Oncorhynchus* sp.	Salmonidae
Pacific cod	*Gadus macrocephalus*	Gadidae
Sablefish	*Anoplopoma fimbria*	Hexagrammidae
Atka mackerel	*Pleurogrammus monopterygius*	Hexagrammidae
Rock greenling	*Hexagrammos lagocephulus*	Hexagrammidae
Kelp greenling	*Hexagrammos decagrammos*	Hexagrammidae
Greenling sp	*Hexagrammmos* sp.	Hexagrammidae
Rockfish sp	*Sebastes* sp.	Scorpaenidae
Yellow Irish lord	*Hemilepidotus jordani*	Cottidae
Red Irish lord	*Hemilepidotus hemilepidotus*	Cottidae
Great-type sculpin	*Myoxocephalus* sp.	Cottidae
Leister sculpin	*Enophrys lucasi*	Cottidae
Antlered sculpin	*Enophrys diceraus*	Cottidae
Bigmouth sculpin	*Hemitripterus bolini*	Hemitripteridae (Cottidae)
Rock sole sp	*Lepidopsetta* sp.	Pleuronectiformes
Halibut	*Hippoglossus stenolepis*	Pleuronectiformes
Longnose lancetfish	*Alepisaurus ferox*	Alepisauridae
In ADK-009 only		
Capelin	*Mallotus villosus*	Osmeridae
Sandlance	*Ammodytes hexapterus*	Ammodytidae
Dolly varden	*Salvelinus malma*	Salmonidae
Kokanee salmon	*Oncorhynchus nerka*	Salmonidae
White-spotted greenling	*Hexagrammos stelleri*	Hexagrammidae
Lingcod	*Ophiodon elongates*	Hexagrammidae
Threaded sculpin	*Gymnocanthus pistilliger*	Cottidae
Ronquil sp	*Bathymasteridae*	Bathymasteridae
Smooth lumpsucker	*Aptocyclus ventricosus*	Cyclopteridae
Snake prickleback	*Lumpenus sagitta*	Stichaeidae
Rock prickleback	*Xiphister mucosus*	Stichaeidae
Gunnel sp	*Pholidae*	Pholidae
Starry flounder	*Platichthys stellatus*	Pleuronectidae
Alaska plaice	*Pleuronectes quadrituberculatus*	Pleuronectidae
Yellowfin sole	*Limanda aspera*	Pleuronectidae
Longhead dab?	cf. *Limanda proboscidea*	Pleuronectidae

salmon (land-locked sockeye), capelin, sandlance, Atka mackerel, white-spotted greenling, threaded sculpin, leister and/or antlered sculpin, an unidentified sculpin, rockfish, ronquils, sablefish, lingcod, pricklebacks, gunnels, longnose lancetfish, smooth lumpsucker, snailfish, Alaska plaice, yellowfin sole, and putative longhead dab. No halibut remains were identified.

Adak (Late)--This site had relatively fewer Pacific cod (perhaps *c.*50%) because rock greenling, yellow Irish lord sculpin and halibut were fairly well represented. The following taxa were relatively rare (*c.* <1% of the sample): small flounders (including rock sole and starry flounder), rockfish, salmon, and great sculpin. Salmon shark, skate, herring, red Irish lord and longnose lancetfish were not identified in this assemblage, but that could be an artifact of the small sample size (a single unit identified).

Adak (Late2)--Pacific cod, virtually all representing fish estimated as greater than or equal to *c.*86cm total length, account for 90% of the 94 pieces of identified fish bone (NISP 85). Irish lord made up 10% of the sample, with definitively identified yellow Irish lord accounting for half this amount. Two pieces attributed to greenling and one piece to an extremely large rockfish comprise the rest of the sample. The single rockfish element (an incomplete but enormous metapterygoid) may represent a shortraker rockfish (*Sebastes borealis*) close to its maximum reported size (Figure 6.6). This species reaches 120cm and is the largest rockfish found in Alaska. It lives on or near the bottom on offshore shelf slopes, usually at depths of 100-600m (Mecklenburg, Mecklenburg and Thorsteinson 2002), which is likely too deep for ancient hook and line fishing. However, shape differences discernable in Figure 6.6 suggest this specimen may be an extraordinarily large

Figure 6.3. ADK-011 lies in a protected cove on the Sitkin Sound side of Zeto Point. (Photo courtesy Dixie West).

rougheye rockfish (*S. aleutianus*), which today attains lengths of 97cm. In addition, while the rougheye is said to prefer waters *c.*100-500m in depth, it has been reported from waters as shallow as 25m. The ADK-084 specimen is much too large to represent Pacific ocean perch (POP), an important commercial species (Logerwell *et al.* 2005) that has a maximum reported length of only 55cm.

Birds (ADK-011, ADK-012)

Details of the Adak bird remains are presented in Tables 6.3, 6.4 and 6.11

Adak (Early)-- In all, 20 species of birds were identified. Ancient murrelet, a small alcid, was the most commonly occurring species (43%), while puffins (23%) were next most common (these could be either tufted or horned puffin, or both). A variety of auklet remains were recovered in relatively low numbers (least, Cassin's, whiskered, and rhinoceros auklet), as were a few pigeon guillemot bones, although remains of common murre are conspicuous by their absence. Cormorant remains were only moderately common (10%), with all three species known to occur in Alaska identified (red-faced, pelagic, and double-crested, although double-crested were the least abundant). Short-tailed albatross was relatively uncommon (3%) and all other species were rare, including ducks, geese and storm petrel. One piece of ptarmigan was recovered: based on present distributions, this probably represents rock ptarmigan, currently a common year-round resident on Adak (Holder, Montgomerie and Friesen 2000; Williams 2006).

Adak (Late)--A total of 34 bird species were identified from this site. Alcids in general were the most commonly occurring family and ancient murrelet by far the most common species (37%). Other alcids occurred at low frequencies, including whiskered auklet, common murre, Cassin's auklet, rhinoceros auklet, pigeon guillemot, marbled murrelet, puffins and crested and/or parakeet auklet. Ducks of various species were fairly well represented, especially eiders (11%), oldsquaw/longtail and harlequin ducks (5% each). The eiders were primarily common eider, although a few pieces of king eider may be present as well. A few other duck species were also identified but were rare, including white-winged scoter, surf scoter, black scoter, greater scaup, goldeneye and bufflehead. Medium-sized geese were fairly well represented and were the only group where unfledged individuals were identified (characterized

Table 6.2 Fish remains comparison, Adak Early and Early 2 NISP, absolute relative frequency and estimated relative abundances vs. Adak Late estimated relative abundance, where P=present only (<2%), x=uncommon (c. 2-9%), xx=common (10-35%), xxx=very common (c. 36-55%), xxxx abundant (55-80%), xxxxx=very abundant (>80%). Adak Early (ADK-012), Adak Early2 (ADK-009), Adak Late (ADK-011).

Common name	Adak Early NISP	Adak Early %	Adak Early2 NISP	Adak Early2 %	Adak Early Estimated relative abundance (from %)	Adak Early2 Estimated relative abundance (from %)	Adak Late Estimated relative abundance (no counts)
Pacific cod	1615	63	7719	31	XXXX	XX	XXX
Greenling sp (cf. rock)	501	20	11905	48	XX	XXX	XX
Red Irish lord	4	<1	1060	4	P	X	-
Yellow Irish lord	5	<1	6	<1	P	P	XX
Irish lord, undet.	42	1.6	199	1	X	P	XX
Great sculpin spp.	64	2.5	759	4	X	X	P
Sculpin, other spp.	4	<1	21	<1	P	P	-
Flatfish undet. (not halibut)	65	2.5	563	2	X	X	P
Starry flounder	0	0	490	2	-	X	-
Rock sole	0	0	76	<1	-	P	-
Halibut	0	0	21	<1	-	P	XX
Rockfish spp.	102	4	103	<1	X	P	P
Atka mackerel	13	<1	185	1	P	P	P
Ronquil spp.	0	0	62	<1	-	P	-
Pacific salmon spp.	104	4	392	2	X	X	P
Herring	33	1	936	4	X	X	-
Capelin	0	0	33	<1	-	P	-
Skate	2	<1	28	<1	P	P	-
Sablefish	0	0	1	<1	-	P	P
Salmon shark	2	<1	0	0	P	-	-
Longnose lancetfish	6	<1	13	<1	P	P	-
Pricklebacks & gunnels	0	0	73	<1	-	P	-
Misc. inshore species†	0	0	36	<1	-	P	-
Total bone count	**2562**	**100**	**24825**	**100**	**2562**	**24825**	**n/a**

† Misc. inshore species, Adak Early2 include (NISP): pollock 10, snailfish 2, lumpsucker 8, lingcod 2, sandlance 14.

by their small size, porous cortex and unfused growth plates (epiphyses) of the major long bones).

For prehistoric Adak, any of the following medium-sized goose species could be represented (Armstrong 1995; Williams 2006): emperor goose (*Chen canagica*), Aleutian cackling goose (*Branta hutchinsii leucopareia*), snow goose (*Chen caerulescens*), brant (*Branta bernicla*), and white-fronted goose (*Anser albifrons*). Note that the Aleutian cackling goose (formerly Aleutian Canada goose) is now classified as a distinct subspecies of cackling goose (*B. hutchinsii*), in part because it is distinguished genetically from large-bodied forms of Canada geese, *B. canadensis* (Banks *et al.* 2004; Scribner *et al.* 2003). Due to a lack of adequate goose comparative material, species identity in this case was resolved by using mitochondrial DNA (mtDNA) analysis, a method increasingly used to answer taxonomic questions for archaeological skeletal material (e.g., Eda *et al.* 2006; Frey *et al.* 2005; Barnes, Dobney and Wong 1998). Out of the 132 specimens of medium-sized goose identified from the four units analyzed, 26 elements (20%) were successfully sequenced. Only emperor goose (Figure 6.7) and Aleutian cackling goose were identified. Four of the 22 adult specimens analyzed for DNA yielded emperor goose sequences (18%), while 18 sequences matched cackling goose (82%). Unfledged nestlings comprised 11% of the total goose sample, while a few (1%) were slightly older, probably fully-fledged juveniles (almost adult sized but with juvenile texture and long bone growth plates fully fused). All four of the unfledged goose specimens analyzed matched cackling goose mtDNA sequences.

Table 6.3 Bird remains data for Adak Early vs. Adak Late, NISP and relative frequency. ADK-12, Units 1-3 and ADK-011, Units D, E, F, G.

Common name	Adak Early NISP	Adak Early frequency %	Adak Late NISP	Adak Late frequency %
Common loon	0	0	4	0.3
Red-necked loon	0	0	1	0.1
Shorttailed albatross	5	2.7	9	0.7
Leach's storm petrel	1	0.5	0	0
Northern fulmar	0	0	1	0.1
Short-tailed shearwater	0	0	6	0.5
Double-crested cormorant	4	2.2	3	0.2
Pelagic cormorant	7	3.8	1	0.1
Red-faced cormorant	8	4.3	27	2.1
Cormorant undetermined	0	0	35	2.7
Goose sp.	2	1.1	132	10.4
Common eider	0	0	79	6.2
King eider	0	0	1	0.1
Common/King eider	3	1.6	59	4.6
Misc. sea ducks	0	0	24	1.9
Oldsquaw (Long-tailed)	0	0	62	4.9
Harlequin	0	0	68	5.3
Duck undetermined	3	1.6	79	6.2
Bald eagle	0	0	2	0.2
Ptarmigan sp.	1	0.5	7	0.5
Sandpiper sp.	0	0	12	0.9
Gull sp. (lg)	3	1.6	14	1.1
Common murre	0	0	41	3.2
Pigeon guillemot	3	1.6	8	0.6
Marbled murrelet	0	0	4	0.3
Ancient murrelet	80	43.2	469	36.8
Whiskered auklet	2	1.1	45	3.5
Crested-type auklet	0	0	8	0.6
Least auklet	10	5.4	0	0
Cassin's auklet	6	3.2	7	0.5
Rhinoceros auklet	1	0.5	3	0.2
Puffin sp.	42	22.7	13	1.0
Alcids (sm only)	2	1.1	36	2.8
Alcids (vsm, med, lg)	0	0	6	0.3
Winter wren	0	0	6	0.3
Med. song bird (robin?)	1	0.5	0	0
Snow bunting	0	0	4	0.3
Gray-crowned rosy finch	0	0	2	0.2
Total NISP/frequency	**185**	**100**	**1273**	**100**

Note: "Misc. sea ducks (Adak Late only, NISP) include surf scoter 2, black scoter 8, white-wing scoter 4, greater scaup 1, goldeneye 1, bufflehead 8.

Table 6.4 Bird remains comparison of major taxa for Adak Early vs. Adak Late, relative frequency of NISP.

Major taxa	*Early Adak %*	*Late Adak %*
Geese	1	10
Eider ducks	2	11
Harlequin & oldsquaw ducks	0	10
Cormorants	10	5
Ancient murrelet	42	37
Puffins	23	1
Other species, including undetermined ducks & alcids	21	8
Total frequency	**100**	**100**
Total NISP per site	***185***	***1273***

Figure 6.4. Photo of a beach off the Sweeper Cove site, showing rocky reef and kelp habitat. Olga Krylovich is collecting comparative samples of invertebrates. (Photo courtesy Arkady Savinetsky).

Cormorant remains are relatively rare in this assemblage, although all three species known to occur in the central Aleutians are represented. Most of the remains appear to be red-faced rather than pelagic cormorant and only a few specimens were definitely double-crested cormorant. Other bird species identified in the assemblage, although rarely, included large gulls, sandpipers, shorttailed albatross, fulmar, common loon, red-throated loon, bald eagle, winter wren, snow bunting, gray-crowned rosy finch and ptarmigan.

Mammals (ADK-011, ADK-012)

Mammal remains are reported in Tables 6.5, 6.6 and 6.12

Adak (Early)--Sea otter comprised 61% of this sample and harbor seal another 20%. Northern sea lion and northern fur seal bones were represented at 4% and 5%, respectively. An unidentified "true" seal (*Phoca* sp.) was represented by two flipper bones, which were most similar to ribbon seal, an ice-breeding arctic species that ranges throughout the

Figure 6.5. Enormous rockfish sp. metapterygoid (L, incomplete) from ADK-084 compared to a modern c. 60cm shortraker rockfish (Sebastes borealis).

Bering Sea during the ice-free summer months (Reeves *et al.* 2002; Wilson and Ruff 1999). A few pieces of porpoise were present, as were a few fragments of vertebra from a small whale, *c.*20 ft or so. There were a few land mammals recovered from this site: remains tentatively identified as caribou are represented by two pieces, each from a small adult animal (perhaps female) and a late-term fetus or newborn (all are from upper levels and presumed to be historic intrusions into the prehistoric layers, as caribou are not native to the island). The single piece of dog identified may also be of historic origin. Note that while domestic dog (*Canis familiaris*) is generally a common human commensal prehistorically and regularly recovered from northwest coast archaeological sites (Crockford 1997a), this was the only bone attributable to dog recovered from any of the Adak sites.

All age categories of sea otter were represented: 49% of the 72 bones for which age estimates could be made were attributed to various juvenile categories from dependent newborns to young subadult animals; 51% were either subadult and/or adult, similar to the pattern apparent in the Late deposits (see Table 6.7). The few sea lion remains recovered were classified as either adult male or juvenile/subadult male, typical of many Aleutian and Northwest Coast sites in that remains of males are most common (Crockford *et al.* 2004; Lyman 1989; McKechnie and Wigen 2011; Yesner 1988). Fur seal remains included bones of one rookery-aged newborn, one somewhat older juvenile (probably < one year of age), and one adult male individual. Harbor seal remains were represented by five individuals of various ages (two adults, one juvenile/subadult, two young juvenile/ newborn).

Adak (Late) Sea otter and harbor seal were the most common taxa, each comprising about one quarter of the sample (26% and 25%, respectively). Northern sea lion and northern fur seal each comprised 19% of the sample. Porpoise and whale remains were present but rare, and some remains could be identified no further than family (e.g., pinnepedia). At least one bone of ringed seal was identified, as was a piece of putative ribbon seal. While both are ice-breeding arctic species, a few individuals venture into the southern Bering Sea during the summer and fall, making their presence in these deposits unremarkable (Reeves *et al.* 2002; Wilson and Ruff 1999).

Several pieces of unidentified small whale may have come

Figure 6.6. Dominant kelp forest vertebrates in the prehistoric Aleutians: rock greenling (upper L), sea otter (upper R), harbor seal (lower L), northern fur seal (lower R., proposed, see text). Photos from Wikipedia.com (greenling), Steve Mlodinow (otter), James Baichtal (harbor seal) and NOAA (fur seal).

from the same species (or even the same individual) as that represented by a large fragment of cranium from the occipital region that was identified through mtDNA analysis as Stejneger's beaked whale, a small whale that reaches lengths of 15-18ft. All specimens of this species that have been collected in recent decades have come from stranded individuals, whales that for reasons unknown, deliberately or accidentally (alone or in groups) swim up onto beaches where they get stuck and die. Between 1972 and 1992, 11 stranding incidents involving 23 individual Stejneger's beaked whales have been recorded for Kuluk Bay on NE Adak, suggesting that this is a relatively common species at this location (Walker and Hanson 1999). It is quite possible this species also stranded at this location in the past and that the prehistoric remains identified here represent utilization of natural mortality events (dispatching dying individuals and/or butchering freshly-dead stranding victims) rather than actively pursued (i.e. hunted) animals.

Two additional pieces of whalebone were identified by mtDNA analysis as minke and humpback whale. It was not possible to say whether these specimens were the remains of scavenged carcasses or hunted individuals, although both species have been known historically to enter Kuluk Bay (Missile Defense Agency 2005).

All age classes of sea otter were well represented in the sample, from newborns to fully mature adults, which was also true for true seals (Table 6.7). In contrast, amongst fur seal remains, young subadults and older juvenile classes were poorly represented, while adults (33.5%) and young-of-the-year juveniles (including unweaned newborns) were well represented (50.5%).

The age profile for sea lion remains at this site closely resembles the profile for fur seals. That is, young juveniles (less than five-six months) and adult/subadult classes (including females) were relatively well represented (34% and 61%, respectively), while older juveniles were poorly represented (5%). A few very poorly developed elements were recovered that almost certainly represent pups less than three months of age (compared to two modern specimens in the reference collection). No young subadult sea lions were identified.

A few rat remains were recovered from Adak Late but these were historic intrusions into the deposits. The Norway rat was introduced early in the historic period and rapidly became a common human commensal throughout the Aleutians (Bailey 1993). There were no dog or caribou remains recovered.

Figure 6.7 Identity of this goose synsacrum from ADK-011 was determined by mtDNA analysis to be emperor goose rather than cackling goose (Wilson et al. 2011).

Evidence for Fur Seal and Sea Lion Rookeries (ADK-011, ADK-012)

At Adak Late, the remains of young juvenile fur seals and sea lions (< four month old unweaned fur seals; < five month old sea lions) provide good evidence that both species maintained rookeries or nursery haulouts (areas utilized by sea lion mother/nursing pup pairs, distinct from birthing/breeding rookeries) accessible from NE Adak during the late prehistoric period. Today, fur seal pups born in the Pribilofs do not leave their natal rookery until they are weaned and abandoned by their mothers at about four months of age, and most of them leave the Bering Sea, via one of the eastern passes, before five months of age (Baker 2007). The high frequency of rookery-age fur seal remains (*c.*32%) recorded at Adak Late is solid evidence that a fur seal rookery was located near Zeto Point. In addition, the relatively strong representation of slightly older juveniles approximately four-six months old (Table 6.7, 'young juvenile/newborn' category) suggests that locally breeding fur seals may have been non-migratory, as this age class of the migratory Pribilof-based population is rare in the Aleutians (Baker 2007). While the sample size of fur seal remains from Adak Early is much smaller, the presence of a single rookery-aged specimen confirms that a local rookery was being accessed. Today, there are no fur seal rookeries in the Aleutians nor were any reported to exist during the historic period (Gibson 1996; Jochelson 1933; Kenyon and Wilke 1953; Roppel 1984). I address this issue again in the Discussion and Conclusions section.

Sea lion pups leave rookery locations with their mothers before they are three months old (well before they are weaned), although mother/nursing pup pairs generally stay close to a haulout area (Recovery Plan for the Steller Sea Lion 2008). There is currently an actively utilized sea lion breeding rookery on the Pacific Ocean side of Adak, at Cape Yakak, and two known haulout sites on the rocky west coast (Call and Loughlin 2005; Kenyon and Rice 1961; Steller Sea Lion Recovery Plan 2008) but it is doubtful if these locations were accessed by Unangan hunters from the Zeto Point region where Adak Late is situated. Sea lion remains are much better represented at Adak Late than at Adak Early. While none of the sea lion remains from Adak Late appear to represent an age class indicative of true rookery harvest, the age profile for sea lion remains still resembles closely the profile for fur seals. That is, young juveniles (less than five-six months) and adult/subadult

Table 6.5 Mammal data from Adak Early and Late Adak, NISP totals and relative frequencies per species or genera for all units analyzed.

Common name	*Adak Early Total NISP*	*Adak Early Frequency %*	*Adak Late Total NISP*	*Adak Late Frequency %*
Norway rat †	0	0	9	1
Dog †	1	< 1	0	0
Caribou	4	3	0	0
Sea otter	88	61	322	26
Fur seal	7	5	237	19
Northern sea lion	6	4	232	19
Sea lion/fur seal	0	0	60	5
Ringed seal	0	0	1	< 1
Ribbon seal	0	0	1	< 1
Harbor & spotted seal	31	21	313	25
Pinnepedia	1	< 1	27	2
Dall's porpoise	0	0	2	< 1
Harbour porpoise	1	< 1	4	< 1
Porpoise sp.	2	2	9	1
Stejneger's beaked whale	0	0	1	< 1
Small/medium whale undet.	4	3	10	
Humpback whale	0	0	1	< 1
Minke whale	0	0	1	< 1
Large whale undet.	0		4	< 1
Total NISP/frequency	**100**	***100***	***1234***	***100***

* "Harbor & spotted seals" are predominantly harbor seal but may include a few spotted seal.
† taxa assumed to be historic intrusions (from upper levels)

Table 6.6 Sea mammal frequencies for major taxa compared from Adak Early vs. Adak Late, relative frequency of NISP.

Major taxa	*Early Adak %*	*Late Adak %*
Sea otter	61	26
True seal*	21	25
Sea lion	4	19
Fur seal	5	19
Porpoise & whale	5	2
Other species plus unidentified seals/sealions	4	9
Total frequency	100	100
Total NISP	***145***	***1234***

* "True seals" are predominantly harbor seal but may include a few spotted, ribbon and/or ringed seal remains.

classes (including females) are well represented, while older juveniles and young subadults are poorly represented. A similar pattern has been described for sea lion material recovered from Buldir Island in the western Aleutians (Lefèvre *et al.* 1997), a known sea lion rookery location (Sease and Gudmundson 2002).

Evidence for Kelp Forest Habitat (ADK-009, ADK-011, ADK-012)

For all Adak sites reported here, the high relative abundance of sea otter plus harbor seal remains (Table 6.6, 6.12), in combination with relatively high frequencies of rock greenling (Table 6.2, 6.10), suggests that one or several mature kelp forests were present in the waters around NE Adak, accessible from both Zeto Point and Sweeper Cove (Figure 6.1). Archaeological and modern ecological studies have demonstrated that sea otters consume primarily kelp-dwelling and nearshore-dwelling fish species rather than urchins or clams when they are at high densities in established kelp forests and that high densities of rock greenling are characteristic of this habitat (Estes 1990; Reisewitz, Estes and Simenstad 2006; Watt, Siniff and Estes 2000). Kelp forests are preferred habitat for Aleutian harbor seals, where they likely feed on a variety of kelp and nearshore fishes (Estes, Simenstad and Palmisano 1978; Small *et al.* 2008). On Amchitka, Simenstad, Estes and Kenyon (1978) proposed that abundant archaeological

Table 6.7 Sea mammals by age class for Adak Late only, relative frequency of total NISP per age-estimated category. See text for age category definitions.

Age-estimated category	*Sea otter %*	*True seal* %*	*Sea lion %*	*Fur seal %*
Adult/subadult	47	45	61	33.5
Subadult/juvenile	35	31	0	8
Juvenile/young juvenile	10	19	5	8
Young juvenile/ newborn	5	4	34	19
Newborn/foetal	2	1	<1	31.5
Total NISP aged	***316***	***258***	***201***	***236***

* "True seals" are predominantly harbor seal but may include a few spotted or ribbon seal remains.

remains of harbor seal were a strong indicator that mature kelp forests existed there prehistorically.

Season of Occupation (ADK-011, ADK-012)

Taxa well represented in the faunal assemblage that have a limited seasonal availability make good indicators of when a site was occupied (Table 6.8). Absence of taxa known to have been seasonally-abundant constitutes moderately good evidence for absence of occupation (but does not rule out occupation in that season), while good representation of a seasonally-abundant taxon is considered relatively strong evidence that the site was occupied in that season. On Adak, relative abundance data of various bird taxa provide the strongest evidence for seasonality of site occupation, and this is one of the best reasons for comprehensive analysis of bird remains that takes identifications to species whenever possible.

Summer/Fall

At Adak Early, the abundance of summer visiting/breeding alcid and cormorant species, particularly ancient murrelet, puffin and red-faced cormorant, suggest strongly that bird hunting was done predominantly during the late spring through late summer. Neither harbor seal nor sea otter remains provide information on seasonal procurement practices. Fishing was likely a summer activity in all localities, as virtually all species are at their highest abundance in shallow, nearshore waters in this season (Mecklenburg, Mecklenburg and Thorsteinson 2002). All of the harbor seal remains represent either adult or older juveniles and while some newborn and nursing-aged sea otter remains were recovered, sea otters do not have a seasonally-constrained birthing period.

The presence of unfledged Aleutian cackling geese at Adak Late indicates summer procurement of this species. No geese of any species currently nest on Adak; nor are there records of geese breeding on Adak in historic times (Byrd, Gibson and Johnson 1974; Taber 1946; US Fish and Wildlife Service 1999; Williams 2006). The Aleutian cackling goose was historically known as a common breeder on a few islands, particularly Attu in the Near Islands (Bailey 1993; Turner 1885), but by the mid-20th century—after foxes had been introduced to many islands—breeding populations existed only on Buldir in the west and Chagulak in the east (Bailey and Trapp 1984; Byrd, Gibson and Johnson 1974; Byrd and Day 1986; US Fish and Wildlife Service 1999). The mtDNA analysis results of ADK-011 goose remains thus provide the first direct evidence that Adak Island was part of the former breeding range of Aleutian cackling geese (Wilson *et al.* in press).

The presence of both cackling goose goslings as well as adults in the prehistoric remains is predictable because geese provide parental care and protection to their unfledged young, which ensures that geese encountered during the summer nesting period are primarily family groups. Harvesting of geese during the breeding season would likely net both adults and juveniles. In contrast, Kristine Bovy (2007) reported that 97% of the cormorant remains (NISP 2420) recovered from a site in coastal British Columbia (situated adjacent to known cormorant breeding colonies) were unfledged newborns, suggesting that nestlings were targeted almost exclusively while adults were away foraging at sea.

The only duck species represented in the Adak Late assemblage that is currently a summer breeder on Adak (other than the greater scaup, of incidental occurrence) is the common eider. Eider bones are relatively well represented, suggesting that eiders were harvested primarily during the summer breeding season, as they are rare in other seasons. Surprisingly, however, no remains of nestling-age chicks of eider were identified.

Fall/Winter

Adak Early contained very few or no strong winter indicators (Table 6.8). The predominance of summer-available resources and lack of winter indicators suggest this site was occupied seasonally during the summer, with hunting activities directed at a limited number of species.

In contrast, as both oldsquaw and harlequin ducks are almost exclusively winter residents on Adak (Byrd, Gibson and Johnson 1974; Williams 2006), their presence in the Late Adak material is a strong indicator of winter occupation. The confirmed presence of emperor geese also indicate

Table 6.8 Seasonal indicators for Adak Early and Adak Late.

Season and Faunal indicator	*Adak Early*	*Adak Late*
Spring/summer		
- unfledged cackling goose	no	yes
- adult cackling goose	?	yes
- ancient murrelet	YES	YES
Summer		
- puffins	YES	few
Summer/fall		
- rookery-age fur seal	yes	YES
- < 6 month old sea lion	no	few
Fall/winter		
- emperor goose	?	few
Winter/spring		
- harlequin duck	no	yes
- oldsquaw/long-tail duck	no	yes

Note: "YES" indicates frequencies of >20% for this taxon or age category, "yes" indicate frequencies < 20%.

winter occupation, as today these geese are common visitors on Adak from about late September to early May (Byrd, Gibson and Johnson 1974; Hupp, Schmutz and Ely 2008; Taber 1946). The presence of winter through summer resources suggest Adak Late was occupied year-round, a conclusion supported by the somewhat greater diversity of bird species at Adak Late compared to Adak Early.

Site Location (ADK-009, ADK-011, ADK-012, ADK-084, ADK-171)

Adak Earliest--Located on the Clam Lagoon side of Zeto Point (Figures 6.1, 6.2), there should have been a wide range of vertebrate resources available (from both the lagoon habitat and the open ocean). This was a protected site with access to both the lagoon (indicated by saffron cod and perhaps flatfishes) and to mature nearshore kelp forests outside (rock greenling, sea otters, harbor seal and fur seal).

Adak Early--This site is located on the barrier beach that separates Clam Lagoon from Sitkin Sound (Figures 6.1, 6.2). The beach location of this site would have been convenient in the summer but very exposed in the winter. The lack of wintering birds and the exposed location suggests this site was occupied from spring through fall at most, and perhaps in summer only. Kelp forest vertebrates were well represented, especially sea otters and harbor seals, suggesting that a mature nearshore kelp forest existed close by. Of all the sites examined, Adak Early had the highest frequency of salmon (4%), which is consistent with a summer fishing site. However, salmon is still very uncommon compared to kelp and nearshore marine fish such as cod. Although today there are a number of salmon streams on Adak (Holmes 1997), salmon were not heavily utilized prehistorically relative to other species and their bones did not enter refuse middens at the analyzed sites. We would expect salmon bones to be found in archaeological sites near Adak salmon streams where salmon processing would occur.

Adak Early2--Located on Sweeper Cove, south of Zeto Point, this location has rocky shores and lots of kelp but no sandy beaches (Figures 6.1, 6.4). Mature kelp forests vertebrates were very well represented at this site, but with more diversity and at higher frequencies than Clam Lagoon area sites. There was an abundance of kelp-associated fish species, including large sculpins such as red Irish lord and great sculpin (which are rare elsewhere). Great sculpin prefer sand and mud bottoms around rocks and are common from intertidal depths to <200m (Burger *et al.*, 2007; Mecklenburg, Mecklenburg and Thorsteinson 2002; Ormseth and TenBrink 2009). Red Irish lords are usually found in rocky-bottomed habitats that are more shallow (intertidal to *c.*48m) than the subtidal to 110m depths preferred by yellow Irish lords (Isakson, Simenstad and Burgner 1971; Mecklenburg, Mecklenburg and Thorsteinson 2002; Simenstad, Estes and Kenyon 1978).

Adak Late--Located on a protected cove on the Sitkin Sound side of Zeto Point, Adak Late inhabitants had easy access to mainland Adak, Clam Lagoon and Sitkin Sound/Kuluk Bay resources (Figure 6.1, 6.3). It was protected enough to be occupied year-round. Terrestrial bird hunting (especially for geese, eiders, and small terrestrial birds) may have taken place more often and further afield than at other locations or time periods. In the fish assemblage, the presence of species found offshore in relatively deeper water (especially halibut and yellow Irish lord, see Adak Late2 below) as well as nearshore kelp forest species (such as rock greenling) suggests access to or control over a number of fish habitats.

Adak Late2--Located around Mt. Adagdak from Clam Lagoon, on a protected cove on north-facing Andrew Bay, this site is very close to relatively deep, offshore resources (Figure 6.1, 6.2). The location of the site is strongly reflected in the marked abundance and consistently-large size of Pacific cod and presence of yellow Irish lord, a large offshore-dwelling sculpin that is most common from subtidal depths to about 110m (Isakson, Simenstad and Burgner 1971; Mecklenburg, Mecklenburg and Thorsteinson 2002). Although cod generally move into shallow water in summer, large cod are found at relatively greater depths than smaller individuals (Orchard and Crockford 2010).

Change Over Time on Adak

The sites included in this comparison, with relevant dates, are: Adak Earliest (Tutiakoff Site, ADK-171, *c.*7000-6000 BP), Adak Early (Dozered Site, ADK-012, *c.*1865-1515 BP), Adak Early2 (Sweeper Cove, ADK-009, *c.*2080-970 BP), Adak Late (Zeto Point, ADK-011, *c.*400-170 BP), Adak Late2 (Andrew Bay, ADK-084, *c.*480-250 BP), see Tables 6.10-6.12.

Table 6.9. Fish remains comparison, central Aleutian Adak sites vs. eastern and western Aleutian sites, by relative frequency of NISP (%) or subjective relative abundance (see Table 6.2, "Est."). UNL-050 (Amaknak Bridge, c. 3500-2500 BP), UNL-055 (Amaknak Spit, c. 600-350 BP), ADK-171 (Adak Earliest, c. 7000-6000 BP), ADK-012 (Adak Early, c. 1865-1515 BP), ADK-009 (Adak Early2, c. 2080-970 BP), ADK-011 (Adak Late, c. 400-170 BP), ADK-084 (Adak Late2, c. 480-250 BP), ATU-061 (Shemya, c. 3300 BP), ATU-003 (Shemya, c. 2500 BP), KIS-008 (Buldir Whalehouse, c. 530-280 BP).

	East	***East***	***Central***	***Central***	***Central***	***Central***	***Central***	***West***	***West***	***West***
Taxa	***UNL-050***	***UNL-055***	***Earliest ADK-171***	***Early ADK-012***	***Early2 ADK-009***	***Late ADK-011***	***Late 2 ADK-084***	***ATU-061***	***ATU-003***	***KIS-008***
	Level %	***Est. †***	***Level%***	***Level %***	***Lev+Mat. %***	***Est.***	***Level%***	***Matrix %***	***Level %***	***Level %***
P. cod &/or pollock§	89	XXXX	35	63	31	XXX	90	38	36	5
Saffron cod	-	-	**5**	-	-	-	-	-	-	-
Greenling (cf. rock)	2	X	49	20	48	XX	*	10	12	1
Atka mackerel	<1		-	<1	1	P	-	6	9	85
Red Irish lord	<1		-	*	4	-	-	<1	2	-
Yellow Irish lord	<1		-	*	<1	XX	5	9	1	-
Banded Irish lord	-		-	-	-	-	-	<1	1	-
Irish lord, undet.	1	XX	-	2	1	XX	5	25	14	1
Great sculpin spp.	1		-	2	4	P	-	1	1	<1
Sculpin, other spp.	*	X	-	*	<1	-	-	2	1	-
Sculpin, undet.	-		1	-	*	-	-	-	<1	-
Rock sole	*		-	*	<1	-	-	1	3	<1
Starry flounder	<1		<1	1	2	-	-	-	*	-
Halibut	3	P	2	-	<1	XX	-	2	<1	-
Flatfish undet.	<1		6	1	2	P	-	1	3	-
Rockfish sp.	1	*	<1	4	<1	P	*	4	16	8
Herring	-		-	1	4	-	-	-	-	-
Pac. salmon sp.	4	*	<1	4	1	P	-	-	*	-
L-N lancetfish	-		-	<1	<1	-	-	*	<1	-
Misc. inshore species ♫ *				*	*	-	-	-	-	-
Total NISP	***15069***	***126 †***	***8414***	***2562***	***24825***	***n/a***	***94***	***3053***	***5952***	***1091***

Notes: § where Pacific cod and pollock were distinguished, pollock comprise <0.1 to <1% of total NISP; ♫ "misc. inshore species" (in trace amounts to <1%) may include: sandlance, capelin, smooth lumpsucker, small sablefish, lingcod, white-spotted greenling, sandfish, snailfish, salmon shark, arrowtooth flounder, petrale sole, dover sole, yellowfin sole, longhead dab, Alaska plaice, pricklebacks and gunnels, skate, ronquil sp., Dolly Varden, kokanee salmon.
†UNL-055 only 126 dentary bones identified to family (total fish NISP 899), except for P. cod and halibut, and as exact frequencies are not comparable to other samples, estimates of general abundance are used instead. "*" indicates trace amounts only (1-4 specimens). "Level" samples may or may not have been screened through 1/4" mesh, "Matrix" samples have been screened through 1/8" mesh. For data sources, see Methods.

Fish

Fishing was almost certainly a late spring through early fall activity at all sites, as virtually all species are found near shore in greatest abundance during the summer (Jochelson 1933; Mecklenburg, Mecklenburg and Thorsteinson 2002). The one exception to this is saffron cod, a species that is available inshore only in winter. Saffron cod was recovered only at Adak Earliest, perhaps the initial occupation site at Clam Lagoon. Although it cannot be determined when this population was established, the lagoon certainly has the type of spawning habitat saffron cod seem to prefer in colder regions. However, this local population appears not to have survived intense aboriginal harvest: saffron cod never appear again in the archaeological record after this period, not even incidentally (although admittedly no sites have yet been found on Adak around Clam Lagoon that were occupied during the cold Neoglacial period).

Adak Earliest fish remains were very similar to Adak Early2. At both of these sites, rock greenling was as common or more so (*c.*50%), than cod remains (which comprised only

Table 6.10 Birds of Adak Island, modern relative abundance from census data vs. prehistoric relative abundance of harvested taxa from faunal data, for Adak Early and Adak Late.

	Census Status 1968-73*	***Census Status 2003-06*****	***Prehistoric Adak Early‡***	***Prehistoric Adak Late§***
S-T albatross	absent	absent	uncommon	incidental
S-T shearwater	very abundant	uncommon	absent	incidental
	spring visitor	spring visitor		
R-F cormorant	moderately common	moderately common	uncommon	uncommon
	summer breeder	summer breeder		
Emperor goose	common	common	incidental?	incidental
	winter visitor	winter visitor		
Aleut. Can. goose	rare spring visitor ('68)	absent	incidental?	common
Black scoter	common	common	absent	incidental
	winter visiter	winter visitor		
W-W scoter	common	uncommon	incidental	incidental
	winter visitor	winter visitor		
Mallard duck	moderately common	moderately common	absent	absent
	resident breeder	resident breeder		
G-W teal	moderately common	moderately common	absent	absent
	summer breeder	summer breeder		
Scaup (greater)	common	common	absent	incidental
	winter visitor	winter visitor		
Bufflehead duck	moderately common	moderately common	absent	incidental
	winter visitor	winter visitor		
Oldsquaw duck	abundant	abundant	absent	moderately
	winter visitor	winter visitor		common
Harlequin duck	abundant	abundant	absent	moderately
	winter visitor	winter visitor		common
Common eider	common	common	incidental	common
	summer breeder	summer breeder		
Common murre	common visitor	rare resident	absent	uncommon
	spring & summer	summer breeder		
Anc. murrelet	moderately common	moderately common	very common	very common
	summer visitor	summer visitor		
Whisk. auklet	occasional, abundant	occasional, abundant	incidental	uncommon
	spring visitor	spring visitor		
Crested auklet	moderately common	moderately common	absent	incidental
	winter & spring	winter & spring		
Puffin spp.	abundant visitor	abundant visitor	common	incidental
	summer breeder	summer breeder		
Pig. guillemot	common resident	common resident	incidental	incidental
	summer breeder	summer breeder		

Notes: * as per Byrd *et al.* 1974; ** as per Byrd *et al.* 2005, Williams 2006; ‡ Units 1-3, bone count 185, goose specimens not identified to species; § Units F & G only, bone count 696, a subsample of goose specimens identified by DNA. Estimated relative abundance for archaeological samples: incidental, <2%; uncommon, 2-4%; moderately common, 5-9%; common, 10-29%; very common, 30-50%, abundant, 50-70%.

*c.*30-40%). In contrast, Adak Early fish remains were most similar to Adak Late: at these sites, cod were the most common (*c.*60% or so) and greenling much less so (*c.*20% of fish NISP). However, relative frequencies of sea otter and harbor seal (major predators of rock greenling and also common residents of mature kelp forests) are present at their highest frequencies at Adak Early and Adak Early2 (where together, sea otter and harbor seal comprise > 80% of mammals). In other words, although relative abundance of greenling went up and down a bit over time, these changes do not track with changes in sea otter and harbor seal abundance over time as might be expected if sea otters and/or greenling are indicators of kelp forest equilibrium, as Simenstad, Estes and Kenyon (1978) have proposed. However, relative frequencies calculated by level might show another pattern to the site totals compared here.

Rockfish and salmon were present at frequencies greater than incidental (i.e. *c.*1-5% of the fish NISP) only at Adak Early, despite the very large sample size at Adak Early2.

Table 6.11 Bird remains comparison across the Aleutians, relative frequency of NISP per dominant species or taxonomic group. Amaknak (UNL-050, c. 3500-2500 BP); Early Adak (ADK-012, c. 1865-1515 BP); Late Adak (ADK-011, c. 400-170 BP); Amchitka (RAT-031, c.2550-890 BP); Shemya (ATU-022, c. 2000 BP); Buldir (KIS-008, c. 530-280 BP), by relative frequencies of NISP ("other species" include ducks other than eiders, additional alcid species and other incidental taxa).

Major bird species or group	***East Amaknak UNL-050 %***	***Central Adak Early ADK-012 %***	***Central Adak Late ADK-011 %***	***Central Amchitka RAT-031 %***	***West Shemya ATU-022 %***	***West Buldir KIS-008 %***
Albatross/fulmar shearwater	**17**	3	2	12	**55**	<1
Cormorants (all spp.)	3	10	5	**33**	3.5	<1
Geese & eiders	5	3	**21**	6	1	1
Murres (lg alcids)	**29**	0	3	<1Φ	2	<1
Puffins (med alcids)	2	**23**	1	1‡	1	3
Ancient murrelet (sm alcid)	<1	**43**	**37**	?†	4.5	5
Whiskered/crested/ least auklets (sm alcids)	**10¶**	6¥	4§	**12†**	2	22
Other species	33	12	27	32	31	68
Total NISP	***10986***	***185***	***1273***	***4723***	***1722***	***5903***

Notes: Note that "albatross/fulmar/shearwater" figures do not include storm petrel. ¶ These are primarily crested ¥ These are primarily least auklet § These are primarily whiskered auklet. † These figures appear to represent "small alcids" not exclusively *Aethia* sp. as stated in Causey *et al.* 2005, and original figures listed by Warter 1970, cited in Funk (2009, 50), suggest that least auklet was the most common. ‡ Original figures listed by Warter 1970, cited in Funk (2009, 50), suggest that ca. 1% of this total were puffins. Φ These figures appear to represent "large alcids" not exclusively *Uria* sp. as stated in Causey *et al.* 2005, but original figures listed by Warter 1970, cited in Funk (2009, 50), suggest that murres were indeed the most common of these. For data sources, see Methods.

The location of these sites and the summer-only occupation of Adak Early likely explain the difference. Similarly, herring are present at frequencies greater than incidental only at Adak Early2, which may be partly due to location differences but are also likely influenced by the fact that matrix samples (screened through 1/8" mesh) were included in the analysis. Lastly, small flatfish species (i.e. species other than halibut) were present at frequencies greater than incidental only at Adak Earliest, where proximity to the sandy mud habitat of Clam Lagoon, and/or the region immediately outside it on Kuluk Bay, likely explains their prevalence.

A few specimens of longnose lancetfish, a large offshore species with very fragile bones and spectacular teeth (Mecklenburg, Mecklenburg and Thorsteinson 2002), were found at both Adak Early and Adak Early 2. As the nearshore fish fauna at these sites are otherwise not similar, it is likely this species was caught incidentally while fishing further offshore for halibut or cod.

Birds

As bird remains were not available for Adak Earliest, Adak Early2 or Adak Late2, the discussion of change over time for Adak birds is confined to Adak Early and Adak Late. Table 6.10 summarizes current and historic bird abundance on Adak and contrasts that to the relative abundance of remains identified from the Early and Late archaeological deposits. From this comparison, some notable differences are apparent: 1) The low representation of any medium-sized goose in Adak Early contrasts with the relative abundance of geese in the Late deposits and the abundance of emperor geese in modern census counts (note that cackling geese are not documented historically); 2) The definitive identification of a breeding population of Aleutian cackling goose in the Late deposits confirms that this species once maintained a breeding population of Adak; 3) The absence or near absence of short-tailed shearwater in the archaeological deposits contrasts with records of very high (if short-term) abundance in modern times; 4) The relative rarity or absence prehistorically of scoters contrasts with the fact that these ducks are considered fairly common today (a pattern also apparent for mallards, green-winged teals, greater scaups and buffleheads) (Taber 1946; Williams 2006); 5) The fact that both oldsquaw/longtail and harlequin ducks are absent from the Early deposit contrasts strongly with their relative abundance in the Late deposits as well

Table 6.12 Marine mammal comparison across the Aleutians, relative frequency of NISP per dominant species and/or family groups ("other seals" are primarily ringed and/or bearded seal). Sites are: MargBay (Margaret Bay, Unalaska, UNL-048, c. 4700-4100 BP); Amak1 (Amaknak Bridge, Unalaska, UNL-050, c. 3500-2500 BP); Amak2 (Amaknak Spit, Unalaska, UNL-055, ca 600-350 BP); Adak Earliest (ADK-171, c. 7000-6000 BP; Adak Early (ADK-012, c. 1865-1515 BP); Adak Early2 (ADK-009, c. 2080-970 BP), Adak Late (ADK-011, c. 400-170 BP); Amchitka (RAT-031, c.2550-890 BP); Shemya (ATU-061, c. 3300 BP); Buldir (KIS-008 Pit 1, c. 530-280 BP).

Species or group	***East MargBay UNL-048 %***	***East Amak01 UNL-050 %***	***East Amak02 UNL-055 %***	***Central Adak Earliest ADK-171 %***	***Central Adak Early ADK-012 %***	***Central Adak Early2 ADK-009 %***	***Central Adak Late ADK-011 %***	***Central Amchitka RAT-031 %***	***West Shemya ATU-061 %***	***West Buldir KIS-008 %***
Sea otter	*2*	*<1*	*1*	*30*	61	45	26	57	<1	0
Harbor &/or spotted seal	49	9	6	16	21	37.5	25	32	<1	0
Other seals	11	37	0	0	<1	-	<1	0	0	0
Sea lion	8	3	9	19	4	12.5	19	6	3	99.6
Fur seal	6	40	70	34	5	5	19	2	72	0
Porpoises & whales	20	11	5	1	5	-	2	?	2	<1
Other spp & unident. pinnipeds	4	<1	9	0	4	-	9	3	22	0
Total NISP	***5392***	***5947***	***4775***	***138***	***145***	***40***	***1234***	***1300***	***518***	***~1100***

Note that Davis (2001) reported material identified to family level (for UNL-048) differently than other authors but percentages appear to be comparable; for KIS-008, exact numbers are not reported for sea lion remains and total mammal NISP - the total presented here has been estimated from the graph provided; for RAT-031, numbers cited are for whole bones only (no fragments). For data sources, see Methods.

Table 6.13 Sea otter and harbor seal relative frequency along the NW coast, as a percent of the NISP totals reported for all pinniped species + sea otter (seals/sea lions/fur seals, including undetermined pinnipeds, sea otters), compared to Early Adak (UNL-012, c. 1815-1515 BP). Sites are: Amchitka (RAT-031, c.2550-890 BP); Margaret Bay, AK (Unalaska, UNL-048, c. 4700-4100 BP); Cape Addington (Noyes Island AK, 49-CRG-188, c. 2000-170 BP); Moresby Island SE (Haida Gwaii BC, 699T, c. 1000-170 BP); Hesquiat Harbour (Vancouver Island BC, DiSo1-IV, c. 780-1220 BP); Ozette Village (Cape Flattery, Olympic Peninsula, WA, 45CA24, c. 400-250 BP; Seal Rock (SRI, OR, c. 400-100 BP).

Species	***Central Aleutians***	***Central Aleutians***	***Eastern Aleutians***	***SE Alaska***	***Haida Gwaii, BC***	***Vancouver Island, BC***	***Olympic Penn. WA***	***Oregon***
	Early Adak	***Amchitka (S. shore)***	***Margaret Bay, AK***	***Cape Addington***	***Moresby Island SE***	***Hesquiat Harbour***	***Ozette Village***	***Seal Rock (SRI)***
Sea otter frequency	**67%**	**57%**	**2%**	**18%**	**52%**	**23%**	**1%**	**11%**
Harbor seal frequency	23%	32%	64%	27%	34%	27.5%	<1%	3%
Pinniped plus sea otter total NISP	**132**	**1300**	**4142**	**357**	**155**	**240**	**49135**	**1287**

For data sources, see Methods

as in modern times, suggesting that Adak Early was not occupied during the winter; 6) The relative abundance of common eider in modern times and in the Late deposits contrasts with their rarity in the Early material; 7) While the diminutive ancient murrelet was by far the most common species identified in both assemblages (Table 6.3), it is considered only a moderately common species around Adak today. This pattern is explained by the severe impact that introduced rats and foxes likely had on this burrow-nesting species (Bailey 1993; Byrd, Renner and Renner 2005); 8) In contrast to ancient murrelets, burrow- and/or crevice-nesting puffins are common around Zeto Point today (where they breed offshore) but puffins were well represented only in the early prehistoric deposits.

Thus, although there are some quite dramatic differences in relative abundance of various bird taxa between these two site assemblages, and between these and modern abundance levels, there does not appear to be a blanket explanation for either. I address these issues again in the Discussion and Conclusions section.

Sea Mammals

Considering only the four most prominent taxa (sea otter, harbor seal, fur seal and sea lion), Adak Earliest and Adak Early2 is very similar to Adak Late (where all four are present at almost equal frequencies) while Adak Early is most similar to Adak Early2 (where sea otter and harbor seal predominate, with low frequencies of fur seal and sea lion). The relatively high rate of removal of sea otters in all periods may have impacted sea otter genetic composition over time at this location (Nishida *et al.*, Chapter 7 this volume), since localized reduction of a sea otter population might have been followed by immigration from nearby areas (Estes 1990).

It is apparent, however, that fairly substantial populations of sea otter and harbor seal were locally available for the entire time that NE Adak was occupied. Although there appears to be some differences in frequency over time, I am not convinced that they are significant. Three out of the four sample sizes are fairly small (and one is very small). The largest sample available, from Adak Late, is likely to be the most representative for the region: virtually identical frequencies of sea otter, harbor seal, fur seal and sea lion.

Lastly, one of the most significant findings is the evidence that local fur seal rookeries were being accessed from all locations and all time periods on Adak, including Adak Earliest. This extends the record for locally breeding fur seals in the Northeast Pacific back another 1000-2000 years, the implications of which I address in more detail in the Discussion and Conclusions section below. As I argue there, I am convinced that fur seal assemblages, which include rookery-aged newborns, may be a strong indicator of mature kelp forests, although these were probably located away from settlement areas rather that directly offshore from site locations.

Change Over Time Across the Aleutians

Figures 6.8-6.11 present a chronological summary of faunal data reported to date across the Aleutians. These figures divide the prehistoric period into four parts defined by their relationship to two well-known cold periods, the Neoglacial (a well-documented mid-Holocene period of cold climate) and the late Holocene Little Ice Age (Crockford and Frederick 2007). Thus, sites with analyzed fauna are grouped into the Early Period (an early, pre-Neoglacial period, *c.*6000-4700 BP), the Neoglacial (the known record for the Neoglacial itself, 4700-2500 BP), the Post-Neoglacial (*c.*2500-1000 BP) and the Late Period (which encompasses the Little Ice Age era within it, *c.*1000-100 BP). Each site summary lists the one most abundant taxon (or two or three most abundant, by relative frequency of total NISP, whether accompanied by actual data or inferred from reports lacking precise data) for fish, birds and sea mammals. Less common and/or noteworthy species are listed in brackets (such as eider, harlequin duck, and geese) and the presence of rookery-aged fur seals is indicated with an asterisk. Holes in the data (e.g., 'no fish report') are readily apparent for specific regions and/or time periods.

Fish Summary (Table 6.9, Figures 6.8-6.11)

Pacific cod was extremely common at Unalaska (UNL-050) during the Neoglacial and at the pre-contact site on Adak (ADK-084), completely eclipsing other taxa. Cod from these sites were primarily very large and associated species included yellow Irish lord and longnose lancetfish, suggesting that fishing was done primarily in relatively deep, offshore areas. Although not included here, ATU-198 in the Near Islands fits this pattern also (West *et al.* 2003). In all other sites, cod was common to very common. One of the few locations where cod was not overwhelmingly the most common taxon was the late period 'Whalehouse' site on Buldir (KIS-008), where Atka mackerel took on the role of dominant fish species, leaving cod frequencies a mere 5% of the total.

Rock greenling frequencies exceeded those of Pacific cod at several sites in the central Aleutians, including Adak Earliest (ADK-171) and Adak Early2 (ADK-009), suggesting that these sites had mature nearshore kelp forest habitats nearby that were being utilized.

Kelp-dwelling and/or rocky bottom fish species (rock greenling, large sculpins and/or rockfish) were found at relatively high frequencies only in the central and western Aleutians. In some areas, the available kelp forests may have been in more exposed locations than in the central region, and in some locations, the kelp forests were perhaps further offshore in somewhat deeper water (e.g., at Shemya, Adak Early2 (ADK-009)).

Rockfish was only moderately common at a single site, on Shemya in the western Aleutians (ATU-003), and was incidental or uncommon elsewhere.

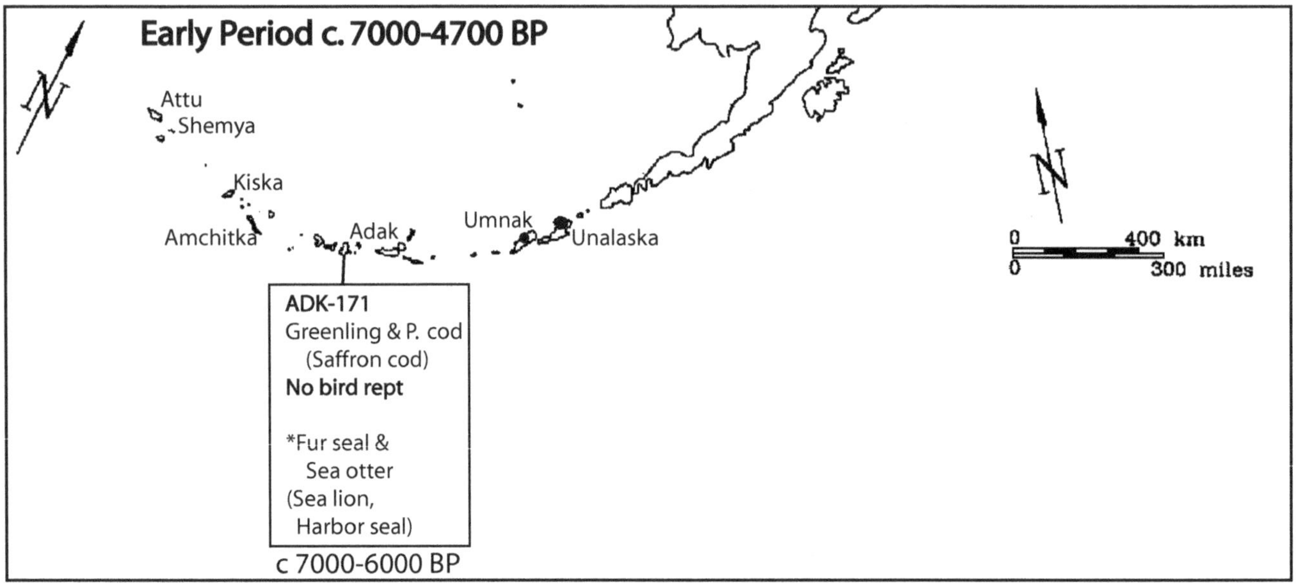

*Figure 6.8 Summary of dominant vertebrate taxa across the Aleutians recorded for the Early Period c. 7000-4700 BP. See Methods for references. Most common taxa only are listed, with less common or noteworthy taxa in brackets, and rookery-aged fur seals are indicated with an asterisk ('*Fur seal'). Note that some major categories have no reported fauna or the analysis report was not available.*

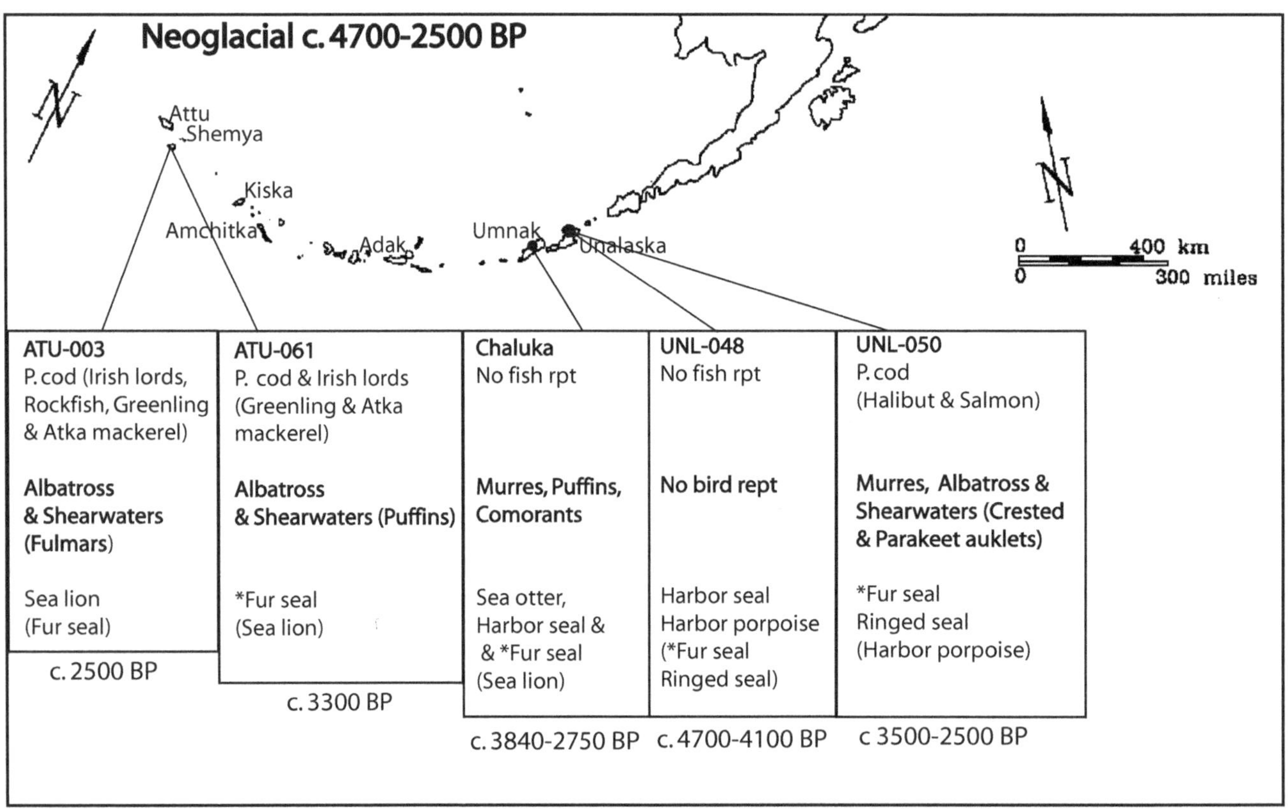

*Figure 6.9. Summary of dominant vertebrate taxa across the Aleutians recorded for the Neoglacial c. 4700-2500 BP. See Methods for references. Most common taxa only are listed, with less common or noteworthy taxa in brackets, and rookery-aged fur seals are indicated with an asterisk ('*Fur seal'). Note that some major categories have no reported fauna or the analysis report was not available.*

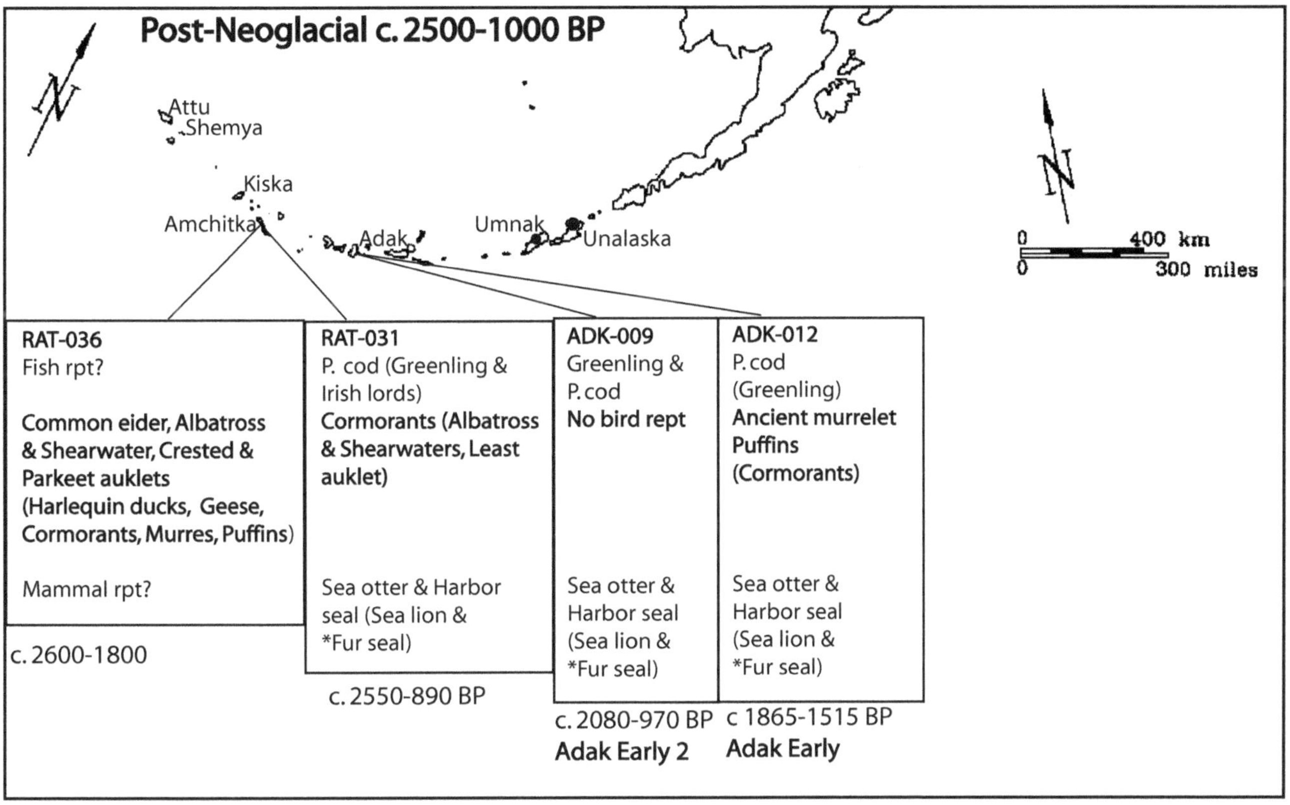

*Figure 6.10. Summary of dominant vertebrate taxa across the Aleutians recorded for the Post-Neoglacial c. 2500-1000 BP. See Methods for references. Most common taxa only are listed, with less common or noteworthy taxa in brackets, and rookery-aged fur seals are indicated with an asterisk ('*Fur seal'). Note that some major categories have no reported fauna or the analysis report was not available.*

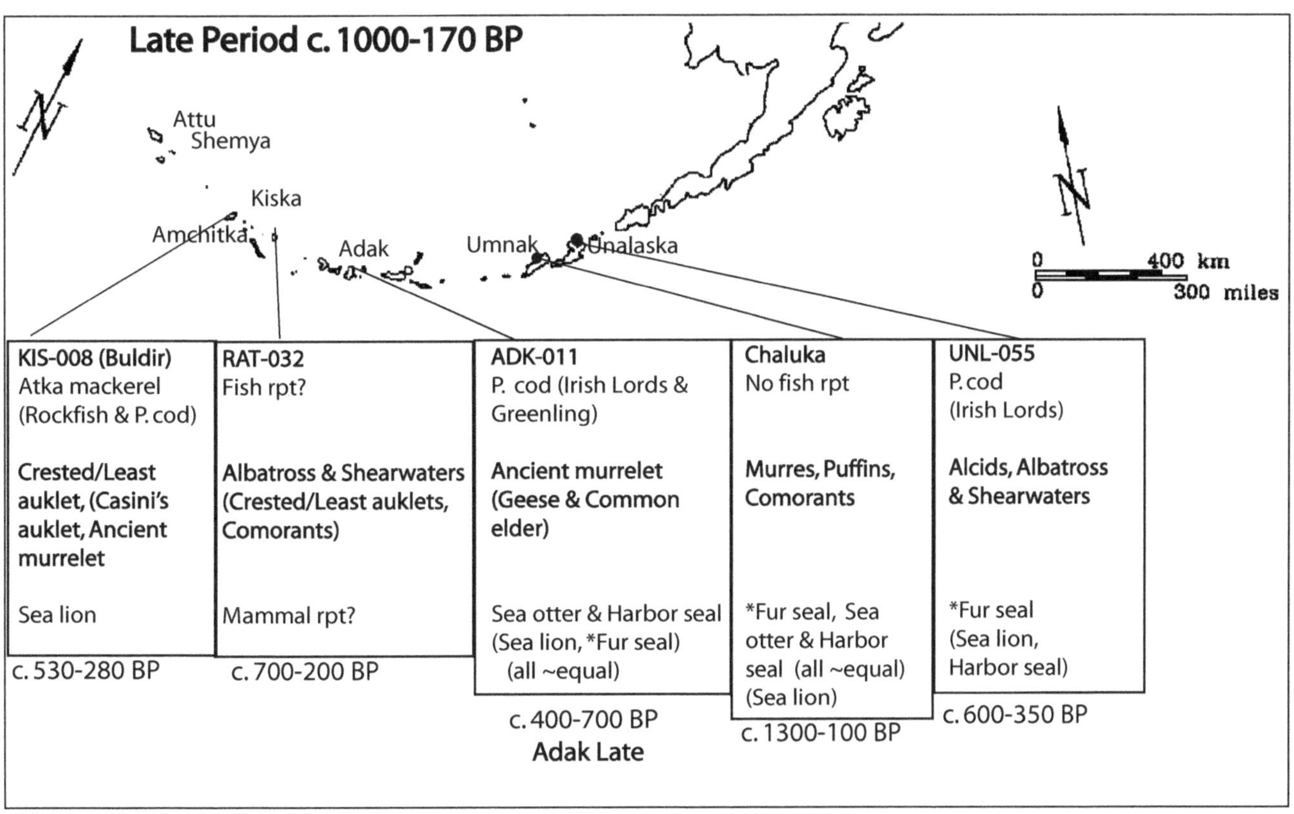

*Figure 6.11. Summary of dominant vertebrate taxa across the Aleutians recorded for the Late Period c. 1000-170 BP. See Methods for references. Most common taxa only are listed, with less common or noteworthy taxa in brackets, and rookery-aged fur seals are indicated with an asterisk ('*Fur seal'). Note that some major categories have no reported fauna or the analysis report was not available.*

The only places where frequencies of salmon were more than incidental (i.e. comprised at least 1-5% of the fish sample) were at Amaknak Bridge during the Neoglacial and Adak Early and Adak Early2 in the Post-Neoglacial period (2500-1000 BP).

The only location where Atka mackerel was most common fish was the late period site on Buldir (KIS-008) in the far western Aleutians, where Atka mackerel dominated the assemblage (85% of the total). Otherwise, the only place where Atka mackerel was even moderately common (6-9% of the fish sample) was on Shemya (ATU-003, ATU-061), also in the far west, during the Neoglacial. Everywhere else prehistorically it was an incidental species (<1%). This supports recent findings by Lauth, McEntire and Zenger Jr. (2007) that most Aleutian Atka mackerel spawn in rocky areas from summer through fall at much greater depths than previously thought and thus may not have been generally available to nearshore Unangan fishers, except at Buldir, where suitable conditions may have been present closer to shore than elsewhere.

The only place that winter-spawning, cold-loving saffron cod was recorded was at Adak Earliest, *c.* 7000-6000 BP, on Clam Lagoon. It was not recorded elsewhere. Note that saffron cod bones are distinctive and the identification of this species by Savinetsky and colleagues (Chapter 5 this volume) is certain.

Surprisingly perhaps, halibut was an incidental species everywhere across the Aleutians except during the Late period on Adak, where it was only moderately common (comprising perhaps 10% of the sample). This may be an artifact of butchering large specimens on the shore, which would result in few skeletal elements making their way into midden material.

Herring remains were much more abundant (4% of the NISP) at Adak Early2 (ADK-009) than at any other site except UNL-050 on Unalaska (Crockford *et al.* 2004) where herring was recovered in matrix samples. Elsewhere, herring was either not recovered or incidental.

Capelin, although it has bones smaller than herring, did not show up in the fine-screened samples from either UNL-050 (Crockford *et al.* 2004) or ATU-061 and was recovered in only small amounts from Adak Early2 (ADK-009): this suggests capelin was either rarely harvested prehistorically or that its bones did not usually enter refuse middens.

Long-nosed lancetfish is a large, conspicuous species with very fragile bones found incidentally only in central and western sites. As it is an offshore species, it may have been an incidental catch to halibut or Pacific cod or may have been taken opportunistically while hunting for sea birds or mammals. It was also recovered from cave burial site ATU-198 in the Near Islands (West *et al.* 2003).

Detailed fish bone analyses and/or published reports are lacking from several Aleutian faunal assemblages, including Shemya (ATU-022), Amchitka (RAT-031, RAT-032, RAT-036), Unalaska (UNL-048, UNL-055), Umnak (Chaluka). However, general summaries or family level identifications were available for some of these sites (see Method section for details).

Fish Discussion

Across the Aleutians, Pacific cod was the most common fish harvested in all time periods and at virtually all locations, except in the far western islands and at several Adak sites (ADK-171 and ADK-009) where cod was somewhat less common and rock greenling and/or Irish lords more abundant. Few if any walleye pollock were recorded among the tens of thousands of cod bones that have been identified from Aleutian Island sites (Table 6.9). This pattern is striking in contrast to modern commercial groundfish (demersal) catches, which are strongly dominated by pollock (Logerwell *et al.* 2005; North Pacific Fisheries Management Council 2007). What could explain this huge discrepancy in the dominant fish taken by prehistoric and modern harvesters?

I suggest that the harvesting disparity may reflect combined differences in preferred habitat between these two species and distinctly different fishing methods. Although adult, commercial-sized pollock and cod are both classified as demersal groundfish, cod occupy the bottom-most layer of this habitat while pollock live somewhat higher up. That is, adult cod are strongly demersal, dwelling within 7-10m of the bottom (Nichol, Honkalehto and Thompson 2006) while pollock are semi-demersal, generally residing within 30m above the ocean floor (Kotwicki *et al.*, 2009; Wilson *et al.*, 2003). As a consequence, weighted hooks on hand-lines lowered rapidly to the bottom and jigged up to the surface by ancient Unangan fishers would have exposed cod to the bait first. Hand-lines would therefore catch predominantly cod rather than pollock even if both were present and abundant, and would also yield an associated catch of other strictly bottom-dwelling species (e.g., skate, flounders, Irish lords, great sculpin, etc.), which is certainly what we recover archaeologically (Tables 6.2 and 6.9). Modern fishers, however, have methods at their disposal that are able to target the semi-demersal zone above the sea floor and thus yield primarily pollock.

Birds Summary (Table 6.11, Figures 6.8-6.11)

Short-tailed albatross and shearwaters (and in some cases fulmar) were the most common taxa at several sites in the western Aleutians and were moderately common in eastern and central Aleutian sites.

The only place where cormorants were the most common taxa was at Amchitka (RAT-031), although they were moderately common elsewhere.

The only place that breeding Aleutian cackling goose was identified was at Adak Late.

Geese were not common anywhere across the Aleutians prehistorically and were recorded in greater than incidental amounts only at the post-Neoglacial Amchitka site of RAT-036 (Funk 2010) and the pre-contact Adak Late site (ADK-011).

The only place where puffins were relatively common was at Adak Early: nowhere across the Aleutians were puffins the most common taxa, although they appear to have been moderately common at Chaluka on Umnak and on Amchitka (RAT-036).

The only places where common and thick-billed murres were the most common taxa were the eastern Aleutians (Unalaska sites UNL-050 and UNL-055, Umnak site Chaluka).

The only place where ancient murrelet was the most common bird was at Adak Early and Adak Late: nowhere else in the Aleutians could this diminutive sea bird have been called a common species prehistorically, although today they are relatively common across the archipelago (Byrd, Renner and Renner 2005).

Crested and/or parakeet auklets (difficult to distinguish osteologically) were uncommon taxa in virtually all sites prehistorically, although they were incidental in a few places.

Detailed bird analyses are lacking for several site assemblages from critical time periods, including Adak Earliest (ADK-171), Margaret Bay (UNL-048), Adak Early2 and Amaknak Spit (UNL-055).

Birds Discussion

Overall, the pattern of bird harvest reflected is markedly different from one time period to the other and one island or island group to the other (Table 6.11, Figures 6.8-6.11). The most common bird in both Adak sites was ancient murrelet and in this respect Adak is distinct from sites on islands to the east and west. While the prehistoric bird assemblages from some other islands had abundant albatross and shearwater remains (e.g., Shemya), others had an abundance of murre (e.g., Unalaska) or cormorant (e.g., Amchitka). Some researchers have attempted to attribute such differences across the Aleutians to changes in climatic conditions (temperature and precipitation) over time (Causey *et al.* 2005; Savinetsky, Kiseleva and Khassanov 2004; Yesner 1981) without much success. However, I suggest that cultural differences in preference for various bird species (Black 1984; Jochelson 1933, 53-54) plus a consideration of the precise ecological preferences and population dynamics of individual species (independent of human influences) almost certainly preclude making a climate change interpretation regarding differences in species abundance in prehistoric deposits over time. Few of the bird species represented in the Adak deposits give any indication of having been harvested in the proportions that they occurred (cf. Corbett, Lefèvre and Siegel-Causey 1997; Yesner 1981). For example, mallard ducks are currently moderately common year round yet appear to be absent from the Adak archaeological record. Similarly, emperor geese are common winter visitors throughout the Aleutians today (Armstrong 1995) and yet appear to have been rarely taken prehistorically. The same appears to have been true for ptarmigan: common now, rare prehistorically. That being said, specific birds could not have been harvested in large numbers prehistorically unless they were at least moderately abundant in the area.

Unless the abundance of various Aleutian bird species was never in the past similar to their abundance today, it seems obvious that ancient Unangan hunters regularly chose specific types of birds out of the landscape for specific reasons other than for food: one sort (or several) for making bird skin clothing, another sort for making bone tools and yet another because their feathers are just right for decorating garments and other items, regardless of how abundant they might have been. Thus if puffins were preferred for clothing over all other species (including eiders) by the occupants of Adak Early, while the occupants of Adak Late preferred eiders over all others (including puffins), both puffins and eiders may have been common and available during both time periods. This does not mean that the differences in prehistoric bird abundance across the Aleutians could not at least partially reflect past climatic change. However, specific local environmental conditions, such as proximity to shelf-edge upwelling or inter-island passes associated with particular food resources, or availability of suitable nesting sites (Byrd, Renner and Renner 2005; Freethy 1987; Hunt and Stabeno 2005; Jahncke, Coyle and Hunt 2005; Piatt *et al.* 2006) almost certainly make one island or island group better habitat for one species over another. While some of these physical features may change over time with changing climatic conditions, others may be remarkably stable: broad-scale difference from region to region and one time period to another may be more practically informative than trying to tease fine-scaled patterns from fauna data with statistical methods.

The presence of unfledged cackling goose remains at Adak is not only unusual in respect to goose remains but is atypical for Aleutian bird remains generally. Although many species represented at Aleutian Island sites are colonial nesters that breed locally (including cormorants, gulls and a variety of alcids), no unfledged juveniles of any of those species were recovered (cf. Bovy 2007) from Adak and the only mention of unfledged chicks of any species in the Aleutian zooarchaeological literature is the report from Buldir site KIS-008 of 'a few bones of Leach's storm petrel chicks in Pit 1 and of chicks and fledglings of several alcid species in Pit 2' (Lefèvre *et al.* 1997, 125). Crockford *et al.* (2004, 45) remark on the lack of unfledged chicks of any species in the UNL-050 bird assemblage, despite its impressive sample size (NISP 10,986). If unfledged chicks were taken regularly prehistorically, their bones cannot have been discarded in the refuse middens. However, sea bird eggs may have been harvested in large numbers without leaving an obvious archaeological signature.

Sea Mammals Summary (Tables 6.12, 6.13; Figures 6.8-6.11)

There is strong evidence for a continuous 6000-7000 year history of fur seal breeding rookeries adjacent to sites across the Aleutians, similar to those documented prehistorically to the south along the west coast of North America (e.g., Crockford and Frederick 2007), which contrasts with the modern pattern where virtually all NFS (except the San Miguel, CA colony) breed and give birth only on the Pribilof and Commander Islands during the summer, after which females and juveniles migrate south (Bigg 1990).

Sites in the central Aleutians (Adak and Amchitka) were the only places that document high frequencies of both sea otter and harbor seal. The only other site found in the literature that has similar frequencies of sea otter and harbor seal as these central Aleutian sites is the Moresby Island (Canada) site of 699T (Table 6.13). However, Cape Addington in SE Alaska and Hesquiat Harbor's DiSo I (level IV), on the west coast of Vancouver Island, British Columbia, both have moderately high frequencies of sea otter and harbor seal.

The only site where sea lion remains were the most common sea mammal was on Buldir, at KIS-008, during the Late Period (*c.*1000-170 BP), where remains of pups < five-six months were also recovered.

Nowhere were porpoise remains more than an incidental component of the faunal assemblage, except at Neoglacial-age sites on Unalaska in the eastern Aleutians (UNL-048, UNL-050) where harbor porpoise were relatively common.

Sea Mammals Discussion

Whale remains are difficult to quantify in a manner comparable to other taxa and while they appear incidental in many site assemblages, this almost certainly belies the importance of whale products to Unangan inhabitants.

The strongest pattern for sea mammals across the Aleutians is the evidence that locally breeding fur seals were present throughout the Aleutians, from at least 6000 years ago until the late pre-contact period. Rookery-aged pups as well as adults of both sexes were harvested, and in many areas fur seal was the most common taxon. Despite the various ethnographically-known uses for sea lion products, including the construction of gut clothing and skin boats (Black 1984; Jochelson 1933; Haynes and Mishler 1991; Hudson 2008), sea lion remains are less common prehistorically than these critical utilitarian uses might suggest. Only at Buldir, in the late pre-contact period, were sea lions remains much more common than fur seal. This is not so surprising when we realize that the fur seal that ancient Unangan hunters were harvesting prehistorically was very likely extinct by the historic period, perhaps as early as 1800 or so. Outside the Aleutians, we see a similar pattern. Sea lions were uncommon prehistorically except in areas where locally-breeding fur seals were not present (McKechnie and Wigen 2011) and the evidence that sea lion rookeries were located close to human settlements (i.e. remains of young or newborn pups) is slim (one known location is Seal Rock, Oregon; Lyman 1989). Unangan hunters during the historic period would have had to shift to hunting sea lions for all uses that formerly were satisfied by a mix of fur seal and sea lion products. I address this issue more fully in the Discussion and Conclusions.

In the central Aleutians, the high proportion of sea otter remains is quite unusual. While the 26% seen at Late Adak is remarkable, the 61% in the Early deposit is so far unprecedented in the Aleutians and perhaps elsewhere as well. Exceptionally good sea otter habitat at Adak Early, plus the summer-only occupation inferred from other data, might explain this pattern. Adak Late sea otter frequencies, for example, contrast strongly with the assemblage from the closely contemporaneous site in the eastern Aleutians, Amaknak Spit (UNL-055), which reports only 1% sea otter out of a total NISP of 4714 (Yesner, in Knecht and Davis 2003). Similarly, a huge sample of sea mammal remains recovered from the Ozette village site on the Olympic Peninsula, WA, had only 1% sea otter (Table 6.13). In contrast, while a few other North Pacific archaeological sites have reported somewhat higher proportions of sea otter remains, such as Cape Addington on Noyes Island, AK (20%), Hesquiat Harbour on Vancouver Island BC (23%), and Seal Rock, OR (11%), none except a site on Haida Gwaii (52%) come close to the sea otter frequency seen at Adak. It is tempting to speculate that sea otters were indeed more abundant around Adak than at other locations but perhaps they were simply highly sought after.

Whales and porpoises have been given little attention so far, in part, because they are relatively rare in most sites. However, the relatively high frequency of harbor porpoise remains (*c.*10-12%) recovered from each of the two Neoglacial sites on Unalaska (UNL-048 and UNL-050) is more than a little curious (Crockford *et al.* 2004; Table 6.12). Dall's porpoise are far more common today than harbor porpoise in the Bering Sea and along the Aleutians (Moore *et al.* 2002). Further south, NE Pacific harbor porpoise have declined noticeably from historic levels in many areas (e.g., Raum-Suryan and Harvey 1998), suggesting they were formerly more common everywhere. Dall's porpoise are a relatively pelagic species that primarily occupy offshore continental shelf regions, where they are gregarious and conspicuous. They will often ride the bow-wave of fast-moving vessels and leap high in the air as they travel (Reeves *et al.* 2002). In contrast, harbor porpoise are shy and elusive occupants of nearshore habitat (the exclusive domain of coastal aboriginal peoples everywhere) and are notoriously difficult to study. They generally travel in small groups and barely break the surface as they come up for air (Reeves *et al.* 2002). As biologists learn more about the mysterious harbor porpoise—and as more faunal analysts take the time and effort to identify to species the archaeological remains of porpoises—we may come to understand why harbor porpoise were more common at some places than others in the past.

Regarding whale remains, it is apparent that real progress has been made over the last few years in the identification of archaeological whale remains from Aleutian sites, due primarily to DNA analysis. We now know for sure that the currently endangered North Pacific right whale (*Eubalaena japonica*) was actively hunted by Unangan people in the eastern Aleutians during the Neoglacial and that the long-finned pilot whale (*Globicephala melas*), a species now considered extinct in the North Pacific, was once a constituent of the Aleutian ecosystem (Crockford and Frederick 2007; Frey *et al.* 2005). Humpback whales were available to Unangan inhabitants of Adak (Garong *et al.* Chapter 8 this volume) and Unalaska (Frey *et al.*, 2005). Stejneger's beaked whale, a seldom-seen species known historically as a regular strander on the beaches of Adak's Kuluk Bay, likely did the same thing hundreds of years ago (Walker and Hanson 1999). It would be fascinating to learn the species identity of the whale remains reported by West, Corbett and Lefèvre (2005), from the whalebone houses on Buldir and Attu: ancient mtDNA analysis is tailor-made for such situations.

Lastly, I must include a few words regarding polar bears and brown bears. The expanded sea ice that characterized the Neoglacial in the Bering Sea, *c.*4700-2500 BP (Crockford and Frederick 2007), made it possible for the occasional polar bear to make its way down to Unalaska Island in the spring and for brown bears to travel west from the Alaska Peninsula (Davis 2001). Whether brown bears established a self-sustaining colony on Unalaska, even if short-lived, has not yet been determined.

Conclusions and Discussion

The diversity and abundance of vertebrate species that now exist in the Aleutian Island ecosystem are a distorted reflection of the one that sustained Unangan populations before the mid 18th century. The zooarchaeological evidence presented here verifies that profound ecological differences existed prior to Russian contact in the mid-1700s. While some of these distinctions have been previously noted and are well documented archaeologically (such as the decimation of sea otters and the severe declines in vulnerable sea bird colonies due to introduced rats and foxes), perhaps the most important constituent of the prehistoric ecosystem is historically invisible. The fur seal that sustained Aleutian Island inhabitants for more than 6000 years was extinct even before its presence could be recorded, for it was not the same animal that today is famous for the enormous herds that gather in the Pribilof and Commander Islands each summer.

Contrary to what Aleutian explorers, ethnographers and traders have told us, we know from the archaeological record reported here that the northern fur seal once maintained local breeding rookeries throughout the Aleutians, from the Rat Islands in the west to Unalaska in the east. These local fur seal rookeries existed from perhaps 7000 years ago until the early contact period (*c.*1741-1800) or just before. None of these local rookeries are mentioned in early historic or ethnographic documents—indeed, the presence of fur seals anywhere in the Aleutians appears to have gone unnoticed by Europeans until the 1786 discovery of the enormous herds using the Pribilof Islands as rookeries (Jochelson 1933, 37-43), some 45 years after Aleutian sea otters had been discovered. This same pattern of widely-distributed, locally-breeding fur seals has also been documented along the west coast of North America (from perhaps SE Alaska south to California), from *c.*1700 to 4500 BP (Crockford and Frederick 2007, 2011; Crockford, Frederick and Wigen 2002; Moss 2004).

Isotope analysis (Burton *et al.* 2001; Gifford-Gonzalez *et al.* 2005) confirms that female and juvenile fur seals in at least some of these extinct locally-breeding populations were essentially non-migratory (but see Moss *et al.* 2006). They moved offshore at mid-latitudes during the winter to feed rather than undertaking extensive north-to-south migrations as Pribilof-based animals do today (Kenyon and Wilke 1953; Bigg 1990). In addition, isotope analysis of young animal remains (< four months old pups and five-twelve month olds) from several prehistoric assemblages (including the Chaluka site on Umnak and the Ozette site on the Olympic Peninsula, WA) suggests that fur seals from prehistoric colonies in the eastern Aleutians as well as the Pacific Northwest weaned their pups at a much older age than do modern seals (Newsome *et al.* 2007, 9711). If nursing was prolonged in these animals and they did not migrate, a number of pups > four months of age would likely have been available to ancient hunters into the winter months. What we see in the Aleutian zooarchaeological record, at least from sites where sample sizes are large enough and fur seal age determination has been done precisely enough (i.e. Adak Late ADK-011 and Amaknak Bridge UNL-050), is a greater-than-expected representation of five-twelve month old pups (*c.*17-19%) than if Pribilof-based animals were being harvested. Today, Pribilof-born pups older than *c.* five months are rare winter visitors in Aleutian nearshore habitats (Baker 2007).

It would appear, therefore, that prehistoric Unangan hunters targeted locally breeding, non-migratory fur seals only, leaving animals from the migrating Pribilof stock largely untouched, as did their southern neighbors. Moreover, data presented here from Adak Island extends previous evidence for these populations back another 1000-2000 years. This moves the earliest date known for a locally-breeding fur seal population to well before the onset of the Neoglacial cold period, an event previously suggested as a cause for the establishment of non-migratory rookeries (Crockford and Frederick 2007, 2010). If tenable, that explanation needs amending, although only slightly. A non-migratory/ locally-breeding strategy may have existed in the Aleutians prior to the Neoglacial as one of two possible life-history approaches available to fur seals or 'alternative ecotypes.' However, the onset of the Neoglacial would have made a Pribilof-based strategy impossible under these conditions; abundance of the non-migratory ecotype would have increased significantly, and the distribution of locally breeding populations would have expanded.

As Crockford and Frederick (2010) have pointed out, the Pribilofs were part of the Bering Land Bridge during the Last Glacial Maximum (*c.*25,000-11,500 BP) and thus inaccessible to fur seals during that time. Moreover, while we cannot be sure that fur seals used the Pribilofs between the early Holocene and the end of the Neoglacial (when there is subfossil evidence for their presence), Crockford and Frederick (2010, 79) state that '...fur seal life history strategies are sufficiently flexible to adjust to fairly major disruptions in their migratory habits and preferred breeding environment. The fact that non-migratory NFS colonies established in southern CA in 1968 are still flourishing today is proof that such life history flexibility still exists.'

Even if both fur seal life history strategies existed simultaneously during most of the Holocene, it would appear that prehistoric hunters everywhere harvested locally breeding fur seals virtually exclusively. However, if locally breeding Aleutian fur seal rookeries did not exist at the beginning of the commercial fur seal extravaganza in the early 1800s, when and why did they disappear? For Northwest Coast rookeries, the blame for exterminated local rookeries has been placed, rightly or not, on late 19th and early 20th century pelagic sealing operations (e.g., Newsome *et al.* 2007) but in the Aleutians that explanation simply does not fit the evidence.

Late prehistoric Aleutian fur seal rookeries are documented in several locations just prior to Russian contact in 1741 (at Adak ADK-011 in the west and Amaknak Spit UNL-055 in the east). It is possible that hunting pressure from large precontact Unangan populations had already reduced local fur seal rookeries to unsustainable levels before Russians arrived. For example, at the Amaknak Spit site in the eastern Aleutians, dated *c.*600-350 BP (Yesner, in Knecht and Davis 2003), fur seal remains (reportedly almost all pups < four months old) comprise fully 70% of the large marine mammal sample (Table 6.12)—from a rookery that had disappeared by the time Russian fur seal traders reached Unalaska less than 200 years later (Jochelson 1933). Adak Late shows a similar pattern for the time immediately prior to Russian contact. While the total sample is much smaller and fur seals comprise only 19% (but where 32% are pups < four months), locally-breeding fur seals were still being harvested at that time. Therefore, while it is possible that some local fur seal rookeries were already gone when the Russians arrived in the Aleutians, it seems implausible that all local rookeries had disappeared, since many were undoubtedly far enough from Unangan settlements to escape harvest.

Was it early Russian fur hunters? Probably not, as it is clear from the historic reports (e.g., Jochelson 1933, Roppel 1984) that sea otters in high numbers were what first captured the mercantile interest of Russian traders. Since sea otter pelts fetched the highest price of any fur-bearing animal, fur seals in relatively small numbers, scattered amongst numerous islands, may not have been noteworthy. Only sea otter skins are mentioned as booty of the returning Bering Expedition (Jochelson 1933, 2 and 37) and it is worth noting that the furs of primary interest to Russians at the newly discovered Pribilof Islands in 1786 were those of sea otters, not fur seals (Atkinson 1988, 112; Roppel 1984, 5). So while something in excess of four million Pribilof fur seals were harvested in the 42 years between 1786 and 1828, it appears to have been the exceptionally large concentrations of seals on the Pribilofs that made hunting them attractive (Roppel 1984, 5). Moreover, during the early years of Russian occupation (*c.*1745-1800), the dramatic declines in many local Unangan populations due to disease, forced labor, relocation and violent conflict (Black 1984; Gibson 1996, 25 and 40), meant that aboriginal subsistence hunting of non-Pribilof fur seals in most areas would have decreased rather than intensified.

Sea otters were Russian gold. The sea otter trade began within years of the Bering Expedition of 1741 (see Garong *et al.* Chapter 8 this volume for details), some 50 years or more before interest in exploitation of fur seals outside the Pribilofs began. We know from a combination of modern ecological studies and comprehensive archaeological analyses, including studies in the Aleutians (e.g., Bodkin, Robertson and Doroff 2004; Dayton 1975; Duggins 1980; Duggins, Simenstad and Estes 1989; Estes 1990; Estes *et al.* 1998; Isakson, Simenstad and Burgner 1971; Estes, Smith and Palmisano 1978; Watt, Siniff and Estes 2000) that removal of sea otters is strongly correlated with reduction of kelp forests. Sea urchins are the preferred food of sea otters but sea urchins eat kelp (however, sea otters also eat clams, crabs, octopus and fish—especially when there are large numbers of otters and sea urchins have been locally depleted (Estes 1990; Reisewitz, Estes and Simenstad 2006; Watt, Siniff and Estes 2000). Therefore, when sea urchin predators (otters) are removed by hunting or other causes, sea urchins rise dramatically in abundance and their voracious kelp consumption soon reduces the density and extent of kelp forests (e.g., Dayton 1975; Duggins 1980; Estes and Duggins 1995; Estes *et al.*, 1998; Simenstad, Estes and Kenyon 1978).

When nearshore kelp forests declined, there would have been little habitat left for the nearshore fish (especially rock greenling) that depended on the lush expanses of kelp for food and protection (Estes 1990; Isakson, Simenstad and Burgner 1971; Reisewitz, Estes and Simenstad 2006; Watt, Siniff and Estes 2000). As kelp forests shrank in extent and thickness, preferred foraging habitat for harbor seals would have diminished (Simenstad, Estes and Kenyon 1978; Small *et al.* 2008) and the dense surface of the kelp canopy, which locally-breeding fur seals may have used as a platform for giving birth and nursing, would have disappeared.

The suggestion that Aleutian fur seals may have used mature kelp forests as rookery habitat derives from ethnographic evidence collected at Neah Bay, WA by James Swan (Swan 1883). Crockford, Frederick and Wigen (2002) discuss in detail the implications of evidence presented by Swan, given by a number of credible informants, that fur seals of that region gave birth during the summer in the kelp as sea otters do today (note that sea otters are known to give birth on land as well as in kelp and that no other living member of the Mustelidae family is known to use marine kelp as essential habitat—fortunately, enough sea otters survived the historic-era slaughter that biologists have seen them use kelp with their own eyes and cannot dismiss it as a hunter's tall tale). If Aleutian fur seals indeed used the innumerable kelp forests scattered across the archipelago as rookery habitat, it would explain their low profile in the early historic period as well as the sudden demise of local rookeries, as I explain below.

The ethnographic information gathered by Swan in Neah Bay does not make clear whether 'kelp beds' refer to kelps attached by their holdfasts within kelp forests or to enormous mats of storm-detached kelp canopy known to drift about offshore waters (e.g., Hobday 2000). While pelagic fur seals are known to use offshore drifting kelp mats as rest stops (Fowler 1987), giving birth and nursing almost certainly require a more stable and enduring platform. Note that mating (a post-partum activity in fur seals usually performed on land) can be accomplished in the water if necessary, as observed and documented by Baker (1989). If fur seals used mature nearshore kelp forests for giving birth and nursing, then the decline of those forests across the Aleutians as a result of sea otter removal would have resulted in most locally-breeding fur seals losing the essential quality of their preferred rookery habitat: there may have been some kelp left but it was likely not thick enough to support a fur seal giving birth.

The decimation of sea otters across the Aleutians in the early Russian occupation period (Kenyon 1969), in combination with reduced human predation on sea urchins due to smaller Unangan populations, would have reduced the formerly extensive kelp forests within decades. Significant reductions in essential rookery habitat for non-migratory fur seals likely occurred before commercial exploitation of Pribilof-based seals had begun in earnest. That being said, it is also possible that very early in the historic period, pregnant fur seals ready to give birth in July may have moved away from kelp beds to avoid the disturbance created by the much intensified, Russian-driven sea otter hunts. If such avoidance caused them to lose their pups and the disturbance took place year after year, fur seal populations would certainly have declined due to disturbance alone, even before kelp forests declined.

Support for this suggestion comes from Lydia Black (1984, 89), who says that in the early Russian occupation period, sea otter hunts for profit involved at least 12 one-man bidarkas surrounding the prey, although larger parties were preferred. Sea otter hunts occurred only in summer. Seals and sea lions were said to have been hunted along the way for food (whether 'seal' in this context means harbor seal or fur seal, or both) is not clear from these accounts).

While the late period of commercial hunting (*c.*1871-1910) is generally considered to have been what brought sea otter populations to the brink of extinction (Johnson 1982), the early period (*c.*1741-1820) appears to have had a similar impact on fur seals. Paul Dayton *et al.* (1998) would call these fur seals 'kelp forest ghosts.' The suggestion that severe declines in sea otter numbers early in the historic period caused, indirectly, the extinction of wide spread populations of non-migratory fur seals is a plausible hypothesis that would not have been possible to formulate without detailed zooarchaeological data.

Testing this hypothesis should be possible. If well-dated early Russian occupation sites in the Aleutians contain few, but some, newborn fur seal pup remains, it would tell us that locally-breeding fur seal populations survived through the first wave of unregulated sea otter harvests and were likely not being subjected to commercial-level hunting pressure at that time. Intriguingly, Michael Etnier (2002) has reported that 28% of the fourteen bones measured for age estimation purposes from the historic (<200 BP) level sod of the Umnak site of Oglodax' were rookery-age pups and that fur seals were less common in the historic level than the late prehistoric one below. Another promising assemblage derives from recent excavations at a site on Rat Island, which lies northwest of Amchitka (RAT-00081). One unit tested had an upper component with Russian trade goods (Corbett, West and Lefèvre 2003), marine mammal remains and radiocarbon dates *c.*1790 (Funk 2008, 2010). Unfortunately, the specific purpose for which the faunal analysis was done resulting in the data reported by MNI rather than NISP and thus it cannot be used to address the 'fur seals in kelp' hypothesis. However, further analysis of this material may prove fruitful. The Reese Bay site on Unalaska (UNL-063) apparently has some early post contact era material that includes faunal remains, and although they have not yet been analyzed (Veltre and McCartney 2001). Future work may reveal useful evidence regarding the fate of the local fur seal rookery that existed during the time that the Amaknak Spit site (UNL-055) was occupied. Finally, Corbett *et al.* (2010, Table 5.1) list a dozen unexcavated sites in the Near Islands with an early Russian component that might generate faunal samples useful for addressing the question of whether the extirpation of local fur seal populations followed the decline of local sea otters or preceded them.

Another line of evidence that might be explored is analysis of ancient molecules extracted from bone. If kelp-rookery fur seal females foraged in kelp while they nursed their pups, the particular fish they fed upon may have left distinctive isotope or fatty acid signatures. Ancient DNA analysis might help as well, although if the prehistoric use of kelp forests as rookery habitat was simply a long-standing, alternative life history option for fur seals (i.e. one of two available ecotypes, see the discussion in Crockford and Frederick, 2011), then the two types might not possess the kind of distinct mtDNA haplotypes that take thousands of years of isolation to develop. As molecular techniques improve, however, so might the potential for applying them to the issue of whether kelp fur seals are distinguishable from migratory seals.

Returning to sea otters themselves, this analysis reveals that there were relatively higher frequencies of sea otter remains found at Adak and Amchitka sites than others to the east or west. Two explanations come to mind that explain this difference. One plausible reason is that otters were always more plentiful in the Rat and Andreanof Island groups because kelp habitat there was exceptionally good and extensive. The high relative abundance of rock greenling in all Adak assemblages reported here (and described by other researchers for sites on Amchitka (Isakson, Simenstad and Burgner 1971) suggests strongly that local kelp forests were at equilibrium between otter predation and sea urchins (Estes 1990; Reisewitz, Estes and Simenstad 2006; Watt, Siniff and Estes 2000).

While Simenstad, Estes and Kenyon (1978) have stated explicitly that the size of sea urchins represented in Aleutian Island middens are too large to have come from kelp forests shared with sea otters, I suggest that modern studies of the otter/urchin relationship lack one essential component that existed prehistorically: the continued onshore presence of Unangan hunters. Knowing that hunters were on the beach, sea otters almost certainly occupied positions within the kelp that put them as far away from shore as efficient benthic foraging permitted (noting that since foraging for kelp-dwelling fish is less restricted by depth, consumption of fish allows a broader occupation zone than benthic feeding alone). Hunter-avoidance by sea otters would have created a nearshore zone with kelp that could be occupied by urchins, where humans would have been the only significant urchin predator. Hence, Aleutian kelp forests at 'equilibrium' prehistorically must have been comprised of abundant deep-dwelling kelp, abundant sea otters, abundant sea urchins, abundant rock greenling and astute Unangan hunter-gathers.

It is possible that ancient Unangan hunters specifically targeted Adak and Amchitka sea otters in relatively large numbers, either because they preferred sea otter skins for their own use or because surplus pelts could be traded. An abundance of otters might naturally lead to surplus harvests that could be used as trade goods with neighbors, in exchange, perhaps, for raw obsidian or obsidian tools from the eastern Aleutians (see Nicolaysen *et al.* Chapter 11 this volume). It may eventually be possible to identify Adak sea otter bones found in other prehistoric Aleutian sites using ancient mtDNA analysis, since there appears to be at least one Adak-specific haplotype (W3, see Nishida *et al.* Chapter 7 this volume) and more might eventually be discovered. However, if skins were processed without foot bones attached, nothing would remain as evidence of their origin.

The above example raises the point that biomolecular techniques, including analysis of ancient DNA, isotopes and fatty acids (as well as others not yet developed or perfected) have great potential for addressing some of the pressing questions that remain about ancient Aleutian and Bering Sea fish, birds and mammals. Aleutian faunal material is generally well preserved and tailor-made for such work: molecular biologists willing and able to appreciate the anthropological and archaeological contexts of such samples have a bountiful array of specimens at their disposal. But that is for the future.

For now, the suggestion that Adak and Amchitka kelp forests were especially well-developed and stable prehistorically perhaps comes as no surprise, given that these locations appear to have harbored remnant populations of sea otters after most others had been exterminated (Estes 1990; Estes *et al.* 1998, 2005; Kenyon 1969). In this regard, Adak and Amchitka appear to have had a relatively unique environment seldom found elsewhere in the Aleutians and beyond. The particular mix of vertebrates recovered from sites on Adak and Amchitka (abundant sea otters, harbor seals, rock greenling, and perhaps fur seals) indicate strongly that a well-developed nearshore kelp forest existed nearby.

It also appears that Clam Lagoon, as a unique habitat immediately adjacent to protected nearshore kelp forests, contributed significantly to the overall attractive attributes of NE Adak that made it a preferred residence location for more than 6000 years. The lagoon provided a source of easily harvested cockles (West *et al.* Chapter 10 this volume) and protected habitat for a number of vertebrate species, including the saffron cod reported from the earliest occupation. Sea otters and harbor seals currently reside in Clam Lagoon or use it for refuge and/or feeding (Bodkin, Robertson and Doroff 2004). However, the relatively high frequencies of kelp-associated fish species in the archaeological record (particularly rock greenling) at all Adak sites suggests that since well-developed kelp forests must have existed nearby, sea otters and harbor seals probably resided in local kelp forest habitats, rather than in the lagoon, at least when ancient Unangan people were settled on its shores. Although today laws protect sea otters and harbor seals from human hunting, it is likely that Clam Lagoon would have been anything but a safe haven from prehistoric human hunters, as otters using the lagoon during the early years of Unangan settlement would have discovered (Savinetsky *et al.* Chapter 5 this volume). Today, the lagoon provides an effective refuge from transient killer whales (Estes *et al.* 1998), predators that might not have been as worrisome in the past, when kelp forests were more extensive and larger prey was perhaps more readily available (Springer *et al.* 2003, but see also Kuker and Barrett-Lennard 2010).

Due exclusively to the analysis of faunal remains generated by the Central Aleutians Project, we now know that a locally breeding population of Aleutian cackling goose existed on Adak before European contact. While this had been surmised before, it is now a certainty. However, it appears this species was more rare prehistorically than has been assumed by ornithologists, as remains of unfledged goose chicks (which should all be cackling goose) have not been found or reported from any other Aleutian site. Is it possible that Adak was the only island with a breeding population of cackling geese? Perhaps, but time and future faunal and genetic analyses from other islands are needed to be sure. The rarity of emperor goose prehistorically across the Aleutians (suggested by the low frequency of adult goose remains generally) is similarly surprising, considering the relative abundance of this species today in nearshore habitats.

Prehistorically, as is true today, winter-specific Aleutian resources would have been relatively impoverished compared to the bounty of summer, so if wintering emperors were available, it is reasonable for us to expect that they would have been hunted. The same could be said for harlequin and oldsquaw/long-tail ducks, which today are very common winter visitors on most islands in nearshore habitats. Both of these small ducks are quite diagnostic osteologically, although some faunal analysts do not bother to distinguish duck remains to species. However, the relative rarity of these species in only one of the two Adak samples reported here (all identified according to the same standard) suggests that the absence of wintering oldsquaws, harlequins and emperor geese in the Adak early sample is real. In this case, I have concluded that the absence of wintering birds at Adak Early is likely due to the site not being occupied in the winter because it was located on a low exposed beach without protection from gales and storm surges.

However, for other taxa, the interpretation of cultural choice and/or avoidance might better explain the differences between sites and/or site occupations compared to modern abundance records. How else but local cultural choice explains a targeted harvest of puffins plus ancient murrelets at Adak Early but nowhere else? Not so far away, we see a very high frequency of cormorants at Amchitka (but few or no puffins), a pattern again not seen elsewhere. Biologists have documented that some Aleutian bird species were severely impacted by the historic introduction of rats and/or arctic foxes and appear to have assumed that all species which are common today or that have rebounded after fox eradication, were relatively common prehistorically in the same areas (e.g., Bailey 1993; Byrd, Renner and Renner 2005). However, zooarchaeological data do not appear to back this up: prehistorically, the abundance of bird taxa is disjunct over time and space (see also Causey *et al.* 2005). Modern sea bird abundance and distribution on short time scales have been shown to be affected by local topography and proximity to major passes, where upwelling increases productivity (Hunt and Stabeno 2005; Jahncke, Coyle and Hunt 2005; Piatt *et al.* 2006). What is unclear is whether or not this relationship translates to distinct enough differences in prehistoric availability of different bird species from one island or island group to another over time: that is, whether these are different enough to be evident from faunal data. The specter of cultural choice makes it less likely that prehistoric Unangan hunters harvested birds in particular in the relative proportions that they existed in the environment (cf. Yesner 1981). I am not convinced that it is valid to assume that ancient Unangan hunters did not make choices about which animal to use for a certain purpose based on a complex mix of local tradition, personal preference, past experience and belief systems (including taboos), in addition to an animal's size, relative abundance and ease of procurement.

A number of bird species are mentioned ethnographically as being used for skin clothing, primarily cormorants, puffins and murres (Black 1984; Jochelson 1933; Turner, in Hudson 2008). Lisenkov (cited in Black 1984, 59) suggested that different types of bird skins were used to indicate rank. It thus seems eminently reasonable to me to suggest that different Unangan local groups had specific preferences for the birds they used for clothing, given that most of these species were probably available in high enough numbers everywhere that most Unangan groups could pick and choose among them. Abundance of all species, in other words, appears to have given local Unangan groups the luxury of choice. Similarly, birds have other uses aside from providing meat and skins. In particular, the bones of certain species or families were used for making bone tools. For example, short-tailed albatross bone was preferred for needles virtually everywhere and cormorant wing bones (including humerus, radius and ulna) were preferred for certain types of awls (Jochelson 1933; Knecht and Davis 2004, 88; West and Hatfield Chapter 15 this volume). The selection of particular kinds of birds for particular tool types (i.e. for their bones) and/or for particular decorative purposes, bedding or regalia rather than basic clothing (feathers/skins) would account for a modest representation of a variety of taxa regardless of their relative natural abundance in the landscape or the location and/or chronological age of the site. There are as yet no clear-cut patterns in prehistoric bird use across time and space in the Aleutians, but that might change with future research.

Lastly, I include a word about dogs, although they are not considered subsistence fauna. Dogs were virtually non-existent prehistorically in the Aleutians, which is very curious. Except for a few historic and late prehistoric specimens (dated stratigraphically to *c.*950 BP) from Akun Island that may have arrived via trade through the Alaska Peninsula (Holland 2004), dogs are archaeologically and ethnographically absent in the Aleutians. They are, however, common at Kodiak and points south (e.g., Allen

1939; Clark 1974, 43-5; Crockford 1997b), within cultures very nearly as marine-focused as ancient Unangan people were.

In summary, I restate my contention that the diversity and abundance of vertebrate species that now exist in the Aleutian Island ecosystem is a distorted reflection of the one that sustained Unangan populations before the mid 18th century. The distribution and abundance of marine and large terrestrial mammals is quite different today than it was prehistorically. Not only have sea otters become much less common, but locally-breeding fur seals have disappeared altogether. Where once ancient Unangan hunters used a mix of fur seal and sea lion products to satisfy their need for waterproof gut clothing and skin boats, since perhaps 1800 or so they would have had to use sea lion exclusively. It is possible that sea lion populations in the Aleutians, now known to be genetically distinct from those in Southeast Alaska and British Columbia, did not increase to modern, pre-recent-decline levels (National Marine Fisheries Service 2008) until locally breeding fur seals had become extinct. Harbor porpoises appear to have been quite common in the eastern Aleutians during the Neoglacial, a time when the now extinct long-finned pilot whale, as well as the currently-endangered right whale and beluga, were common enough to be taken fairly regularly by Unangan hunters (Crockford and Frederick 2011). The expanded sea ice of the Neoglacial also brought polar bears and brown bears to the eastern Aleutians, where they no longer exist. Curiously, the expanded faunal data set confirms that dogs were virtually as non-existent prehistorically in the Aleutians (over the entire 6000-7000 years of the record), as they were reported to be ethnographically, even though dogs were common prehistorically just about everywhere else on the planet (Crockford 2009)—which is not, by the way, an exaggeration.

While a wide variety of sea bird species (and a few terrestrial ones) were universally utilized by prehistoric Unangan people it is difficult to ascertain from archaeological remains any distinct pattern in either chronology or spatial distribution of specific taxa because comprehensive data for several critical areas are still unavailable. We need more bird remains analyzed, identified to species and reported by NISP so that they are comparable. In contrast, while Aleutian prehistoric fish data have similar holes, the abundance and distribution of most nearshore fish taxa so far appear to have been minimally impacted by either traditional Unangan procurement methods, Neoglacial cold or modern commercial fishing practices—with the exception of rock greenling and saffron cod. As large populations of rock greenling depend so strongly on a well-developed kelp forest, the overall abundance of this species across the Aleutians is much diminished from prehistoric times: kelp and masked greenling (*Hexagrammos octogrammus*) appear to be far more common today (Dean *et al*. 2000; Mecklenburg, Mecklenburg and Thorsteinson 2002). However, without a doubt, the archaeological record of past Unangan resource procurement described here demonstrates that the post-contact decline of lush prehistoric kelp forests—above all else—has had a lasting impact on the ecology of the Aleutians in ways we never could have imagined. One of the species most critical to the survival of ancient Unangan populations, the non-migratory Aleutian fur seal, is now but a kelp forest ghost.

Acknowledgements

This research was supported by NSF Grant # 0353065 (Office of Polar Programs), with permission from the Aleut Corporation. Special thanks are due to Dixie West for enthusiastic feedback and editorial comment, and to Vernon Byrd (Alaska Maritime National Wildlife Refuge, Homer AK) for comments on a previous version of this paper and to Doug Veltre and Debra Corbett for useful citations and reference material. Pacific Identifications colleagues Gay Frederick and Rebecca Wigen provided essential intellectual support. That being said, I bear total responsibility for any errors.

References

Allen, Glover M. 1939. Dog Skulls from Uyak Bay, Kodiak Island. *Journal of Mammalogy* 20, 336-340.

Armstrong, Robert H. 1995. *Guide to the Birds of Alaska* (4th ed.). Anchorage, Alaska Northwest Books.

Atkinson, Clinton E. 1988. Fisheries Management: An Historical Overview. *Marine Fisheries Review* 50, 112-123.

Bailey, Edgar P. 1993. Introduction of Foxes to Alaskan Islands – History, Effects on Avifauna, and Eradication. *U S Fish and Wildlife Service Resource Publication* 193.

Bailey, Edgar P. and John L. Trapp. 1984. A Second Wild Breeding Population of the Aleutian Canada Goose. *American Birds* 38, 284-286.

Baker, Jason D. 1989. Aquatic Copulation in the Northern Fur Seal, *C. ursinus. Northwest Nature* 70, 33-36.

Baker, Jason D. 2007. Post-weaning Migration of Northern Fur Seal *Callorhinus ursinus* Pups from the Pribilof Islands, Alaska. *Marine Ecology Progress Series* 341, 243- 255.

Banks, Richard C., Carla Cicero, John L. Dunn, Andrew W Kratter, Pamela C. Rasmussen, J. V. Remsen Jr., James D. Rising and Douglas F. Stotz. 2004. Forty-fifth Supplement to the American Ornithologists' Union *Checklist of North American Birds. The Auk* 121, 985-995.

Barnes, Ian, Keith M. Dobney and J. Peter W. Wong. 1998. The Molecular Palaeoecology of Geese: Identification of Archaeological Goose Remains Using Ancient DNA analysis. *International Journal of Osteoarchaeology* 8, 280-287.

Bigg, Michael A. 1990. Migration of Northern Fur Seals. (*Callorhinus ursinus*) off Western North America. *Canadian Technical Report of Fisheries and Aquatic Sciences* 1764. Nanaimo, Fisheries and Oceans Canada.

Black, Lydia. 1984. *Atka: An Ethnohistory of the Western Aleutians*. Kingston, Ontario, Limestone Press.

Bodkin, James L., Morgan A. Robertson and Angela Doroff. 2004. Sea Otter Foraging Behavior and Population Status in the Aleutian Islands and the Alaska Peninsula. Unpublished report on file, Anchorage, U S Fish and Wildlife.

Bovy, Kristine M. 2007. Prehistoric Human Impacts on Waterbirds at Watmough Bay, Washington, USA. *Journal of Island and Coast Archaeology* 2, 210-230.

Burger, Joanna, Michael Gochfeld, Christian Jeitner, Sean Burke and Timothy Stamm. 2007. Metal Levels in Flathead Sole (*Hippoglossoides*) and Great Sculpin (*Myoxocephalus polyacanthocephalus*) from Adak Island, Alaska: Potential Risk to Predators and Fishermen. *Environmental Research* 103, 62-69.

Burton, Robert K., J. Josh Snodgrass, Diane Gifford-Gonzalez, Tom Guilderson, Tom Brown and Paul L. Koch. 2001. Holocene Changes in the Ecology of Northern Fur Seals: Insights from Stable Isotopes and Archaeofauna. *Oecologia* 128, 107-115.

Byrd, G. Vernon and Robert H. Day. 1986. The Avifauna of Buldir Island, Aleutian Islands, Alaska. *Arctic* 39, 109-118.

Byrd, G. Vernon, Daniel D. Gibson and David L. Johnson. 1974. The Birds of Adak Island, Alaska. *The Condor* 76, 288-300.

Byrd, G. Vernon, Heather M. Renner and Martin Renner. 2005. Distribution Patterns and Population Trends of Breeding Seabirds in the Aleutian Islands. *Fisheries Oceanography* 14 (Suppl. 1), 139-159.

Calvert, S. Gay. 1980. A Cultural Analysis of Faunal Remains from Three Archaeological Sites in Hesquiat Harbour, BC. Unpublished Ph.D. thesis, University of British Columbia.

Causey, Douglas, Debra G. Corbett, Christine Lefèvre, Dixie L. West, Arkady Savinetsky, Nina K. Kiseleva, and Bulat F. Khassanov. 2005. The Palaeoenvironment of Humans and Marine Birds of the Aleutian Islands: Three Millennia of Change. *Fisheries Oceanography* 14 (Suppl. 1), 259-276.

Clark, Donald W. 1974. *Koniag Prehistory: Archaeological Investigations at Late Prehistoric Sites on Kodiak Island, Alaska*. Stuttgart, Verlag W. Kohlhammer.

Corbett, Debra G., Christine Lefèvre, Dixie West and Douglas Siegel-Causey. 1997. Excavations at KIS-008, Buldir Island: Evaluation and Potential. *Arctic Anthropology* 34, 110–117.

Corbett, Debra G., Christine Lefèvre and Douglas Siegel-Causey. 1997. The Western Aleutians: Cultural Isolation and Environmental Change. *Human Ecology* 25, 459-479.

Corbett, Debra, Douglas Causey, Mark Clementz, Paul Koch, Angela Doroff, Christine Lefèvre and Dixie West. 2008. Aleut Hunters, Sea Otters, and Sea Cows: Three Thousand Years of Intereactions in the Western Aleutian Islands, Alaska, in R. Torben and J. Erlandson (eds.), *Human Impacts on Ancient Marine Ecosystems: A Global Perspective*, 43-76. Berkeley, University of California Press.

Corbett, D., West, D. and Lefèvre, C. 2003. 2003 Progress Report: The Western Aleutians Archaeological and Paleobiological Project: Rat Island. Unpublished report on file, Anchorage, US Fish and Wildlife Service.

Corbett, Debra, Dixie, West and Christine Lefèvre (eds.) 2010. *The People at the End of the World: The Western Aleutians Project and the Archaeology of Shemya Island.* Anchorage, Aurora Monograph Series.

Crockford, Susan J. 1997a. *Osteometry of Makah and Coast Salish Dogs*. Publication No. 22, Burnaby, Simon Fraser University Archaeology Press.

Crockford, Susan J. 1997b. Fish Bone Analysis from Shemya Island Site ATU-061. Unpublished report on file, Anchorage, US Fish and Wildlife Service.

Crockford, Susan J. 1999. Fish Bone Analysis from Shemya Island Site ATU-003. Unpublished report on file, US Fish and Wildlife Service, Anchorage, AK.

Crockford, Susan J. 2009. *A Practical Guide to In Situ Dog Remains for the Field Archaeologist*. Victoria, Pacific Identifications Inc.

Crockford, Susan J. and Gay Frederick. 2011. Neoglacial Distribution of North Pacific Pinnipeds Addresses Fundamental Questions About Ringed Seal and Fur Seal Life History. In T. Braje and T. Rick (eds.), *Human and Marine Ecosystems: Archaeology and Historical Ecology of Northeastern Pacific Seals, Sea Lions, and Sea Otters*, 65-91. Los Angeles, University of California Press.

Crockford, Susan J. and Gay Frederick. 2007. Sea Ice Expansion in the Bering Sea During the Neoglacial: Evidence from Archaeozoology. *The Holocene* 17, 699-706.

Crockford, Susan J., Gay Frederick and Rebecca Wigen. 1997. A Humerus Story: Albatross Element Distribution from Two Northwest Coast sites, North America. *International Journal of Osteoarchaeology* 7, 378.1-5.

Crockford, Susan J., Gay Frederick and Rebecca Wigen. 2002. The Cape Flattery Fur Seal: An Extinct Species of *Callorhinus* in the Eastern North Pacific? *Canadian Journal of Archaeology* 26, 152-174.

Crockford, Susan, Gay Frederick, Rebecca Wigen and Iain McKechnie. 2004. Final Report on the Analysis of the Vertebrate Fauna from Amaknak Bridge, Unalaska, AK, UNL050, Supplement to R. Knecht and R. Davis, *The Amaknak Bridge Site Data Recovery Project Final Report*, Project # MGS-STP-BR-0310(S)/52930, Unpublished report on file, Anchorage, Alaska Department of Transportation.

Dayton, Paul K. 1975. Experimental Studies of Algal Canopy Interactions in a Sea Otter-dominated Kelp Community at Amchitka Island, Alaska. *Fishery Bulletin* 73, 230-237.

Dayton, Paul K., Mia J. Tegner, Peter B. Edwards and Kristin L. Riser. 1998. Sliding Baselines, Ghosts, and Reduced Expectations in Kelp Forest Communities. *Ecological Applications* 8, 309-322.

Dean, Thomas A., Lewis Haldorson, David R. Laur, Stephen C. Jewett and Arny Blanchard. 2000. The Distribution of Nearshore Fishes in Kelp and Eelgrass Communities in Prince William Sound, Alaska: Associations with

Vegetation and Physical Habitat Characteristics. *Environmental Biology of Fishes* 57, 271-287.

Duggins, David, O. 1980. Kelp Beds and Sea Otters: An Experimental Approach. *Ecology* 61, 447-453.

Duggins, David O., Charles A. Simenstad, and James A. Estes. 1989. Magnificant of Secondary Production by Kelp Detritus in Coastal Marine Ecosystems. *Science* 245, 170-173.

Eda, Masaki, Yoshiuki Baba, Hiroko Koike and Hiroyoshi Higuchi. 2006. Do Temporal Size Differences Influence Species Identification of Archaeological Albatross Remains when Using Modern Reference Samples? *Journal of Archaeological Science* 33, 349-359.

Estes, James A., 1990. Growth and Equilibrium in Sea Otter Populations. *Journal of Animal Ecology* 59, 385-401.

Estes, James A. and David O. Duggins. 1995. Sea Otters and Kelp Forests in Alaska: Generality and Variation in a Community Ecological Paradigm. *Ecological Monographs* 65, 75-100.

Estes, James A., Norman S. Smith and John F. Palmisano. 1978. Sea Otter Predation and Community Organization in the Western Aleutian Islands, Alaska. *Ecology* 59, 822-833.

Estes, James A., M. Tim Tinker, Terrie M. Williams and Daniel F. Doak. 1998. Killer Whale Predation on Sea Otters Linking Oceanic and Nearshore Ecosystems. *Science* 282, 473-477.

Etnier, Michael A. 2002. The Effect of Human Hunting on Northern Fur Seal (*Callorhinus ursinus*) Migration and Breeding Distribution in the Late Holocene. Unpublished PhD. thesis, Seattle, University of Washington.

Fowler, Charles W. 1987. Marine Debris and Northern Fur Seals: A Case Study. *Marine Pollution Bulletin* 18, 326-335.

Freethy, Ron. 1987. *Auks: An Ornithologists Guide*. Poole, Blandford Press.

Frey, Amy, Susan J. Crockford, Matthias Meyer and Gregory O'Corry-Crowe. 2005. Genetic Analysis of Prehistoric Marine Mammal Bones from an Ancient Aleut Village in the Southeastern Bering Sea. Conference poster, 16th Biennial Conference on the Biology of Marine Mammals, San Diego.

Funk, Caroline L. 2008. Birds and Mammals from the 2003 Test of Features 2 and 6 at Rat-00081 on Rat Island, Alaska: Part 2 - Species Level Analysis. Unpublished report on file, Anchorage, US Fish and Wildlife Service.

Funk, Caroline L. 2010. Rats and Birds: Tracking Ecological Change with Evidence from Prehistoric to Historic Aleut Village Midden Test Excavations, Rat Island, Alaska. Unpublished report on file, Anchorage, US Fish and Wildlife Service.

Garong, Ame, Chiozuru Takashima, Shin Nishida, Akihiro Kano and Hiroko Koike. 2011. Carbon, Nitrogen and Oxygen Isotopic Compositions of Bone Collagen and Tooth Enamel Carbonate of the Sea Otter *Enhydra lutris* from Archaeological Sites at Adak Island, Aleutian Archipelago. Chapter 8 in D. West, V. Hatfield, E. Wilmerding, C. Lefèvre and L. Gualtieri (eds.), *The People Before: The Geology, Paleoecology and Archaeology of Adak Island, Alaska*. Oxford, British Archaeological Reports.

Gibson, James R. 1996. Russian Dependence on the Natives of Alaska. In S. W. Haycox and M. C. Mangusso (eds.), *An Alaskan Anthology: Interpreting the Past*, 21-42. Seattle, University of Washington Press.

Gifford-Gonzalez, Diane, Seth D. Newsome, Paul L. Koch, Tom P. Guilderson, J. Josh Snodgrass, and Robert K. Burton. 2005. Archaeofaunal Insights on Pinniped-human Interactions in the Northeastern Pacific. In G. Monks (ed.), *The Exploitation and Cultural Importance of Sea Mammals*, 19-38. Oxford, Oxbow.

Haynes, Terry L., and Craig Mishler. 1991, The Subsistence Harvest and Use of Steller Sea Lions in Alaska. Technical Paper 198. Juneau, Alaska Department of Fish and Game.

Hobday, Alistair J. 2000. Abundance and Dispersal of Drifting Kelp *Macrocystis pyrifera* Raft in the Southern California Bight. *Marine Ecology Progress Series* 195, 101-116.

Holder, Karen, Robert Montgomerie and Vicki L. Friesen. 2000. Glacial Vicariance and Historical Biogeography of Rock Ptarmigan (*Lagopus mutus*) in the Bering Region. *Molecular Ecology* 9, 1265-1278.

Holland, Katheryn M. 2004. A Brief Note on the Significance of Prehistoric Dogs from the Eastern Aleutian Islands. *Arctic Anthropology* 41, 50-54.

Holmes, Patrick B. 1997. Aleutian Islands Salmon: 1982 Stock Assessment Survey and Current Status. Regional Information Report 4K97-6. Kodiak, Alaska Department of Fish and Game.

Hudson, Raymond L. (ed.) 2008. *An Aleutian Ethnography by Lucien M. Turner*. Fairbanks, University of Alaska Press.

Huelsbeck, David R. 1983. Mammals and Fish in the Subsistence Economy of Ozette. Unpublished Ph.D. thesis. Pullman, Washington State University.

Hunt Jr., George L. and Phyllis J. Stabeno. 2005. Oceanography and Ecology of the Aleutian Archipelago: Spatial and Temporal Variation. *Fisheries Oceanography* 14 (Suppl.1), 292-306.

Hupp, Jerry W., Joel A. Schmutz and Craig R. Ely. 2008. The Annual Migration Cycle of Emperor Geese in Western Alaska. *Arctic* 61, 23-34.

Isakson, John S., Simenstad, Charles, A. and Robert L. Burgner. 1971. Fish Communities and Food Chains in the Amchitka area. *BioScience* 21, 666-670.

Jahncke, Jaime, Kenneth O. Coyle and George L. Hunt Jr. 2005. Seabird Distribution, Abundance and Diets in the Eastern and Central Aleutian Islands. *Fisheries Oceanography* 14 (Suppl.1), 160-177.

Jochelson, Waldemar. 1933. *History, Ethnology and Anthropology of the Aleut*. 2002. [Reprint] Foreword by K. L. Reedy-Maschner and H. D. G. Maschner. Salt Lake City, University of Utah Press.

Kenyon, Karl W. 1969. The Sea Otter in the Eastern Pacific Ocean. *North American Fauna* 68, 1-352.

Kenyon, Karl W. and Dale W. Rice. 1961. Abundance and

Distribution of Steller Sea Lion. *Journal of Mammalogy* 42, 223-234.

Kenyon, Karl W. and Ford Wilke. 1953. Migration of the Northern Fur Seal, *Callorhinus ursinus*. *Journal of Mammalogy* 34, 86-98.

Kotwicki, Stan, Alex De Robertis, Paul von Szalay and Rick Towler. 2009. The Effect of Light Intensity on the Availability of Walleye Pollock (*Theragra chalcogramma*) to Bottom Trawl and Acoustic Surveys. *Canadian Journal of Fisheries and Aquatic Sciences* 66, 983-994.

Knecht, Richard A. and Richard S. Davis. 2003. Archaeological Evaluation of Tanaxtaxak, the Amaknak Spit Site (UNL-055) Final Report. Unpublished report on file, Unalaska, Museum of the Aleutians.

Knecht, Richard A. and Richard S. Davis. 2004. The Amaknak Bridge Site Data Recovery Project Final Report, Project # MGS-STP-BR-0310(S)/52930. Unpublished report on file. Anchorage, Alaska Department of Transportation.

Knecht, Richard A. and Richard S. Davis. 2008. The Amaknak Bridge Site: Cultural Change and the Neoglacial in the Eastern Aleutians. *Arctic Anthropology* 45, 61-78.

Kuker, Katie and Lance Barrett-Lennard. 2010. A Re-evaluation of the Role of Killer Whales *Orcinus orca* in a Population Decline of Sea Otters *Enhydra lutris* in the Aleutian Islands and a Review of Alternative Hypotheses. *Mammal Review* 40, 103-124.

Laughlin, William S. and William G. Reeder. 1962. Revision of Aleutian Prehistory. *Science* 137, 856-857.

Lauth, Robert R., Scott W. McEntire and Harold H. Zenger Jr. 2007. Geographic Distribution, Depth Range, and Description of Atka Mackerel *Pleurogrammus monpterygius* Nesting Habitat in Alaska. *Alaska Fishery Research Bulletin* 12, 165-186.

Lefèvre, Christine, Debra G. Corbett, Dixie West and Douglas Siegel-Causey. 1997. A Zooarchaeological Study at Buldir Island, Western Aleutians, Alaska. *Arctic Anthropology* 34, 118-131.

Lippold, Lois K. 1966. Chaluka: The Economic Base. *Arctic Anthropology* 3, 125-131.

Logerwell, Elizabeth A., Kerim Aydin, Steven Barbeaux, Eric Brown, M. Elizabeth Conners, Sandra Lowe, Jay W. Orr, Ivonne Ortiz, Rebecca Reuter and Paul Spencer. 2005. Geographic Patterns in the Demersal Ichthyofauna of the Aleutian Islands. *Fisheries Oceanography* 14 (Suppl.1), 93-112.

Lyman, R. Lee. 1989. Seal and Sea Lion Hunting: A Zooarchaeological Study from the Southern Northwest Coast of North America. *Journal of Anthropological Archaeology* 8, 68-99.

McKechnie, Iain and Rebecca J. Wigen. 2011. Towards a Historical Ecology of Pinniped and Sea Otter Hunting Traditions on the Coast of Southern British Columbia. In T. Braje and T. Rick (eds.), *Human and Marine Ecosystems: Archaeology and Historical Ecology of Northeastern Pacific Seals, Sea Lions, and Sea Otters*, 129-166. Los Angeles, University of California Press.

Mecklenburg, Catherine W., T. Anthony Mecklenburg and Lyman K. Thorsteinson. 2002. *Fishes of Alaska*. Bethesda, American Fisheries Society.

Moore, Sue E., Janice M. Waite, Nancy A. Friday and Taina Honkalehto. 2002. Cetacean Distribution and Relative Abundance on the Central-eastern and the Southeastern Bering Sea Shelf with Reference to Oceanographic Domains. *Progress In Oceanography* 55, 249-261.

Moss, Madonna L. 2004. Archaeological Investigation of Cape Addington Rockshelter: Human Occupation of the Rugged Seacoast on the Outer Prince of Wales Archipelago, Alaska. *University of Oregon Anthropological Paper* 63. Eugene, University of Oregon.

Moss, Madonna L., Dongya Y. Yang, Seth D. Newsome, Camilla F. Speller, Iain McKechnie, Alan D. McMillan, Robert J. Losey and Paul L. Koch. 2006. Historical Ecology and Biogeography of North Pacific Pinnipeds: Isotopes and Ancient DNA from Three Archaeological Assemblages. *Journal of Island and Coastal Archaeology* 1, 165-190.

National Marine Fisheries Service. 2008. Recovery Plan for the Steller Sea Lion (*Eumetopias jubatus*), Revision. Silver Springs, MD, National Marine Fisheries Service.

Newsome, Seth D., Michael A. Etnier, Diane Gifford-Gonzalez, Donald L. Phillips, Marcel van Tuinen, Elizabeth A. Hadly, Daniel P.Costa, Douglas J. Kennett, Tom P. Guilderson and Paul L. Koch. 2007. The Shifting Baseline of Northern Fur Seal Ecology in the Northeast Pacific Ocean. *Proceedings of the National Academy of Sciences USA* 104, 9709-9714.

Nichol, Daniel G., Taina Honkalehto and Grant G. Thompson. 2006. Proximity of Pacific Cod to the Sea Floor: Using Archival Tags to Estimate Fish Availability to Research Bottom Trawls. *Fisheries Research* 86, 129-135.

Nicolaysen, Kirsten, Taylor Johnson, Elizabeth Wilmerding, Virginia Hatfield, Dixie West, and Robert G. McGimsey. 2011. Provenance of Obsidian Fragments Recovered from Adak Island, Central Aleutian Islands: Evidence for Long-distance Transport of Raw Lithic Material. Chapter 11 in D. West, V. Hatfield, E. Wilmerding, C. Lefèvre and L. Gualtieri (eds.), *The People Before: The Geology, Paleoecology and Archaeology of Adak Island, Alaska*. Oxford, British Archaeological Reports.

Nishida, Shin, Dixie West, Susan Crockford and Hiroko Koike. 2011. Ancient DNA analysis of the Sea Otter (*Enhydra lutris*) from Archaeological Sites on Adak, Aleutian Islands, Chapter 7 in D. West, V. Hatfield, E. Wilmerding, C. Lefèvre and L. Gualtieri (eds.), *The People Before: The Geology, Paleoecology and Archaeology of Adak Island, Alaska*. Oxford, British Archaeological Reports.

North Pacific Fisheries Management Council. 2007. Aleutian Islands Fishery Ecosystem Plan, May 17 Draft. Anchorage, North Pacific Fisheries Management Council.

Orchard, T. J. 2003. *An Application of the Linear Regression Technique for Determining Length and Weight of Six*

Fish Taxa: The Role of Selected Fish Species in Aleut Paleodiet. Oxford, British Archaeological Reports, International Series 1172.

Orchard, T.J. 2007 Otters and Urchins: Continuity and Change in Haida Economy During the Late Holocene and Maritime Fur Trader Periods. Unpublished PhD thesis, University of Toronto.

Orchard, Trevor J. and Susan J. Crockford. 2010. Size Estimation of Pacific Cod (*Gadus macrocephalus*) remains, in D. Corbett, D. West, and C. Lefèvre (eds.), *The People at the End of the World: The Western Aleutians Project and the Archaeology of Shemya Island*, 93-98. Anchorage, Aurora Monograph Series.

Ormseth, Olav, A. and Todd T. Tenbrink. 2009. Bering Sea/Aleutian Islands Sculpins. In *Stock Assessment and Fishery Evaluation Report for the Groundfish Resources of the Bering Sea and Aleutian Islands*, pp. 1267-1304. Anchorage, North Pacific Fisheries Management Council.

Piatt, John F., Jennifer Wetzel, Kevin Bell, Anthony R. DeGange, Gregory R. Balogh, Gary Drew, Tracee Geernaert, Carol Ladd and G. Vernon Byrd. 2006. Predictable Hotspots and Foraging Habitat of the Endangered Short-tailed Albatross (*Phoebastria albatrus*) in the North Pacific: Implications for Conservation. *Deep Sea Research Part II: Topical Studies in Oceanography* 53, 387-398.

Raum-Suryan, Kimberly L. and James T. Harvey. 1998. Distribution and Abundance of and Habitat Use by Harbor Porpoise, *Phooena phocoena*, off the Northern San Juan Islands, Washingon. *Fishery Bulletin* 96, 808-822.

Reeves, Randall R., Brent S. Stewart, Phillip J. Clapham and James A. Powell. 2002. *National Audobon Society's Guide to Marine Mammals of the World*. New York, Alfred A. Knopf.

Reisewitz, Shauna, James A. Estes and Charles A. Simenstad. 2006. Indirect Food Web Interactions: Sea Otters and Kelp Forest Fishes in the Aleutian Archipelago. *Oecologia* 146, 623-631.

Roppel, Alton Y. 1984. Management of Northern Fur Seals on the Pribilof Islands, Alaska, 1786-1981. Seattle, National Oceanic and Atmospheric Administration Technical Report National Marine Fisheries Service 4.

Savinetsky, Arkady B., Dixie L. West, Zhanna A. Antipushina, Bulat F. Khassanov, Nina K. Kiseleva, Olga A. Krylovich and Andrei M. Pereladov. 2011. The Reconstruction of Ecosystem History of Adak Island (Aleutian Islands) During the Holocene, Chapter 5 in D. West, V. Hatfield, E. Wilmerding, C. Lefèvre and L. Gualtieri (eds.), *The People Before: The Geology, Paleoecology and Archaeology of Adak Island, Alaska*. Oxford, British Archaeological Reports.

Savinetsky, Arkady B., Nina K. Kiseleva and Bulat F. Khassanov. 2004. Dynamics of Sea Mammal and Bird Populations of the Bering Sea Region Over the Last Several Millennia. *Palaeogeography, Palaeoclimatology, Palaeoecology* 209, 335-352.

Scribner, Kim T., Sandra L. Talbot, John M. Pearce, Barbara J. Pierson, Karen S. Bollinger and Dirk V. Derksen. 2003. Phylogeography of Canada Geese (*Branta canadensis*) in Western North America. *The Auk* 120, 889-907.

Sease, John L. and Carolyne J. Gudmundson. 2002. Aerial and Land-based Surveys of Steller Sea Lions (*Eumetopias jubatus*) from the Western Stock in Alaska, June and July 2001 and 2002. Seattle, National Oceanic and Atmospheric Administration Technical Memorandum NMFS-AFSC-131.

Simenstad, Charles A., James A. Estes and Karl W. Kenyon. 1978. Aleuts, Sea Otters, and Alternate Stable-state Communities. *Science* 200, 403-411.

Small, Robert J., Peter L. Boveng, G. Vernon Byrd and David E. Withrow. 2008. Harbor Seal Population Decline in the Aleutian Archipelago. *Marine Mammal Science* 24, 845-863.

Springer, Alan M., James A. Estes, Gus B. van Vleit, Terrie M. Williams, Daniel F. Doak, Eric M. Danner, Karen A. Forney and B. Pfister. 2003. Sequential Megafaunal Collapse in the North Pacific Ocean: An Ongoing Legacy of Industrial Whaling? *Proceedings of the National Academy of Science USA* 100, 12223-12228.

Taber, Richard D. 1946. The Winter Birds of Adak, Alaska. *The Condor* 48, 272-277.

Turner, Lucien M. 1885. Notes on the Birds of the Nearer Islands, Alaska. *Auk* 2, 154-159.

US Fish and Wildlife Service. 1999. Threatened and Endangered Species: Aleutian Canada Goose (*Branta canadensis leucopareia*). US Fish and Wildlife Fact Sheet, On file at http://alaska. fws.gov/media/acg.htm

Veltre, Douglas W. 2011. One Hundred Forty years of Archaeology in the Central Aleutian Islands, Alaska, Chapter 2 in D. West, V. Hatfield, E. Wilmerding, C. Lefèvre, L. Gualtieri (eds.), *The People Before: The Geology, Paleoecology and Archaeology of Adak Island, Alaska*. Oxford, British Archaeological Reports.

Veltre, Douglas W. and Allen P. McCartney. 2001. Ethnohistorical Archaeology at the Reese Bay Site, Unalaska Island, Alaska. In D. E. Dumond (ed.), *Anthropological Papers of the University of Oregon* 58, 87-104. Eugene, University of Oregon.

Walker, William A. and M. Bradley Hanson. 1999. Biological Observations on Stejneger's Beaked Whale, *Mesoplodon stejnegeri*, from Strandings on Adak Island, Alaska. *Marine Mammal Science* 15, 1314-1329.

Watt, Jon, Donald B. Siniff and James A. Estes. 2000. Inter-decadal Patterns of Population and Dietary Change in Sea Otters at Amchitka Island, Alaska. *Oecologia* 124, 289-298.

West, Dixie L., Debra Corbett and Christine Lefèvre. 1999. 1999 NSF Grant Report: The Western Aleutians Archaeological and Paleobiological Project. Unpublished report on file, Washington, DC, National Science Foundation.

West, Dixie L., Debra Corbett and Christine Lefèvre. 2005. 2005 NSF Grant Report: The Western Aleutians Archaeological and Paleobiological Project. Unpublished report on file, Washington, DC, National Science Foundation.

West, Dixie, Christine Lefèvre, Debra Corbett and Susan Crockford. 2003. A Burial Cave in the Western Aleutian Islands, Alaska. *Arctic Anthropology* 40, 70-86.

West, Dixie and Virginia Hatfield. 2011. Bone Artifacts and Debris from Three Archaeological Sites on Adak Island, Alaska, Chapter 15 in D. West, V. Hatfield, E. Wilmerding, C. Lefèvre and L. Gualtieri (eds.), *The People Before: The Geology, Paleoecology and Archaeology of Adak Island, Alaska*. Oxford, British Archaeological Reports.

West, Dixie, Zhanna Antipushina, Arkady Savinetsky, and Olga Krylovich. 2011. Invertebrate Remains: A Pan Aleutian Comparison, Chapter 10 in D. West, V. Hatfield, E. Wilmerding, C. Lefèvre and L. Gualtieri (eds.), *The People Before: The Geology, Paleoecology and Archaeology of Adak Island, Alaska*. Oxford, British Archaeological Reports.

Wilson, Bryan, Crockford, Susan J., Johnson, Jesse W., Malhi, Ripan S., and Brian M. Kemp. 2011. Genetic and Archaeological Evidence of a Breeding Population of Formerly Endangered Aleutian Cackling Goose, *Branta hutchinsii leucopareia*, on Adak Island in the Central Aleutians, Alaska. *Canadian Journal of Zoology* 89, 732-743.

Wilson, Christopher D., Anne B. Hollowed, Michiyo Shima, Paul Walline and Sarah Stienessen. 2003. Interactions Between Commercial Fishing and Walleye Pollock. *Alaska Fishery Research Bulletin* 10, 61-77.

Wilson, Don E., and Sue Ruff (eds.) 1999. *The Smithsonian Book of North American Mammals*. Vancouver, University of British Columbia Press.

Williams, Jeff. 2006. *Birds of Adak Island*. US Fish and Wildlife Service Bulletin. Homer, Alaska Maritime National Wildlife Refuge.

Yesner, David R. 1976. Aleutian Island Albatrosses: A Population History. *Auk* 93, 91-112.

Yesner, David R. 1981. Archaeological Applications of Optimal Foraging Theory, in B. Winterhalder and E. Smith (eds.), *Harvest Strategies of Aleut Hunter-Gatherer Foraging Strategies*, 148-170. Chicago, University of Chicago Press.

Yesner, David R. 1988. Effects of Prehistoric Human Exploitation on Aleutian Sea Mammal Populations. *Arctic Anthropology* 25, 28-43.

Chapter 7
Ancient DNA Analysis for the Sea Otter (*Enhydra lutris*) from Archaeological Sites on Adak, Aleutian Islands

Shin Nishida, Dixie West, Susan Crockford, and Hiroko Koike

Abstract: *We analyzed 36 bone samples of the sea otter (Enhydra lutris) excavated from archaeological sites on Adak Island, Alaska, in the central Aleutians. The otter bones represent four different time periods: the recent (170-285 years BP), the middle (285-415 years BP), the old (c. 2200 years BP), and the oldest (6000 years BP). We discovered five mitochondrial DNA (mtDNA) haplotypes: B, B2, B3, B4 and W3 in the Adak samples. Four of these—B2, B3, B4, and W3-- are new sea otter haplotypes for this region. Haplotypes B and B3 are dominant in Adak otters and are observed in all periods except the oldest. Conversely, haplotypes B2, B4, and W3 were present in all periods prior to the most recent period. Additionally, the haplotype diversity of the old period (h = 0.75) was higher than either the middle (h=0.29) or the recent (h=0.39) periods. Genetic diversity of sea otter mtDNA may have also decreased between the middle and the oldest periods. Our evidence suggests that genetic 'bottleneck' events may have occurred multiple times in the past, either due to human hunting and/or ecological and environmental changes in sea otter habitat. All Aleutian Islands sea otter populations are assigned to the subspecies E. l. kenyoni. Although the contemporary sea otter population of Amchitka in the western Aleutian Islands possesses haplotypes A (dominant) and B (Larson et al. 2002a), the ancient Adak sea otter population had no haplotype A. This suggests that gene flow has been restricted between these neighboring islands in the Aleutian archipelago.*

Keywords: *sea otters, ancient DNA, Aleutian Islands*

Introduction

The sea otter (*Enhydra lutris*), *one of the smallest of all marine mammals,* belongs to the family Mustelidae, which also includes minks, weasels, badgers, wolverines, and other otter species (Berta *et al.* 2006). Three sea otter subspecies have been recognized based on: 1) their original distribution, 2) morphology of their skulls (Wilson *et al.* 1991) and, 3) genetic differences (e.g., Cronin *et al.* 1996). *E. l. lutris* lives from the Kurile Islands north to the Commander Islands in the western North Pacific; *E. l. kenyoni* occurs throughout the Aleutian Islands and southward in the eastern Pacific to Washington; and *E. l. nereis* is found from central California to northern Mexico (Cronin *et al.* 1996).

Historically, sea otters numbered between 150,000 and 300,000 (Doroff and Burdin 2008) with colonies located throughout the circum Pacific area from southern California up the west coast of North America; the Aleutian, Kurile, and Commander Islands; and the Kamchatka coast (Nowak 1999). Soon after Vitus Bering explored the Aleutian archipelago in 1741 for Czar Peter the Great, eager Russian entrepreneurs sailed into the Near Islands to exploit the virgin territory for what appeared to be an almost unending supply of sea otter pelts. By 1830, Russian traders regularly visited Adak, conscripting local inhabitants to hunt sea otters. On March 30, 1867, the United States purchased Alaska from Russia. During the American Period commercial harvesting intensified, nearly exterminating Aleutian otter populations by the beginning of the 20th century. By 1910 the sea otter population numbered only 1000- 2000 individuals in 13 colonies in central California, western Alaska, and the Commander and Kurile Islands. To save these remnant populations, the first international treaty to protect fur-bearers was signed in 1911. Adak and most other islands in the archipelago were declared a national wildlife refuge to ensure a successful sea otter recovery. Under protective covenants, sea otter populations slowly grew and sufficiently recovered reaching a stable population of 107,000 distributed on the Russian east coast, Alaska, British Columbia, Washington, and California, and expanded to Mexico and Japan. However, disturbing population statistics emerged during the last quarter of the 20th century. Despite general sea otter recovery, local populations of Alaskan sea otters dramatically dropped. In 1973, the Alaskan sea otter population was estimated between 100,000 and 125,000. By 2006, the population declined to 73,000. Estes *et al.* (1998) recorded that sea otter populations on north Adak declined approximately 25% per year throughout the 1990's. This massive depopulation in the Aleutian Islands accounts for most of the change in Alaskan sea otter populations. The cause of recent sea otter reduction is unknown, although killer whale predation is suspected (Estes *et al.* 1998; Schrope 2007). The sea otter is listed in the International Union for Conservation of Nature's Red List of Threatened Species as 'Endangered' (Doroff and Burdin 2008).

The sea otter inhabits a relatively narrow ribbon of coast within a kilometer of the shore. Its territory is generally bounded by the shoreline and by water over 100m deep (Kenyon 1969). Although adult male sea otters establish territories and patrol territorial boundaries to exclude other adult males from the area, females freely move between and among male territories. Sea otter annual home ranges are relatively small, up to 0.8km^2 (80ha) and can extend along 16km of coastline (Doroff and Burdin 2008; Kenyon 1969; Loughlin 1980). For unknown reasons, both sexes occasionally move long distances, more than several hundred kilometers (Estes *et al.* 2009). Females generally occupy larger home ranges than males, but have lifetime home ranges considerably smaller than adult males (Ralls *et al.* 1988).

Currently, nearly all sea otter populations, including remnant and translocated groups, display relatively low levels of genetic variation. This low genetic diversity has been documented using a variety of analyses including: 1) allozymes (Lidicker and McCollum 1997); 2) restriction fragment length polymorphism (RFLP) of mitochondrial DNA (mtDNA) (Bodkin *et al.* 1999; Cronin *et al.* 1996); 3) mtDNA sequence variation (Larson *et al.* 2002a); 4) microsatellites (Aguilar *et al.* 2008; Larson *et al.* 2002a); and 5) MHC sequence (Aguilar *et al.* 2008). Only six haplotypes of the control region on mtDNA have been reported in contemporary and ancient sea otter populations: A to D (Larson *et al.* 2002a, 2002b), and X and W (Valentine *et al.* 2007). Larson *et al.* (2002b) compared a pre-fur harvest population (ca. 100-450 years BP) from Washington to a modern population (see Table 7.1). Their results suggested significant losses of microsatellite alleles and heterozigosity, but a bottleneck effect on mtDNA sequences was not detected. Valentine *et al.* (2007) found ancient haplotypes of mtDNA sequences representing X and W from archaeological samples in Oregon (ca 175–2000 years BP). These haplotypes are absent in modern otter populations.

The Aleuts are aboriginal inhabitants of the Aleutian archipelago who shared common ancestry with Chukchi and Siberian Eskimos of Chukotka (West *et al.* 2007). Archaeological evidence suggests that the Aleutians were first settled in the east approximately 9000 years ago (Laughlin 1963; McCartney 1984; West *et al.* 2007). Aleuts migrated westward across the island chain, reaching the central Aleutians including Adak Island approximately 6000 years ago (O'Leary 2001). Aboriginal maritime hunters developed sophisticated methods to hunt sea otters and may have caused local depletions around village sites, as evidenced from faunal remains in Aleut kitchen middens (Estes *et al.* 2009; Simenstad *et al.* 1978). Sea otter remains in archaeological sites indicate that sea otters were hunted from 6000 years BP to the 19th century on Adak Island (see Crockford Chapter 6 this volume and Savinetsky *et al.* Chapter 5 this volume).

At the beginning of the 20th century the sea otter population was grossly over hunted and genetic diversity underwent a severe bottleneck. Under federal protection the population subsequently recovered. However, we know very little about the diversity of sea otter populations in deep time. Specimens collected from archaeological sites as well as those stored in museum collections potentially provide useful information about modern sea otter populations as well as those from the near and remote past (Hattori *et al.* 2003; Scribner *et al.* 1997). Using ancient DNA methods, we analyzed sea otter bones and teeth excavated from archaeological sites on Adak Island; we estimated the genetic structure and diversity of natural populations.

Materials and Methods

To avoid contamination, analyses were conducted in a room specified only for ancient DNA of the laboratory at Kyushu University, Fukuoka, Japan. A closed grove box was used for three steps: sampling, isolation of DNA, and preparation of PCR. Standard precautions were used to avoid contamination (Adachi *et al.* 2004), including the separation of pre- and post-PCR experimental areas, the use of disposable laboratory wares and filter-plunged pipette tips, treatment with DNA contamination removal solution (DNA-OFF™, TaKaRa) and UV ray irradiation to the equipment and grove box, negative extraction controls, and negative PCR controls.

We did not have the opportunity to analyze modern individuals from Adak; as protected species, extant sea otter remains are difficult to access. We analyzed 36 bone samples (Table 7.1, Appendix 7.A) of the sea otter (*Enhydra lutris*) excavated from archaeological sites representing four different time periods on Adak Island. Eleven samples, recovered from archaeological site ADK-011, upper layers date between 170 and 285 years BP. This most recent period coincides with an era of sea otter overexploitation. Thirteen samples recovered from archaeological site ADK-011 comprise the middle period (285-415 years BP). Ten otter samples, found at site ADK-011 represent the old period, (ca. 2200 years BP). The oldest period (ca. 6000 years BP) is represented by two otter samples from archaeological site ADK-171.

Many cleaning steps were employed to remove contaminant DNA and other pollutants such as soil humic acids; these often interfere with PCR. In molecular biology, the polymerase chain reaction (PCR) is a technique to amplify a single or few copies of a piece of DNA across several orders of magnitude, generating millions or more copies of a particular DNA sequence. Bones and teeth were washed with DNA-OFF solution and super distilled water (SDW), and air-dried. Two sides of each sample were exposed to UV radiation for 10 to 15 minutes. Approximately100mg to 400mg of powder were collected from each bone specimen using a dental drill. After freezing sea otter teeth in liquid nitrogen, the teeth were crushed using a Cryo-Press CP-100 (Microtech Nichion). Nearly all powdered samples were placed in a 15ml conical tube and dipped in a bleach

Table 7.1. Distribution and frequency of sea otter haplotypes from Adak Island compared with other populations/periods.

Region/site	***sub-species***	***Period (YBP)***	***N***	***(bone)***	***(teeth)***	***A***	***A2***	***B***	***B3***	***B4***	***B2***	***C***	***D***	***X***	***W***	***W3***	***h****	***Source***
Adak/ADK011: layer1	*E. l. kenyoni*	recent (170-285)**	11	11				8	3								0,44	This study
Adak/ADK011: layer2 and 3		middle (285-415+)	13	13				11	1							1	0,29	
Adak/ADK012: layer1-3		old (ca 2200)	10	8	2			4	4	1	1						0,73	
Adak/ADK171		oldest (ca 6000)	2		2						2						-	
Amchitka (Aleutian Is.): AM	*E. l. kenyoni*	contemporary	36			24		12									0,46	Larson *et al.* 2002a, b
Prince William Sound	*E. l. kenyoni*	contemporary	33			31						2					0,12	
Southeast Alaska	*E. l. kenyoni*	contemporary	24			13		11									0,52	
Washington (translocated AM)	*E. l. kenyoni*	contemporary	27			13		14									0,52	
California	*E. l. nereis*	contemporary	54			4						40	10				0,42	
Washington	*E. l. kenyoni*	100-ca 450	9			7		2									0,39	
Seal Rock, Oregon	*E. l. nereis*	170-400	6									5		1			0,33	Valentine *et al.* 2007
Nah-So-Mah, Oregon	*E. l. nereis*	695-1060	6									4			2		0,53	
		980-1430	3			1						2					-	
		1800-2000	1			*1*											-	

**: including the last hunting period: 1911-1741(97-267 ybp)
*: haplotype diversity

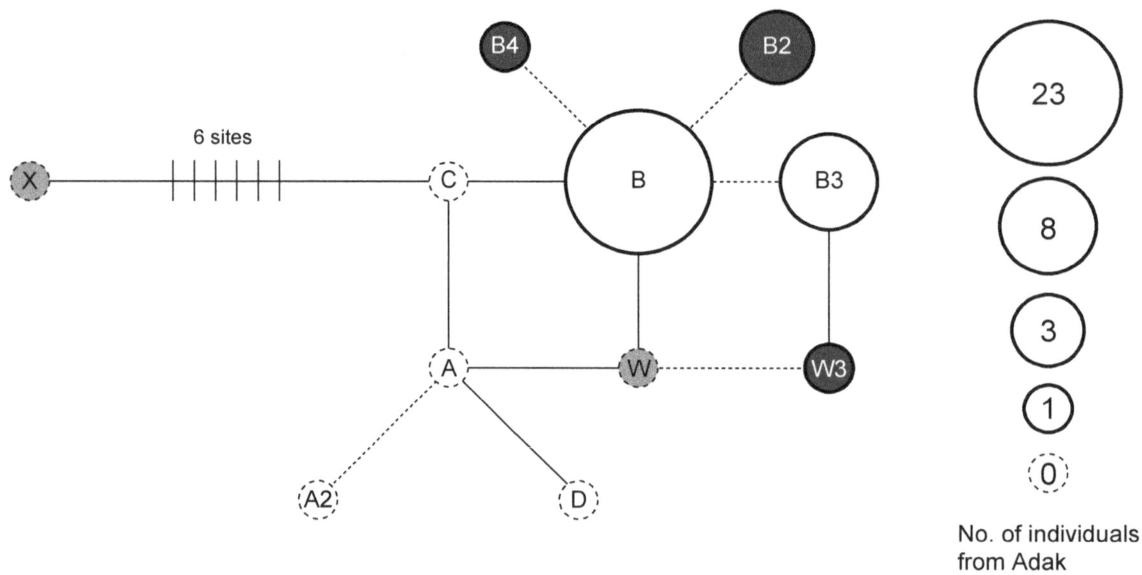

Figure 7.1. The MJ network for mtDNA control region haplotypes from Adak (B, B2, B3, B4, and W3 and others. (Circles with dot line: A, A2, C, D, X, and W). The area of the circles is proportional to the frequency of the haplotype. Each line between haplotypes represents a mutational step, and the dotted line shows indels. Filled circles indicate ancient haplotypes.

solution (low concentration of sodium hypochlorite; NaOCl) for over 15 minutes and then washed several times with SDW after centrifuging. This bleaching method effectively destroys contaminant DNA (Kemp and Smith 2005). Total DNA was extracted using a GeneClean kit for ancient DNA (Bio101/MP biomedical) following Shinoda *et al.* (2006). All samples were decalcified and digested proteins by 5ml of 0.5 M EDTA (pH 8.0), 200 µl of 20mg/ml proteinase K and 200 µl of 10 % SDS (Sodium Dodecyl Sulfate) at 37°C for over 16 hours with rotate. The supernatant was used for DNA extraction with the kit. We extracted approximately 80 µl of DNA solution and measured the total volume of DNA after isolation using Qubit™ fluorometer with Quant-iT™ dsDNAHS Assays (Invitrogen Corp.).

New primer pairs of tRNApro-F.enh (5' –TCAGTA CCCAAAACTGACATTCTAACTA–3') and Hcon1. enh (5'–TGGCCCTGAAGTAAGAACCAGATGC–3') amplified at ca 350 bp of mtDNA including 283–285 bp of the left domain of the control region. These primers were designed based on reported haplotypes of this species (Larson *et al.* 2002a; Valentine *et al.* 2007), and applied to all PCR's and sequencing. To amplify the volume of DNA from our small amount of ancient DNA, we used a second PCR. The first PCR reactions (20 µl) contained 2.5 µl of total DNA, 1 × PCR buffer, 0.2 mM each dNTP mixture, 0.5mg/ml BSA, 0.2 µM each primer, and 0.938 units of TaKaRa Ex *Taq* Hot Start Version DNA polymerase (TaKaRa Bio Inc.), which has proofreading activity. PCR parameters consisted of 30 cycles of denaturation at 94°C for 30 seconds, annealing at 57°C for 45 seconds, and extension at 72°C for 45 seconds. The second PCR reactions (25µl) contained 0.5-0.8 µl of product of 1st PCR, 1 × PCR buffer, 0.2 mM each dNTP mixture, 0.2mg/ml BSA, 0.2 µM each primer, and 0.625 units of TaKaRa Ex *Taq* Hot Start Version DNA polymerase. The second PCR parameters were the same as those for the first PCR reactions.

The sum of the PCR products were subcloned using the TOPO TA Cloning Kit (Invitrogen Corp.) into the PCR 2.1 TOPO-vector and then transformed into the TOP 10F' strain of *E. coli*. Positive colonies were selected and screened for the correct product insert. After the initial screen, the PCR products from each clone were sequenced. PCR products were enzymatically purified (PCR Product Pre-Sequencing kit, USB Corporation), and direct sequencing was carried out using a Dye Terminator Cycle Sequencing Kit (Beckman Coulter Inc.), according to the manufacturer's protocol. Cycle sequencing consisted of 30 cycles of 96°C for 20 seconds, 50°C for 20 seconds, 60°C for 150 seconds. Capillary sequencing was conducted using the CEQ2000XL DNA sequencer (Beckman Coulter Inc.), and data were analyzed using CEQ Software, Sequence Analysis ver. 2 (Beckman Coulter Inc.). Alignment of the sequence data was performed manually using MEGA 4.02 (Tamura *et al.* 2007). A network tree was constructed by the median joining (MJ) method (Bandelt *et al.* 1999) using Network (ver. 4.510; http://www.fluxus-engineering.com). Genetic variation in populations, expressed as gene diversity (*h*, Nei 1987), was estimated from the frequency distribution of haplotypes using Arlequin (ver.3.11; Excoffier *et al.* 2005).

Results

DNA was successfully extracted from all 36 sea otter samples (Appendix 7.A) including samples from the oldest period (6000 years BP), and sequenced 295–297 bp of the

Table 7.2. Haplotypes table of control region for mtDNA for sea otter.

									1	*1*	*1*	*1*	
site/	***5***	***5***	***5***	***7***	***8***	***8***	***8***	***8***	***4***	***4***	***6***	***6***	
haplotype	***1***	***6***	***7***	***0***	***2***	***5***	***6***	***7***	***7***	***8***	***1***	***4***	***GenBank***
A	C	C	C	C	C	C	-	T	G	T	C	C	AF333029
A2	.	.	.	.	.	.	C	.	.	.	.	.	AF333031
B	.	.	.	.	.	.	-	.	A	.	.	T	AF333030
B2	.	.	.	.	.	.	C	.	A	.	.	T	AB497070
B3	.	.	.	.	.	.	-	-	A	.	.	T	AB497071
B4	.	.	.	.	.	-	-	.	A	.	.	T	AB497072
C	.	.	.	.	.	.	-	.	.	.	.	T	AF333032
D	.	.	.	.	.	.	-	.	.	C	.	.	AF333033
W	.	.	.	.	.	.	-	.	A	.	.	.	EF640383
W3	.	.	.	.	.	.	-	-	A	.	.	.	AB497073
X	T	T	T	T	T	.	-	.	.	.	T	T	EF640381

Numbers as site potisions are from the top of control region. Dots indicate same nucleotide as haplotype A, and "-" are shown indels.

left domain of the control region with 12 bp of partial tRNA-Pro. All samples, except for one, were successfully amplified by the first PCR. Volumes of total, isolated DNA (Appendix 7.A) ranged from 0.04 to 2.53 μg/mil (average: 0.44 μg/ml) for recent period, 0.07 to 0.87 μg/ml (average: 0.33 μg/ml) for middle period, 0.05 to 0.95 μg/ml (average: 0.33 μg/ml) for old period, 0.14 and 1.55 μg/ml (average: 0.39 μg/ml) for oldest period, and 0.04 to 2.53 μg/ml (average: 0.39 μg/ml) for all. The amount of total DNA in each specimen was not correlated with sample age ($p>0.05$, Mann-Whitney U test) probably, but with the level of preservation.

Five mtDNA haplotypes were found in the Adak samples: B, B2, B3, B4 and W3; four: B2, B3, B4 and W3 (Table 7.2, Figure 7.1) are new haplotypes based on comparisons with 221 bp of the consensus region of other haplotypes reported. The description of each haplotype as 2, 3 and 4 were defined by indel(s) based on numbers of poly C with indel of the adjacent T from site 85 in the sequence from the top of control region (none of description; C×6+T, 2; C7+T, 3: C6 without T and 4; C5+T). Haplotypes B and B3, dominant in the Adak sea otter populations, occurred during modern, middle and old periods (Table 7.1); these haplotypes were not found in otter samples from the oldest (6000 years BP) period. Our most recent sample (08Ad003) also was represented by haplotype B. On the other hand, W was only found in the middle period, B4 in old period, and B2 in old and oldest periods. As a result (Table 7.1), the sea otter population representing the recent period consisted of haplotype B (n = 8, frequency in this population is 0.73) and B3 (n = 3, 0.27). Sea otter populations of the middle period are represented by haplotypes B (n = 11, 0.85), B3 (n = 1, 0.08), and W3 (n = 1, 0.08). The sea otters of the old period are represented by haplotypes B (n = 4, 0.50), B3 (n = 2, 0.25), B4 (n = 1, 0.125) and B2 (n = 1, 0.125). The sea otters of the oldest period are represented only by haplotype B2 (n = 2, 1.00). Haplotype diversities (h) were 0.44, 0.29, and 0.73 for the recent, middle, and old periods, respectively.

Discussion

The concentration of total DNA isolated in this study was low when compared with DNA levels of modern specimens. Nevertheless, due to the cool, damp environment of the Aleutians, DNA preservation of the archaeological samples, including those from the 6000-year-old site, was remarkably good. In the Aleutian Islands the average summer high temperatures are around 10°C, while average winter lows are 0°C (Armstrong 1977). Low temperatures slow the rate of chemical processes that degrade DNA (e.g., Höss *et al.* 1996; Hofreiter *et al.* 2001). In fact, the most successful ancient DNA studies have been based on samples collected from higher latitudes (Karanth and Yoder In press).

Using archaeological remains, we have recovered lost genetic information about past, natural sea otter populations. At least three (B2, B4 and W3) ancient haplotypes in sea otter remains from Adak archaeological sites were detected; these haplotypes have not been found in any modern sea otter populations. Haplotype W group (W and W3), discovered in the middle period (285 to 415 years BP) sequences, is an ancient type found previously only in prehistoric sea otter populations at Nah-So-Mah, Oregon (Valentine *et al.* 2007). Haplotypes B2 and B4 were only found in otter bones dated between 2000 and 6000 years ago (old and oldest periods). Interestingly, otters from the oldest period (6000 years BP) possessed only haplotype B2; however, only two samples were analyzed. The above ancient haplotypes have disappeared from sea otter populations; the reasons for this loss may be due to overexploitation by past human hunters or unknown climatic/ecological causes.

Although the old period had higher haplotype diversity (h = 0.73) than later periods (middle, 0.29; recent, 0.44), only four haplotypes including sub-haplotypes were detected from this period. Sea otter populations of the middle and recent periods are represented by three and two haplotypes, respectively. Essentially, our research shows that genetic diversity of mtDNA for Adak sea otter populations has been relatively low for the past 6000 years; only 5 haplotypes, counting sub-haplotypes, have been found thus far, and each period possessed only a few haplotypes (Table 7.1). Sub-haplotypes are defined by numbers of indels. Here, 'indel' is used to mean an insertion or a deletion and 'indels' simply refers to the mutation class that includes both insertions, deletions, and the combination thereof. In general, such indels occurred more frequently than other substitutions. Sea otters of the old period had four haplotypes, but three of these were sub-haplotypes. This suggests that the ancient sea otter population had a low genetic diversity as far back as 2200 years ago. Statistical tests for estimating historical demography using microsatellite data largely failed to find a signal of a recent population size reduction (within the past 200 years), but a Baysian method found a signal of a strong reduction in California sea otter populations over a longer time scale--up to 500 years (Aguilar *et al.* 2008). Simenstad *et al.* (1978) reported that Aleut prey species dramatically changed during the 2500 years of Aleut occupation at Amchitka Island, Alaska, and these changes were probably due to local overexploitation of sea otters by prehistoric Aleuts.

Long-term ecological and environmental changes/events might also impact population dynamics of this species. Currently, sea otter populations are again decreasing after recovering from 19th and early 20th century exploitation (Doroff *et al.* 2003; Estes *et al.* 1998; 2005). Estes *et al.* (1998) suggested that, through some, possibly human caused, disruption of the ecosystem, killer whales have increased predation on sea otters, causing this decline. Geologically, the Aleutian Islands are volcanic in origin and owe their existence to the dynamics of plate tectonics. This is a highly volatile region where volcanic eruptions, earthquakes and tsunamis can cause mass mortality of marine fauna (Black 1981; McCartney and Veltre 1999). Sea otters inhabiting inshore areas may have been directly or indirectly affected by such events. Our research suggests that Adak sea otter genetic diversity, represented by mtDNA, declined not only during the last overexploitation (1741–1911), but also between the middle (285-415 years BP) and the oldest period (6000 years BP). Such 'bottleneck' events probably occurred multiple times either by human predation and/or ecological and environmental changes of otter habitat.

The Aleutians sea otter populations belong to subspecies *E. lutris kenyoni* (e.g., Cronin *et al.* 1996). Using restriction fragment analysis of mtDNA, Cronin *et al.* (1996) detected genetic differentiation between sea otters living in Prince William Sound, Alaska, and the western Aleutians. Our research shows a genetic differentiation in the western populations of this subspecies. Although the modern sea otter population of Amchitka in Aleutian Islands has haplotype A (dominant) and B (Larson *et al.* 2002a), the ancient Adak population had no haplotype A. This evidence suggests that gene flow has been restricted between these two islands in Aleutian archipelago. However, it is necessary to analyze modern sea otters from the Adak population. Gene flow among sea otter populations is limited to linear, shallow-water coastal habitats or to relatively short distances of open water between suitable habitats (Scribner *et al.* 1997). Moreover, each population has experienced decreases in population size due to overexploitation and/or environmental changes over a long time. As a result, such small populations have been greatly affected by genetic drift. In this case, a new population/stock may be formed by a relatively small number of individuals from a nearby stock, and could be influenced by the founder effect. These ecological and historical events should lead to increased genetic differences among populations and decreased genetic diversity in this species.

Acknowledgements

Sea otter elements used in this study were recovered during 2005 and 2006 with a grant from the National Science Foundation, Office of Polar Programs, Arctic Social Sciences (Award # 0353092). We are grateful to Dr Géraldine Véron at Departement Systématique et Evolution, Muséum National d'Histoire Naturelle and anonymous reviewers who provided helpful suggestions for this paper.

References

Armstrong, Robert H. 1977. Weather and Climate. In (M.L. Merritt and R.G. Fuller, eds.), *Environment of Amchitka Island, Alaska,* 53–58. Springfield, Virginia, National Technical Information Center.

Adachi, Noboru. Kazuo Umetsu, Wataru Takigawa and Kazuhiro Sakaue. 2004. Phylogenetic Analysis of the Human Ancient Mitochondrial DNA. *Journal of Archaeological Science* 31(10), 1339–1348.

Aguilar, Andres, David A. Jessup, James A. Estes and John C. Garza. 2008. The Distribution of Nuclear Genetic Variation and Historical Demography of Sea Otters. *Animal Conservation* 11(1), 35-45

Bandelt, Hans-Jürgen, Peter Forster and Arne Röhl. 1999. Median-joining Networks for Inferring Intraspecific Phylogenies. *Molecular Biology and Evolution* 16(1), 37–48.

Berta, A., J.L. Suchach, and K.M. Kovacs. 2006. *Marine Mammals: Evolutionary Biology* (2nd edition). San Diego, Academic Press.

Black, Lydia T. 1981. Volcanism as a Factor in Human Ecology: The Aleutian Case. *Ethnohistory* 28(4), 313-340.

Bodkin, James L., Brenda E. Ballachey, Matthew A. Cronin and Kim T. Scribner. 1999. Population Demographics and Genetic Diversity in Remnant and Translocated Populations of Sea Otters. *Conservation Biology* 13(6), 1378–1385.

Cronin, Matthew A., James Bodkin, Brenda Ballachey,

James A Estes and John. C. Patton. 1996. Mitochondrial-DNA Variation Among Subspecies and Populations of Sea Otters (*Enhydra lutris*). *Journal of Mammalogy* 77(2), 546–557.

Crockford, Susan. 2011. Archaeozoology of Adak Island: 6000 Years of Subsistence History in the Central Aleutians. Chapter 6. In D. West, V. Hatfield, E. Wilmerding, C. Lefèvre, and L. Gualtieri (eds.), *The People Before: The Geology, Paleoecology and Archaeology of Adak Island, Alaska*. Oxford, British Archaeological Reports.

Doroff, Angela M., James A. Estes, M. Tim Tinker, Douglas M. Burn and Thomas J. Evans. 2003. Sea Otter Population Declines in the Aleutian Archipelago. *Journal of Mammalogy*. 84(1), 55-64.

Doroff, Angela M. and Alexander M. Burdin. 2008. *Enhydra lutris*. In 2008 IUCN Red List of Threatened Species, International Union for Conservation of Nature 2008 (http://www.iucnredlist.org).

Estes, James A., James Bodkin L. and M. Ben-David. 2009. Otters, Marine. In W. F. Perrin, B. Würsig and J. G. M. Thewissen (eds.), *Encyclopedia of Marine Mammals*, Second Edition, 807–816. Amsterdam, Academic Press–Elsevier.

Estes, James A., M. Tim Tinker, Terrie M. Williams and Daniel F. Doak. 1998. Killer Whale Predation on Sea Otters Linking Oceanic and Nearshore Ecosystems. *Science* 282(5388), 473–476.

Excoffier, Laurent, Guillaume Laval and Stefan Schneider. 2005. Arlequin ver. 3.0: An Integrated Software Package for Population Genetics Data Analysis. *Evolutionary Bioinformatics Online* 1, 47–50.

Hofreiter, Michael, David Serre, Hendrik N. Poinar, Melanie Kuch and Svante Pääbo. 2001. Ancient DNA. *Nature Reviews Genetics* 2(5), 353-359.

Höss, Matthias, Pawel Jaruga, Tomasz H. Zastawny, Miral Dizdaroglu and Svante Pääbo. 1996. DNA Damage and DNA Sequence Retrieval from Ancient Tissues. *Nucleic Acids Research*. 24(7), 1304–1307.

Hattori, Kaoru, Alexander M. Burdin, Manabu Onuma, Masatsugu Suzuki and Noriyuki Ohtaishi. 2003. Sex Determination in the Sea Otter (*Enhydra lutris*) from Tissue and Dental Pulp using PCR Amplification. *Canadian Journal of Zoology* 81(1), 52-56.

Karanth, K. Praveen and Anne D. Yoder. In press Geographic Origins of aDNA Samples and Prospects for aDNA Studies in the Tropics. In G. Dorado (ed.), *Molecular Markers, PCR, Bioinformatics and Ancient DNA - Technology, Troubleshooting and Applications*. New York, Science Publishers.

Kemp, Brian M and David G. Smith. 2005. Use of Bleach to Eliminate Contaminating DNA from the Surface of Bones and Teeth. *Forensic Science International* 154(1), 53–61.

Kenyon, Karl W. 1969. The Sea Otter in the Eastern Pacific Ocean. In *North American Fauna* 68. Washington DC, US Fish and Wildlife Service.

Larson, Shawn, Ronald Jameson, James Bodkin, Michelle Staedler and Paul Bentzen. 2002a, Microsatellite DNA and Mitochondrial DNA Variation in Remnant and Translocated Sea Otter (*Enhydra lutris*) Populations. *Journal of Mammology* 83(3), 893–906.

Larson, Shawn, Ronald Jameson, Michael Etnier, Melissa Fleming and Paul Bentzen. 2002b. Loss of Genetic Diversity in Sea Otters (*Enhydra lutris*) Associated with the Fur Trade of the 18th and 19th Centuries. *Molecular Ecology* 11(10), 1899–1903.

Laughlin, William S. 1963. The Earliest Aleuts. *Anthropological Papers of the University of Alaska* 10, 73–91.

Lidicker, William Z, Jr. and F. C. McCollum. 1997. Allozymic Variation in California Sea Otters. *Journal of Mammalogy* 78(2), 417–425.

Loughlin, Thomas R. 1980. Home Range and Territoriality of Sea Otters near Monterey, California. *Journal of Wildlife Management* 44(3), 576-582.

Nei, Masatoshi. 1987. *Molecular Evolutionary Genetics*. New York, Columbia University Press.

Nowak, Ronald M. 1999. *Walker's Mammals of the World*. Baltimore and London, The Johns Hopkins University Press.

McCartney, Allen P. 1984. Prehistory of the Aleutian Region. In D. Damas (ed.), *Handbook of North American Indians,* Vol. 5: Arctic, 119-135. Washington, DC, Smithsonian Institution Press.

McCartney, Allen P. and Douglas W. Veltre. 1999. Aleutian Island Prehistory: Living in Insular Extremes. *World Archaeology* 30, 503-515.

O'Leary, Matthew. 2001. Volcanic Ash Stratigraphy for Adak Island, Central Aleutian Archipelago. In Don Dumond (ed.), *Archaeology in the Aleut Zone of Alaska, Some Recent Research. University of Oregon Anthropological Papers* 58, 215–234. Eugene, University of Oregon Press.

Ralls, Katherine, Eagle Thomas C. and Siniff Donald. B. 1988. Movement Patterns and Spatial Use of California Sea Otters. In D. B. Siniff and K. Ralls (eds.), *Population Status of California Sea Otters*, 33-63. Final Report on Contract No. 14-12-001-3003.Los Angeles, California, Minerals Management Service.

Riedman, Marianne L. and James A. Estes. 1990. The Sea Otter (*Enhydra lutris*): Behavior, Ecology, and Natural History. *Biological Report* 90, Washington, DC, US Fish and Wildlife Service.

Savinetsky, Arkady B., Dixie L. West, Zhanna A. Antipushina, Bulat F. Khassanov, Nina K. Kiseleva, Olga Krylovich, and Andrei M. Pereladov. 2011. The Reconstruction of Ecosystems History of Adak Island (Aleutian Islands) During the Holocene. Chapter 5. In D. West, V. Hatfield, E. Wilmerding, C. Lefèvre, and L. Gualtieri (eds.), *The People Before: The Geology, Paleoecology and Archaeology of Adak Island, Alaska*. Oxford, British Archaeological Reports

Schrope, Mark. 2007. Food Chains: Killer in the Kelp. *Nature* 445(7129), 703–705.

Scribner, Kim T., James Bodkin, Brenda Ballachey, Steven R. Fain, Matthew A. Cronin and Maria Sanchez. 1997. Population Genetic Studies of the Sea Otter (*Enhydra lutris*): A Review and Interpretation of Available Data. In A.E. Dizon, S.J. Chivers, and W.F. Perrin (eds.), *Molecular Genetics of Marine Mammals* 197–208. Special Publication No. 3. Lawrence, KS, Society for Marine Mammals.

Shinoda, Ken-ichi, Noboru Adachi, Sonia Guillen and Izumi Shimada. 2006 Mitochondrial DNA Analysis of Ancient Peruvian Highlanders. *American Journal of Physical Anthropology* 131(1), 98–107.

Simenstad, Charles A., James A. Estes and Karl W. Kenyon. 1978. Aleuts, Sea Otters, and Alternate Stable-State Communities. *Science* 200(4340), 403-411.

Tamura, Koichiro, Joel Dudley, Masatoshi Nei and Sudhir Kumar. 2007. MEGA4: Molecular Evolutionary Genetics Analysis (MEGA) software version 4.0. *Molecular Biology and Evolution* 24(8), 1596–1599.

Valentine, Kim, Deborah A. Duffield, Lorelei E. Patrick, David R. Hatch, Virginia L. Butler, Roberta L. Hall and Niles Lehman. 2007. Ancient DNA Reveals Genotypic Relationships Among Oregon Populations of the Sea Otter *(Enhydra lutris*). *Conservation Genetics* 9(4), 933–938.

West, Dixie, Michael Crawford and Arkady B. Savinetsky. 2007. Genetics, Prehistory and the Colonisation of the Aleutian Islands. *Earth and Environmental Science Transactions of the Royal Society of Edinburgh* 98(1), 47-57.

Wilson, Don E., Michael A. Bogan, Robert L. Brownell, Jr., Alexander M. Burdin and M.K. Maminov. 1991. Geographic Variation in Sea Otters, *Enhydra lutris*. *Journal of Mammalogy* 72(1), 22-36.

Appendix 7A. Adak sea otter samples

Lab.No.	*Haplotype*	*Catalogue No.*	*Specimen No.*	*Date*	*Site*	*House*	*Unit*	*layer*	*level*	*Period (YBP)*	*part of sample*	*volume of sample**	*consentration of DNA***	*age (sex)*
08Ad002	B	#0017		06/16/06	ADK011	House 1	Unit G	layer 1		recent (170-285)	bone	-	-	adult
08Ad003	B	#8		05/18/07	ADK-SW		Unit D	-	level 1	recent (170-285)	bone	285 mg	0.42 μg/ml	juvenile
08Ad004	B	#0323		06/21/06	ADK011	House 1	Unit D	layer 2		middle (285-415+)	bone	-	0.67 μg/ml	subadult/ juvenile
08Ad005	B	#11		06/16/06	ADK011	House 1	Unit E	layer 1		recent (170-285)	bone	-	2.53 μg/ml	adult
08Ad006	B	#701		06/26/06	ADK011	House 1	Unit C	layer 2	level 4	middle (285-415+)	bone	263 mg	0.07 μg/ml	adult
08Ad007	B	#268		06/20/06	ADK011	House 1	Unit H	layer 2		middle (285-415+)	bone	283 mg	0.54 μg/ml	
08Ad008	B	#0908		07/01/06	ADK011	House 1	Unit J	layer 1	level 1	recent (170-285)	bone	269 mg	0.67 μg/ml	
08Ad009	B	#0232		06/20/06	ADK011	House 1	Unit D	layer 2		middle (285-415+)	bone	230 mg	0.11 μg/ml	
08Ad010	B	#185		06/19/06	ADK011	House 1	Unit H	layer 2	level 2	middle (285-415+)	bone	254 mg	0.20 μg/ml	juvenile
08Ad011	B3	#0966		07/02/06	ADK011	House 1	Unit K	layer 1	level 2	recent (170-285)	bone	251 mg	0.07 μg/ml	adult
08Ad012	W3	#0578		06/25/06	ADK011	House 1	Unit E	Layer 2	level 3	middle (285-415+)	bone	239 mg	0.22 μg/ml	adult
08Ad013	B	#150		06/18/06	ADK011	House 1	Unit C	layer 2		middle (285-415+)	bone	245 mg	0.12 μg/ml	
08Ad014	B	#23		06/17/06	ADK011	House 1	Unit F	layer 1		recent (170-285)	bone	253 mg	0.07 μg/ml	adult/male
08Ad015	B	#592		06/25/06	ADK011	House 1	Unit H	layer 3 east	level 4	middle (285-415+)	bone	262 mg	0.24 μg/ml	adult
08Ad016	B	#0679			ADK011	House 1	Unit D	layer 2	level 3	middle (285-415+)	bone	302 mg	0.30 μg/ml	adult
08Ad017	B3	#047		06/17/06	ADK011	House 1	Unit G	layer 1	level 1	recent (170-285)	bone	274 mg	0.12 μg/ml	adult
08Ad018	B3	#011		06/25/06	ADK011	House 1	Unit B	layer 2	level 3	middle (285-415+)	bone	262 mg	0.87 μg/ml	subadult
08Ad019	B	#0323		06/21/06	ADK011	House 1	Unit D	layer 2		middle (285-415+)	bone	277 mg	0.39 μg/ml	adult
08Ad020	B	#861		06/27/06	ADK011	House 1	Unit D	layer 2	level 6	middle (285-415+)	bone	353 mg	0.35 μg/ml	

Lab.No.	*Haplotype*	*Catalogue No.*	*Specimen No.*	*Date*	*Site*	*House*	*Unit*	*layer*	*level*	*Period (YBP)*	*part of sample*	*volume of sample**	*consentration of DNA***	*age (sex)*
08Ad021	B	#0629		06/25/06	ADK011	House 1	Unit C	layer 2	level 3	middle (285-415+)	bone	397 mg	0.19 μg/ml	adult
08Ad022	B3	#65	#049		ADK012		EU3	layer 1		old (ca 2200)	bone	290 mg	0.05 μg/ml	adult
08Ad023	B2	#5	#010		ADK012		EU1	layer 2		old (ca 2200)	bone	297 mg	0.13 μg/ml	adult/ subadult
08Ad024	B	#39	#027		ADK012		EU2	layer 1		old (ca 2200)	bone	258 mg	0.41 μg/ml	juvenile
08Ad025	B	#75	#073		ADK012		EU3	layer 2		old (ca 2200)	bone	365 mg	0.50 μg/ml	adult/ subadult
08Ad026	B3	#39	#025		ADK012		EU2	layer 1		old (ca 2200)	bone	311 mg	0.39 μg/ml	adult
08Ad027	B	#84	#077		ADK012		EU3	layer 3		old (ca 2200)	bone	408 mg	0.08 μg/ml	juvenile/ subadult
08Ad028	B	#5	#008		ADK012		EU1	layer 2		old (ca 2200)	bone	284 mg	0.09 μg/ml	juvenile/ subadult
08Ad029	B4	#65	#046		ADK012		EU3	layer 1		old (ca 2200)	bone	326 mg	0.09 μg/ml	juvenile
08Ad030	B3	#0094		06/18/06	ADK011	House 1	Unit G	layer 1		recent (170-285)	bone	292 mg	0.16 μg/ml	
08Ad031	B	#452		06/22/06	ADK011	House 1	Unit D	layer 1	level 2	recent (170-285)	bone	238 mg	0.04 μg/ml	adult
08Ad032	B	-		06/25/06	ADK011	House 1	Unit F	layer 1	level 3	recent (170-285)	bone	270 mg	0.09 μg/ml	juvenile
08Ad033	B	#0494			ADK011	House 1	Unit D	layer 1	level 2	recent (170-285)	bone	266 mg	0.26 μg/ml	junior
08Ad034	B3	#61	#038		ADK012		EU2	layer 2		old (ca 2200)	tooth	< 200 mg	0.63 μg/ml	juvenile
08Ad035	B3	#39	#029		ADK012		EU2	layer 2		old (ca 2200)	tooth	410 mg	0.95 μg/ml	adult
08Ad036	B4			1999	ADK171				level 1	oldest (ca 6000)	tooth	101 mg	0.14 μg/ml	
08Ad037	B4			2005	ADK171				level 3	oldest (ca 6000)	tooth	393 mg	1.55 μg/ml	

*volume of sample for DNA isolation
**consentration of total DNA isolated

Region/site	***sub-species***	***Period (YBP)***
Adak/ADK011: layer1	*E. l. kenyoni*	recent (170-285)**
Adak/ADK011: layer2 and 3		middle (285-415+)
Adak/ADK012: layer1-3		old (ca 2200)
Adak/ADK171		oldest (ca 6000)

Chapter 8
Carbon, Nitrogen, and Oxygen Isotopic Composition of Bone Collagen and Tooth Enamel Carbonate of the Sea Otter (*Enhydra lutris*) from Archaeological Sites on Adak Island, Aleutian Archipelago

Ame Garong, Chizuru Takashima, Shin Nishida, Akihiro Kano and Hiroko Koike

Abstract: *Isotopic compositions measured for bone collagen and tooth enamel carbonate were conducted on sea otter (Enhydra lutris) bones collected from four archaeological sites on Adak Island, Aleutian Islands, Alaska: ADK-011, Layer 1 (170-285 yr BP), ADK-011, Layer 2 (285-415 yr BP), ADK-012 (circa 2200 yr BP), and ADK-171 (circa 6000 yr BP). Three bones representing minke, humpback and beaked whales, were used for comparison. Bone collagen $\delta^{13}C$ and $\delta^{15}N$ cross-plots suggest that the archeological sea otters can be divided into two groups independent of archaeological age and location. The first group represents a domain of -11 to -13‰ in $\delta^{13}C$ and 8.5-12‰ in $\delta^{15}N$. The second group falls within the domain of –12.5 to –14.5‰ in $\delta^{13}C$ and 10-14‰ in $\delta^{15}N$, a range of values that includes minke and humpback whales. Higher $\delta^{13}C$ values of the first group correspond to a typical range of marine mammals dependent upon the kelp ecosystem. The second group consists of individuals that, along with minke and humpback whales, are adapted to an oceanic ecosystem where ^{13}C-depleted collagen is synthesized. The tooth enamel carbonate $\delta^{13}C$ and $\delta^{18}O$ cross-plots also suggest that the samples can be divided into two groups. The first group is in a domain of -8 to -12‰ in δ13C and around -5‰ in $\delta^{18}O$. The second group is in –12.5 to –10‰ in $\delta^{13}C$ and around -6‰ in $\delta^{18}O$. Again, these two groups can be attributed to food differences between the kelp and oceanic ecosystems.*

Keywords: *sea otter, bone collagen, tooth enamel carbonate, $\delta^{13}C$, $\delta^{15}N$, $\delta^{18}O$*

Introduction

Prehistoric Aleut hunters targeted sea otters for their valuable pelts; consequently, their remains are frequently recovered in coastal archaeological sites across the archipelago (see Crockford Chapter 6 this volume). Since the Russian discovery of sea otters in the 18th century, human hunting pressure has largely controlled sea otter distribution and population. Prior to commercial exploitation, the sea otter population is estimated to have ranged between 150,000 and 300,000 (Doroff and Burdin 2008; Kenyon 1969) and encompassed the circum-Pacific region, from southern California, the west coast of North America, Aleutian Islands, Kurile-Commander Islands, and the coast of Kamchatka, an overall range of 10,000km (Nowak 1999). Both Russian and American commercial otter hunting for valuable, lush pelts, beginning circa 1741 and continuing for the next 170 years, nearly exterminated the species. By the beginning of the 20th century, the world sea otter population ranged between 1000 and 2000 individuals in 13 colonies in central California, western Alaska, and the Commander and Kurile Islands (Estes 1990; Kenyon 1969). In 1911, under the International Fur Seal Treaty, the otter was declared a protected species, and 11 sea otter colonies recovered, reaching a stable population within 50 years. In 1965, the Aleutian sea otter population was estimated at 9700 (Kenyon 1969). In 1973 the US Fish and Wildlife Service, under the authority of the Endangered Species Act, declared 'endangered status' for this species (Doroff and Burdin 2008), but sea otter populations continued to decrease. By 1992 the Aleutian population had dropped to 8048 (Evans *et al.* 1997), and by 2000 the sea otter saw a drastic reduction to 2442 individuals (Doroff *et al.* 2003). In sum, Alaskan sea otter populations declined 55-67% between the mid 1980's and 2005 in Alaska, the center of the sea otter's range. The cause of this decline remains unknown, although killer whale predation is suspected (Estes *et al.* 1998; Schrope 2007).

The sea otter (*Enhydra lutris*) usually lives within a kilometer of the shoreline in relatively shallow waters less than 100m deep with most foraging occurring at depths less than 30m (Bodkin *et al.* 2004; Riedman and Estes 1990). Individuals generally form colonies that inhabit a few kilometers of coastline year-round; wide, deep-water passes are barriers to sea otter movements (Kenyon 1969). Habitat determines otter diets. Sea otters can inhabit rocky substrates between shoreline and the seaward limits of the kelp canopy (Riedman and Estes 1990) where they maintain kelp forest viability (Love 1992) by feeding almost exclusively on sea urchins, along with a variety of mollusks, crustaceans, snails, and crabs. Otters also inhabit soft sediment substrates where they generally rely on clams. A third habitat is the open ocean where otters feed on fish, octopus and squid (Estes *et al.* 1982; Estes 1990).

For decades, carbon, nitrogen, and oxygen stable isotopic ratios have been used to study various types of ecosystems and applied intensively to reconstruct environmental conditions or food web pathways (Dunton *et al.* 1989; Hobson and Welch 1992; Jennings *et al.* 2002; Peterson and Fry 1987; Wada *et al.* 1991; Walker and Macko 1999). Because variations in stable isotope ratios are so small, the isotopic ratio (R) is expressed as a per mil (‰) deviation from international standards, using the following formula:

$$d\,(‰) = [R_{sample} / R_{standard} - 1] \times 1000$$

where R is ($^{13}C/^{12}C$), ($^{15}N/^{14}N$) or ($^{18}O/^{16}O$). Conventional standards for carbon and nitrogen are PDB (Pee Dee Belemnite) and atmospheric N_2, respectively. Oxygen has two types of standards: SMOW (standard mean oceanic water) and PDB. In this study we use $\delta^{18}O_{PDB}$ for calcite, and $\delta^{18}O_{SMOW}$ for water The two expressions are in the following relation; $\delta^{18}O_{PDB} = 0.97006 * \delta^{18}O_{SMOW} - 29.94(‰)$ (Friedman and O'Neil 1977).

Marine plants and phytoplankton assimilate carbon using the C_3 cycle from dissolved inorganic carbon that has an isotope value of approximately 0‰. Photosynthetic fractionation of $\delta^{13}C$ values was shown to be about –19.5‰ (Degens *et al.* 1968; Sackett *et al.* 1965). The ratio of stable isotopes can change between diet and consumer based on differential digestion or fractionation during dissimilation and metabolic processes. Trophic enrichment of $\delta^{13}C$ is small, mostly within 1‰ for each trophic level (DeNiro and Epstein 1978; Fry and Sherr 1984; Peterson *et al.* 1985). Trophic enrichment of $\delta^{15}N$ is usually positive, with larger amplitude of 3 to 4‰ (DeNiro and Epstein 1981; Hobson and Welch 1992; Minagawa and Wada 1984; Vander Zanden and Rasmussen 1999). The nitrogen isotope ratio of an analyzed animal has been used as an index for trophic level in various ecosystems (Fry 1988; Hobson and Welch 1992; Schoeninger and DeNiro 1984; Sealy *et al.* 1987; Wada *et al.* 1987).

Dietary studies using $\delta^{13}C$ and $\delta^{15}N$ have been performed for marine mammals including the sea otter (Fry 1988; Hobson and Welch, 1992; Walker and Macko 1999). Bone collagen from sea otters on the California coast (Walker and Macko 1999) gave $\delta^{13}C$ values of -11.1± 0.5‰, and $\delta^{15}N$ values of 7.8 ± 0.8‰ (N=10). The $\delta^{13}C$ and $\delta^{15}N$ signatures of those sea otters apparently reflected their diets in the kelp ecosystem, which provides a habitat for diverse marine animals. Newsome *et al.* (2009) analyzed the isotopic composition of sea otter vibrissae (whiskers) to characterize inter- and intra-individual variation in sea otter diets at Monterey Bay, California. The results showed substantial variation in both $\delta^{13}C$ and $\delta^{15}N$ values, which Newsome observes resulted from differences in the values of diets.

Thus the isotopic signatures of collagen can be used to identify feeding strategies of the sea otter. However, isotopic variations can occur in the kelp plant, complicating the analyses. According to Fredriksen (2003), kelp *Laminaria hyperborea* on the Norwegian coast showed a significant variation between the basal ($\delta^{13}C$: –16.65‰, $\delta^{15}N$: 5.54‰) and distal ($\delta^{13}C$: –18.67‰, $\delta^{13}C$: 3.54‰) parts of the lamina (leaf).

Along with collagen, animal bones and teeth also provide isotopic signatures that record ecological and environmental characteristics. In the case of enamel carbonate, $\delta^{18}O$ value reflects the source of body water, while the $\delta^{13}C$ value is linked to the diet (Lee-Thorp *et al.* 1989). The oxygen isotope composition ($\delta^{18}O$) of biogenetic apatite is strongly correlated with that of body water, and is offset from temperature-dependent fractionation that is constant for homeothermic mammals (Longinelli 1984; Luz *et al.* 1984). External sources of oxygen, e.g., ambient water and water from food, provide the baseline from which body water $\delta^{18}O$ deviates. Regional and temporal variations in these sources range from high values of ambient seawater, and food water to relatively low values of coastal freshwater and snowmelt water (Craig and Gordon 1965). Oxygen isotopes have been assessed as an indicator of habitats that can be grouped into four categories: aquatic-marine, aquatic-estuarine, aquatic-fresh-water, and terrestrial (Clementz and Koch 2001). Isotopic analysis was also applied to tooth enamel calcite (Clementz and Koch 2001). Roe *et al.* (1998) analyzed the $\delta^{18}O$ of phosphate in tooth enamel and bone, which are about 8.8‰ depleted in ^{18}O relative to its calcite value (Bryant *et al.* 1996). Sea otters from Central California measured 27.3 ± 0.6‰ (-3.5‰ vs. PDB), and river otters from Washington 25.8 ± 0.9‰ (-5.3‰ vs. PDB) and river otters from Oregon 23.0 ± 0.3‰ (-7.6‰ vs. PDB) (Clementz and Koch 2001).

Tooth enamel calcite basically reflects carbon isotopic composition from dissolved inorganic carbon (DIC) in the mouth, and involves some ecological information because the DIC largely consists of metabolic CO_2. The $\delta^{13}C$ value of tooth enamel carbonate is however expected to be higher than the value of collagen due to two processes: 1) contribution of seawater DIC (~O‰) and, 2) degassing of ^{13}C depleted CO_2 in the mouth.

In this study, we measured $\delta^{13}C$ and $\delta^{15}N$ values using bone collagen from archaeological specimens recovered from Adak Island to determine prehistoric sea otter diet over time. Samples of sea otter tooth enamel were also used for $\delta^{18}O$ and $\delta^{13}C$ values in order to determine environmental conditions during the periods when prehistoric sites were occupied at Adak Island.

Materials and Methods

To determine ancient sea otter diet using $\delta^{13}C$ and $\delta^{15}N$ analysis we used 28 bones previously identified as sea otter, by ancient DNA analysis (Nishida *et al.* Chapter 7 this volume), representing three archaeological periods on Adak Island: (1) ADK-011, Layer 1 (170-285 yrs BP: N=7), (2) ADK-011, Layer 2 (285-415 yrs BP; N=13); and (3) ADK-012 (circa 2200 yrs BP; N=8). Three bones representing minke, humpback and Stejneger's beaked whales were similarly analyzed; results of these analyses were compared with results obtained from sea otters (Table 8.1).

Using a dental drill, approximately 200mg of each bone sample was surface cleaned, powdered, collected and cleaned by sonication in double distilled water (DDW). Alkaline treatment with 0.1 N NaOH to remove soil acids followed. This step was repeated until the solutions were colorless. The samples were then washed with DDW and placed in cellulose dialysis tubes for acid treatment to remove inorganic carbon. Each sample tube was placed in a beaker containing 0.1 N HCl with a stirrer (Horai *et al.* 1989). The HCl solutions were changed daily for five days to achieve complete dissolution. The organic residues were centrifuged to pellets, and rinsed with DDW three times. The samples were again washed with 0.01N NaOH and then rinsed with DDW and centrifuged to pellets. The samples were frozen and then freeze-dried overnight using an Eyela freeze dryer (Tokyo-Rika, FDU 506). Aliquots of 0.8 ± 0.05mg of each sample were placed in tin capsules (two capsules per sample) for isotopic measurement.

Two sets of $\delta^{13}C$ and $\delta^{15}N$ values were obtained for each sample, using a continuous-flow mass-spectrometer (ANCA-mass 20-20, Europa Scientific Instruments, UK). Repeated measurements of a running standard (glycine) produced measurement errors of less than 0.1‰ for $\delta^{13}C$ and 0.3‰ for $\delta^{15}N$. If the difference in results between the two analyses per sample was larger than the measurement error, that sample was re-measured.

For $\delta^{18}O$ and $\delta^{13}C$ analysis, a total of 13 sea otter teeth (ADK-011, Layer 1, N=2; ADK-011, Layer 2, N=4; ADK-012, N=6; and ADK-171, N=1) were used (Table 8.2). These teeth were cleaned ultrasonically with DDW to remove adhering soil. Enamel between cervical lines was sampled using a diamond-coated dental drill. Organic components of the enamel were removed by soaking in 3% sodium hypochlorite for 24 hours. The samples were rinsed four times with distilled water and freeze-dried.

Oxygen and carbon isotopes were measured by gas mass spectrometry (Finnigan Delta Plus) using a gas separation system (Thermo Gas Bench). In this system, the samples were reacted with purified H_3PO_4 at 55°C for 3 hours within a sealed glass vial filled with 1 atm He. The CO_2 emitted from the carbonate fraction was, together with He, first introduced to the gas chromatograph and separated from other components, such as H_2O. The separated CO_2 was then introduced with He to the mass spectrometer. The measured carbon and oxygen isotopic ratios were normalized to the laboratory standard (the Solnhofen Limestone; $\delta^{13}C$=-0.47‰, $\delta^{18}O$=-5.04‰) that was calibrated against NBS19 ($\delta^{13}C$=+1.95‰, $\delta^{18}O$=-2.20‰ vs. VPDB). Repeated measurements of the laboratory standard (N=22) showed that reproducibility for $\delta^{13}C$ and $\delta^{18}O$ was within ±0.20 and ±0.27‰, respectively (Hori *et al.* 2009).

Results

$\delta^{13}C$ and $\delta^{15}N$ Analyses of Bone Collagen

Extracted collagen samples from sea otter yielded acceptable carbon and nitrogen quantities. The C/N ratio ranged between 2.7 and 3.0 (Table 8.1) indicating that the dialysis extraction was sufficient and that gelatinization was not needed. Collagen $\delta^{13}C$ values for otter bones from ADK-011, Layer 1 (170-285 yrs BP) averaged -12.3 ± 0.8‰, ranging from -11.7 to 13.8‰. The values from ADK-011 Layer 2 (285-415 yrs BP) averaged -11.9 ± 0.6‰, ranging between -10.7 and 12.9‰. The samples from ADK-012 (circa 2200 yrs BP) gave a $\delta^{13}C$ average value of -12.8 ± 1.2‰ and ranged between -11.6 and 15.4‰. No significant difference was observed in the $\delta^{13}C$ values among these sites. These $\delta^{13}C$ values were higher than the values we obtained for the minke whale (-14.0‰ from the ADK-011 Layer 1), humpback whale (-14.5‰ from ADK-011 Layer 2) and beaked whale (-16.3‰ from ADK-011 Layer 2; Table 8.1).

The collagen $\delta^{15}N$ values of the otter samples from ADK-011, Layer 1 averaged 11.0 ± 1.0‰, ranging between 9.9 and 12.4‰. The values from ADK-011 Layer 2 averaged 10.6 ± 0.7‰, ranging between 9.9 and 11.9‰. The samples from ADK-012 yielded average $\delta^{15}N$ values of 11.9 ± 1.7‰, ranging between 8.8 and 13.6‰. There is no significant difference in collagen $\delta^{15}N$ value due to either age or geographical location from which the samples were taken. The archeological whale samples clearly yielded higher $\delta^{15}N$ values; the minke, humpback, and Stejneger's beaked whales recorded $\delta^{15}N$ values of 12.7‰, 13.1‰, and 17.8‰, respectively.

The $\delta^{13}C$ and $\delta^{15}N$ cross-plots (Figure 8.1) suggest that the archeological sea otters can be divided into two groups. The first group is in a domain ranging from -11 to -13‰ for $\delta^{13}C$ and from 8.5-12‰ for $\delta^{15}N$. The second group ranges from –12.5 to –14.5‰ for $\delta^{13}C$ and 10-14‰ for $\delta^{15}N$. The stable isotope values of this latter group are similar to the values seen in the archaeological minke and humpback whales.

$\delta^{18}O$ and $\delta^{13}C$ Using Tooth Enamel Carbonate

As with collagen samples, we observed no significant differences in isotopic compositions of the tooth calcite based on either the location or age of the archaeological sites from which samples were taken. The $\delta^{13}C$ value of sea otter teeth from ADK-011 Layer 1 averaged -11.5 ±

Table 8.1. $\delta^{13}C$ and $\delta^{15}N$ values using bone collagen from the sea otter and whales.

Lab. No.	Cat.No.	Species	Site and Layer	%C	%N	C/N	$\delta^{13}C$	$\delta^{15}N$	Average ±SD
08Ad02	#0017	sea otter	ADK011-layer 1	44.90	15.85	2.83	-11.80	10.44	$\delta^{13}C$
08Ad03	#8	sea otter	ADK011-layer 1	44.11	15.61	2.83	-13.76	12.22	$\delta^{15}N$
08Ad05	#11	sea otter	ADK011-layer 1	46.14	16.20	2.85	-11.89	9.94	
08Ad08	#0908	sea otter	ADK011-layer 1	44.32	15.59	2.84	-11.93	10.07	
08Ad14	#23	sea otter	ADK011-layer 1	44.95	16.11	2.79	-12.09	10.68	
08Ad17	#47	sea otter	ADK011-layer 1	42.70	15.39	2.78	-11.73	10.98	-12.15 ±0.72
08Ad31	#452	sea otter	ADK011-layer 1	50.68	19.01	2.67	-11.84	9.92	10.61 ±0.81
08Ad04	#0323	sea otter	ADK011-layer 2	46.24	15.99	2.89	-12.34	9.89	
08Ad06	#701	sea otter	ADK011-layer 2	45.36	16.26	2.79	-12.12	10.06	
08Ad07	#268	sea otter	ADK011-layer 2	43.94	15.40	2.85	-12.25	10.53	
08Ad09	#0232	sea otter	ADK011-layer 2	43.82	15.43	2.84	-11.71	10.11	
08Ad10	#158	sea otter	ADK011-layer 2	32.88	11.56	2.84	-11.64	9.79	
08Ad12	#0578	sea otter	ADK011-layer 2	41.35	14.86	2.78	-11.81	9.95	
08Ad13	#150	sea otter	ADK011-layer 2	46.65	16.54	2.82	-10.74	10.82	
08Ad16	#679	sea otter	ADK011-layer 2	44.02	15.32	2.87	-11.97	11.17	
08Ad18	#011	sea otter	ADK011-layer 2	42.57	14.98	2.84	-11.89	11.14	
08Ad19	#0323	sea otter	ADK011-layer 2	45.35	16.06	2.82	-11.71	9.94	
08Ad20	#681	sea otter	ADK011-layer 2	43.59	15.47	2.82	-11.17	11.85	
08Ad21	#0629	sea otter	ADK011-layer 2	42.82	14.83	2.89	-12.89	10.07	-11.83 ±0.53
08Ad15	#592	sea otter	ADK011-layer 3	43.16	15.49	2.79	-11.60	11.25	10.50 ±0.67
08Ad22	#49	sea otter	ADK012-layer 1	44.37	16.01	2.77	-12.61	9.97	
08Ad24	#27	sea otter	ADK012-layer 1	29.59	9.84	3.01	-12.82	8.78	
08Ad26	#25	sea otter	ADK012-layer 1	28.52	10.29	2.77	-12.39	13.60	
08Ad29	#46	sea otter	ADK012-layer 1	41.90	14.81	2.83	-11.57	11.67	
08Ad23	#10	sea otter	ADK012-layer 2	42.30	14.91	2.84	-12.45	13.86	
08Ad25	#73	sea otter	ADK012-layer 2	39.34	13.30	2.96	-12.78	10.82	
08Ad28	#08	sea otter	ADK012-layer 2	38.40	13.72	2.80	-11.61	12.30	-12.60 ±0.93
08Ad27	#77	sea otter	ADK012-layer 3	29.63	9.96	2.98	-14.57	10.14	11.39 ±1.80
08Ad11	#0966	sea otter	unknown	41.76	14.96	2.79	-12.04	11.03	
07Ad01	recent	sea otter	Kuluku Bay, Adak	41.13	14.44	2.85	-10.16	13.19	
07Ad02	#52	minke whale	ADK011-layer 1	39.27	13.01	3.02	-14.07	12.70	
07Ad03	#1088	humpback whale	ADK011-layer 2	41.52	14.65	2.83	-14.49	13.14	
08Ad01	#0922	a beaked whale	ADK011-layer 2	39.40	11.96	3.29	-16.04	16.33	

Table 8.2. $\delta^{13}C$ and $\delta^{18}O$ using tooth enamel carbonate from the sea otter.

O^{18} Sample No.	Lab No.	Cat. No.	Site	Layer	Age	Sampled tooth	$\delta^{13}C$ True PBD	$\delta^{18}O$ True PDB	Average ±SD
S013	08Ad008	#0908	ADK011	1		molar	-10.61	-5.33	-11.49 ± 1/24
S016		#0052	ADK011	1	adult/subadult	upper M1	-12.37	-5.87	-5.60 ± 0.38
S09	08Ad011	#0096	ADK011		adult	molar	-11.29	-5.12	
S011		#1729	ADK011	2	adult	molar	-9.48	-4.87	
S012		#5669	ADK011	2	adult/subadult	upper left PM2	-9.75	-5.16	-10.39 ± 0.90
S015		#0788	ADK011	2	adult	R. canine upper tooth	-11.03	-6.30	-5.36 ± 0.63
S01		#5-#11	ADK012	2	subadult	upper canine	-11.85	-5.87	
S02		#15 1/8"-#13	ADK012	2	adult	lower M2	-10.22	-6.09	
S03		#39-029	ADK012	2	adult	upper canine (right)	-10.25	-5.15	
S04		#61-#038	ADK012	2	juvenile	lower canine (right)	-10.50	-5.99	
S05		#6 1/8-#85	ADK012	1	juvenile	lower M1 (deciduous)	-10.47	-5.04	-10.62 ± 0.62
S06		#76 1/8-#091	ADK012	2	juvenile	lower M1 (deciduous left)	-10.40	-6.27	-5.74 ± 0.51
S07			ADK177 (1999)	1	juvenile?	incisor	-10.08	-5.45	

1.2‰, ranging from -10.6 to 12.4‰. The samples from ADK-011 Layer 2 gave average $\delta^{13}C$ values of -9.8 ± 1.2‰ ranging from -9.5 to 11.3‰, while sea otter tooth enamel from ADK-012 averaged -10.6 ± 0.6‰ with a range from -10.3 to 11.9‰. The $\delta^{13}C$ value of one sample from ADK-171 (6000 yrs BP) was -10.8‰.

The samples from ADK-011 Layer 1 gave an average $\delta^{18}O$ value of -5.6 ± 0.4‰ with values ranging from -5.3 to -5.9‰. The samples from the ADK-011 Layer 2 averaged –5.3 ± 1.0‰ and ranged from -4.9 to -6.3‰, and those from ADK-012 averaged -5.7 ± 0.5‰ with values ranging from -5.0 to -6.3‰. The $\delta^{18}O$ value of one sample from ADK-171 was -5.5‰.

The $\delta^{13}C$ and $\delta^{18}O$ cross-plots (Figure 8.2) suggest that the tooth samples, like the collagen samples, can be divided into two groups. The first group has a domain of -8 to -12‰ for $\delta^{13}C$ and around -5‰ for $\delta^{18}O$. The second group's domain is from –12.5 to –10‰ for $\delta^{13}C$ and around -6‰ for $\delta^{18}O$.

Discussion

In marine mammal diet analysis using tooth dentine (e.g., Walker and Macko 1999), a ^{13}C depleted signal has been typically documented for harp seals (mean -15.1‰), common dolphins (mean -14.5‰), and bottlenose dolphins (mean -13.9‰) in the western North Atlantic. The walrus and killer whales from the Pacific Ocean have $\delta^{13}C$ values of -13.3‰ and -12.9‰, respectively. The teeth of sea otters, as well as pygmy sperm whales and manatees, are more enriched in ^{13}C relative to the other marine mammals. Manatees consume large amounts of sea grasses, known to be enriched in ^{13}C (Harrigan *et al.* 1989) relative to phytoplankton. Thus, the enriched values, of around -11.9 to -9.5‰, for the Adak sea otter teeth could result from otters feeding in the kelp ecosystem. In comparison, the ^{13}C-depleted values of the harp seals and the dolphins reflect the signal of the fish that they eat, which ultimately originates from phytoplankton.

Our analyses of bone collagen show that the prehistoric sea otters on Adak Island represent two specific groups. The higher $\delta^{13}C$ values of the first group (around -11 to -13‰) correspond to a typical range for sea mammals depending on the kelp ecosystem, as discussed by Walker and Macko (1999). The second group has isotopic signatures similar to the oceanic minke and humpback whales, suggesting that they also thrived on small fish within the oceanic ecosystem (Table 8.1).

Differences in food preference are also reflected in the $\delta^{15}N$ compositions for marine mammals. According to Walker and Macko (1999), the herbivore manatees, found stranded in Florida, had the lowest $\delta^{15}N$ values (mean of 7.8‰). Walker and Macko (1999) found that the average $\delta^{15}N$ values for common dolphins and pygmy sperm whales in the Atlantic and Pacific Oceans were 12.4‰. The Pacific Ocean sea otters (14.5‰), and walruses (14.6 ‰), and

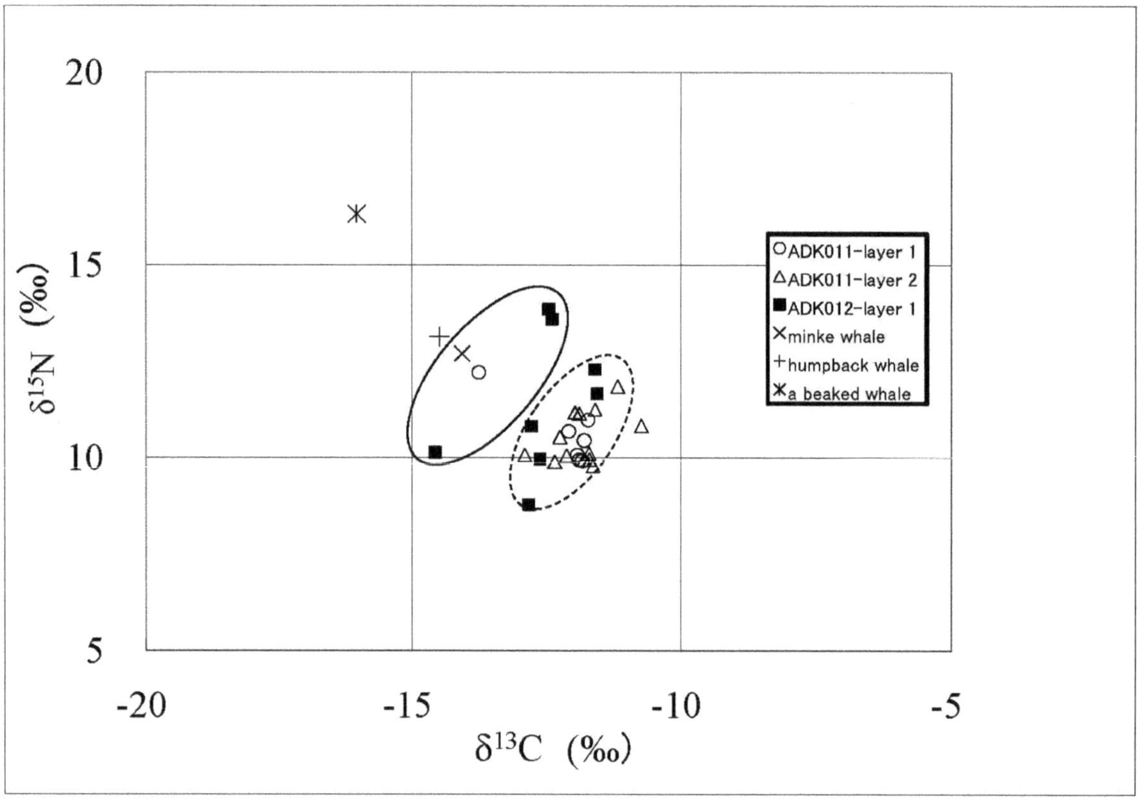

Figure 8.1. $\delta^{13}C$ and $\delta^{15}N$ cross plots of bone collagen from the sea otter and whales.

the Atlantic Ocean bottlenose dolphins were 14.8 ‰; all had $\delta^{15}N$ values typical of carnivores at a higher trophic level. As we suspected from their collagen $\delta^{13}C$ values, it is most likely that our first group of sea otters, with the lower values, depended on the kelp ecosystem or a similar ecosystem, while individuals of the second group were fish consumers. This interpretation is supported by our collagen $\delta^{15}N$ values, values that are known to increase with trophic levels. The second group has slightly higher $\delta^{15}N$ values, suggesting that they ate more fish than the first group members. In comparison, a high $\delta^{15}N$ value (17.6‰) was reported for killer whales that eat carnivorous marine mammals (Walker and Macko 1999), and we found a comparable value for the beaked whale whose main food is deep-sea squid.

Newsome *et al.* (2009) analyzed keratin $\delta^{13}C$ and $\delta^{15}N$ values of vibrissae of California sea otters. The isotopic values were within a narrow range for many analyzed individuals, and nearly constant along the length of a single whisker; this suggests that the relative proportion of different diet items of these animals did not vary substantially over time. However, seven of the 31 analyzed individuals showed large intra-vibrissa variability, which indicates that these individuals had changed their diets temporally. Four animals showed a shift from sea urchin/mussels to kelp crab (Newsome *et al.* 2009). We also found a wide variation in $\delta^{13}C$ and $\delta^{15}N$ value, but for canine teeth, which is associated not only with trophic level differences, but also represents dietary variation within the ecosystem, likely between sea urchin/mussels in kelp and oceanic, small fish.

Clementz and Koch (2001) reported that marine cetaceans exhibit remarkably little $\delta^{18}O$ variability (±0.2 to ±0.4‰) for tooth enamel carbonate, which is consistent with their expectation that aquatic mammals in isotopically homogeneous bodies of water should show low variability. Pinnipeds (±0.3 to ±0.4‰), sea otters (±0.6‰) and river otters (±0.3‰) also exhibited small $\delta^{18}O$ variability in similar situations, while estuarine otters showed a larger variability (±0.9‰). Our results show the sea otters had relatively high variability (±0.4 to ±0.6‰), perhaps due to reservoir temperature fluctuations, that would have affected dietary species values, over a number of archaeological periods. We can evaluate the fluctuations using a calculation of the expected $\delta^{18}O$ value of tooth calcite of sea otters. Fractionation of oxygen isotope between water ($\delta^{18}O_{SMOW}$) and calcite ($\delta^{18}O_{PDB}$) is principally controlled by temperature according to the following equation (Clayton *et al.* 1975):

$$T(^{\circ}C) = 15.7 - 4.36(\delta^{18}O_{PDB} - \delta^{18}O_{SMOW}) + 0.12(\delta^{18}O_{PDB} - \delta^{18}O_{SMOW})^2$$

Assuming that body temperatures and $\delta^{18}O_{SMOW}$ were 38°C and 0‰, the expected $\delta^{18}O_{PDB}$ value would be –4.55‰, roughly corresponding to the highest value among our analyzed specimens (-4.87‰). The mixing of freshwater from river or snowmelt might cause such a small difference.

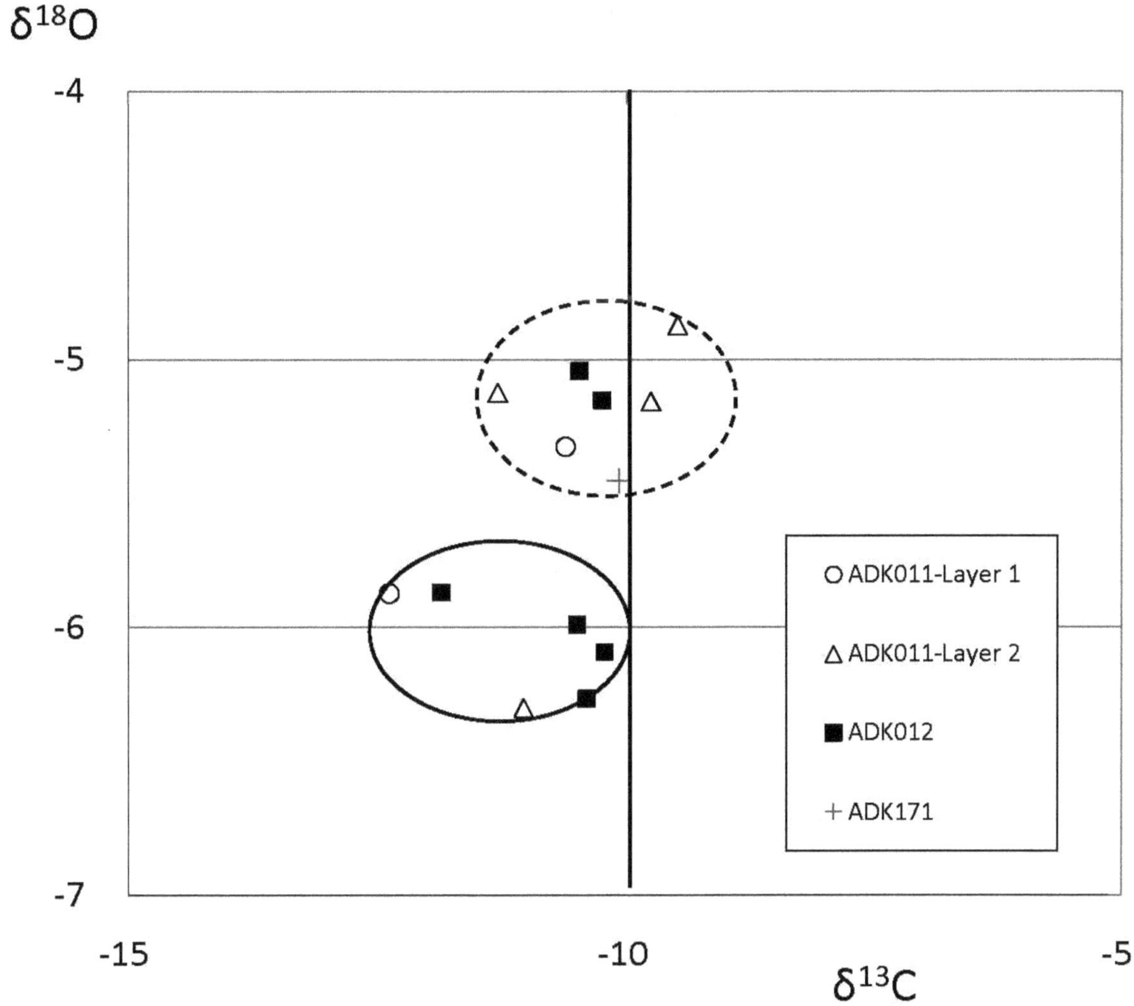

Figure 8.2. $\delta^{13}C$ and $\delta^{18}O$ cross plots of tooth enamel from the sea otters.

Our $\delta^{18}O$–$\delta^{13}C$ cross-plots of tooth carbonate represent two groups (Figure 8.2). As we suspected in the collagen $\delta^{15}N$-$\delta^{13}C$ cross plots (Figure 8.1), two groups can be ascribed to food differences between the kelp ecosystem and the oceanic ecosystem. If so, the groupings appearing in Figures 8.1 and 8.2 correspond to each other. Thus, we suggest that a variation in $\delta^{13}C$ values of tooth calcite basically inherits a variation in the value of metabolic CO_2 and therefore involves ecological information on the food for sea otters.

An Aleutian refuge for sea otter populations is found at Adak Island in the Andreanof group of the Aleutian Islands, Alaska. Although sea otter populations on the outer coasts of Adak have seen an appreciable decline (Estes *et al.* 1998), the sea otter population in Clam Lagoon, a body of water on northeast Adak Island, nearly closed off from the ocean, appears to be in equilibrium. The closed lagoon system may protect sea otters from killer whale predation. According to Bodkin *et al.* (2004), who compared foraging data for current sea otters: 1) within Clam Lagoon and 2) on the outer coastline of Adak, the two groups' prey compositions were different. At Adak the preferred food items for both lagoon and outer coast otters were bivalves. Clam Lagoon otters fed on butter clams, and outer coast otters fed on razor clams. Outer coast sea otters regularly ate green sea urchins and periodically consumed very large, high caloric prey items including flatfishes and octopi. Our stable isotope evidence suggests that some prehistoric Adak sea otters fed in the kelp forest, but others also ate different foods, in some cases, oceanic fish.

Acknowledgements

We are deeply grateful to Dr. Dixie West at the University of Kansas, Dr. Susan Crockford at the University of Victoria for their kindness in providing the samples used in this study, and Dr. Brian Chisholm at the University of British Columbia for his valuable suggestions on our isotope analysis. Vernon Byrd, Alaska Maritime National Refuge provided useful USFWS reports and published data for this paper.

References

Bodkin James L., Morgan A. Robertson, and Angela Doroff. 2004. Sea Otter Foraging Behavior and Population Status in the Aleutian Islands and the Alaska Peninsula. Report on file, US Geological Survey, Anchorage, Alaska.

Bryant Daniel J., Paul L. Koch, Philip N. Froelich, William J. Showers and Bernard J. Genna. 1996. Oxygen Isotope Partitioning Between Phosphate and Carbonate in Mammalian Apatite. *Geochim Cosmochim Acta* 60, 5145–5148.

Clayton Robert N., Julian R. Goldsmith, Karin J. Karel, Toshiko K. Mayeda and Robert C. Newton. 1975. Limits on the Effect of Pressure on Isotopic Fractionation. *Geochimica et Cosmochimica Act* 39, 1197–1201.

Clementz, Mark T. and Paul L. Koch. 2001. Differentiating Aquatic Mammal Habit and Foraging Ecology with Stable Isotopes in Tooth Enamel. *Oecologia* 129, 461–472.

Craig, Harmon and L. I. Gordon. 1965. Deuterium and Oxygen-18 Variations in Ocean and the Marine Atmosphere. In E. Tongiorgi (ed.), *Proceedings of a Conference on Stable Isotopes in Oceanographic Studies and Paleotemperatures*, 277–374. Kingston, University of Rhode Island.

Crockford, Susan. 2011. Archaeozoology of Adak Island: 6000 Years of Subsistence History in the Central Aleutians, Chapter 6 in D. West, V. Hatfield, E. Wilmerding, C. Lefèvre and L. Gualtieri (eds), *The People Before: The Geology, Paleoecology and Archaeology of Adak Island, Alaska*. Oxford, British Archaeological Reports.

Degens, Egon T., Robert R. L. Guillard, William M. Sackett and Johan A. Hellebust. 1968. Metabolic Fractionation of Carbon Isotopes in Marine Plankton. I. Temperature and Respiration Experiments. *Deep-Sea Research* 15, 1–9. http://www.iucnredlist. org/apps/redlist/details/7750/0

DeNiro, Michael J, and Samuel Epstein. 1978. Influence of Diet on the Distribution of Carbon Isotopes in Animals. *Geochimica et Cosmochimica Acta* 42, 495–506.

DeNiro, Michael J, and Samuel Epstein. 1981. Influence of Diet on the Distribution of Nitrogen Isotopes in Animals. *Geochimica et Cosmochemica Acta* 45(5), 657–660.

Doroff, Angela M., James Estes, M. Tim Tinker, Douglas M. Burn, and Thomas J. Evans. 2003. Sea Otter Population in the Aleutian Archipelago. *Journal of Mammalogy* 84, 55–64.

Doroff, Angela and Alexander M. Burdin. 2008. *Enhydra lutris*. In 2008 IUCN Red List of Threatened Species, IUCN 2008.

Dunton, Kenneth H., Suan M. Saupea, Alexander N. Golikov, Donald M. Schell, and Susan V. Schonberg. 1989. Trophic Relationships and Isotopic Gradients Among Arctic and Subarctic Marine Fauna. *Marine Ecology Progress Series* 56, 89–97.

Estes, James A. 1990. Growth and Equilibrium in Sea Otter Populations. *Journal of Animal Ecology* 59, 384–401.

Estes, James A., R. J. Jameson, and E. B. Rhode. 1982. Activity and Prey Selection in the Sea Otter: Influence of Population Status on Community Structure. *American Naturalist* 120, 242–258.

Estes, James A., M. Tim Tinker, Terrie M. Williams, and Daniel F. Doak. 1998. Killer Whale Predation on Sea Otters Linking Oceanic and Near Shore Ecosystems. *Science* 282 (5388), 473–476.

Evans, T. J., D. M. Burn, and A. R. DeGange. 1997. Distribution and Relative Abundance of Sea Otters in the Aleutian Archipelago. *US Fish and Wildlife Service, Marine Mammals Management Technical Report MMM* 97-5, 1–29.

Fredriksen, Stein. 2003. Food Web Studies in a Norwegian Kelp Forest Based on Stable Isotope ($\delta^{13}C$ and $\delta^{15}N$) Analysis. *Marine Ecology Progress Series* 260, 71–81.

Friedman, Irving and James R. O'Neil. 1977. Compilation of Stable Isotope Fractionation Factors of Geochemical Interest. In M. Fleischer (ed.), *Data of Geochemistry*, Sixth Edition,1–12. US Geological Survey Professional Paper 440-KK, Washington, DC, US Government Printing Office.

Fry, Brian. 1988. Food Web Structure on Georges Bank from Stable C, N, and S Isotopic Compositions. *Limnology Oceanography* 33 (5), 1182–1190.

Fry, Brian and Evelyn B. Sherr. 1984. $\delta^{13}C$ Measurements as Indicators of Carbon Flow in Marine and Freshwater Ecosystems. *Contributions in Marine Science* 27, 13–47.

Harrigan, P., J. C. Zieman, and Stephen A. Macko. 1989. The Base of Nutritional Support for the Gray Snapper (*Lutjanus griseus*): An Evaluation Based on a Combined Stomach Content and Stable Isotope Analysis. *Bulletin of Marine Science* 44, 65–77.

Hobson, Keith and Harold E. Welch. 1992. Determination of Trophic Relationships Within a High Arctic Marine Food Web Using $\delta^{13}C$ and $\delta^{15}N$ Analysis. *Marine Ecology Progress Series* 84, 49–18.

Horai, Satoshi, Kenji Hayasaka, Kumiko Murayama, Noriyuki Wate, Hiroko Koike, and Nouyuki Nakai. 1989. DNA Amplification from Ancient Human Skeletal Remains and Their Sequence Analysis. *Proceedings of the Japan Academy* 65 (B) No. 10, 229–233.

Hori, Masako, Chizuru Takashima, Jun Matsuoka, and Akihiko Kano. 2009. *Carbon and Oxygen Stable Isotopic Measurements of Carbonate and Water Samples Using Mass Spectrometer with Gas Bench.* Bulletin of the Graduate School of Social and Cultural Studies, Kyushu University 15, 51–57.

Jennings, Simon, Karema J. Warr, and Steve Mackinson. 2002. Use of Size-based Production and Stable Isotope Analyses to Predict Trophic Transfer Efficiencies and Predator-prey Body Mass Ratios in Food Webs. *Marine Ecology Progress Series* 240, 11–20.

Kenyon, K. W. 1969. The Sea Otter in the Eastern Pacific Ocean. *North American Fauna* 68, 1–352.

Lee-Thorp, J., J. Sealy, J. and N. van der Merwe, N. 1989. Stable Carbon Isotope Ratio Differences Between Bone Collagen and Bone Apatite, and Their Relationship to Diet. *Journal of Archaeological Science* 16, 585–599.

Longinelli, Antonio. 1984. Oxygen Isotopes in Mammal Bone Phosphate: a New Tool for Paleohydrological and Paleoclimatological Research? *Geochimica et Cosmochimica Acta* 48, 385–390.

Love, John A. 1992. *Sea Otters*. Golden, Colorado, Fulcrum Publishing.

Luz, Boaz, Yehoshua Kolodny and Michal Horowitz. 1984. Fractionation of Oxygen Isotopes Between Mammalian Bone-phosphate and Environmental Drinking Water. *Geochimica et Cosmochimica Acta* 48(8), 1689–1693.

Minagawa, Masao and Eitaro Wada. 1984. Stepwise Enrichment of ^{15}N Along Food Chains: Further Evidence and the Relation Between $\delta^{15}N$ and Animal Age. *Geochimica et Cosmochimica Acta* 48 (5), 1135–1140.

Newsome, Seth D., M. Tim Tinker, Daniel H. Monson, Olav T. Oftedal, Katherine Ralls, Michelle M. Staedler, Marilyn L. Fogel, and James A. Estes. 2009. Using Stable Isotopes to Investigate Individual Diet Specialization in California Sea Otters (*Enhydra lutris nereis*). *Ecology* 90(4), 961–974.

Nowak, Ronald M. 1999. *Walker's Mammals of the World*. Baltimore and London, The Johns Hopkins University Press.

Peterson, Bruce J. and Brian Fry. 1987. Stable Isotopes in Ecosystem Studies. *Annual Review of Ecology and Systematics* 18, 293–320.

Peterson, Bruce J., Robert W. Howarth, and Robert H. Garritt. 1985 Multiple Stable Isotopes Used to Trace the Flow of Organic Matter in Estuarine Food Webs. *Science* 227(4692), 1361–1363.

Reidman, M. L. and J. M. Estes. 1990. The Sea Otter ((*Enhydra lutris*): Behavior, Ecology and Natural History. United States Fish and Wildlife Service Biological Report 90, 1–126.

Roe, Lois J, J. G. M. Thewissen, Jay Quade, James R. O'Neil, Sunil Bajpal, Ashok Sahni, and S. Taseer Hussain. 1998. Isotopic Approaches to Understanding the Terrestrial-to-marine Transition of the Earliest Cetaceans. In J. G. M. Tewissen (ed.), *The Emergence of Whales*, 399-422. New York, Plenum Press.

Sackett, William M., Walter R. Eckelmann, Michael L. Bender, and Allan W. H. Bé. 1965. Temperature Dependence of Carbon Isotope Composition in Marine Plankton and Sediments. *Science* 143(3667), 235–237.

Schoeninger, Margaret J., and Michael J. DeNiro. 1984. Nitrogen and Carbon Isotopic Composition of Bone Collagen from Marine and Terrestrial Animals. *Geochimica et Cosmochimica Acta* 48(4), 625–639.

Schrope, Mark. 2007. Food Chains: Killer in the Kelp. *Nature* 445(7129), 703–705.

Sealy, Judith C., Nikolaas J. van der Merwe, Julia A. Lee Thorp, and John L. Lanham. 1987. Nitrogen Isotopic Ecology in Southern Africa: Implications for Environmental and Dietary Tracing. *Geochimica et Cosmochimica Acta* 51(10), 707–2717.

Wada, Eitaro, Masao Minagawa, Hiroshi Mizutani, Takashi Tsuji, Reiko Imaizumi and Kyoko Karasawa. 1987. Biogeochemical Studies on the Transport of Organic Matter along the Otsuchi River Watershed, Japan. *Estuarine, Coastal and Shelf Science* 25(3), 321–336.

Wada Eitaro, Hiroshi Mizutani, and Masao Minagawa. 1991. The Use of Stable Isotopes for Food Web Analysis. *Critical Reviews in Food Science and Nutrition* 30(4), 361–371.

Walker, Jane L. and Stephen A. Macko. 1999. Dietary Studies of Marine Mammals Using Stable Carbon and Nitrogen Isotopic Ratios of Teeth. *Marine Mammal Science* 15(2), 314–334.

Van der Zanden, M. Jake and Joseph B. Rasmussen. 1999. Primary Consumer $\delta^{15}N$ and $\delta^{13}C$ and the Trophic Position of Aquatic Consumers. *Ecology* 80 (4) 1395–1404.

Chapter 9
The Aleut Cockle, *Clinocardium nuttallii* (Conrad, 1837) from ADK-011, Adak Island, Alaska

Hiroko Koike, Chizuru Takashima and Akihiro Kano

Abstract: *Cockles from archaeological site ADK-011, Component 2 (170-400 years BP) are compared with modern cockles at Clam Lagoon, Adak Island, Alaska. Growth ring and stable isotope analyses reveal that prehistoric cockles, in nutrient rich environment, grew faster than cockles in the region today. Prehistoric cockles are larger and older than extant cockles, which could infer a comparatively rich ecosystem or that human and sea otter predation pressure on cockles was low when ADK-011 was occupied. Seasonality studies indicate that Aleuts at ADK-011 collected cockles from early spring through early autumn.*

Keywords: *Clinocardium, stable isotope analysis, growth rates, seasonality*

Introduction

Clinocardium nuttallii, known as the basket cockle, heart cockle, Nuttall's cockle, or Aleut cockle, is distributed along the northern Pacific coast from the Bering Sea to San Diego, California. It is also found in northern Japan. This bivalve inhabits shallow, intertidal zones or subtidal zones to 200m deep. Its habitat is usually fine sediment, especially muddy fine sand, where it is shallowly buried. The cockle has a thick shell, with more than 30 (often around 35) distinct radial ribs covering the entire valve. This fast growing cockle can reach over 15cm in diameter.

Cockle Shells from ADK-011

Located on the northeastern side of Adak Island, Clam Lagoon currently opens into Kuluk Bay through a narrow inlet at Candlestick Bridge. We sampled cockle shells, representing recently deceased individuals, swept onto the beach by waves just outside the Candlestick Bridge inlet. For comparison we also sampled ancient cockles from strata dated between 170 and 400 years old at archaeological site ADK-011, a relatively large Aleut village situated in low lying hills overlooking the Bering Sea on the northeast side of Clam Lagoon. Archaeological cockles came from Layers 1 and 2 in units D, E, F, G, H and I at ADK-011. The shell layer represents a shellfish and bone midden (Units D-F) approximately 40cm deep at its maximum located adjacent to a prehistoric Aleut house pit (Units G-1).

During 2006, Dave Brewer, teacher at the Adak elementary school sided, measured, weighed, and recorded all of the cockle shells (Table 9.1). A total of 1156 cockle shells were represented by 96 complete shells, 603 shells identified as either right or left valves, and 406 nearly complete valves. Cockle shell height from Layer 1 (N=158) and Layer 2 (N=24) averaged 107.6±12.2mm and 105.9±19.2mm respectively. Although no significant statistical differences exist between cockles from the two layers, cockles from Layer 1 were slightly larger than those from Layer 2. Shell length measurements were also slightly larger in layer 1 (110.9±12.0mm) than layer 2 (108.7±17.6mm). Shell weights were also similar between Layer 1 (110.8±2.4g) and Layer 2 (109.7±8.3g). Statistics suggest that the Layer 1 and Layer 2 cockles represent shell debris from a single human occupation component.

Oxygen Isotope Analysis

Clear winter bands were observed through growth line analysis (Figure 9.1). Such winter bands were also observed on the shell surface. To cross check between the growth signature and its season, oxygen and carbon isotopes ($\delta^{18}O$ and $\delta^{13}C$) were measured by gas mass spectrometry (Finnigan Delta Plus) attached to a gas separation system (Thermo Gas Bench. Detailed methods are described in Chapter 8 (Garong *et al.* Chapter 8 this volume).

Cockle shell is composed of aragonite, a form of calcium carbonate. The $\delta^{18}O$ values in the carbonate reflect seawater temperatures (Epstein *et al.* 1953), according to the aragonite sea temperature scale (Horibe and Oba 1972). The $\delta^{13}C$ value also changes with temperature, but this change is not as large as $\delta^{18}O$. The $\delta^{13}C$ reflects salinity or density of seawater (Spero *et al.* 1997).

Samples for the oxygen isotope measurements were collected with a 1mm diamond-coated ball-type dental drill. Winter bands appear thick and swollen. Summer bands are thinner due to rapid growth. Figure 9.2 shows three winter and two summer growth bands on a cockle shell. The first winter band, in general, is not as clear as the later winter band.

Total # of Clams:	
1) Whole Clams	96
a) minus missing whole clams (1/2 shells)	-3
2) Identified Clams(L or R)	627
a) minus abs(R - L)/2	-24
3) Est.# clams (Pieces)	460
TTL # of Clams	**1156**

(This does not account for the pieces missing from the clams identified as Left or Right but were missing part of the 'whole clam')

Greatest Length - Layer 1	
Mean	10.77
Standard Error	0.11
Median	10.90
Mode	11.00
Standard Deviation	1.22
Sample Variance	1.49
Kurtosis	12.40
Skewness	-2.43
Range	9.50
Minimum	3.70
Maximum	13.20
Sum	1432.10
Count	133.00
Confidence Level(95.0%)	0.21

Greatest Length - Layer 2	
Mean	10.59
Standard Error	0.39
Median	11.00
Mode	11.00
Standard Deviation	1.92
Sample Variance	3.67
Kurtosis	14.70
Skewness	-3.42
Range	10.30
Minimum	2.50
Maximum	12.80
Sum	254.20
Count	24.00
Confidence Level(95.0%)	0.81

z-Test: Two Sample for Means
Greatest Length

	Layer 1	*Layer 2*
Mean	10.82785	10.59167
Known Variance	1.486598	3.673841
Observations	158	24
Hypothesized Mean Diffe	0.176	
z	0.149298	
P(Z<=z) one-tail	0.440659	
z Critical one-tail	1.644853	
P(Z<=z) two-tail	0.881318	
z Critical two-tail	1.959961	

Table 9.1. Total number of Clinocardium from archaeological site ADK-011 Layers 1 and 2 used in this study.

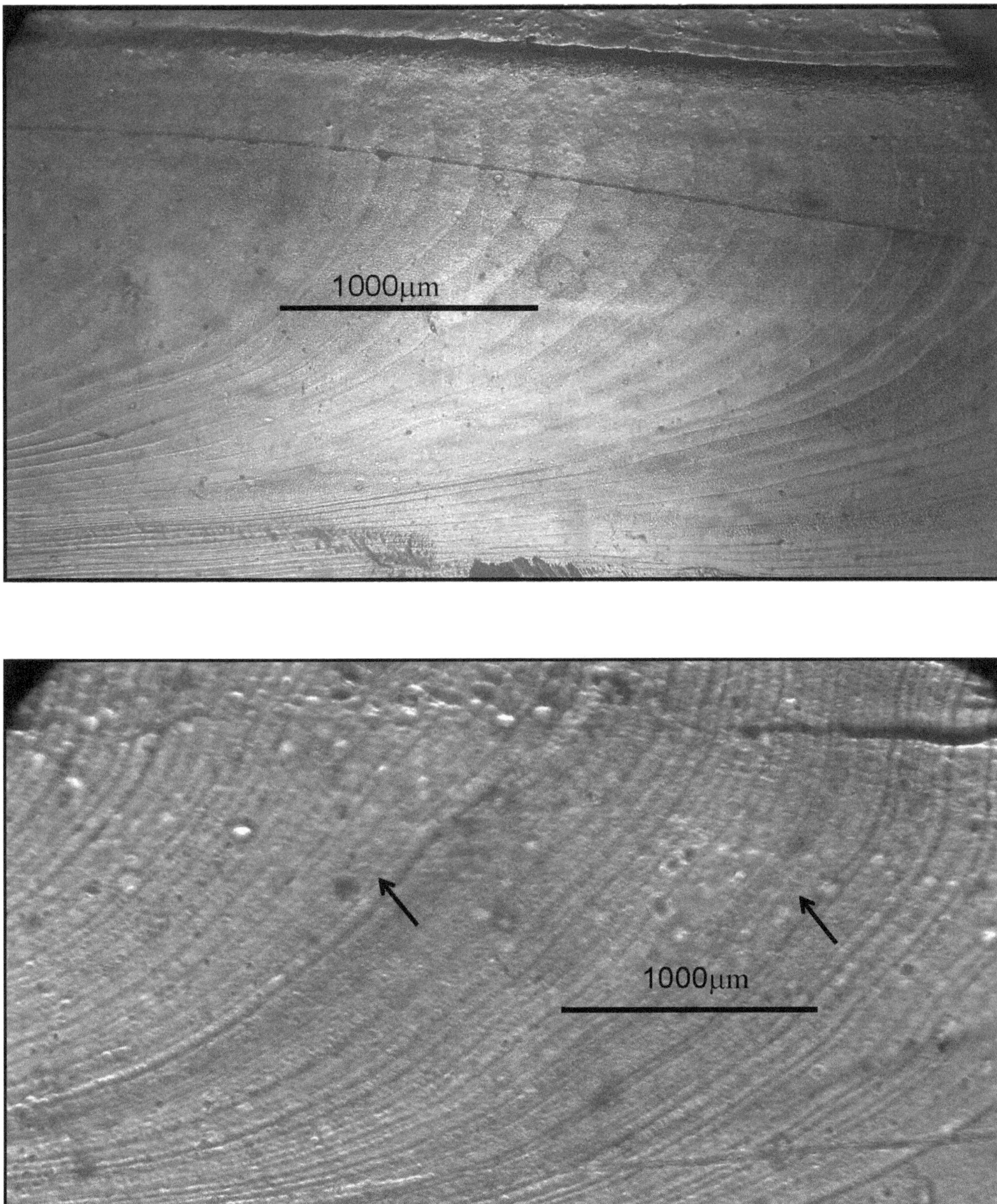

Figure 9.1. Growth bands in Clinocardium. Top photo illustrates summer growth. Bottom photo illustrates winter growth.

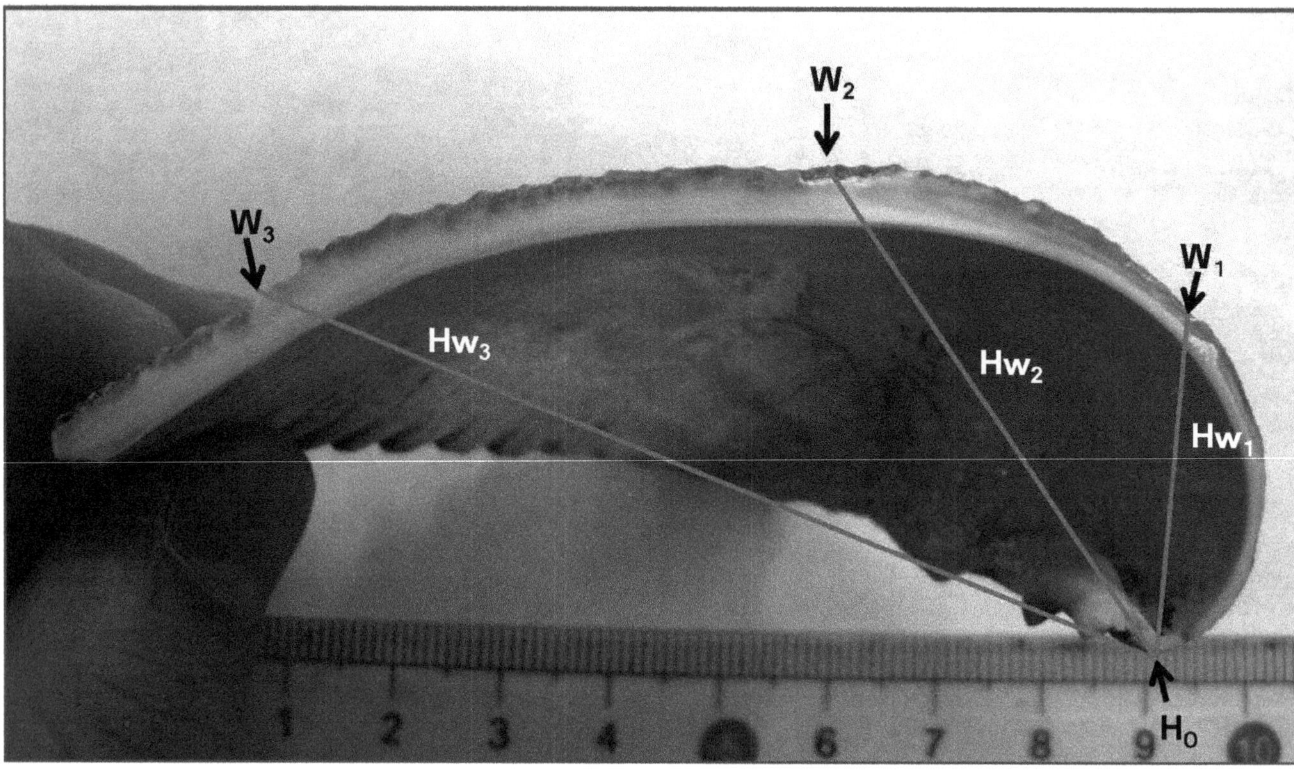

Figure 9.2. Measurements taken on ADK-011 Clinocardium showing shell growth. W1, W2 and W3 illustrate winter bands.

The $\delta^{18}O$ and $\delta^{13}C$ curves of modern and ADK-011 cockles show a clear correlation with growth line analysis (Figure 9.3). $\delta^{18}O$ ranged from -0.42‰ during the second summer and 0.51‰ for the first summer to +1.8‰ during the second winter. The first winter and third winter did not show peaks, yet their values were +1.4‰ and +1.1‰, respectively. This corresponds to seawater temperatures between +8° C to -17° C. The $\delta^{13}C$ showed no correlation with winter bands, but had a gradual trend from 1.0% during the first winter near the umbo, to around 0% after the second winter. This suggests that cockles might migrate from shallow water in estuaries where fresh water slightly mixed with seawater to deeper zones off shore.

Shell Growth Analysis

Most of the ADK-011 cockleshells represent large and relatively old individuals. Shell growth on seven complete shells from ADK-011 and eight individuals collected from the Candlestick Bridge inlet were analyzed.

After shell heights and lengths were measured, shells were cut using a dental mortar with a diamond disk. Cut shells were embedded in polyester resin to avoid destruction of fragile parts. The cut surface of the shell was polished mechanically with carborandum (silicon carbide) sheets, and then etched with 0.1N HCl solution for three minutes. Imprin, or replica film of the etched surface was made using 0.034mm thick cellulose acetate film (Oken Bioden Film). Imprint, or replica film of the etched surface was made using 0.034 mm thick cellulose acetate film (Oken Bioden Film). The imprint films were observed under a microscope at a magnification of 100 to 200 (Koike 1973).

The Adak cockle shell growth appears to be divided into three stages. Growth lines unfortunately are difficult to observe because the outer layer within 5-10mm from the umbo point is usually eroded in older shells. Some of the cockles from ADK-011 exhibited clear burn traces from the middle of shell to around the umbo. Burning could suggest that the Aleuts did not boil cockles, but directly placed the cockles into fires in order to open the valves.

Shell growth in the vicinity of 5mm from the umbo point (Ho) (Figure 9.2) was usually slow; shell growth gradually accelerated and reached a rapid growth stage at about 30-70mm in shell height. Shell growth in this stage is the most rapid in the life history of the cockle, and daily growth increments of 200mm are recorded in the summer growth of this stage. Rapid growth may be due to the rich nutrients in Aleutian waters. Such rapid growth of 200mm/day was also reported in Meterix lamarckii (Chinzei *et al.* 1987). After the stage of maximum growth, the shell growth rate decreased with age. Decrease in growth with age may be due to energy expenditure during reproduction.

In adults, the daily shell growth increment becomes very slow, particularly during winter. This cessation of growth is reflected in narrow growth bands (Figure 9.1 bottom). Declining growth can be recognized as prominent growth ridges on the exterior shell surface. In the Bering Sea, the northern range of the cockle, *Clinocardium* nearly quit

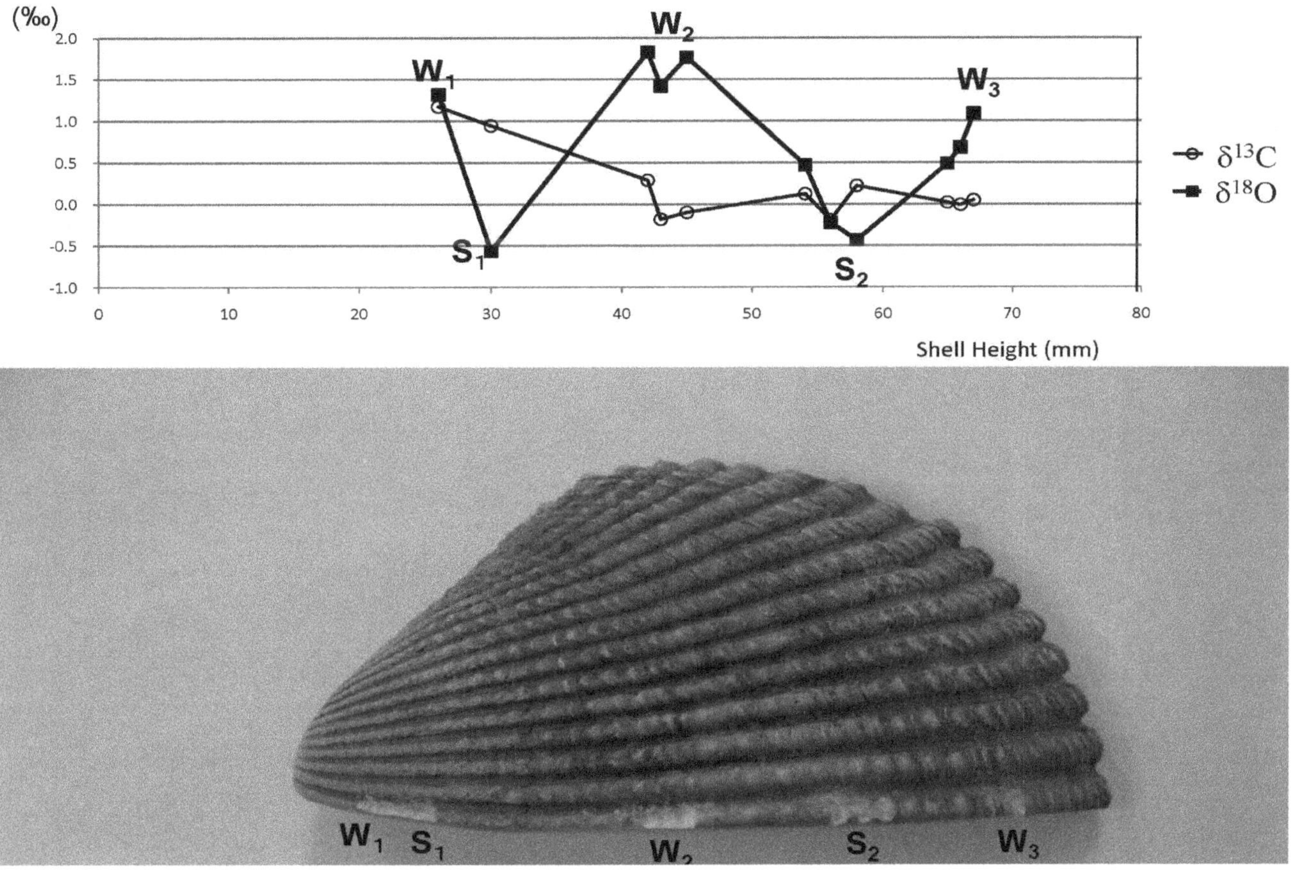

Figure 9.3. $\delta^{18}O$ and $\delta^{13}C$ curves of modern and ADK-011 cockles showing a correlation with growth lines.

feeding during winter, which results in the formation of narrow winter bands. Using a caliper, the shell height at the each winter band was measured from the umbo point to the each winter band (represented as Hwn height at winter band).

For the mathematical treatment of the growth curve, Bertalanffy's (1938) growth formula is widely used in fishery studies. When plotted in natural logarithms against age, the differences between the actual lengths at given ages and the initial length estimated with this formula results in a straight line. In 1946, Walford developed a new graphic method based on the exponential growth formula to obtain the shell length; the annual increase was represented by the differences in the successive shell heights at the annual growth minimum (Hwn). If the annual growth minimum is exponential, Hw_n on the X axis and Hwn-1 on the Y axis are plotted, the regression line will show a straight line. Based on this method, we can directly calculate full age length (eg. 1-year-old length, 2-year-old length, etc.) and length at time of death as intersection points between growth regression line and the Y=X line. The shell age or growth rate was used in comparisons among local groups to observe the shell growth variations in detail (Koike 1982).

Modern Cockles

The annual bands on shell surfaces for the eight modern cockles were calculated using Walford's formula based on the paired data of Hwn and Hwn+1 (Figure 9.4). Estimated one year-old shell height (initial shell height) was 28.5mm, and maximum shell height was estimated as 116.3mm. Growth rate was 0.775; this is very rapid among bivalves. Estimated heights for the 1, 2, 3, 4 and 5 year-olds were 28.5, 50.0, 66.2, 78.5, 87.8mm, respectively.

Archaeological Specimens

Using Walford's formula, annual bands of seven cockle specimens collected from ADK-011 were analyzed (Figure 9.5). Results showed that initial shell height was 39.46 mm; this is astonishingly larger than heights observed in modern Adak cockles. Maximum height (estimated final shell height) was 115.7mm, which was almost the same as modern specimens. Growth rate was 0.66, smaller than modern specimens. Estimated shell heights for one-year-old to five-year-old were 39.46, 65.50, 82.69, 94.04, 101.52mm, respectively. Consequently, shell growth environments for younger phases represented by archaeological specimens were far better than the modern specimens, although both

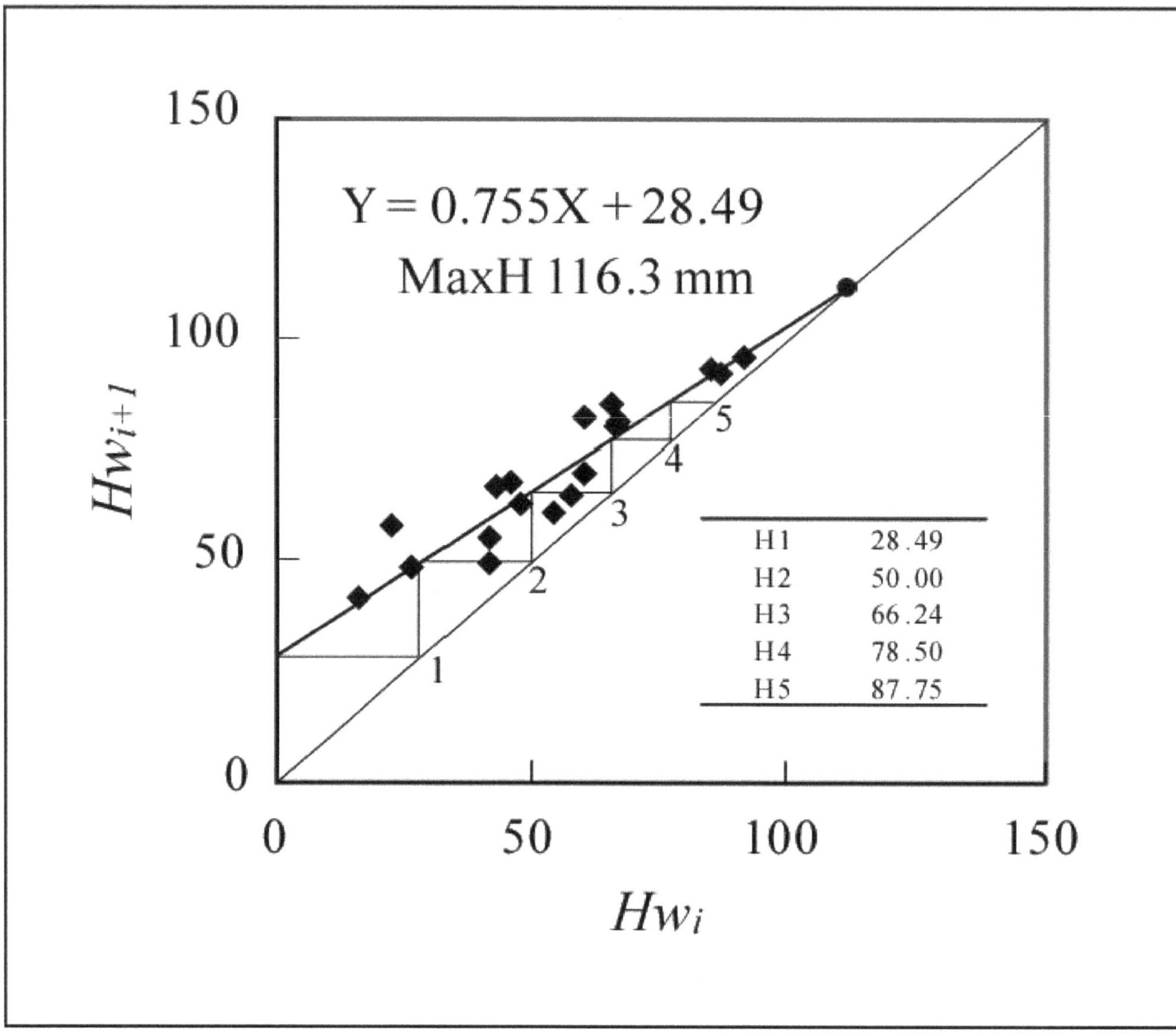

Figure 9.4. Growth of modern Clam Lagoon Clinocardium using Walford (1946).

phases of shells had nearly the same final growth size. The shell height histogram for modern specimens are smaller than that for the archaeological specimens; however, the final shell height calculated by Walford's formula was similar for both the modern and archaeological specimens.

Investigation of shell growth variation showed that a growth formula using natural logarithms is useful for observing details of shell growth patterns. Shells with a high growth rate at a younger age tended to eventually decrease in growth rate, while those exhibiting an early low growth rate experienced a slow increase, approaching normal shell length during later growth stages. The former pattern seems to correspond with specimens that grow in more southern regions and the latter with specimens from northern regions (Koike 1982).

Age Composition

Complete, unbroken cockles were used to estimate age by counting annual growth disturbances. Ages of cockle shells from Layer 1 ranged from four to 16 years old with a mode at seven years old. In Layer 2 cockle ages ranged between four and 15 years old with a mode at eight years old (Figure 9.6). Average and standard deviation of cockle age composition from Layer 1 and Layer 2 were 8.07±2.12 and 8.86±2.41, respectively. Layer 1 produced more than 140 shells, and Layer 2 contained approximately 40 shells. However, age difference of these cockle shells from Layer 1 and Layer 2 were not statistically significant. This suggests that deposition of Layers 1 and 2 occurred nearly simultaneously (Figure 9.6).

Modern shells collected on the shore near Clam Lagoon ranged between four and seven years old with an average of 5.13±1.06 years. Large *Clinocardium* shells are conspicuously absent at Clam Lagoon. Perhaps the oldest clams inhabit deeper ocean levels and are not easily swept on shore. Or, perhaps the largest *Clinocardium* currently no longer grow in the Clam Lagoon area.

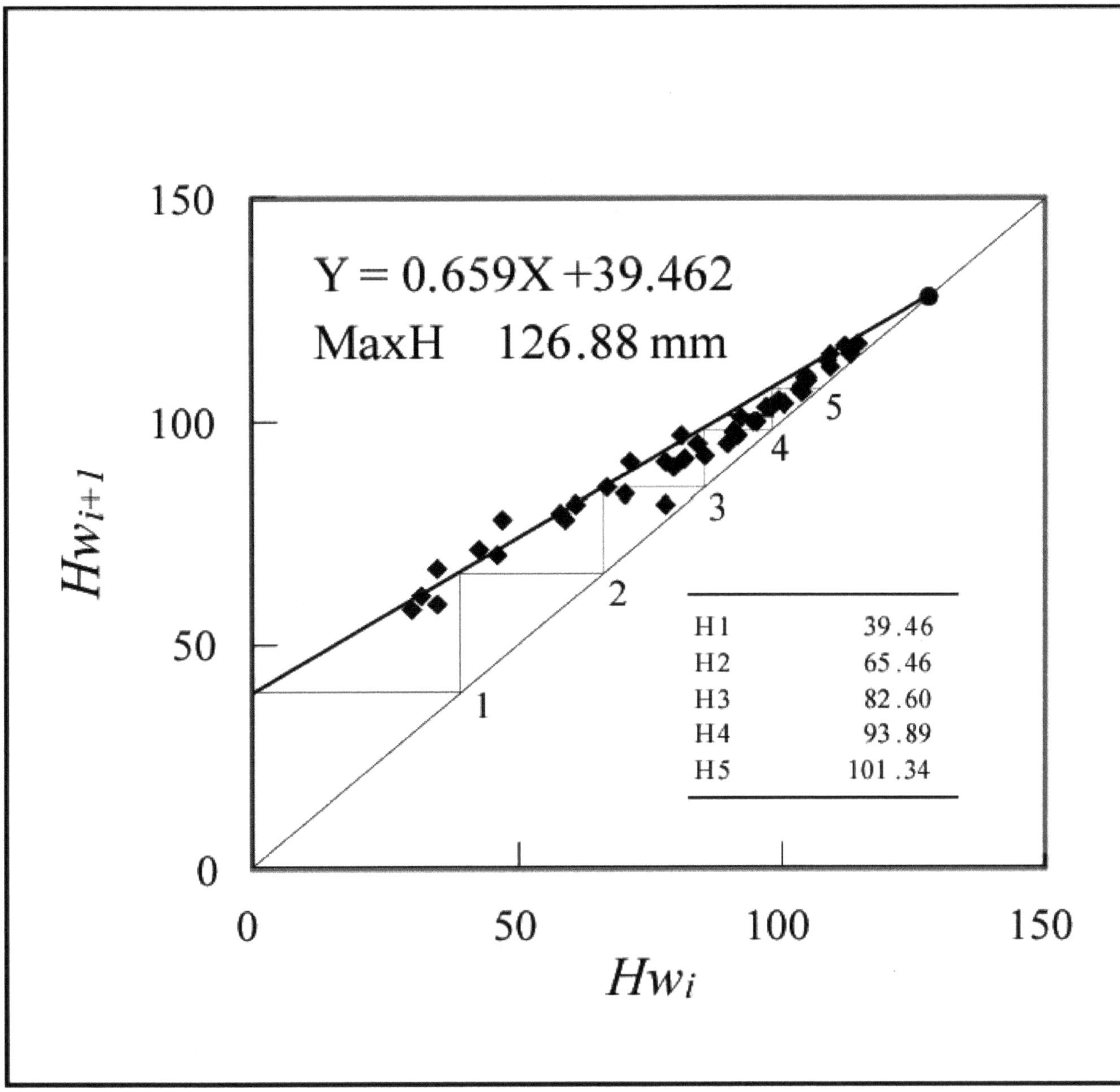

Figure 9.5 Growth of ADK-011 Clinocardium using Walford (1946).

Age composition is an important index to estimate collecting pressure on cockle biomass (Koike 1986). Archaeological specimens exhibited a standard age distribution between 7 and 10 years old, suggesting that when Aleuts inhabited ADK-011 cockles were under a low collecting pressure. Savinetsky *et al.* (Chapter 5 this volume) and Crockford (Chapter 6 this volume) also suggest that Aleuts heavily hunted sea otters, predators of *Clinocardium*. The removal of sea otters from the local ecosystem could have enhanced cockle survival and longevity.

Shell Collecting Season

Cockle shells from ADK-011 were too old and shell margins too eroded to estimate the collecting season by counting growth lines. The shell collecting season was estimated by dividing shell growth into eight categories from the late mid-winter band to the shell edge. The Jomon people in Japan (circa 6000-2500 years BP) collected shellfish most intensively during spring and early summer. During these seasons the large tidal range enabled prehistoric gatherers to collect large shells offshore (Koike 1980). However, cockles from ADK-011 suggest that Aleuts collected *Clinocardium* between early summer and early autumn (Figure 9.7). Perhaps Aleuts ventured into deeper waters during warmer months to collect where the largest *Clinocardium* live.

Discussion

Aleutian ethnographic literature reveals that hunter-gatherers collected and ate shellfish and sea urchins in great quantities (Veniaminov 1984, 279). Some Aleutian

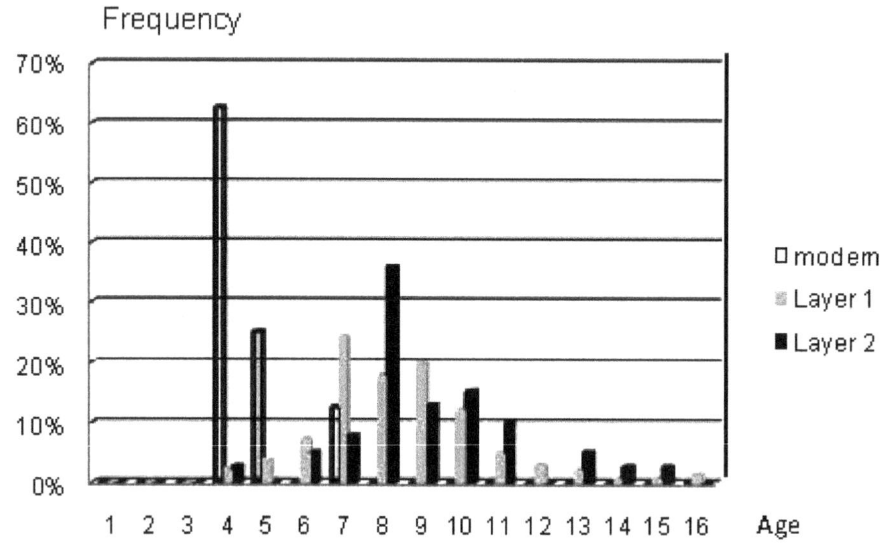

Figure 9.6 Age of modern Clam Lagoon Clinocardium compared with those from ADK-011 Layers 1 and 2.

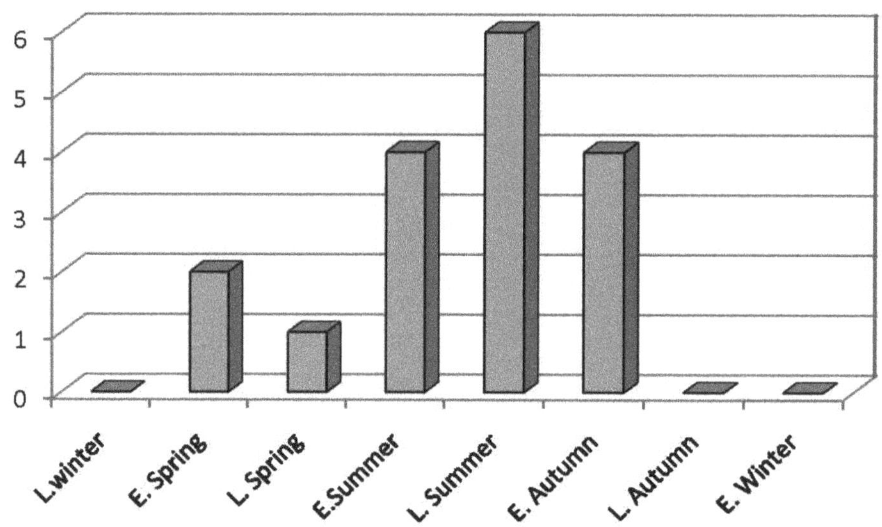

Figure 9.7. Estimation of season of death for Clinocardium at ADK-011.

shell middens measure 6.5m deep (Jochelson 1925, 118) and indicate a continued and great reliance on shellfish in the Aleut economy. With a simple probe to find buried *Clinocardium* and a container for transport, Aleuts could potentially gather many cockles in a comparatively short time with little effort.

Gleaning of the largest cockles from shellfish beds apparently did not reduce the overall size of individual specimens. However, the accumulation at ADK-011 may constitute a single warm season of collection. Savinetsky *et al.* (Chapter 5 this volume) determined that human collecting pressure over time at ADK-009, (archaeological layers that date between 970 to 2080 year BP) did not impact invertebrate biodiversity. Numerous, large cockles could also imply that sea otter populations in the vicinity of ADK-011 were low at the time the site was occupied and when cockles were collected.

Some ADK-011 *Clinocardium* shells were charred, possibly inferring that prehistoric Aleuts opened these bivalves by placing them in a fire. Once heated, the strong muscle of dead *Clinocardium* would have relaxed and the shell could be easily opened. Stone boiling, observed in eastern Aleut villages (Veniaminov 1984, 278), is a possible way to extract meat from large cockles. Stone boiling is labor intensive and requires large amounts of fuel—a rare commodity in the Aleutians. It probably would have been far more efficient to directly heat the live *Clinocardium* to relax the muscle and open the shell rather than immerse it in hot water. Cockle meat also could have been extracted by simply breaking thick shells with cobbles.

Acknowledgements

The authors are grateful to Mr. Dave Brewer, a teacher of Adak Elementary School for his help in measuring cockles. This work was made possible with a grant from the National Science Foundation, Office of Polar Programs, Arctic Social Sciences Division (Award # 0353065)

References

Bertalanffy, L. von. 1938. A Quantitative Theory of Organic Growth (Inquiries on Growth Laws. II). *Human Biology* 10, 181–213.

Chinsei, K., H. Koike, T. Ohba, H. Kitazato and Y. Matsushima. 1987. Secular Changes in the Oxygen Isotope Ratios of Molluscs Shells During the Holocene of the Central Japan. *Paleogeography, Palaeoclimatology, Paleoecology* 61, 155–166

Crockford, Susan. 2011. Archaeozoology of Adak Island: 6000 Years of Subsistence History in the Central Aleutians. Chapter 6, in D. West, V. Hatfield, E. Wilmerding, C. Lefèvre and L. Gualtieri (eds), *The People Before: The Geology, Paleoecology and Archaeology of Adak Island, Alaska.* Oxford, British Archaeological Reports.

Epstein, S., R. Buchsbaum, H. A. Lowenstam, and H. C. Urey. 1953. Revised Carbonate-water Isotopic Temperature Scale. *Geological Society of America Bulletin* 64, 1315–1326.

Garong, Ame, Chiozuru Takashima, Shin Nishida, Akihiro Kano, and Hiroko Koike. 2011. Carbon, Nitrogen, and Oxygen Isotopic Compositions of Bone Collagen and Tooth Enamel Carbonate of the Sea Otter *Enhydra lutris* Archaeological Sites at Adak Island, Aleutian Archipelago. Chapter 8 in D. West, V. Hatfield, E. Wilmerding, C. Lefèvre and L. Gualtieri (eds), *The People Before: The Geology, Paleoecology and Archaeology of Adak Island, Alaska.* Oxford, British Archaeological Reports.

Horibe, H. and T. Ob. 1972. Temperature Scales of Aragonite-water and Calcite-water Systems. *Fossils* 23/24, 69–79.

Jochelson, Waldemar. 1925. *Archaeological Investigations in the Aleutian Islands.* Carnegie Institute of Washington Publication 367. Washington, DC.

Koike, Hiroko. 1973. The Use of Daily and Annual Growth Lines of the Clam Meretrix lusoria in Estimating Seasons of Jomon Period Shell Gathering. In *Quaternary Studies, The Royal Society of New Zealand Bulletin* 13, 189–193.

Koike, Hiroko. 1980. *Seasonal Dating by Growth-line Counting of the Clam, Meretrix lusoria.*The University of Tokyo Bulletin 18. Tokyo, The University Museum.

Koike, Hiroko. 1982. Shell Growth Line Analysis Using *Meretrix* from Hokuriku Area in Japan Sea. *Quaternary Research* 21, 273–282 (in Japanese with English summary).

Koike, Hiroko. 1986. Prehistoric Hunting Pressure and Paleobiomass: An Environmental Reconstruction and Archaeozoologicical Analysis of a Jomon Shellmound Area. In *Prehistoric Hunter- gatherers in Japan.* The University Museum, The University of Tokyo Bulletin No.27, 27–53.

Savinetsky, Arkady B., Dixie West, Zhanna A. Antipushina, Bulat F. Khassanov, Nina K. Kiseleva, Olga A Krylovich, and Andrei M. Perelodov. 2011. The Reconstruction of Ecosystems History of Adak Island (Aleutian Islands) During the Holocene, Chapter 5 in D. West, V. Hatfield, E. Wilmerding, C. Lefèvre and L. Gualtieri (eds), *The People Before:The Geology, Paleoecology and Archaeology of Adak Island, Alaska.* Oxford, British Archaeological Reports.

Spero, H. J., J. Bijma, D. W. Lea, and B. E. Bemis. 1997. Effect of Sea Water Carbonate Concentration on Foraminiferal Carbon and Oxygen Isotopes. *Nature* 390, 497–500.

Veniaminov, Ivan. 1984. *Notes on the Islands of the Unalashka District* [Original Russian edition, St. Petersburg, 1840]. [L. T. Black, and R. H. Geoghegan, trans.]. Ontario, Limestone Press.

Walford L. A. 1946. A New Graphic Method of Describing the Growth of Animals. *Biological Bulletin* (Woods Hole) 90, 141–147.

Chapter 10
Invertebrate Remains: A Pan Aleutian Comparison

Dixie West, Zhanna Antipushina, Arkady B. Savinetsky and Olga Krylovich

Abstract: *Invertebrates from four archaeological sites on Adak Island, Alaska, are described and quantified. These are compared with other mollusk and urchin assemblages across the Aleutian Islands. Dissimilar proportions of recovered invertebrate taxa at different sites do not suggest that Aleuts changed littoral collecting practices over time or across the archipelago. Rather, differences in the percentages of invertebrate species among Aleutian sites reflect the structure of the littoral zone immediately adjacent to each Aleut settlement. It is often accepted that children, women, and old people gathered shellfish. Stable isotope analysis seems to confirm this age and gender based intertidal practice. Aleuts, anticipating late winter food shortages, probably positioned themselves adjacent to highly productive intertidal zones to exploit invertebrate communities when winter stores were depleted.*

Keywords: *Invertebrate analysis, Adak Island, stable isotope analysis*

Introduction

As part of the US Coast Survey in 1871-1874, William Healey Dall (1877) collected extant invertebrates from around, and conducted archaeological excavations on, Atka, Adak, Amchitka, Kiska, Little Kiska, Attu and other islands in the Aleutian chain. Later researchers including Jochelson (1925) and Hrdlička (1945) shoveled through decaying shellfish and sea urchin layers in Aleutian middens to collect the rich inventory of stone, bone and wooden artifacts that were ultimately packed and transported back to museums for study and display. Invertebrate remains, rich in $CaCO_3$, played a critical role in preserving bone and ivory artifacts by buffering acidic volcanic soils at Aleutian sites. Gleaning the best artifacts, early scientists discarded invertebrates in excavation backfill, sometimes only making scant notations of species identified. Their immense quantity and ubiquity in Aleutian archaeological sites are the invertebrates' undoing.

Aleutian shell middens primarily consist of sea urchins, barnacles, and chitons, along with a variety of bivalves (cockles, butter, little neck and razor clams) and gastropod mollusks (whelks, limpets, periwinkles, dogwinkles). Their unassuming, and often shattered, remnants provide important information not only about past Aleut subsistence strategies but also about ecosystem dynamics (Corbett *et al.* 2008; Savinetsky *et al.* 2004; Savinetsky *et al.* Chapter 5 this volume).

As r-selected species (MacArthur and Wilson 1967; Pianka 1970) invertebrates have not undergone population bottlenecks currently plaguing endangered Aleutian marine mammals (e.g., sea otters, northern fur seals, and Steller sea lions) or birds (short tailed albatross, Aleutian cackling goose). Consequently, prehistoric invertebrates are given neither the study nor press of their Aleutian K-selected compatriots. Nevertheless, it is this r-selection adaptation that helps regulate marine ecosystems, and invertebrates comprise a major component of marine food webs. In unpredictable environments, r-selected invertebrates mature early and widely disperse millions of offspring each year. Following a catastrophic environmental event, r-selected species play a pivotal role in regenerating ravaged ecosystems. Opportunists with high reproductive rates, invertebrates are the first to recolonize a devastated marine landscape. Over time, higher trophic K-strategists encroach, feeding on or competing with r-selected species, until a new equilibrium, or climax community, is established. During periodic Aleutian catastrophic events that result in local subsidence or uplift, including volcanoes, earthquakes, or even the underground nuclear tests at Amchitka (O'Clair 1977), invertebrates recover quickly. If uplift or subsidence occurs, and if the appropriate substrate and tidal conditions are available, invertebrates quickly colonize new habitats by moving vertically. Species displacement creates an empty niche that other invertebrate species quickly fill.

This high recovery rate also helps offsets faunal and human predation pressure on invertebrate communities. Savinetsky *et al.* (Chapter 5 this volume) determined that the Aleut traditional use of the littoral resources did not alter or decrease the biodiversity of invertebrate species at ADK-009 on Adak Island. Koike *et al.* (Chapter 9 this volume) proposed that the Aleuts at ADK-011 (component 2) collected extremely large and long-lived *Clinocardium*. Heavy human predation pressure did not apparently impact the bivalve population. Corbett *et al.* (2008, 62) reported that, by and large, the Aleutian ecosystem 'was stable over the entire prehistoric span of human occupation....' The large number of shells and sea urchin remains at many archaeological sites across the Aleutian archipelago demonstrates that invertebrate resilience offered a dependable, although calorie poor, subsistence option for Aleuts for millennia.

Methods

Preservation is irregular among invertebrate taxa. The large, heavy umbo of Nuttall's cockle (*Clinocardium*) remained relatively intact when the bivalve was collected, transported to the place of consumption, opened to extract meat, and discarded in the kitchen refuse pile. Following deposition the thick cockle valve survived sediment acidity and overburden pressure. Under similar taphonomic processes, the more fragile mussel (*Mytilus*) was often pulverized, making its quantification sketchy. Gastropod shell survives well in archaeological sites. Sea urchins are never recovered intact, and researchers must sort mouthparts (Aristotle's lanterns) from tests in order to count sea urchin numbers. Quantification of shell from Aleut kitchen middens is a challenging, time-consuming task. Researchers quantify invertebrates differently. Some calculate number of identified specimens (NISP); others calculate minimum number of individuals (MNI). Some use both NISP and MNI to quantify remains. Some simply report invertebrate taxa occurrence at sites. Level of expertise and adequate comparative collections play an important role in invertebrate identification.

Researchers collected invertebrates by bulk and in water screens using either 1/4 or 1/8 inch mesh in all Adak excavations. Arkady Savinetsky, Zhanna Antipushina, and Olga Krylovich identified invertebrates from ADK-171, ADK-009, and ADK-012. Number of identified specimens (NISP) was used to calculate the number of chitons. Bivalves were quantified using the total number of left and right umbos, barnacles using the number of plates, and brachiopods using the number of valves (Antipushina 2008). The number of sea urchins was determined by counting the total number of Aristotle's lantern, genital plates, and madreporits. Gastropods were quantified by counting shells of which more than half was present (Bird *et al.* 2002). The Russian team aged cockles (*Clinocardium nuttallii*) at ADK-171 by counting growth lines on the anterior adductor muscle scar and on the hinge plate (Figure 10.1). Cockle valve lengths and widths were measured to determine *Clinocardium* sizes (Figure 10.2). Measurements of the complete shells are useful because they provide information on cockle growth rates and longevity of exploited *Clinocardium* as well as cockle season of collection.

Dixie West identified invertebrates from ADK-011. NISP was calculated by counting invertebrate pieces to the lowest identifiable taxon. West calculated minimum number of individuals (MNI) by counting identifiable gastropod apexes, bivalve umbos and/or hinges, chiton plates, and sea urchin mouth parts. Left and right valves of medium and large sized bivalves (e.g., *Clinocardium*, *Mytulis*) were easily distinguished. For these taxa, MNI was based on the

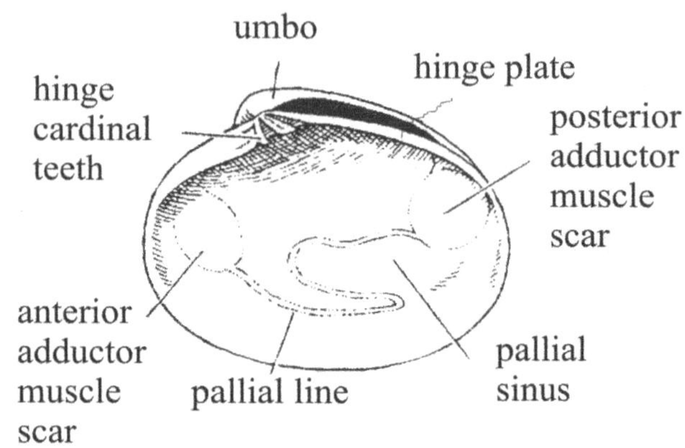

Figure 10.1 Structure of a bivalve shell (after Harbo 2001).

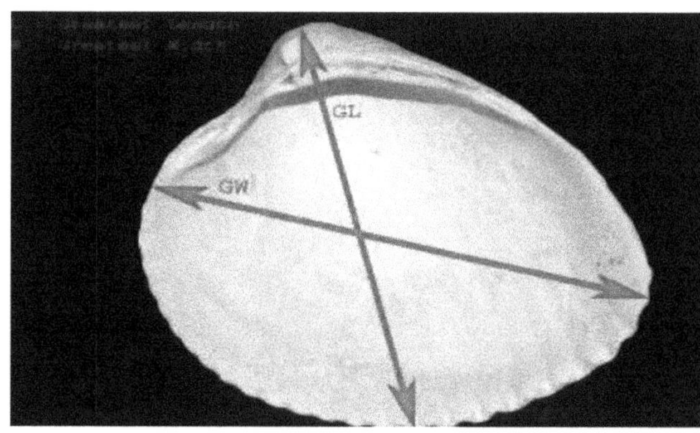

Figure 10.2. Shell measurements taken on Adak Clinocardium. GL = greatest length; GW = greatest width.

higher number of left or right valves or hinge fragments. Small *Mytilus* hinge fragments generally could not be easily separated to left or right, so MNI of small shells was estimated by simply dividing the count of valves containing umbos in half. Chiton MNI was determined by counting the plates and dividing by eight, the number of plates per individual. Urchin (*Strongylocentrotus* sp.) MNI is based on counting jaw parts and dividing by five, the number per individual. Sea urchin test fragments were not counted. Dave Brewer, local Adak school teacher, measured lengths, widths, and heights of complete cockles from ADK-011. Using Brewer measurements, Koike *et al.* (Chapter 9 this volume) determined size and age of ADK-011 cockles. Koike also estimated growth rates and collection season using thin sections and statistical techniques.

Description of Investigated Sites

ADK-171 is located on a bluff overlooking the east side of Clam Lagoon. The prehistoric kitchen refuse area consists of two distinct cultural layers—a loosely packed *Clinocardium* layer containing animal bones and human artifacts separated by a thin lens of redeposited Intermediate volcanic ash from a more densely packed and highly deteriorated mussel shell layer below. Intermediate Ash lies immediately below the midden. A series of radiocarbon dates on midden organics and the Intermediate Ash suggest the ADK-171 midden formed between 6000 and 7000 years ago (Okuno *et al.* Chapter 4 this volume; O'Leary 1998, 2001; Savinetsky *et al.* Chapter 5 this volume).

ADK-012, the Dozered Site, is located at the northern end of a barrier beach separating Clam Lagoon from the open water of Sitkin Sound (Luttrell and Corbett 2000). During 1943 the American military pushed the prehistoric site into a berm while building a road around the lagoon. USFWS archaeologists described the site as an 80m by 20m mound on a 'badly disturbed linear ridge of disturbance vegetation with indistinct features' running parallel to the coastline (Luttrell and Corbett 2000, 32-34). During 2005 Central Aleutians project archeologists excavated a 110cm thick profile in a grassy knoll at the north end of the site. Military bulldozing had displaced all cultural remains above 70cm. These bulldozed layers contained a rich inventory of animal bones, shell, tools and tool making debris. Mitsuru Okuno identified a distinct band of 40 Year Ash 70cm below datum; beneath the volcanic ash, dark brown, peaty strata were intact. Archaeologists opened a 1m X 1m test pit immediately in front of the profile and excavated 30cm below the 40 Year Ash layer. Cultural materials were rare in the 1m X 1m unit. Excavations essentially revealed a 70cm thick deposit of cultural and natural lenses bulldozed into a knoll covering a 30cm thick intact stratum largely devoid of cultural material. Charcoal from the lower, undisturbed zone dated at 1515±15 years BP (UCIAMS #22158). All other dates were taken on organics from the disturbed midden zone. A passerine leg bone recovered from Layer 2 of the upper zone dated to 1865±15 years BP (UCIAMS #22161). Dates from the site taken on other bone range between 2085±15 (UCIAMS #22159) and 2470±15 years BP (UCIAMS #22273). Based on the bird bone, a terrestrial based organism, ADK-012 dates to approximately1865 years BP.

Archaeological site ADK-009 is located near the head of Sweeper Cove on a bedrock shoreline with a gravel beach. ADK-009 is a small site with seventeen features. Fourteen are on an 'elevated mound with abundant midden' overlooking Sweeper Cove and three are on a rocky islet, accessible intertidally, 20m from the shoreline (Luttrell and Corbett 2000, 16). In 1999, Russian historical ecologists excavated a 1m X 1 m test pit, 1.92m deep, in Feature 12 at ADK-009. Cultural and faunal remains were recovered from the 166cm thick cultural deposit. The ADK-009 profile contained five distinctive, cultural layers that accumulated from the beginning of the 6th to the beginning of 17th century (Savinetsky *et al.* Chapter 5 this volume).

Located at Zeto Point, ADK-011 is a comparatively large, multicomponent site with over 100 pit features. The US Bureau of Indian Affairs (O'Leary 1998), US Fish and Wildlife Service (Luttrell and Corbett 2000), and the Western Aleutians Archaeological and Paleobiological Project (West *et al.* 1999) previously investigated the site. CAAPP archaeologists excavated ten 1m X 1m units at the site. Five excavation units (B-F) transected a kitchen midden and five units (G-J) transected a house interior. The midden area (Layers 1 and 2) was dominated by a dense shell and bone deposit containing many stone and bone artifacts. House layers contained a thinner scatter of invertebrate remains. CAAPP archaeologists excavated nine archaeological levels that span from 170-2500 years BP. Based on radiocarbon dates, the ADK-011 kitchen midden and Layers 1-3 house floors formed between 170 and 400 years ago.

Results

ADK-171 Clinocardium Midden Invertebrates

The ADK-171 samples, recovered from a 1m X 1.5m test pit and a cleared profile, represent two cultural layers dating within 30 years of each other. Savinetsky *et al.* (Chapter 5 this volume) provides detailed analyses of invertebrates from each cultural layer at the site. Here, the taxa counts from both cultural layers are combined to provide a broad overview of the ADK-171 invertebrate community. Historical ecologists identified five taxonomic groups: bivalves, gastropods, chitons, barnacles, and sea urchins. A total of 4207 invertebrate remains were analyzed and are presented in Table 10.1. Bivalves dominated among invertebrates in both numbers and weight. By NISP, bivalves constitute 66.5% (N=2799) of all invertebrates, and by weight they represent 99.8% (95.4 kg). *Clinocardium* shells are much heavier than other shells; hence, the high weight number for bivalves at the site. Sea urchins represent 16.3% (N=685) and barnacles represent 14% (N=588) of all invertebrates. Chitons and gastropods are rare, representing only 2.1% (N=87) and 1.1% (N=48) of all ADK-171 invertebrates. Olga Krylovich, (Russian Academy of Sciences) recorded the dimensions and estimated ages of unbroken *Clinocardium* valves from ADK-171. *Clinocardium* ages ranged between 2.5 and 13 years and averaged 6.4 years (N=1123). Shells measured 88-131mm in length with an average of 108mm (N=186). Shell width ranged between 91mm and 128mm with an average of 112mm (N=145).

ADK-012 Invertebrates (1865 BP)

CAAPP researchers identified 7487 invertebrate remains from ADK-012 (Table 10.2). Bivalves dominated among invertebrates, representing 38.3% (N=2867) of all shellfish remains. *Mytilus trossulus* comprises the largest percentage of bivalves, representing 36.7% (N=2749) of all invertebrates and 95.9% of all bivalves. *Clinocardium* (N=105) are comparatively rare. Barnacles (28.9%; N=2166) and echinoderms (25.4%; N=1898) are also found in high numbers. Chitons (5%) and gastropods (3%) are quite rare in the deposits.

Table 10.1 ADK-171 Clinocardium Midden invertebrates and their relative percentages (modified from Savinetsky et al. Chapter 5 Appendix 5.H this volume).

Taxonomic group	***Scientific name***	***N***	***% of total invertebrates***	***% within taxonomic group***
Chitons		87	**2.1**	
	Tonicella submarmorea	22		25.3
	Katharina tunicata	65		74.7
Gastropods		48	**1.1**	
	Littorina sitkana	19		39.6
	Lottia pelta	23		47.9
	Nucella lima	6		12.5
Bivalves		2799	**66.5**	
	Mytilus trossulus	1022		36.5
	Hiatella arctica	15		0.5
	Clinocardium nuttallii	1452		51.9
	Saxidomus giganteus	295		10.5
	Protothaca staminea	1		<0.1
	Cardiidae gen. et sp. *indet*	1		<0.1
	Mactromeris polynima	11		0.4
	Mya truncata	2		0.1
Barnacles		588	**14**	
	Balanus crenatus	473		80.4
	Semibalanus balanoides	17		2.9
	S. cariosus	61		10.4
	Hesperibalanus hesperius	37		6.3
Sea urchins		685	**16.3**	
	Strongylocentrotus polyacanthus	677		98.8
	Echinarachnius parma	8		1.2
Total		4207		

Table 10.2. ADK-012 invertebrates and their relative percentages. (identified by Antipushina).

Taxonomic group	***Scientific name***	***N***	***% of total invertebrates***	***% within the group***
Bivalves		2867	**38.3**	38.3
	Mytilus trossulus	2749		95.9
	Clinocardium nuttallii	105		0.4
	Saxidomus giganteus	11		<0.1
	Hiatella arctica	2		<0.1
Barnacles		2166	**28.9**	28.9
	Balanus crenatus	1906		87.9
	Semibalanus cariosus	260		12.1
Echinoderms		1898	**25.4**	25.4
	Strongylocentrotus polyacanthus	1883		99.2
	Echinarachnius parma	15		0.8
Chitons		360	**5**	5
	Katharina tunicata	360		100
Gastropods		196	**3**	
	Littorina sitkana	127		64.7
	Lottia pelta	61		31.3
	Nucella sp.	8		0.4
Total		7487		

ADK-009 Invertebrates (2080-970 years BP)

Historical ecologists recorded bryozoan, polychaetes, and decapod genera at ADK-009 (See Savinetsky *et al.* Chapter V; Appendix 5H this volume). Because these invertebrates do not constitute human food, they are not shown in Table 10.3 nor are they discussed here. The ADK-009 invertebrate assemblage was large (N=34,961) and varied, consisting of six chiton, 23 gastropod, 11 bivalve, one brachiopod, and six barnacle species. Ecologists also recovered sea urchins and two types of decapods. Mussels, barnacles and sea urchins are well represented over time at ADK-009. *Clinocardium* is rare. Although many gastropod and chiton species were identified, their numbers are comparatively few. *Katharina tunicata* (Katy chiton) and *Lottia pelta* (shield limpet) are the most common chitons and gastropods at the site over time.

ADK-011, Component 2 Invertebrates (400-170 years BP)

ADK-011 Invertebrates were recovered from Layers 1 and 2 across the excavated units. The largest concentration of invertebrates at the site occurred in Units D-F in the midden. Koike *et al.* (Chapter 9 this volume) sized, aged, calculated growth rates, and determined season of collection for complete or nearly complete *Clinocardium* from Layers 1 and 2 at the site. Koike's team calculated that the ADK-011, Layer 1 *Clinocardium* ranged between four and 16 years old at time of death with an average age of seven. Complete *Clinocardium* valves in Layer 2 aged between four and 15 years old at time of death with an average age of 8. Layer 1 cockle bivalve lengths averaged 110.9 ± 12.0 mm (N=158). Cockles in Layer 2 were slightly smaller averaging 108.7 ± 17.6 mm in length (N=24). Differences in Layer 1 and Layer 2 *Clinocardium* sizes are statistically insignificant (Koike *et al.* Chapter 9 this volume). Only complete *Clinocardium* bivalves were measured and aged. However, archaeologists recovered many broken *Clinocardium* shells at ADK-011. These broken shells constitute both large and smaller cockles.

Because the ADK-011 invertebrate assemblage is large, West analyzed broken *Clinocardium* valves and all other invertebrates from (1) Unit F, a 1m X 1m unit representing the midden and (2) Unit H, a 1m X 1m unit representing the house feature. Complete *Clinocardium* valves from Units F and H analyzed by Koike were added to the invertebrate inventory (Table 10.4).

Bivalves overwhelm the ADK-011 assemblage, representing 97% of all invertebrates recovered in both the midden and the house. *Clinocardium* represent 60% of the bivalve assemblage. *Mytilus* (40%) is also strongly represented. Chitons, barnacles and echinoderms constitute 0.1% or less of the invertebrate assemblage. Gastropods are exceedingly rare. Bivalves both visually and statistically overwhelm the midden in the most recent occupation at ADK-011.

Adak Invertebrate Comparisons

Invertebrates in Adak sites are taxonomically diverse and frequently quite abundant. *Clinocardium* dominate at ADK-171 and ADK-011. *Mytilus* overwhelms the assemblage at ADK-012. Mussels, barnacles and sea urchin, and some gastropods, are well represented through time at ADK-009. Savinetsky *et al.* (Chapter 5 this volume) estimated that ADK-009 blue mussel density averaged 15,000/m^3 in the cultural layer. At ADK-171 *Clinocardium* averaged 2,500/m^3.

The Adak study sites span nearly 7000 years; however, the dissimilar proportions of recovered invertebrate taxa at different sites do not suggest that Aleuts changed littoral collecting practices over time. As Savinetsky *et al.* (Chapter 5 this volume) emphasizes, differences in the percentages of invertebrate species among Adak sites depends on the structure of the littoral zone adjacent to each Aleut settlement. ADK-009 is located on a rocky, high-energy shoreline where chitons, mussels, and gastropods aggregate today. Sites ADK-171 and ADK-011 are situated along or close to sandy mud shores and once contained vast quantities of *Clinocardium* that prefer low energy environments. ADK-012 is on a barrier beach sandwiched between the sandy coastline of Clam Lagoon and the rocky, high energy shore of Sitkin Sound. Archaeologists recovered comparatively large numbers of *Mytilus*, barnacles, and urchins from this site. Corbett *et al.* (2008) reported that prey species were taken most intensively within 4km of villages. If the Aleuts living at ADK-009 at Sweeper Cove collected *Clinocardium* at Clam Lagoon, they probably ate the mollusk at the lagoon and did not transport the shellfish to their village at Sweeper Cove. Invertebrates at north Adak sites represent prey collected in the immediate occupation vicinity. If invertebrates were collected further afield their shells, carapaces, and tests were left at the collection site. The invertebrate evidence also suggests that Adak littoral zones probably have not significantly changed during the last 6000 years.

Aleuts collected extremely large *Clinocardium* from the low energy, sandy mud, intertidal zone at Clam Lagoon (Koike *et al.* Chapter 9 this volume; Savinetsky *et al.* Chapter 5 this volume). However, extant *Clinocardium* do not grow to the large sizes observed in prehistory (Koike *et al.* Chapter 9 this volume). Currently Clam Lagoon is a sanctuary for sea otters, shellfish predators. Closed off from the open ocean except by a narrow inlet, the lagoon currently protects sea otters from potential killer whale predation (Estes *et al.* 1998). Consequently, protected sea otter populations are comparatively large in the lagoon today. We suggest that prehistoric *Clinocardium* grew to large sizes at Clam Lagoon because sea otters did not feed there. Threatened by prehistoric human predation, sea otters steered clear of the lagoon where escape routes were limited. Stable isotope analysis (Garong *et al.* Chapter 8 this volume) suggests prehistoric Adak sea otters preferred kelp forest and open ocean habitats. Bodkin (2004, 2) reported

Table 10.3. Invertebrate number and percentages from ADK-009, Adak Island, Alaska (modified from Savinetsky et al. *Chapter 5 Appendix 5H this volume).*

Number of layer, age of formation, yr AD	***V* *710-1060 AD***	***IV* *1060-1260 AD***	***III* *1260-1420 AD***	***II* *1420-1560 AD***	***I* *1560-1810 AD***
Taxonomic group					
Chitons	12[1] 0.7	24 1.0	58 1.0	142 0.9	104 1.2
Lepidozona sp.	-	-	-	2 1.4	5 4.8
Boreochiton beringensis beringensis	-	-	3 5.2	6 4.2	10 9.6
Tonicella submarmorea	-	-	-	1 0.7	-
Mopalia wosnessenskii	-	-	-	2 1.4	1 1.0
Katharina tunicata	12 100	24 100	54 93.1	131 92.3	88 84.6
Cryptochiton stelleri	-	-	1 1.7	-	-
Gastropods	82 4.6	98 4.2	183 3.1	1024 6.2	476 5.7
Problacmaea sybaritica	-	-	1 0.5	-	-
Lottia digitalis	5 6.1	-	4 2.2	6 0.6	-
Lottia pelta	8 9.8	30 30.6	70 38.3	256 25.0	214 45.0
Testudinalia scutum	3 3.7	-	9 4.9	12 1.2	10 2.1
Puncturella longifissa	-	-	2 1.1	5 0.5	-
Crepidula sp.	1 1.2	-	-	-	-
Ariadnaria insignis	-	-	1 0.5	-	1 0.2
Trichotropis bicarinata bicarinata	-	2 2.0	1 0.5	-	-
Littorina aleutica	6 7.3	5 5.1	21 11.5	73 7.1	19 4.0
Littorina sitkana	51 62.2	53 54.1	27 14.8	613 59.9	161 33.8
Cryptonatica sp.	-	-	1 0.5	-	1 0.2
Onoba aurivillii	-	-	13 7.1	22 2.1	6 1.3
Onoba sp.	-	-	2 1.1	-	-
Fusitriton oregonensis	-	-	-	-	1 0.2
Boreotrophon cepula	-	-	2 1.1	9 0.9	5 1.1
Nucella heyseana	-	1 1.0	-	-	1 0.2
Nucella lima	7 8.5	2 2.0	13 7.1	22 2.1	44 9.2
Buccinum percrassum	-	-	1 0.5	-	-
Aulacofusus periscelidus	-	2 2.0	7 3.8	-	5 1.1
Colus sp.	-	-	-	-	1 0.2
Neptunea sp.	-	-	1 0.5	-	-

Number of layer, age of formation, yr AD	*V* *710-1060 AD*	*IV* *1060-1260 AD*	*III* *1260-1420 AD*	*II* *1420-1560 AD*	*I* *1560-1810 AD*
Taxonomic group					
Volutopsius castaneus	-	-	-	-	1 0.2
Astyris sp.	1 1.2	3 3.1	7 3.8	6 0.6	6 1.3
Bivalves	449 25.4	200 8.5	520 8.9	1941 11.7	1803 21.4
Modiolus modiolus	-	-	1 0.2	2 0.1	-
Mytilus trossulus	445 99.1	195 97.5	508 97.7	1920 98.9	1774 98.4
Monia macrochisma	1 0.2	2 1.0	3 0.6	-	-
Hiatella arctica	3 0.7	-	4 0.8	16 0.8	10 0.6
Kellia laperousii	-	1 0.5	1 0.2	1 <0.1	-
Cyclocardia ventricosa	-	-	1 0.2	-	-
Clinocardium nuttallii	-	-	1 0.2	1 <0.1	9 0,5
Serripes groenlandicus	-	1 0.5	-	-	-
Macoma sp.	-	-	1 0.2	-	-
Mactromeris polynima	-	-	-	1 <0.1	10 0.6
Mya truncata	-	1 0.5	-	-	-
Barnacles	356 20.1	382 16.3	1205 20.6	4055 24.4	1419 16.9
Balanus balanus	-	7 1.8	8 0.7	9 0.2	-
Balanus crenatus	3 0.8	19 5.0	1 0.1	167 4.1	68 4.8
Balanus rostratus	-	-	7 0.6	-	7 0.5
Semibalanus cariosus	306 86.0	310 81.2	1069 88.7	2434 60.0	1018 71.7
Semibalanus balanoides	19 5.3	10 2.6	58 4.8	513 12.7	75 5.3
Hesperibalanus hesperius	28 7.9	36 9.4	62 5.1	932 23.0	251 17.7
Decapods					
Decapoda gen. et. sp. indet.	-	-	-	+	-
Crustacea gen. et. sp. indet.	-	-	-	+	-
Brachiopods					
Hemithyris psittacea	1 0.1	1 <0.1	-	8 <0.1	2 <0.1
Sea urchins					
Strongylocentrotus polyacanthus	868 49.1	1644 70.0	3884 66.4	9425 56.8	4607 54.8
Total N	***1767***	***2348***	***5850***	***16587***	***8409***

[1]N / %

Table 10.4. ADK-011, Component 2 invertebrates and percentages.

Taxonomic group	*Unit F midden N/MNI*	*Unit H house N/MNI*	*F + H NISP*	*% of total invertebrates*
Bivalves			**14387**	**97**
Clinocardium nuttallii	7562/183	1120/31	8682	
Mytilus trossulus	5024/123	681/28	5705	
Barnacle			**208**	**0.1**
Balanus crenatus	187/29	21/7	208	
Chiton			**76**	**<0.1**
Katharina tunicata	60/23	16/6	76	
Gastropods			**9**	**<0.1**
Lottia pelta	0	1/1	1	
periwinkle	4/4	2/2	6	
whelk	2/2	0	2	
ECHINODERM			**69**	**<0.1**
Strongylocentrotus polyacanthus (mouthparts)	157/51	8	69	
Total			***14845***	

that reduced Bering Sea sea otter populations result in increased densities and sizes of preferred sea otter prey. The extraordinarily large size of prehistoric Clam Lagoon *Clinocardium* supports this claim.

Pan Aleutian Invertebrate Studies

At the Amaknak Spit Site (UNL-00055), Nora Foster identified 14 gastropod, 17 bivalve, two chiton, and one barnacle species (Knecht and Davis 2003). Table 10.5 shows the MNI of invertebrates from 6 levels at this site. Among gastropods dog winkles (*Nucella lima*) and Neptune snails (*Neptunea*) dominate. Butter clam (*Saxidomus gigantius*) and mussel (*Mytilus*) comprise the largest minimum number of individuals for bivalves. The Amaknak Spit site assemblage is clearly a mixed bag. Aleuts at this site exploited both high and low energy environments. This site has a barrier beach extending into Iliuliuk Bay with a protected cove on one side of the site and a more exposed bay between Amaknak and Unalaska Island.

The green urchin (*Strongylocentrotus drobachiensis*), followed by limpets (*Testudinalia scutum, Lottia pelta*) occurred in 'great heaps' in Amchitka archaeological sites (Desautels *et al.* 1971, 324, 326). In descending order of frequency in Amchitka middens were the mussel (*Mytilus trossulus*), pearly monia (*Monia macrochisma*), black katy (*Katharina tunicata*), hairy triton ((*Fusitriton oregonens)*, the whelk (species unidentified), Sitka periwinkle (*Littorina sitkana*), dogwinkle (*Nucella canaliculata*), island scallop (*Chlamys islandicus*), and the barnacle (*Balanus* sp.) (Desautels *et al.* 1971, 336).

West analyzed invertebrates collected during 1997 from Pits 4 and 6 at site KIS-008 on Buldir Island (Table 10.6). Invertebrate remains were sparse and included sea urchins (NISP=820; MNI=95), foolish mussel (NISP=48; MNI=11), and periwinkle (NISP=1; MNI=1). The paucity of recovered invertebrates suggests that echinoderms and mollusks were not readily accessible and occasionally washed ashore. Invertebrate remains may also have found their way to occupation sites as 'riders' in the stomachs of killed prey or attached to kelp. Much of the Buldir coastline is precipitous and the littoral zone near the excavated Aleut settlement is high energy but has no rocks to serve as holdfasts.

Remains of sea urchins, limpets, periwinkles, Northern striped dogwinkles, channeled dogwinkle, rock oyster, butter clam, and *Ostrea* oyster comprised the shell middens of Agattu Island (Spaulding 1962). These invertebrate assemblages infer a high energy, rocky environment close to the sites sampled.

Debra Corbett and Dixie West analyzed invertebrates from three Shemya sites: ATU-021, ATU-003, and ATU-061 (Lefèvre *et al.* 2010; Tables 11.13-11.15). Shemya invertebrate assemblages included sea urchins, chitons, gastropods, barnacles, and bivalves (Table 10.7). Unfortunately urchins were not quantified at ATU-061 although they comprised a large percentage of the assemblage. At ATU-003 chitons (47.1%) and gastropods—periwinkles, tritons, and limpets (40.5%) comprised most of the assemblage. Bivalves (mussels, rock oyster, and butter clam) were few and represented 7.5% of the assemblage. Sea urchins and barnacles represent 4.6% and 0.2% of the entire invertebrate assemblage at the ATU-003. At ATU-021 gastropods (periwinkles, dogwinkles, limpets and a snail) constitute 84.2% of the assemblage. Chiton represents 15.7%. Only one bivalve, a large clam, and ten sea urchin pieces were recorded for ATU-021. At ATU-061 periwinkles and dogwinkles comprise 86.4% of the assemblage. Chitons represent 13.2%. Sea urchins from ATU-061were not included in the count. The comparatively high number of gastropods and chitons recovered from Shemya infers that Aleuts collected from high energy rocky reefs surrounding the island. (Lefèvre *et al.* 2010).

Table 10.5 Invertebrates from Amaknak Spit Site by level (modified from Knecht and Davis 2003, Tables 1-7)

	(bhc) [1]	***L.2*** [2]	***L.2b***	***L.3***	***L.4b***	***L.5***	***L.6***
Taxon							
Gastropods							
Buccinum baerii							
Fusitriton oregonensis		5	1				
Littorina sitkana		6	3			2	
Lottia/Tectura		8	6				
Lottia pelta			8				
Cryptonatica affinis			4				
Naticidae		6	2				
Neptunea lyrata	1	7	5				
Neptunea ventricosa	4	4	5				
Neptunea?	11	18	3			1	
Nucella canaliculata			1				
Nucella lamellosa	1	3					
Nucella lima	1	95	58	5	1	1	1
Tectura persona							
Tectura scutum		2					
Unidentified gastropods		35	35	6		12	7
Bivalves							
Astarte sp.		2					
Cardiidae		6	4			1	
Clinocardium nuttallii	3	8	15	4			
Clinocardium sp.		3	4				
Hiatella arctica		6	10	1			
Macoma inquinata			1				
Mactromeris polymima		1	3			1	
Mya truncata?	4	20	1				
Mya sp.			10			1	1
Mytilidae			1				
Mytilus trossulus		50	39	19		19	12
Pododesmus macroschisma		2	2			1	
Protothaca staminea	5	76	44	3	1		
Saxidomus giganteus	9	138	115	7		9	13
Serripes?		4	2				
Siliqua sp.		1	5				
Unidentified bivalve		17	20	8			
Veneridae	1	41	30	9		4	
Chitons							
Cryptochiton stelleri		1					
Katharina tunicata		12	11	13	1	2	1
Total Mollusk taxa	10	28		78	3	12	6
Total Mollusk MNI	40	577	448		3	54	35
Barnacles							
Semibalanus cariosus	1	70	82	19	1	7	2
Echinoderms							
Strongylocentrotus sp.		2					3

(bhc) backhoe cut
(L.) layer

Palmisano and Estes (1977) reported high numbers of extant urchins along Attu coasts. Western Aleutians archaeologists excavated a midden and house feature ATU-019 at Austin Cove, Attu (West *et al.* 2002). The feature and midden dated 170 BC to AD 710 (Corbett *et al.* 2008). The entire invertebrate assemblage has yet to be reported, but 2002 excavations revealed a dense sea urchin lens that overwhelms the invertebrate assemblage. Future analysis could reflect an invertebrate assemblage much like that found on Amchitka.

Shellfish as Resources

Shellfish, per unit weight, contain approximately the same

amount of protein as sea mammals (Table 10.8). The daily minimum protein requirement for adults is 40 grams, the amount of protein found in 494 mussels (Osborn 1977). In the Aleutians, ground berries, high in carbohydrates and vitamins, are only seasonally available. Shellfish can provide at least some carbohydrates when plant carbohydrates are inaccessible (Table 10.9). Shellfish also provide necessary trace minerals (Table 10.10). Shellfish supply ascorbic acid, thiamin, potassium and phosphorous, all imperative in maintaining body temperature, heat production and cell regulation (Piantadosi 2003). Without these key nutrients, hunter-gatherers can face nutritional bottlenecks.

Although invertebrates contain protein, some carbohydrates, and valuable micronutrients, they possess relatively little fat (Table 10.9). Extreme high protein, low fat diets lead to starvation. To balance the ratio of protein extracted from shellfish, humans must consume fat to metabolize protein (Piantadosi 2003; Speth 1983, 1987). In the 19th century Veniaminov (1984, 277) reported the critical need for fat in the Aleut diet:

'The main food of the Aleuts is fat.... However many fish an Aleut puts up, if, at the same time he has no fat, one can say with certainty that he is going to suffer either from actual hunger, or illness, because with long usage yukola, or dried fish, without fat, produces a bloody flux. Even the freshest fish without fat is not too nourishing. On the other hand, even if an Aleut does not have a single dried fish, if fat is sufficient, he will not experience hunger. Because with fat, he can use everything--roots, sea cabbage [kelp], which, without fat, especially the latter, are very harmful, and mollusks [rakushki], and the driest yukola and straps, in brief, everything which the stomach can possibly digest.'

Found in large numbers at archaeological sites, invertebrates are frequently considered a marginal resource used either as an additional diet item or as a last resort when other foods are unavailable (Bailey 1975; Parmalee and Klippel 1974; Veniaminov 1984; Waselkov 1987). Fatty fish and sea mammals overwhelmingly provided the fat necessary in the Aleut diet. The common seal (*Phoca vitulina*) provides 143kcal/100g of meat while the mussel (*Mytilus* sp.) provides 95kcal/100g of meat (Erlandson 1988). The common seal weighs 74kg and the weight of edible meat of mussel is 1g. Because of size differences, one seal is equivalent to approximately 110,000 mussels. King salmon flesh provides 15 times as much fat as the same quantity of clam meat (Table 10.9). And any type of fish oil is 100% fat.

Shellfish Seasonality

Whereas sea mammal and migratory bird hunting is confined to warmer months, shellfish can be collected during the entire year. Techniques that analyze seasonal variations in periods of fast and slow growth of shell deposition have helped archaeologists determine the season of harvest (Claassen 1986; Ham and Irvine 1975; Rhoads and Lutz 1980). Koike *et al.* (Chapter 9 this volume) using thin section techniques on *Clinocardium* shells from ADK-011, determined that the largest specimens of these bivalves were collected from spring through early fall. This is an interesting finding in that paralytic shellfish poisoning (PSP), generally a warm month event, apparently did not influence summer *Clinocardium* collecting. Bruce Wright reported a pattern of positive results for PSP in the eastern Aleutians and less PSP in the western Aleutians and Commanders. PSP may be an increasing phenomenon associated with global warming and maritime industrial waste (Wright *et al.* 2006, 2008). However, Black (2004, 133), using the ethnographic literature, recorded that paralytic shellfish poisoning killed 140 Aleut men in 1799.

Shellfish Acquisition and Processing

Littoral collection is a low risk pursuit that requires minimal technology. Wading in shallow water and hand collecting shellfish from rocks exposed during low tide is a frequent method of gathering species that live on rock surfaces (Waselkov 1987). Identifying siphon holes is a convenient way to spot species that dig into sand or mud bottoms. Either method requires relatively simple tools including the fingers, a digging stick, or a pry bar (Harbo 2001; Jochelson 1925; Waselkov 1987). Relatively high investment tools, including harpoons, spears, nets, weirs and traps, are not required to harvest mollusks or sea urchins.

At ADK-011 some *Clinocardium* shells were charred, possibly indicating that Aleuts directly placed these bivalves in a fire. An abundance of griddle stones and charcoal at ADK-011 suggests that intense heat—perhaps steaming—could have been employed to open shells. Once dead, the strong muscle of each *Clinocardium* would have relaxed and the shell could be opened with little effort. Stone boiling, evident as early as the European Upper Paleolithic (West 1997) and observed in eastern Aleut villages (Veniaminov 1984, 278), is a possible way to extract meat from large, tightly closed shells. But stone boiling is an extremely labor intensive chore that requires large amounts of fuel, and wood is in short supply in the Aleutians. It would have been far more efficient to directly heat and kill the *Clinocardium* to relax the muscle and open the shell rather than submerge it in hot water. Or cockle meat may have been extracted by simply breaking thick shells with cobbles or prying shells open with a lever. Burned *Clinocardium* valves at ADK-011 occurred on both exterior and interior shell surfaces as well as on broken edges. This evidence suggests that opened cockle shells were accidentally or intentionally discarded in hearths or near hearth areas, and the burning associated with shells might not be directly related to processing for meat extraction.

Social Organization of Shellfishing

The basic political unit in prehistoric Aleut society was the household, a group of often-related individuals who in some cases lived in a single house. Aleutian households often

Table 10.6. Buldir invertebrates: KIS-008.

KIS-008; Pits 4 and 6		***NISP***		***MNI***	
		#	%	#	%
Echinoderms					
Strongylocentrotus dröbachiensis	Sea Urchin	820	91.6	82	87.2
Mollusks					
Gastropods					
Littorina sp.	Periwinkle	1	>.01	1	1
Bivalves					
Mytilus sp.	Blue mussel	48	5	11	11.7
Unidentified		26	2.9		
Total Invertebrates		***895***	**000*(1)***	***94***	**000*(1)***

Table 10. 7. NISP and % of Shemya Invertebrates (modified from Lefèvre et al. *2010; Tables 11. 13, 11.14, 11.15.)*

	ATU-021		***ATU-003***		***ATU-061***	
TAXON	***NISP***	*%*	***NISP***	*%*	***NISP***	*%*
Echinoderms						
Strongylocentrotus drobachiensis	10	6	100[(1)]	4.6		
Barnacles						
Balanus crenatus			5	0.2		
Chitons						
Katharina tunicata	949	58.6	1,032	47.1	4,700	13.2
Gastropods						
Littorina sp.	496	30.6	445	20.3	18,820	52.7
Thais [(2)] *lima*	76	4.7			12,140	34
Colisella sp.	86	5.3	84	3.8		
Acmaea limatula					20	0.1
Fusitriton oregonensis			359	16.4		
Unidentified snail	1	0.1				
Bivalves						
Mytilus edulis [(3)]			16	0.7		
Mytilus californacus?			22	1.0		
Pododesmus [(4)] *macrochisma*			7	0.3		
Saxidomus giganteus			2	0.1		
Large clam	1	0.1				
Unidentified Invertebrates	118	5.4	118	5.4	10,380	22.5 (1)
Grand Total Invertebrates	***1,619***		***2,190***		***46,060***	

[(1)] mouth parts counted by Debra Corbett; [(2)] *Thais= Nucella*; [(3)] *edulis=trossulus*; [(4)] *Pododesmus=Monia*

Table 10.8. Ratio of protein to edible meat for different marine fauna. (Modified from Osbourne 1977, 175. Table 4.3).

Food Resource	***Protein/Edible Meat***
Fur seal	.10
Hair seal	.10
Sea lion	.11
Whale (including blubber)	.10
Marine fish	.17
Marine birds	.15
Clams	.13
Oysters	.10
Scallops	.15

Table 10.9. Composition of Foods. (modified from Watt, Bernice and Annabelle Merrill 1963. Composition of Foods. Agricultural Handbook No. 8. United States Department of Agriculture, Washington, DC). Some fauna and plants not indigenous to the Aleutians are included here to provide a conceptual framework for comparison. For example, the pheasant is a gallinaceous bird like the rock ptarmagin, and so on.

Group	*Kcal*	*Protein*	*Fat*	*Carbs*
Mammals				
Whale meat, raw	156	20.6	7.5	0
Birds:				
Pheasant	151	24.3	5.2	0
Quail, raw	168	25	6.8	0
Pigeon, flesh and skin	294	18.5	23.8	0
Wild duck	233	21.1	15.8	0
Duck egg	191	13.3	14.5	.7
Goose egg	185	13.9	13.3	1.3
Fish:				
Roe	130	24.4	2.3	1.5
Cod, raw	78	17.6	.3	0
Cod, dried, salted	130	29	.7	0
Cod liver oil	902	0	100	-
Flounder, baked	202	30	8.2	0
Hake	74	16.5	.4	0
Halibut, raw, Pacific	100	20.9	1.2	0
Mackeral, Pacific, raw	159	21.9	7.3	0
Ocean Perch, Pacific	95	19	1.5	0
Pollock	95	20.4	.9	0
Snapper, raw	93	19.2	.9	0
Redhorse	98	18	2.3	0
Rockfish, raw	97	18.9	1.8	0
Salmon, king	222	19.1	15.6	0
Salmon, pink	119	20	3.7	0
Salmon oil	902	0	100	-
Sardine, Pacific, raw	160	19.2	8.6	0
Sucker	104	20.6	1.8	0
Herring oil	902	0	100	-
Shellfish				
Abalone	98	187	0.5	3.4
Clam meat, soft	82	14	1.9	1.3
Clam meat, hard	80	11.1	.9	5.9
Clam meat, hard/soft	76	12.6	1.6	2
Crab, steamed	93	17.3	1.9	.5
Lobster, northern, whole	91	16.9	1.9	.5
Mussels, Pacific, meat	95	14.4	2.2	3.3
Octopus	73	15.3	.8	0
Oyster, raw meat	91	10.6	2.2	6.4
Scallop	81	15.3	.2	3.3
Snail, raw	90	16.1	1.4	2.0
Squid	94	16.4	.9	1.9
Berries				
Blackberries	58	1.2	.9	12.9
Blueberries	62	.7	.5	15.3
Cranberries	46	.4	.7	10.8
Gooseberries	39	.8	.2	11.2
Ground cherries	53	1.9	.7	11.2
Loganberries	62	1	.6	14.9
Greens				
Chickory greens	20	1.8	.3	3.8
Collards	45	4.8	.8	7.5
Garden cress	32	2.6	.7	5.5
Dandelion greens	45	2.7	.7	9.2

Group	*Kcal*	*Protein*	*Fat*	*Carbs*
Dock	28	2.1	.3	5.6
Lambsquarters	43	4.2	.8	7.3
Seaweed: agar	-	-	.3	-
Seaweed: dulse	-	-	3.2	-
Seaweed: irishmoss	-	-	1.8	-
Seaweed; kelp	-	-	1.1	-
Laver	-	-	.6	-
Turnip greens	28	3	.3	5

Table 10.10: Composition of Foods. (modified from Watt, Bernice and Annabelle Merrill 1963. Composition of Foods. Agricultural Handbook No. 8. US Department of Agriculture, Washington, D.C.). The data presented here provide a way to compare micronutrients provided by fish, shellfish, berries and greens.

Group	*Ca*	*P*	*Fe*	*Na*	*K*	*A*	*Th*	*Rib*	*Ni*	*C*
Fish:										
Roe	-	-	.6	-	-	-	.10	.76	1.4	14
Cod, raw	10	194	.4	70	382	0	.06	.07	2.2	2
Cod liver oil	-	-	-	-	-	-	-	-	-	-
Flounder, bakedHake	23	349	1.4	237	587	-	.07	.08	2.5	2
Halibut, raw, Pacific	13	211	.7	54	449	440	.07	.07	8.3	-
Halibut, smoked, Pacific	-	210	-	-	-	-	.01	-	-	-
Mackeral, raw, Pacific	260	288	2.2	-	-	3-	.03	.33	8.8	-
Mackeral, smoked	-	-	-	-	-	-	-	-	-	-
Ocean Perch, Pacific	-	-	-	63	390	-	-	-	-	-
Pollock	-	-	-	48	350	-	.05	.10	1.6	-
Snapper, raw	16	214	.8	67	323	-	.17	.02	-	-
Redhorse	-	-	-	-	-	-	-	-	-	-
Rockfish, raw	-	-	-	60	388	-	.06	.12	-	-
Salmon, king	-	301	-	45	399	310	.10	.23	-	-
Salmon, pink	-	-	-	64	306	-	.14	.05	-	-
Salmon oil	-	-	-	-	-	-	-	-	-	-
Sardine, raw, Pacific	33	215	1.8	-	-	-	-	-	-	-
Sucker	-	220	-	56	336	-	trace	-	1.2	-
Herring oil	-	-	-	-	-	-	-	-	-	-
Shellfish:										
Clam meat, soft	-	183	3.4	36	235	-	-	-	-	-
Clam meat, hard	-	69	151	7.5	205	311	-	-	-	-
Clam meat, hard/soft	69	163	6.1	120	181	100	.10	.18	1.3	10
Crab, steamed	43	175	.8	-	-	2170	.16	.08	2.8	2
Lobster, northern, whole	29	183	.6	-	-	-	.4	.05	1.5	-
Mussels, Pacific, meat	88	236	3.4	289	315	-	.16	.21	-	-
Octopus	29	173	-	-	-	-	.02	.06	1.8	-
Oyster, raw meat	85	153	7.2	-	-	-	.12	-	1.3	30
Scallop	26	208	1.8	255	396	-	-	.06	1.3	-
Shrimp, raw	6.3	166	1.6	140	220	-	.02	.03	3.2	-
Snail, raw	-	-	3.5	-	-	-	-	-	-	-
Squid	12	119	.5	-	-	-	.02	.12	-	-
Berries:										
Blackberries	32	19	.9	1	170	200	.03	.04	.4	21
Blueberries	15	13	1	1	81	100	.03	.06	.5	14
Cranberries	14	10	.5	2	82	40	.03	.02	.1	11
Gooseberries	18	15	.5	1	155	290	-	-	-	33
Ground cherries	9	40	1	-	-	720	.11	.04	2.8	11
Loganberries	35	17	1.2	1	170	200	.03	.24	.4	24
Greens:										
Chickory greens	86	40	.9	-	420	4000	.06	.10	.5	22

Group	Ca	P	Fe	Na	K	A	Th	Rib	Ni	C
Collards	250	82	1.5	-	450	9300	.16	.31	1.7	152
Garden cress	81	76	1.3	14	606	9200	.08	.261	1	69
Dandelion greens	187	66	3.1	76	397	14000	.19	.26	-	35
Dock	66	41	1.6	5	338	12900	.09	.22	.5	119
Lambsquarters	309	72	1.2	1	170	200	.03	.33	8.8	-
Seaweed: agar	567	22	6.3							
Seaweed: dulse	296	267	-	2085	8060					
Seaweed: irishmoss	885	157	8.9	2892	2844					
Seaweed: kelp	1093	240	-	3007	5273					
Seaweed: laver	-	-	-	-	-					

Ca=calcium; P=Phosphorous; Fe=iron; Na=sodium; K=Potassium; A=vitamin A; Th=thiamin; Rib=riboflavin; Ni=niacin; C=ascorbic acid.

consisted of single or several families. It is often accepted that children, women, and old people gathered shellfish (Laughlin 1980). Joan Coltrain (West *et al.* Chapter 16 this volume) conducted stable carbon and nitrogen isotope analyses on a rib of a Aleut child exhumed in the Andreanof Islands. The Andreanof child's stable isotope ratios ($\delta^{13}C$ °/oo=-13.9; $\delta^{15}N$ °/oo = +17.0) compare favorably with values ($\delta^{13}C$ °/oo=-13.4; $\delta^{15}N$ °/oo = +17.8) of an elderly adult recovered in the Near Islands (West *et al.* 2003; West *et al.* Chapter 16 this volume). Aleut skeletons sampled at Kagamil, Shiprock, and Umnak in the eastern Aleutians (N=36) had $\delta^{13}C$ °/oo levels ranging between -12.0$\pm$0.3 and -12.9$\pm$0.1 and $\delta^{15}N$ °/oo levels ranging between +19.5$\pm$0.5 and 20.1$\pm$0.5 (West *et al.* Chapter 16 this volume). Sample size is small, but preliminary stable isotope analyses seem to indicate that central and western Aleuts consumed lower trophic level marine mammal tissue and more sea birds and invertebrates. Stable isotope ratios might also be related to the ages of the humans sampled. The central and western Aleutian samples represent a child and an elderly adult. Laughlin (1980) reported that children, women and the elderly contributed to the greater community by foraging in reefs. Stable isotope analysis could confirm this age and gender based intertidal practice. However, many more Aleutian samples, representing both sexes and all age groups, must be isotopically analyzed to test this idea.

Conclusions

Thick shellfish layers at Aleutian archaeological sites and in ethnographic literature (Collins *et al.* 1945; Jochelson 1933; Veniaminov 1984) confirm that the Aleuts collected and ate mollusks and sea urchins in staggering numbers. The depth of some Aleutian shell middens, as much at 6.5m (Jochelson 1925), attests to the continued accessibility and reliance on invertebrates throughout Aleutian prehistory. The old adage, 'When the tide was out, the table was set' rang true for Aleuts across the archipelago. Intertidal zones were the 'corner grocery stores' where food could be conveniently acquired year round. These intertidal zones were critical during lean months when migratory species were not available and when stored foods had run low. Liebing's Law of the Minimum states: '...an organism is no stronger than the weakest link in its ecological chain of requirement' (cited in Odum 1971, 106). It made no difference if sea lions, fish, or migratory birds appeared by the thousands for a time, because it was the poorest and leanest seasons that set the limit to long-term Aleut sustainability. Burch (1972) pointed out that it only takes several weeks to starve to death. Under such stress situations, humans have three alternatives: 1) population decrease through starvation and/or migration, 2) exploit different resources, or 3) a combination of the above two (Slobodkin 1961). If invertebrates had not existed in the North Pacific Rim, one wonders if human populations could have survived yearly, late winter bottlenecks.

Ancient Aleuts were in tune with their maritime environment. Characteristics of prey species distribution, density, aggregate behavior, condition, and seasonality influenced how specific resources were chosen and procured. As food resources either decreased and/or became dispersed and increasingly unpredictable, Aleut groups faced increased cost and risk. Possible increased costs associated with lower faunal abundance and increased dispersal throughout the year could have resulted in: 1) more frequent movements of a camp during the course of a year, 2) traveling farther from a given camp for resources, 3) increasing search time for a given resource, 4) possible procurement of resources of lesser quality and/or size, or 5) a combination of two or more of the above. These increased costs could be potentially offset by: 1) selecting prey species which provide the most technological/food return for least effort or risk, 2) scheduling seasonal activities in order to be at the right spot at the right time to take advantage of prey, 3) organizing procurement strategies so that resources of lesser quality could be efficiently exploited, or 4) efficiently transporting and processing foods to extract maximum returns (Binford 1980; Kelly 1992; Winterhalder and Smith 1981).

In arctic environments, carnivores and omnivores require substantial quantities of fat to survive. During resource stress, acquisition of fat, rather than protein, becomes critical (Speth 1987; Speth and Spielmann 1983). If fat is in short supply, large quantities of meat must be consumed. If fat is totally absent in the diet, very lean meat should be avoided as it takes a higher metabolism to digest purely lean meat. Invertebrates became the prey choice of Aleuts during lean months when other resources were unavailable and when stores ran low. These low trophic resources provided enough protein, carbohydrates, and micronutrients to sustain people for a short time. Aleuts, anticipating late winter food shortages, probably positioned themselves adjacent to highly productive intertidal zones to exploit invertebrate communities when winter stores were depleted.

Acknowledgements

This research was supported under the National Science Foundation's, Office of Polar Programs, Arctic Social Sciences Program, Award # 0353065. We thank Dave Brewer, local Adak mathematics teacher, for counting, measuring and providing statistical information on ADK-011 *Clinocardium* in the Adak school science laboratory. Special recognition goes to the Adak school children who helped Brewer sort and quantify.

References

Antipushina, Zhanna. 2008. Quantitative Account of Invertebrate Remains in Archaeological Sites. Current Problems of Ecology and Evolution Addressed by Young Scientists, *Proceedings of the Conference Young Scientists and Postgraduates of the Severtsov Institute of Ecology and Evolution,* 11-14. Moscow, Russian Academy of Sciences.

Bailey, Geoffrey N. 1975. The Role of Molluscs in Coastal Economies: the Results of Midden Analysis in Australia. *Journal of Archaeological Science* 2, 45-62.

Binford, Lewis. 1980. Willow Smoke and Dogs' Tails: Hunter-gatherer Settlement Systems and Archaeological Site Formation. *American Antiquity* 45, 4-20.

Bird, Douglas W., Jennifer L. Richardson, Peter M. Veth, and Anthony J. Barham. 2002. Explaining Shellfish Variability in Middens on the Meriam Islands, Torres Strait, Australia. *Journal of Archaeological Science* 29, 457-469.

Black, Lydia. 2004. *Russians in Alaska, 1732-1867.* Fairbanks, University of Alaska Press.

Bodkin, James L. 2004. Sea Otter Foraging Behavior and Population Status in the Aleutian Islands and the Alaska Peninsula. Final Report to the Alaska Maritime National Wildlife Refuge, Homer, Alaska.

Burch, Ernest S., Jr. 1972. The Caribou-Wild Reindeer as a Human Resource. *American Antiquity* 37, 339-368.

Claassen, Cheryl. 1986. Shellfishing Seasons in the Prehistoric Southeastern United States. *American Antiquity* 51(1), 21-37.

Collins Henry B., Austin H. Clark, Egbert H. Walker. 1945. *The Aleutian Islands: Their People and Natural History.* Washington, DC, Smithsonian Institution.

Corbett, Debra G., Douglas Causey, Mark Clementz, Paul L. Koch, Angela Doroff, Christine Lefèvre, and Dixie West. 2008. Aleut Hunters, Sea Otters, and Sea Cows. In C. Rick Torben and Jon M. Erlandson (eds.), *Human Impacts on Ancient Marine Ecosystems: A Global Perspective*, 43-75. Berkeley, California. University of California Press.

Dall, William H. 1877. On Succession in the Shell-Heaps of the Aleutian Islands. In *Tribes of the Extreme Northwest, Contributions to North American Ethnology* Vol 1., 41-91. Washington, DC, US Government Printing Office.

Desautels, Roger J., Allen J. McCurdy, James D. Flynn and Robert R. Ellis. 1971. *Archaeological Report. Amchitka Island, Alaska 1969-1970.* Report by Archaeological Research Inc. to the US Atomic Energy Commission, TID-25481. Las Vegas, Nevada, Holmes and Narver, Inc.

Erlandson Jon M. 1988. The Role of Shellfish in Prehistoric Economies: A Protein Perspective. *American Antiquity* 53(1), 102-109.

Estes, James A. and J. F. Palmisano. 1974. Sea Otters: Their Role in Structuring Nearshore Communities. *Science* 185, 1058-1060.

Estes, James, M. T. Tinker, T. M. Williams, and D. F. Doak. 1998. Killer Whale Predation on Sea Otters Linking Oceanic and Nearshore Ecosystems. *Science* 282, 473-476.

Garong, Ame, Chiozuru Takashima, Shin Nishida, Akihiro Kano, and Hiroko Koike. 2011. Carbon, Nitrogen, and Oxygen Isotopic Compositions of Bone Collagen and Tooth Enamel Carbonate of the Sea Otter *Enhydra lutris* from Archaeological Sites at Adak Island, Aleutian Archipelago. Chapter 8 In D. West, V. Hatfield, E. Wilmerding, C. Lefèvre and L. Gualtieri (eds.), *The People Before: The Geology, Paleoecology and Archaeology of Adak Island, Alaska.* Oxford, British Archaeologial Reports.

Ham, L. C. and M. Irvine. 1975. Techniques for Determining Seasonality of Shell Middens from Marine Mollusk Remains. *Syesis* 8, 363-373.

Harbo Rick. 2001. *Shells and Shellfish of the Pacific Northwest.* Hong Kong, Prolong Press Limited.

Hrdlička, Aleš. 1945. *The Aleutian and Commander Islands and Their Inhabitants.* Philadelphia, The Wistar Institute of Anatomy and Biology.

Jochelson, Waldemar. 1925. Archaeological Investigations in the Aleutian Islands. *Carnegie Institute of Washington Publication* 367. Washington, DC.

Jochelson, Waldemar. 1933. History, Ethnology and Anthropology of the Aleut. *Carnegie Institution of Washington, Publication* 432. Washington, D.C. [Reprinted 1968 by Anthropological Publications, Oosterhout, N.B., The Netherlands].

Kelly, Robert. 1995. *The Foraging Spectrum: Diversity in Hunter-Gatherer Lifeways.* Washington, DC, Smithsonian Institution Press.

Knecht, Richard and Richard Davis. 2003. Archaeological Evaluation of Tanaxtaxak, The Amaknak Spit Site (UNL-00055). Final Report. Prepared for the US Army Corps of Engineers.(With contributions by Nora Foster,

David Yesner, Melia Knechit, Amy Hirons, and Nicole Misarti.)

Koike, Hiroko, Chizuru Takashima and Akihiro Kano. 2011. The Aleut Cockle, *Clinocardium nuttallii* (Conrad, 1837) from ADK-011, Adak Island, Alaska Chapter 9. In D. West, V. Hatfield, E. Wilmerding, C. Lefèvre, and L. Gualtieri (eds.), *The People Before: The Geology, Paleoecology and Archaeology of Adak Island, Alaska*. Oxford, British Archaeologial Reports.

Laughlin, William. 1980. *Aleuts: Survivors of the Bering Land Bridge*. New York, Holt, Rinehart and Winston.

Lefèvre, Christine, Debra Corbett, Susan Crockford, Jessica Czederpiltz, Megan Partlow, and Dixie West. 2010. Faunal Remains and Intersite Comparsions, In Debra Corbett, Dixie West, and Christine Lefèvre (eds.) *The People at the End of the World: The Western Aleutians Project and the Archaeology of Shemya Island*, 121-156. Aurora, Alaska Anthropological Association Monograph Series-VIII. Anchorage, Alaska.

Luttrell Mark, and Debra Corbett. 2000. Archaeological Investigations on Northeast Adak Island, Alaska July, 1999. Unpublished Manuscript, on file Archives, US Fish and Wildlife Service, Anchorage, Alaska.

MacArthur, Robert H., and Edward O. Wilson. 1967. *The Theory of Island Biogeography*. Princeton, New Jersey, Princeton University Press.

O'Clair, Charles E. 1977. Marine Invertebrates in Rocky Intertidal Communities. In M. L. Merritt and R. G. Fuller, (eds.), the *Environment of Amchitka Island, Alaska*, 395-449. Washington, DC, US Energy Research and Development Administration.

Odum, Eugene P. 1971 *Fundamentals of Ecology*. Philadelphia, Pennsylvania, W. B. Saunders Company.

Okuno, Mitsuru, Keiji Wada, Toshio Nakamura, Lyn Gualtieri, Brenn Sarata, Dixie West, Masayuki Torii. 2011. Holocene Tephra Layers on the Northern Half of Adak Island in the West-central Aleutians, Alaska. Chapter 3. In D. West, V. Hatfield, E. Wilmerding, C. Lefèvre and L. Gualtieri (eds.), *The People Before: The Geology, Paleoecology and Archaeology of Adak Island, Alaska*. Oxford, British Archaeologial Reports.

O'Leary, Matthew. 1998. New Site No. 2 ("Tutikoff Site"). Unpublished report on file at US Department of the Interior, Bureau of Indian Affairs, ANCSA Projects Office, Anchorage, Alaska.

O'Leary, Matthew. 2001. Volcanic Ash Stratigraphy for Adak Island, Central Aleutian Archipelago. In Don Dumond, (ed.), *Archaeology in the Aleut Zone of Alaska, Some Recent Research*, 215-234. University of Oregon Anthropological Papers No. 58. Eugene, Oregon, University of Oregon Press.

Osborn, Alan J. 1977. Strandloopers, Mermaids, and Other Fairy Tales: Ecological Determinants of Marine Resource Utilization. The Peruvian Case. In Lewis R. Binford, (ed.), *For Theory Building in Archaeology:Essays on Faunal Remains, Aquatic Resources, Spatial Analysis, and Systemic Modeling*, 157–205. New York, Academic Press.

Palmisano, J. F and James A. Estes. 1977. Ecological Interactions Involving the Sea Otter. In M. L. Merritt and R. G. Fuller, (eds.), *The Environment of Amchitka Island*, 527-567. Washington, DC, Alaska US Energy Research and Development Administration.

Parmalee, Paul W. and Walter E. Klippel. 1974. Freshwater Mussels as a Prehistoric Food Resource. *American Antiquity* 39(3), 421-434.

Pianka, Eric R. 1970. On r and K selection. *American Naturalist* 104, 592-597.

Piantadosi, Claude A. 2003. *The Biology of Human Survival: Life and Death in Extreme Environments*. New York, Oxford University Press.

Rhoads, D. C. and R. A. Lutz (eds.) 1980. *Skeletal Growth of Aquatic Organisms*. New York, Plenum Press.

Savinetsky, Arkady B., Nina K. Kiseleva, and Bulat F. Khassanov. 2004. Dynamics of Sea Mammal and Bird Populations of the Bering Sea Region Over the Last Several Millennia. *Palaeogeography, Palaeoclimatology, Palaeoecology* 209, 335-352.

Savinetsky, Arkady B., Dixie Lee West, Zhanna Antipushina, Bulat F. Khassanov, Nina K. Kiseleva, Olga B. Krylovich, Andrei M. Pereladov. 2011. The Reconstruction of Ecosystems History of Adak Island (Aleutian Islands) During the Holocene. Chapter 5. In D. West, V. Hatfield, E. Wilmerding, C. Lefèvre, and L. Gualtieri (eds.), *The People Before: The Geology, Paleoecology and Archaeology of Adak Island, Alaska*. Oxford, British Archaeologial Reports.

Slobodkin, Lawrence B. 1961. *Growth and Regulation of Animal Populations*. New York, Holt, Rinehart, and Winston.

Spaulding Albert C. 1962. *Archaeological Investigations on Agattu, Aleutian Islands*. Ann Arbor, The University of Michigan.

Speth, John D. 1983. *Bison Kills and Bone Counts: Decision Making by Ancient Hunters*. Chicago, University of Chicago Press.

Speth, John D. 1987. Early Hominid Subsistence Strategies in Seasonal Habitats. *Journal of Archaeological Science* 14, 13-29.

Speth, John, D. and K. A. Spielmann. 1983. Energy Source, Protein Metabolism and Hunter-Gatherer Subsistence Strategies. *Journal of Anthropological Archaeology* 2, 1-31.

Veniaminov, Ivan. 1984. *Notes on the Islands of the Unalashka District*. [Lydia Black and R. H. Geoghegan, trans.]. Kingston, Ontario, The Limestone Press.

Waselkov, Gregory A. 1987. Shellfish Gathering and Shell Midden Archaeology. *Advances in Archaeological Method and Theory* 10, 93-210.

Watt, Bernice and Annabelle Merrill. 1963. Composition of Foods. *Agricultural Handbook* No. 8. Washington, DC, United States Department of Agriculture.

West, Dixie. 1997. *Hunting Strategies in Central Europe During the Last Glacial Maximum*. British Archaeological Reports International Series 672. Oxford.

West, Dixie L., Debra Corbett and Christine Lefèvre. 1999. 1999 NSF Grant Report: the Western Aleutians Archaeological and Paleobiological Project. Unpublished report on file, National Science Foundation, Washington, DC.

West, Dixie L., Debra Corbett and Christine Lefèvre. 2002. 2002 NSF Grant Report: the Western Aleutians Archaeological and Paleobiological Project. Unpublished report on file, Washington, DC, National Science Foundation.

West, Dixie, Joan Coltrain, M. Geoffry Hayes, Dennis H. O'Rourke, Christine Lefèvre, Debra Corbett, Françoise Bouchet, Arkady Savinetsky. 2011. A Human Burial from the Andreanof Islands, Alaska. Chapter 16. In D. West, V. Hatfield, E. Wilmerding, C. Lefèvre, and L. Gualtieri (eds.), *The People Before: The Geology, Paleoecology and Archaeology of Adak Island, Alaska.* Oxford, British Archaeologial Reports.

Winterhalder, Bruce, and Eric A. Smith. 1981. *Hunter-gatherer Foraging Strategies: Ethnographic and Archaeological Analyses.* Chicago, Illinois, University of Chicago Press.

Wright, Bruce, Victoria Gofman, and Ray RaLonde. 2006. Response and Intervention System for Climate Change Induced Paralytic Shellfish Poisoning in Aleut Communities. Project #644. (http://doc. nprb.org/web/06_prjs/644_pr_jul-sept06.pdf)

Wright, Bruce, Ray RaLonde, Victoria Gofman, Michele Harmon, and Kabor Khanna. 2008. North Pacific Research Board Project Final Report: Response and Intervention System for Climate Change Induced Paralytic Shellfish Poisoning (PSP) in Aleut Communities. North Pacific Research Board Project 644 Final Report.

CHAPTER 11
PROVENANCE OF OBSIDIAN ARTIFACTS RECOVERED FROM ADAK ISLAND, CENTRAL ALEUTIAN ISLANDS: EVIDENCE FOR LONG-DISTANCE TRANSPORT OF LITHIC MATERIAL

Kirsten Nicolaysen, Taylor Johnson, Elizabeth Wilmerding, Virginia Hatfield, Dixie West and Robert G. McGimsey

Abstract: *Excavations of archaeological sites ADK-171 and ADK-011 have recovered obsidian artifacts, yet obsidian does not outcrop on Adak Island. The only well-known Aleutian obsidian source is from Okmok volcano on Umnak Island, approximately 650km to the east of Adak Island. This paper describes and characterizes the geochemistry of an obsidian-bearing deposit on Akutan Island, approximately 800km east of Adak; this Akutan locality is a second potential source of obsidian for Aleut lithic tools prior to European contact. To determine whether the Okmok or Akutan obsidian localities provided the raw material for the artifacts on Adak Island, analysis by Inductively Coupled Plasma Mass Spectrometry measured abundances of 38 chemical elements from four obsidian flakes from Adak archaeological sites, a glassy dacitic fragment, and samples from the Akutan and Okmok source obsidian exposures. Bivariate plots and Principle Components Analysis indicate that the Adak obsidian artifacts are chemically indistinguishable and came from the same source, though they were excavated from two different sites and from occupation layers that differ significantly in age. This analysis supports the hypothesis that the obsidian for these Adak artifacts originated from Okmok volcano. The closest potential source for the dacitic fragment is a lava flow from Atka Island, located approximately 160km east of Adak. These results indicate that Aleuts conveyed high-quality lithic materials over long distances through trade, direct transport, or raiding.*

Keywords: *Obsidian, inductively couple plasma mass spectrometry, Adak Island, Okmok, Akutan*

Introduction

Obsidian is a highly prized lithic material for manufacturing tools including projectile points, knives, and scrapers. Controlled flaking of obsidian by percussion or pressure yields extremely thin, sharp edges that are nearly unrivaled for keenness of cutting, but this response depends on the habit of obsidian to break by conchoidal fracture. This characteristic derives from the isotropic silica bonding in obsidian—i.e., it is similar in any direction throughout the glass and lacks a crystalline lattice – and from the small size of any minerals present within the glass. For example, Castro and Mercer (2004) documented that microlites in obsidian are typically less than 12 microns in longest dimension. Although glassy volcanic lavas appear an attractive resource for lithic tools, they contain larger minerals in a microcrystalline matrix and thus do not break to create razor thin, fine edges, though they are more robust for hammer tools (e.g., adzes). The basalt and andesite lavas that commonly comprise the volcanic islands of the central Aleutians are unsuitable for the finest bladed tools. Thus obsidian, though erupted in isolated locations in Kamchatka, Alaska, and the Aleutian Islands, proves a resource worth obtaining despite difficult access (e.g., Cook 1995; Reuther *et al.* 2011; Kuzmin and Glascock 2010 and papers therein).

As documented by Wilmerding and Hatfield (Chapter 12, this volume) obsidian was recovered from three sites excavated on Adak. To summarize based on Okuno *et al.* and Wilmerding and Hatfield (Chapters 4 and 12, respectively, this volume), Test Pit 55-1 of the Tutiakoff site (ADK-171) contained the highest proportion of obsidian debitage from the Late Anangula Phase (7000-4000 years BP) though both the Dozered site (ADK-012) and Component 1 of the Zeto Point site (ADK-011) contained minor quantities of obsidian. In the younger Component 2 of ADK-011 (160-415 years BP, corresponding to Late Aleutian Phase), the only recovered obsidian of note comprised single fragments of two different tools. Not only is obsidian rare in the Adak excavations, its use appears to decrease with time. Three possibilities may explain the scarcity and decreasing abundance of archaeologically-significant obsidian on Adak: a) the material itself was difficult to obtain, b) sampling of Adak obsidian is limited by the size of the excavated sites, c) competition for obsidian increased with an increase in population pressure, or d) technology of lithic tools or lithic resources changed such that obsidian as a resource was devalued, though we find this the least likely hypothesis.

We propose to evaluate the relative difficulty for the Aleuts to obtain obsidian by comparing the geochemical signatures

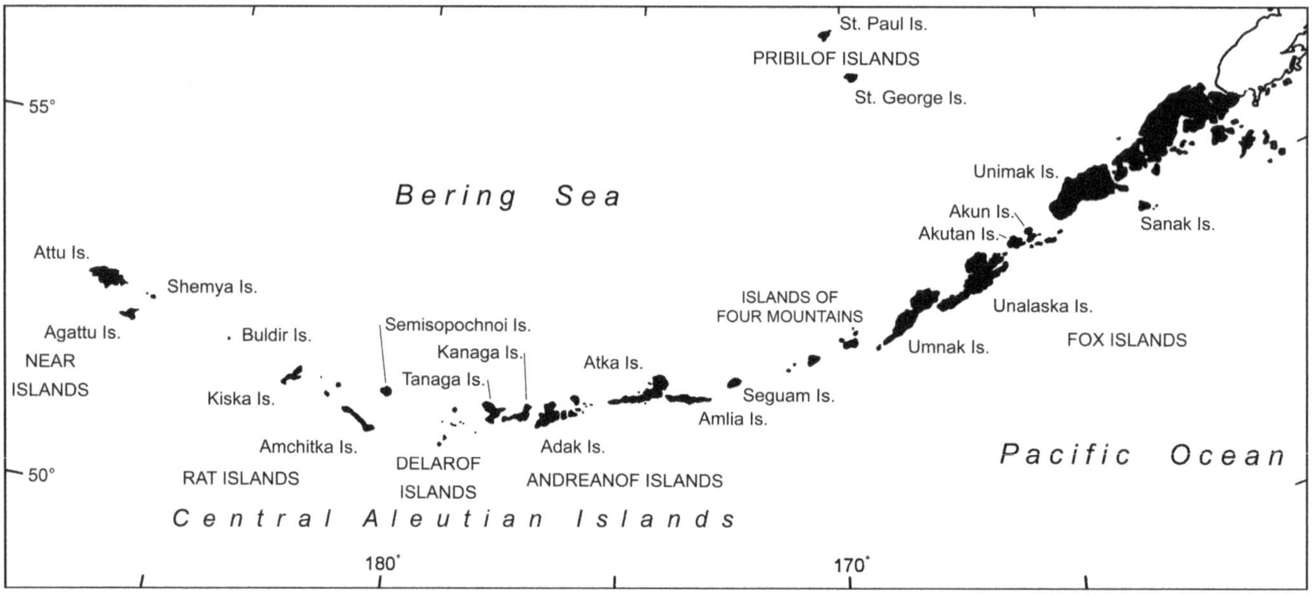

Figure 11.1. Locations of Umnak and Akutan Islands in the Aleutians.

of trace elements in selected obsidian debitage from Adak with the geochemical signatures of obsidian exposed in two locations in the eastern Aleutian Islands (Figure 11.1). Occam's razor and personal experience suggest that people on Adak would have sought the closest possible resources to replenish the supply of stone needed for various tools. Is it possible then that either of the Okmok and Akutan sources provided the obsidian from which Adak flaked tools were created and used when these islands are respectively 650 and 800km east of Adak and the people travelled by hand-paddled kayaks and umiaqs? The hypothesis we test in this paper is that the paucity of obsidian, the small size of recovered flakes, and the lack of known nearby sources requires that the material was transported great distances. By examining the chemical compositions of selected recovered flakes and of known obsidian sources, the data persuade us that Adak obsidian came from a single source and that Okmok is the probable source for the obsidian used on Adak. Furthermore, our results have implications for resource practices and social interactions among Aleuts of distinct cultural traditions.

Aleutian Obsidian Sources

In order for obsidian to erupt as lava, the magma prior to eruption must have extremely high silica content, usually greater than 70% SiO_2 by weight, and low water content, usually less than 2% by weight (e.g., Castro and Mercer 2004). In addition to lava flows, infrequent occurrences of obsidian may be exposed in near-surface vents (Tuffen and Castro 2009), in hyaloclastite deposits formed when water or ice quenches explosively erupt silica-rich magma (Scutter 1998; Tuffen and Castro 2009) or, when hot, thick ash deposits remelt into glass due to their own heat trapped as near-vent deposits cool (e.g., Boyd 1961; Henry and Wolff 1992). Obsidian lavas are most frequently found erupted in volcanic arcs constructed on margins of continental, lithospheric plates because, in the subsurface, magma can preferentially, partially incorporate minerals rich in silica that have low melting temperatures; quartz is the best-known and most abundant, but alkali feldspar minerals also fulfill these criteria. Continental crust tends to have these minerals in much greater abundance than oceanic crust, and the greater thickness of continental crust that the magma traverses to reach the surface provides more opportunity for this partial assimilation process. Accordingly, the identified geologic localities of obsidian used by the peoples of eastern and western Beringia are preferentially located in Kamchatka, Chukotka, and mainland Alaska and Canada (Cook 1995; Grebennikov *et al.* 2010; Kuzmin, 2010; Phillips 2010; Phillips and Speakman 2009; Reuther *et al.* 2011). Thus, the conditions for obsidian eruption in the Aleutians seem to have been met in only two currently known locations: Akutan Island and Umnak Island in the eastern Aleutians (Figure 11.1).

Two known sources of obsidian exist in the eastern Aleutian chain: on Umnak Island and Akutan Island, which are 650 and 800 km, respectively, from Adak Island. (Note, distances are given as the crow flies and do not reflect actual distances traveled by seafaring voyage.) Currently, geochemical analyses of the obsidian from the caldera of Okmok volcano on Umnak Island are in the Alaska Archaeological Obsidian Database (AAOD; Reuther *et al.* 2011), and Cook (1995) surmised that his Group I obsidian came from Okmok. We describe the geologic setting of a second obsidian source, located on Akutan Island, and present the first geochemical characterization of this source in the archaeological literature, though two analyses exist in the geologic literature (George *et al.* 2004; Richter *et al.* 1998). Characterization of the Akutan source is being added to the AAOD based on this work and that of others (J. Rasic, personal communication).

The Okmok obsidian source is located on the northern part of Umnak Island, approximately 100km west of Dutch Harbor. The following geologic description of the Okmok source is derived largely from Larsen *et al.* (2007 and sources therein). The eastern portion of Umnak is roughly 30km in diameter and roughly elliptical in shape due to its formation from the eruption of Okmok volcano over the last ~2 million years. Two large, catastrophic eruptions of Okmok occurred 12,000 ^{14}C BP (Okmok I) and 2050 ^{14}C BP (Begét *et al.* 2005; Byers 1959; Larsen *et al.* 2007; Miller *et al.* 1998), creating two nested calderas, the younger of which (Okmok II) has a diameter of ~10km. Crater Creek drains surface water from the calderas, running approximately 10km from the breached caldera wall to the northeastern coast of Umnak. Larsen *et al.* (2007) report that a Pleistocene age 'glassy rhyolite lava flow [is] exposed high along the outer wall of the North Arcuate rim' of the older caldera. This lava flow is approximately 7km from the coast, as the crow flies. Although this source has been exposed due to the caldera-forming eruption 12,000 years BP and possibly to glaciation, pre-dating the arrival of humans in the Aleutian Islands, more than 16 significant eruptions subsequent to the formation of Okmok I and prior to European contact, including the massive Okmok II eruption of 2050 years BP, would have made this highly-valued resource sometimes dangerous to access.

Akutan Island is located just east of Unalaska Island, approximately 50km from Dutch Harbor, and has two exposed sources of obsidian. The first obsidian locality is at the southeast tip, and we describe this source in some geologic detail. MacDonald and others (1992) reported a second Akutan source at the southwest part of the island and this will be evaluated in 2011 (J. Rasic, pers. comm.). Based on reconnaissance field geology, the first Akutan obsidian is not present as a lava flow or dome but rather crops out as cobble-sized clasts in a thick debris flow exposed in a sea-level beach cliff and adjacent stack at the east end of Green Bight at the southeast tip of Akutan Island. This deposit dips ~3° NNW toward the island indicating its origin from an ancestral off-shore volcano that has eroded away (Richter *et al.* 1998). George *et al.* (2004) provided one analysis of a clast, presumably from the same deposit. We seek to correct the descriptions by Richter *et al.* (1998) and George *et al.* (2004) that report this deposit as a pyroclastic flow by affirming that geological descriptions by two field geologists note this deposit was laid down as a subaqueous volcanic lahar. Major element data indicate that this obsidian is a true rhyolite in composition (72.49 wt. % SiO_2; George *et al.* 2004). Possibly these obsidian fragments initially formed in a near-vent lava dome or satellite lava flow, and the lava eroded and emplaced as a volcanic debris flow. The Akutan deposit is located near a modern sea lion rookery and probably would have been visible to Aleut hunters and local villagers on, or traveling near, the shore. This then provides an opportunity for fortuitous collection of lithic material during hunting forays, an example of an embedded procurement strategy (Binford 1979).

Geology of Adak Island

To provide context for what lithic materials would have been available to the Aleuts on Adak, we briefly summarize the geology near the ADK-171 and ADK-011 archaeological sites based on previous work of Coats (1956), Kay *et al.* (1983), Myers and Marsh (1987), Waythomas (1995), and Gualtieri *et al.* (Chapter 3 this volume). Three major Quaternary volcanoes (Mt. Moffett, Mt. Adagdak and Andrew Bay volcano) and the extensive pre-Quaternary Finger Bay Formation provide the geologic substrate for habitation on northern Adak. ADK-171 and ADK-011 are located within Quaternary unconsolidated deposits overlying the Finger Bay Formation. The Finger Bay Formation consists of basalt flows, associated pyroclastic and epiclastic deposits, and intrusive andesite porphyries and a gabbroic and dioritic intrusions at Finger Bay and Sweeper cove. Jicha *et al.* (2006, 662) show that the basalts and intrusions range from 34-38 Ma in age. Grassy plateaus of unconsolidated sediment drape the lower, southern slopes of the volcanoes of Mts. Moffett and Adagdak. That these lahars, or volcanic debris flows, have not been glaciated suggests that these volcanoes erupted significantly during the Holocene, though the degree of erosion at the volcanic summits indicates that much of the basalt to andesite lava likely erupted in the Pleistocene (also see Okuno *et al.* Chapter 4 this volume). Clam Lagoon as well as nearby Andrew Lake, formed when glaciers descending Mt. Moffett bulldozed shallow basins during the Pleistocene (but see Okuno Chapter 4 this volume).

Most importantly, none of the geologic reports of Adak have described the presence of obsidian on the island. Our observations and conversations with numerous geologic colleagues confirm the absence of exposures of obsidian on Adak or adjacent central Aleutian Islands. Furthermore, prior to our research, archaeologists conducting fieldwork on Adak did not observe obsidian either in natural outcrops or in an archaeological context (Bank 1950; Frolich and Kopjansky 1975; Hrdlička 1945; Luttrell and Corbett 2000; McCartney 1972; O'Leary 1998; Veltre 1997). Only O'Leary (2001, 222) reported, 'a small unifacial scraper made on a fine grained basalt or obsidian flake' at ADK-171 on north Adak Island that was determined in 2006 to have originated from the Okmok source (R. J. Speakman personal communication). More than 60 years of geologic and archaeological study in the central Aleutians documents the absence of geologic obsidian exposures in these islands.

Obsidian Recovered from Adak Archaeological Sites: Sample Locations and Descriptions

The Adak lithic artifacts used in this analysis originated from two archaeological sites, ADK-171 and ADK-011. In general these flakes are small (less than 1cm^2 in greatest dimension) and scarce within the excavated areas,

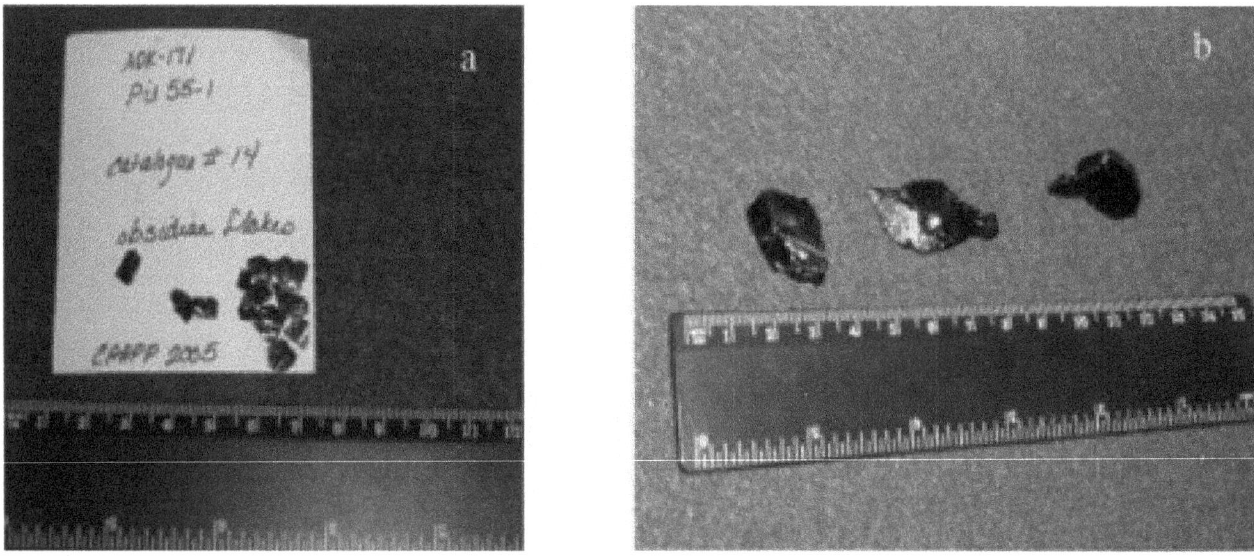

Figure 11.2. Photos of obsidian recovered from archaeological site ADK-171 (a) and from Akutan cobble (b). Scale in centimeters.

particularly in the 8m long trench excavated at ADK-011. Only the debitage of Pit 55-1 of the Tutiakoff site (ADK-171) had a substantial amount of obsidian (~56%). Recovery of these flakes was greatly aided by use of water screening using 1/8" mesh sieves. Some flakes were easily identified as smoky black obsidian with typical glassy luster and conchoidal fracture (Figure 11.2). We provide brief site descriptions and direct interested readers to the extensive descriptions of the sites and the recovered lithic material in West *et al.* and Wilmerding and Hatfield (Chapters 1 and 12, respectively, this volume)

Sample Location 1

ADK-171, Test Pit 55-1 is a 1.3m X 1m unit located on the edge of a 7m X 4m depression flanking the western margin of a low swale overlooking Clam Lagoon. A charcoal lens 75cm below surface, and dating to 6041 $\pm$ 30 years BP (Okuno AMS date; 6040$\pm$30 BP for NUTA2-10967), contained chipped and ground stone tools and flaking debris including obsidian (see Wilmerding and Hatfield Chapter 12 this volume). A total of 128 flakes and five tools were recovered from this cultural zone. Of this total, a little more than 56% represents debris associated with resharpening or final shaping of obsidian bifacial tools. No obsidian tools were recovered at the site; this suggests that Aleuts carried obsidian tools away from the site. ADK-171 represents the earliest known Aleut occupation in the Central Aleutian Islands and correlates with the Late Anangula Phase described for the eastern Aleutians (Knecht and Davis 2001) with only the Amaknak Quarry site dating to around the same time as ADK-171 (Hatfield 2010; Rogers *et al.* 2009).

Sample Location 2

ADK-011 is one of the largest, richest and best-preserved multicomponent sites in the Clam Lagoon vicinity. In 2006 archaeologists excavated ten 1m X 1m units (Units B-K) situated to transect a house feature and an adjacent bone and shell midden at ADK-011. Excavated in 10cm arbitrary levels, Units B-F represented the kitchen midden and Units G-K represented the house feature. Archaeologists excavated nine archaeological levels, in 10cm arbitrary increments, that span from 170-2400 years BP (See West *et al.* Chapter 1 this volume). The kitchen midden accumulated between 170 and 400 years BP (layer 1: 170 years BP; layer 2 level 2: 285 years BP; layer 2 levels 3, 4,5, 6,and 7: 400 years BP). In the house feature, Layers 1-3 date 335-395 years BP. The age of the midden Layer 2, levels 3, 4, 5, 6, and 7 overlaps with the younger levels of the house. Layers dating between 170 and 400 years BP represent Component 2. A series of charcoal lenses (Layer 4 and below) interpreted as house floors, date to approximately 2500 years BP and comprise Component 1. Table 11.1 lists the provenience of obsidian tools recovered from ADK-011.

ADK-011 debitage included 23 very small and two larger obsidian flakes in Layers 1 and 2 (Component 2) and three very small and three larger obsidian flakes in Layer 4 (Component 1). The obsidian debitage was largely undiagnostic and could have resulted during any stage of tool manufacture or curation. Two pressure flakes and bifacial shaping flakes represent the residue of shaping bifaces (either projectile points or knives.)

We selected five stone samples for analysis primarily based on their size and composition; all were obsidian with one exception (Table 11.2). Sample 825 is a piece of aphyric siliceous lava that breaks conchoidally but does not have the glassy luster typical of obsidian and is likely dacitic in composition. Because nearly all flakes were too small for analysis by X-Ray Fluorescence or Neutron Activation, we analyzed individual fragments by Inductively Coupled Plasma Mass Spectrometry (ICPMS), which requires a minimum of approximately 100 mg. In addition to the

Table 11.1. Obsidian tools from ADK-011.

Catalogue #	***Unit***	***Layer: Level***	***Artifact type***	***Age years BP***
814	G	4; 10b	Retouched blade tool	2500
554J	C	2	Projectile point tip	285-400
629L	C	2	Flake tool	285-400

Table 11.2. Provenience and mass weight of analyzed Adak samples.

Catalogue #	***Provenience***	***Unit, Layer***	***Identification***	***Mass (grams)***	***Age, years BP***
749c (a)	ADK-011	G, 4	Obsidian	0.096	2500
664	ADK-011	C, 2	Obsidian	0.146	285-400
14	ADK-171	Pit 55-1	Obsidian	0.843	6000
12	ADK-171	Pit-55-1	Obsidian	0.253	6000
825	ADK-011	D,3,4	Dacite	17.375	170-400

archaeological samples, co-author R. G. McGimsey of the Alaska Volcano Observatory provided five samples of obsidian from cobbles in the Akutan Island debris flow. An analysis of a comparative sample from Okmok Island was generously shared by N. Slobodina (personal communication).

Methods

At the Pacific Centre for Isotopic and Geochemical Research (PCIGR) at the University of British Columbia, the five lithic artifact samples and five Akutan comparative samples were dissolved prior to analysis on a *Thermo Finnigan Element 2* inductively coupled plasma mass spectrometer (ICPMS). Samples were dissolved using concentrated nitric and hydrofluoric acids using the procedure reported by Pretorius *et al.* (2006). A brief description of the methods is outlined here. After evaporation of the HF-HNO_3 solution, samples were converted to chlorides using 6 M HCl, dried, and reconstituted in HNO_3. Samples were diluted to a solution of 2% HNO_3 and spiked with 10 ppb of indium (In; the internal standard). Comparison of the measured abundances of In for each analysis to the known 10 ppb quantity allows one to correct measured abundances of all trace elements for instrumental drift during analysis. Samples were split into two aliquots and diluted appropriately. Analysis of the first aliquot measured primarily high field strength element abundances (Li, Nb, Mo, Cd, Sn, Sb, Cs, Hf, Ta, W, Bi, V, Cu, Zn, Ga, Rb, Sr, Zr, Ba) and analysis of the second focused on the measurement of rare earth (REE) and selected other element compositions (REE, Sc, Y, Th, U. Tm, Pb) (Table 11.3).

Both accuracy and precision of the sample dissolution method and the instrumental analysis were evaluated. Accuracy of the instrumental analysis was controlled by dissolving and analyzing an aliquot of US Geological Survey standard RGM-1, a standard reference material used to ensure quality analysis. We analyzed RGM-1 because this rhyolitic obsidian from Glass Mountain, California is expected to have similar abundances of major and minor elements that create the solution matrix within which the analyzed trace elements are dissolved. Table 11.3 shows the composition of RGM-1 measured at the time of the sample analyses and expected compositions of RGM-1 based on published USGS values and on the long-term average values obtained at PCIGR. With the exception of seven elements, analyses at PCIGR match published values of RGM-1 within 10%, and for 15 elements our analyses match published values at 6% or better. We also evaluated the reproducibility of the method and the instrumental analysis by performing a second digestion of sample 12a and by performing a replicate analysis of the same digested aliquot (12a-1) on the ICPMS. In terms of reproducibility, the duplicate and replicate analyses are within the error defined by the evaluation of RGM-1 (accuracy analysis) except for Ba, Sc, Ce, Nd, Th, U, and V, but in general these differ by a few percent.

Results

Because ICPMS cannot measure major element compositions easily without substantial dilution of sample aliquots, we are unable to classify the recovered fragments as to their lava composition based on silica content. Identifying the variation in abundances of 38 trace element for the samples was prioritized because this variation provides a more reliable chemical fingerprint for provenance determination than the ~8 elements returned by major element analysis. Here we use the term provenance to indicate the relationship between a lithic artifact and its ultimate source in a surface outcrop and provenience to designate the location of a fragment within its archaeological context.

Akutan Source Obsidian

Our analysis of the five different chips of obsidian from Akutan yields a very consistent trace element signature (Table 11.3, Figures 11.3-11.5). For the 38 analyzed trace elements, the standard deviation among the five samples is 3% or less for all elements except Ba, Sr, Zn, and Zr.

Table 11.3: Compositions of lithic flakes from ADK-011 obtained by ICP-MS

	RGM-1	USGS value	PCIGR value	% diff. std	12a-1	12a-2	12a average (duplicate, original)	duplicate %RSD	12a-1 replicate analysis	12a-1 average (replicate, original)	replicate %RSD	14a	749c	664	825	28a	28b	28c	28d	28e
					obsidian ADK-171	duplicate digestion						obsidian ADK-171	obsidian ADK-11, G,4	obsidian ADK-11, C, 2	dacite ADK-11, D, 3 & 4	Akutan source	Akutan source	Akutan source	Akutan source	Akutan source
Sample mass (g)					0,253							0,843	0,010	0,146	17,35					
Li	61,9	57,0	75,0	9	42,1	45,3	43,7	5	42,7	42,4	1	47	45	47	25	40	40	53	41	38
Nb	8,3	8,9	8,0	7	12,9	13,6	13,2	4	12,0	12,5	5	14	13	13	5	13	12	10	13	12
Mo	2,4	2,3	2,4	4	4,1	4,2	4,1	1	4,0	4,1	1	4,3	4,2	4,1	2,9	4,3	4,5	4,0	4,4	4,2
Cd	0,2		0,1		0,4	0,4	0,4	1	0,4	0,4	3	0,5	0,4	0,4	0,4	0,6	0,6	0,6	0,5	0,6
Sn	4,4	4,1	3,8	7	4,8	4,9	4,8	2	4,3	4,5	8	4,9	4,8	4,5	2,2	3,9	3,9	3,7	3,7	3,7
Sb	1,2	1,3	1,0	8	2,1	2,2	2,1	3	2,0	2,0	4	2,1	2,0	2,1	0,3	0,8	0,8	0,7	0,9	0,8
Cs	10,0	9,6	10,6	4	7,1	7,5	7,3	4	7,0	7,1	1	7,6	7,3	7,4	5,8	5,9	6,1	5,8	6,0	5,7
Hf	5,9		5,8		10,7	11,2	11,0	3	9,7	10,2	7	11,2	10,8	10,5	7,3	11,8	12,2	13,6	11,8	11,6
Ta	0,7	1,0	0,9	26	0,7	0,8	0,7	6	0,8	0,7	9	0,8	0,7	0,5	0,3	0,9	0,8	0,8	0,9	0,9
W	1,3	1,5	1,8	13	2,0	2,1	2,1	3	2,0	2,0	3	2,4	2,0	2,1	0,9	1,3	1,3	1,2	1,5	1,4
Bi	0,3		0,2		0,1	0,1	0,1	1	0,1	0,9	3	0,1	0,1	0,1	0,1	0,0	0,0	0,0	0,1	0,0
V	11,0	13,0	10,0	15	2,5	4,2	3,3	35	2,7	2,6	6	1,9	1,6	2,2	19,5	0,6	0,6	0,7	0,6	0,5
Cu	9,4	12,0	9,3	22	17,6	18,6	18,1	4	17,6	17,6	0	18	18	18	4,5	24	25	30	25	24
Zn	43,8	32,0	32,0	37	132,1	166,3	149,2	16	113,0	122,6	11	124	220	173	111	93	71	80	82	71
Ga	15,7	15,0	17,0	5	20,5	21,2	20,9	3	20,4	20,4	0	22	21	21	21	17	17	18	17	17
Rb	143,9	150,0	152,0	4	95,3	99,6	97,4	3	88,0	91,6	6	101	97	95	104	112	112	107	110	107
Sr	100,3	110,0	118,0	9	84,3	89,7	87,0	4	83,0	83,7	1	89	87	88	183	46	47	64	48	43
Zr	210,4	220,0	215,0	4	361,0	377,0	369,0	3	354,3	357,7	1	387	367	366	287	466	468	543	471	446
Ba	813,8	810,0	911,0	0	1125,2	1195,5	1160,4	4	1134,0	1129,6	1	1217	1169	1212	1503	1095	1095	1079	1115	1037
Sc	4,3	4,4	4,5	2	5,7	6,3	6,0	8	5,9	5,8	3	6,2	6,0	6,2	7,0	6,0	6,1	7,3	6,2	5,7
Y	21,2	25,0	20,0	15	68,7	75,7	72,2	7	68,2	68,5	1	76	76	75	30	53	53	56	51	48
La	22,8	24,0	23,4	5	33,8	33,7	33,7	0	34,5	34,1	1	34	34	35	27	33	33	34	35	32
Ce	47,2	47,0	47,0	0	81,4	86,1	83,7	4	79,3	80,4	2	87	85	84	58	76	77	80	77	72

Pr	5,2		5,3		10,2	10,9	10,5	4	9,9	10,1	2	10,9	10,9	10,9	7,3	9,9	10,0	10,5	9,8	9,2
Nd	18,8	19,0	18,8	1	42,6	46,2	44,4	6	42,6	42,6	0	47	46	47	30	41	41	43	41	38
Sm	3,9	4,3	3,9	9	10,3	11,1	10,7	5	10,4	10,4	0	11,2	11,1	11,2	6,3	9,3	9,3	9,9	9,3	8,8
Eu	0,5	0,7	0,6	29	1,5	1,5	1,5	3	1,5	1,5	1	2	2	2	1	1	1	1	1	1
Gd	3,5	3,7	3,5	5	11,4	11,7	11,6	2	10,9	11,2	3	11,8	12,0	12,0	5,6	8,8	9,3	9,8	8,9	8,7
Tb	0,6		0,6		1,8	2,0	1,9	5	1,8	1,8	2	1,9	1,9	1,9	0,8	1,4	1,5	1,6	1,5	1,4
Dy	3,7	4,1	3,5	10	13,1	13,5	13,3	2	11,8	12,5	8	14,0	13,4	13,4	5,3	9,3	9,6	10,3	9,1	8,7
Ho	0,8		0,8		2,8	2,9	2,8	4	2,6	2,7	5	2,9	2,9	2,9	1,2	2,1	2,1	2,2	2,0	2,0
Er	2,3		2,3		8,2	8,8	8,5	5	8,0	8,1	1	8,7	8,7	8,7	3,5	6,2	6,2	6,5	6,2	5,8
Tm	0,3		0,3		1,2	1,3	1,3	5	1,2	1,2	1	1,3	1,3	1,4	0,6	1,0	1,0	1,0	0,9	0,9
Yb	2,4	2,6	2,3	8	8,4	9,1	8,7	6	8,2	8,3	1	9,1	9,0	9,1	3,9	6,6	6,7	6,9	6,5	6,2
Lu	0,4	0,4	0,4	0	1,3	1,4	1,4	7	1,3	1,3	3	1,4	1,4	1,4	0,6	1,0	1,0	1,1	1,0	1,0
Pb	23,5	24,0	20,0	2	32,7	33,7	33,2	2	32,2	32,5	1	33,9	33,3	33,5	17,0	20,3	20,4	19,9	19,9	19,0
Th	14,9	15,0	15,1	1	9,4	10,1	9,8	5	9,1	9,3	2	10,2	9,9	10,1	11,2	13,7	13,8	13,4	13,3	12,9
U	5,7	5,8	5,5	2	4,6	4,7	4,7	2	4,3	4,5	5	4,7	4,7	4,7	4,7	6,0	6,0	5,9	5,7	5,6

Notably one chip has significantly higher Zr than the other four, perhaps indicating some heterogeneity in the source material. In general however, the low standard deviation among the five Akutan source samples testifies both to the homogeneity of the material and the reproducibility of the ICPMS data.

George *et al.* (2004) reported major and trace element data for a single sample of a rhyolitic obsidian clast, presumably taken from the same deposit exposed in the southern beach cliffs of Akutan as our samples. This glass sample has 72.5% SiO_2 and 2.60% FeO (total Fe) as well as significant amounts of sodium and potassium (4.83 wt. % Na_2O and 4.97 wt. % K_2O; George *et al.* 2004). Thirty different trace elements are common to both our analysis and George's reported dataset (Table 11.3) and of these, 23 elements are within 10% of our reported average and 17 of the element abundances are within 5% (e.g., Figure 11.3). We believe it is reasonable to infer that their sample should be added to the characterization of the Akutan obsidian source given that 10% is a reasonable difference between two different ICPMS labs.

Geochemical Characteristics of Obsidian from ADK-171 and ADK-011

Inductively Coupled Plasma Mass Spectrometry (ICPMS) measured abundances of 38 elements from four obsidian flakes and a dacitic fragment from the Adak sites. The standard deviation of the compositions (35 of 38 trace elements) for the archaeological Adak obsidian flakes is less than the error in the analytical accuracy for these elements. Essentially the compositions of the analyzed obsidian flakes suggest they originated from a single source although our archaeological samples were recovered from two different sites (ADK-171 and ADK-011) and from occupation layers that differ in age as much as 5000 years (Figures 11.3-11.5). A comparison of Sr and Zr abundances in samples recovered from ADK-171 and ADK-011 with Okmok and Akutan sources suggests that Okmok is the geologic source for these artifacts (Figure 11.3). Rather than making numerous binary comparisons, we use Principle Components Analysis to compare suites of trace element analyses.

The logic behind using Principle Components Analysis (PCA), a statistical comparison of the similarity among samples, involves reducing the various elemental values within the extensive database and the Adak and Akutan sample values and thus to focus only on the variance for each sample. Because of variation in the data sets available for Aleutian lavas, the two source areas and the analyses of the artifacts, our principle components analysis used data for 13 trace elements (Sr, Ba, REE, Hf, Ta, Th). The greatest variance among the values is associated with the first principle component, while the second greatest variance is associated with the second principle (the second coordinate), etc. In our analysis, the first three

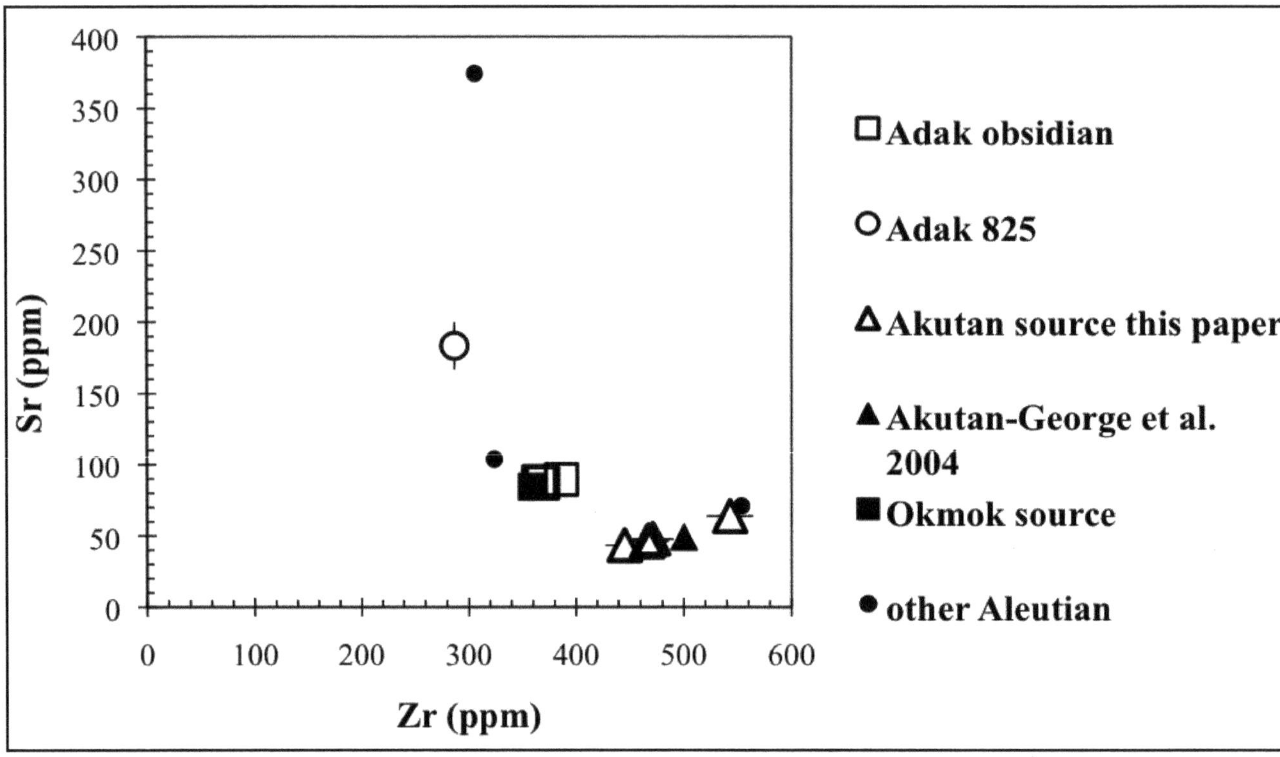

Figure 11.3. Binary comparison of Sr and Zr abundances in samples recovered from ADK-171 and ADK-011 with Okmok and Akutan sources. Sample 825 is a glassy lava, probably dacite in composition. Okmok source data were provided by N. Slobodina (pers. comm.). Other Aleutian points represent three obsidian groups present in samples from the archive of samples excavated in the Aleutians and are distinct geochemical groups in the Alaska Archaeological Obsidian Database as reported by Reuther et al. (2001). Unless otherwise indicated, analytical error is smaller than symbol size for our data.

principle components values explained over 95% of the compositional variance.

Using *SAS 8.0* program and the *PROC PRINCOMP* procedure to run the operation, each principle component produces a column of values that corresponds to the original element input value. The principle component is best represented by elements whose values display a high absolute value. Furthermore, the different principle components are represented by subsets of different elements. These principle component values can be subsequently plotted against one another on bivariate graphs in order to derive a comparative relationship among the Alaskan glassy lava database relative to the Adak and Akutan samples (Figures 11.4, 11.5).

Compared to the compositional variation among all dacitic and rhyolitic lavas, comparative samples from Akutan (open triangles) cluster tightly in both PCA 1 and 2. The slight variation in this cluster indicates the differences from either inhomogeneity within the obsidian itself and/or slight analytical differences. Note that the obsidian from ADK-011 and ADK-171 cluster together; this suggests that the same obsidian source was used although these samples are found at different sites and separated by a significant time gap in the deposition of the flakes (~5,000 years). Furthermore comparison of PC 1 and 2 shows the Okmok obsidian source best matches the compositions of the Adak flakes. The slight difference in PC3 between the Okmok sample and the Adak flakes is linked to the difference in Rb and Ba between the Adak flakes (Rb= 94.9-100.6; Ba=1160-1217ppm) and the Okmok comparative sample (Rb= 107.2 Ba=1545ppm). This difference in Rb and Ba is significant, yet these elements are mobile in the presence of groundwater. Two explanations are possible. First the Adak flakes did come from Okmok originally but long burial in a wet environment leached some Ba from the small flakes. Alternatively the obsidian flakes came from another, as yet unidentified and unsampled location in the Aleutians.

The glassy lithic artifact, sample 825, is geochemically distinct in composition from other samples because this is not a true obsidian but instead a fine-grained lava from an unknown site. Based on the PCA, the closest known potential source for the sample 825 is a lava flow from Atka Island, located ~160km east of Adak. This match was found by using the database of lavas with dacitic or rhyolitic compositions as compiled from Kelemen *et al.* (2003) and Nicolaysen and Nicolaysen and Myers, unpublished. That the recovered dacitic fragment potentially originated from Atka Island is not surprising. Atka is part of the Andreanof Island group and prehistoric contact and communication undoubtedly occurred among these islands. We cannot rule out that another source closer to Adak may ultimately prove a closer match, but this is still a compelling result.

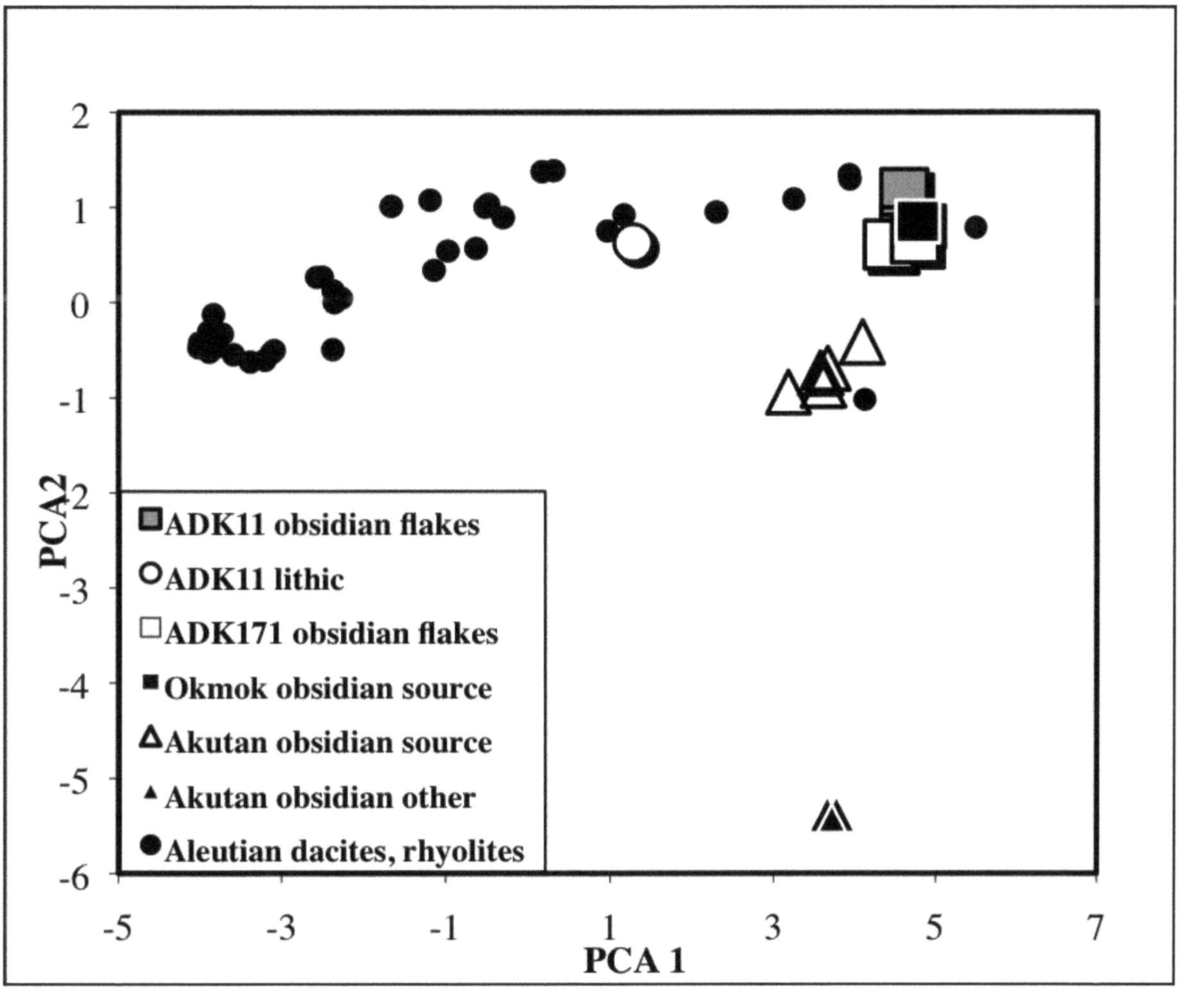

Figure 11.4. Results of Principle Components Analysis. Compared to the compositional variation among all dacitic and rhyolitic lavas, comparative samples from Akutan (open triangles) cluster tightly in both PCA 1 and 2. ADK11 lithic is sample 825. Akutan obsidian other is the analysis from George et al. 2004. Aleutian dacites and rhyolites are compiled from Kelemen et al. 2003 and unpublished data of Nicolaysen and Nicolaysen and J. D. Myers and were selected by screening for lavas whose silica (SiO2) content exceeded 65 wt. %.

Discussion and Conclusions

Obsidian, a rare geologic material in the Aleutians, has been identified in numerous eastern Aleutian archaeological sites and two sites in the central Aleutians. Yet, we know little about how it was quarried and transported from site of origin to place of use. Two possible Aleutian obsidian sources are quite distant and to the east of Adak. A debris flow deposit with obsidian clasts outcrops on Akutan Island, approximately 800km east. A second obsidian deposit from Okmok volcano, Umnak Island is slightly closer to Adak (650km east). We organize our discussion by addressing several questions. What is the source of obsidian debitage recovered from Adak Island Central Aleutians? What is the archaeological record of obsidian use in the eastern Aleutian Islands? Does obsidian appear important in ethnographic accounts of Aleuts? How could obsidian be conveyed to Adak Island? Why does obsidian abundance decrease with time in Adak sites?

What is the source of obsidian debitage recovered from Adak Island Central Aleutians?

The compositional match between the Adak obsidian flakes and the Okmok obsidian source is very compelling (Figure 11.3) with few elements such as Rb and Ba differing significantly among the samples (Ba=1160-1217ppm) and the possible source (Ba=1545ppm). The difference in Ba is possibly explained: 1) heterogeneity in the obsidian, 2) leaching of Ba from the obsidian flakes within the archaeological sites, or 3) another, closer Aleutian obsidian source for the Adak tools has not yet been identified. Based on Neutron Activation Analysis of 74 Okmok source samples, Ba varies from 1056-1322ppm (R.J. Speakman, personal communication), closely overlapping with values for the Adak artifacts, suggesting that source heterogeneity is a reality. Nonetheless Principal Component Analysis (PCA) supports the hypothesis that the Adak obsidian originated at Okmok volcano and demonstrates that the

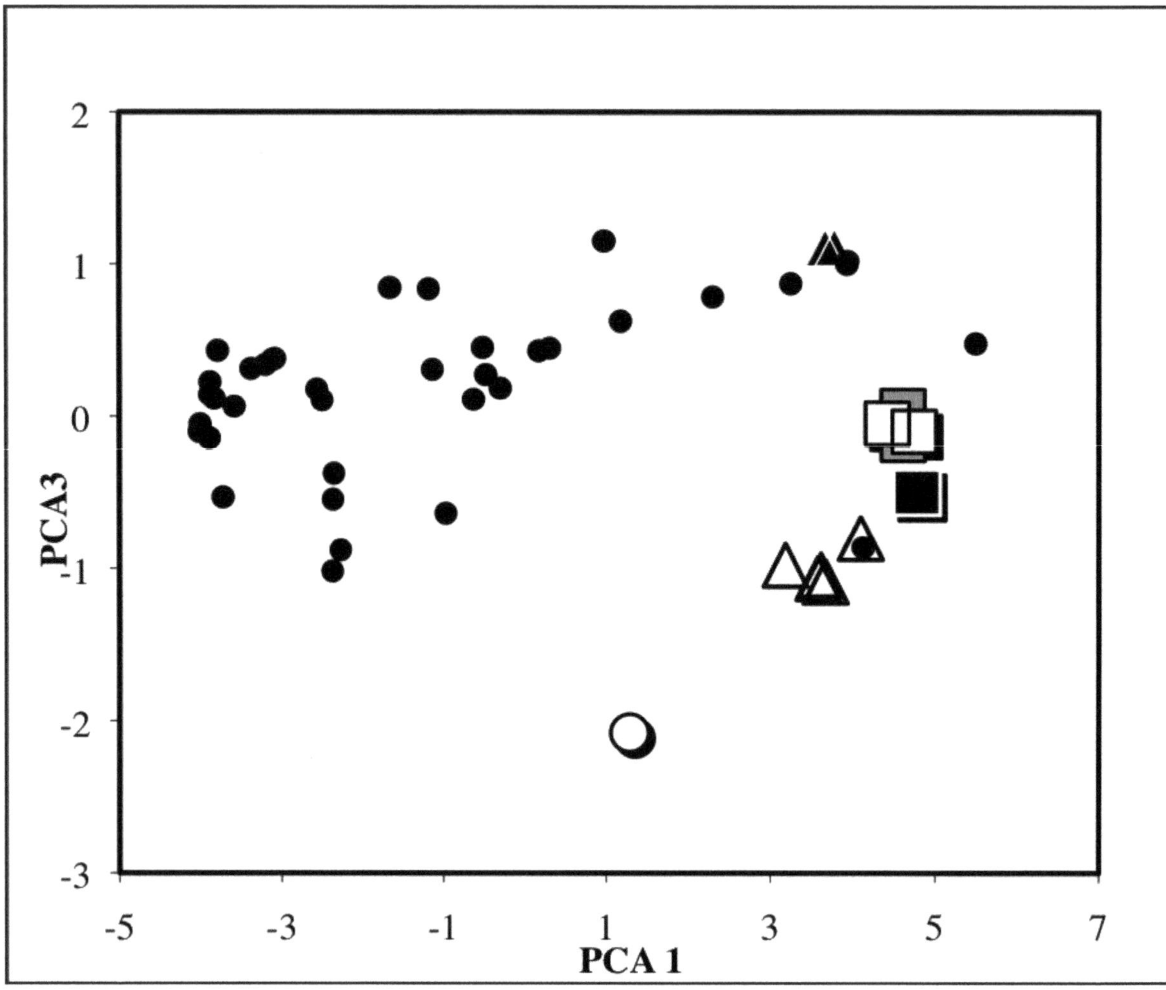

Figure 11.5. Comparison of PCA 1 and PCA 3. Again the ADK-011 and ADK-171 flakes are very similar in composition to each other – essentially identical. ADK-011 and ADK-171 are again closest in composition to the Okmok comparative sample. The slight difference in PCA3 between the Okmok sample and the Adak flakes is linked to the difference in Rb and Ba abundance (see text).

Akutan Island obsidian is not as likely a match as the Okmok source (Figures 11.4, 11.5).

What is the archaeological record of obsidian use in the eastern Aleutian Islands?

Current understanding of the pattern and timing of excavated Aleut settlements indicates that their ancestors crossed the exposed Bering Land Bridge, migrating southward before turning westward and migrating down the Alaska Peninsula and across Aleutian chain from east to west (Liapunova 1996; McCartney 1984; West *et al.* 1999; West *et al.* 2007). The earliest evidence of human occupation in the Aleutians is found at Anangula, north of Umnak Island, and at two sites on Hog Island; all three sites date to circa 8000 years BP. (Knecht and Davis 2001; Laughlin and Marsh 1954). The oldest archaeological site on Adak Island (ADK-171), in the Andreanof Islands, dates to circa 6000 yr. BP. (Luttrell and Corbett 2000; O'Leary 2001; Savinetsky *et al.* Chapter 5 this volume). Archaeologists investigating Amchitka infer that Aleuts reached the Rat Island group at least by 2550 ± 95 yr. BP (Desautels *et al.* 1971, 384); however, prior to nuclear bomb testing, the US Bureau of Indian Affairs examined sites on Amchitka that date to between 4750 and 4300 yr. BP. (Corbett *et al.* 1997). In the Near Islands group, organic material from Krugloi Point on Agattu dated to 2500±300 and 2630 yr. BP. (Spaulding 1962, 12). In the same island group, deposits from Shemya Island dated to 3540±60 yr. BP. and 3096±155 yr. BP. (Corbett 1990; Siegel-Causey *et al.* 1995; West *et al.* 1999).

This westward migration vector provides the possibility that the obsidian resources on Akutan and Okmok were discovered concurrently with settlement of the eastern Aleutians. The archaeological record supports this premise. Archaeologists recovered obsidian at Hog Island, Unalaska (8003±60 RCY BP), one of the earliest sites in the Aleutian archipelago (Dumond and Knecht 2001). At the Margaret Bay Site on Amaknak Island (5500-3000 RCY BP), obsidian comprises as much as 25% of the lithic materials used for chipped stone tools (Knecht *et al.* 2001, Figure 2.4). The only known sources for the Anangula and Hog Island artifacts are believed to come from the

Akutan Island shoreline or Okmok volcano on Umnak (Knecht *et al.* 2001), though Reuther *et al.* (2011) have documented additional geochemical groups of obsidian by analyzing archaeological artifacts whose geographic source is unknown. Similarly, at the South Channel Bridge site, obsidian artifacts were recovered that represented all tool categories; here obsidian was used predominantly for projectile points, as 24% of the projectile points are obsidian (Knecht and Davis 2005). Hoffman (2002, 132) reported small quantities of obsidian tools and debitage for Agayadan Village on Unimak Island. Obsidian is also represented at the Tanaxtaxak (Amaknak Spit) site, UNL-55 (Knecht and Davis 2003) and the Amaknax site UNL-54 (Knecht and Davis 2008). At the Chulka Site, Akun Island, obsidian comprised approximately 4.2% of the total lithic material (Turner and Turner 1974, 49). McCartney (1974, 72) described both obsidian bifaces and debitage among a predominantly basalt assemblage recovered from the Izembek Lagoon whalebone houses on the western end of the Alaskan Peninsula. At Sandy Beach Bay, obsidian was present but comprised less than 1% of material recovered from five excavated house floors (Aigner *et al.* 1976). At Chaluka, Aigner (1978, 21, 23) described ten obsidian bifaces or fragments and noted that obsidian accounted for slightly more than 40% of tools and 37-42% of debitage. Aigner and others (1976, 85) noted that obsidian accounted for 50% of the lithic assemblage in basal Chaluka layers. The Nunik Site on Chernabura Island in the Shumagin Island group contained 23 obsidian artifacts and roughly 300 pieces of debitage in historic levels (110±50 yr. BP), 20 obsidian artifacts and 156 flakes in the Russian contact period house, and a handful of obsidian debitage in archaeological levels dating 3200-400 yr. BP (Wilmerding 2005). Other eastern sites containing obsidian include Ashishik Point, Umnak Island, (Denniston 1972) and Reese Bay, Unalaska Island (McCartney and Veltre 2002, 259). In the central Aleutians, a single obsidian flake tool was recovered from a site on Amchitka (McCartney 1977). Desautels *et al.* (1971, 109) mention a broken obsidian point from the upper level of RAT-31 on Amchitka Island. Evidently a provenance study of these two isolated Amchitka artifacts was not undertaken at the time. In summary, McCartney's generalization (1971, 104) that obsidian has not been reported for the Near Islands but was 'wide-spread in the central and eastern islands' seems to hold true, though the recovery of obsidian from central Aleutian archaeological sites is arguably sparse at the current time.

Does obsidian appear important in ethnographic accounts of Aleuts?

A compelling yet tantalizingly brief ethnographic account in Veniaminov (1984) makes it clear that the Okmok obsidian was known and used. After describing Tulikskoi [Okmok] as a 'lofty and precipitous headland on the easternmost part of the island' [Umnak], the account goes on to say that 'On the NNW side of Tulikskaia sopka [Okmok volcano], black and greenish obsidian is found in the sand and among the ordinary rocks, in chunks of from 20 to 100 pounds, and around this same place porphyry occurs, from which the ancient Aleuts made hand axes and other implements' (Veniaminov 1984, 78). This suggests that the Aleuts collected the obsidian from secondary beach deposits rather than trekking inland to the source described in Larsen *et al.* (2007). It seems possible that the deposit described in Veniaminov might be a glacial till, volcanic debris flow, or stream flood deposit that contains clasts of the inland rhyolitic lava described by Larsen *et al.* (2007). Alternatively, it is possible that the obsidian clasts within the beach deposit are not derived from the inland lava, and archaeologists have yet to sample and characterize this obsidian deposit.

Given that Veniaminov (1984, 79) reports 20 settlements on Umnak Island prior to Russian arrival and that the Okmok obsidian features as a valued resource for tools in his collected ethnography, it seems quite conceivable that obsidian was utilized and shared among the Aleuts on Umnak. Archaeological evidence supports the widespread use of obsidian on Umnak as is seen at Ashishik Point located 6km west of Crater Creek and seen at Anangula Blade and Chaluka about 90km away from Crater Creek as the crow flies (Aigner 1978; Denniston 1972; Knecht and Davis 2001). Interaction among villagers on Umnak and islands to the west would provide a means for obsidian transport to the central Aleutians.

Obsidian also appears in ethnographic accounts regarding Akutan Island, where seven villages existed prior to Russian contact (Liapunova 1996, 165). For example, Veniaminov (1984, 99-100) states 'From the middle of a stratified cliff on the southeast side of the island [Akutan], three streams flow. In the same spot obsidian is found, which breaks off in chunks from above.'

How would obsidian be conveyed to Adak Island?

Correlating the source obsidians with those recovered from Aleutian archaeological sites has major implications for travel and trade patterns among Aleut groups, many of whom were in conflict perhaps due to resource competition in a marginal habitat. Our results indicate that Aleuts transported high quality, lithic materials long distances either through exchange and trade networks, direct transport, or raiding. Archaeological theory presents examples wherein higher value of items is correlated directly with distance from which the item originated and effort required in tool manufacture (e.g., Dillian and White 2010; Earle 2010 and references therein). The literature abounds with examples of high quality obsidian being found at sites long distances from their geologic sources (Dillian 2010; Glascock 2002; Shackley 2005), and this is true in spite of the challenging arctic and sub-arctic conditions of eastern and western Beringia as well (Kuzmin 2010; Kuzmin and Glascock 2010; Reuther *et al.* 2011). However, in most of these documented instances of obsidian transport, the resource network – whether actuated by exchange, trade or raiding – is an overland network. We do not seek to minimize the importance of the danger to prehistoric peoples in crossing mountain ranges, deserts and substantial rivers in effecting obsidian transport, but the use of baidarkas and umiaqs by individuals or small groups to cross open straits between island groups is a formidable undertaking. Nonetheless, documented obsidian transport to and among eight Kuril Islands from source localities in Kamchatka and Japan shows that important resources spurred extraordinary efforts over long periods of time (e.g., 1700 years; Phillips and Speakman 2009) even when sea travel was required (Phillips and Speakman 2009; Phillips 2010).

The most formidable channel that Aleuts would traverse in transporting obsidian from the Okmok source on Umnak is Samalga Pass. Samalga Pass is an approximately 40km-wide strait between the southwestern tip of Umnak Island and the eastern shore of Chuginadak in the Islands of Four Mountains group. As detailed by Veltre (Chapter 2 this volume), this strait marks not only the boundary between the eastern and central Aleutians geographically, but is the probable boundary between eastern and central Aleutian cultural and linguistic traditions. Veniaminov (1984, 12) particularly mentions this strait [Chetyrekhsopochnyi] as one of the largest for circulation of ocean currents in the Aleutians.

In spite of the challenges of navigating currents created where the Pacific Ocean and Bering Sea mingle, ethnographic and archaeological records provide information about Aleut interisland communication and contact. It is unlikely that hunters from Adak retrieved the obsidian directly from Umnak because 'each settlement had its own separate hunting and fishing grounds… The leaders of the settlement were obliged to defend them and … not permit outsiders to hunt or fish within another's boundaries…' (Veniaminov in Liapunova 1996, 137). Instead one or more settlements on Umnak may have controlled access to the Okmok obsidian or used it as a commodity, and this collective ownership of land may have been a source of wealth for the toion who controlled it (e.g., Liapunova 1996, 137). How then did obsidian move from the settlement that controlled it to Adak? Several possibilities include: 1) the group that colonized Adak brought the obsidian (direct procurement), 2) inter-island trade or gift exchanges replenished supplies of important resources, and/or 3) raiding among settlements allowed opportunistic retrieval of obsidian cores or tools.

The ethnographic evidence related to initial colonization of islands is necessarily vague, but Liapunova (1996, 138) notes that oral history suggested that those who were forced westward due to resource competition were 'the weakest and poorest' and 'exiles'. If so, they may have not had many material resources to bring with them. Alternatively establishing a new settlement allowed an opportunity to establish a lineage and become a toion, or chief, who might control fishing ground, slaves and other material wealth (Liapunova 1996). Possibly this would appeal to a nephew of an important lineage who may have had considerable resources. Nonetheless, obsidian flakes are present concurrently with the earliest inhabitation of Adak (see Site ADK-171 above) perhaps suggesting that early colonists initially brought obsidian with them.

Inter-island barter was evidently important both during pre-contact time and during the Russian contact period (Veniaminov 1984; Liapunova 1996), though sometimes this exchange took place under threat of armed conflict (Ziber 1937 in Liapunova 1996). In addition to exchange of game animals such as sea otter skins from the western islands for walrus ivory from the eastern islands, some animal hides used by Aleuts originated on the Alaska Peninsula as did amber and other material used for personal ornamentation, even when these were obtained at the risk of their lives (Jacobi 1937 in Liapunova 1996, 150). Not all exchange of goods required that both parties benefit as potlatch ceremonies redistributed wealth among individuals through gift exchanges (Veniaminov in Liapunova 1996). Marriage between members of different islands brings an opportunity for gift exchange or for a bride to bring a coveted supply of a useful stone (Veniaminov 1984, 204). Clearly, individual belongings were highly valued as Veniaminov (1984, 204) documents further that personal possessions could be cause for internecine strife.

In fact, interisland raids were quite common. Black (1984, 57), citing Russian reports, noted that at time of Russian contact, 'There is no doubt that Atkans raided the Attuan and Rat Islanders…The Atkans themselves, in turn, were subject to raids by easterners.' Black (1984, 73) derived from Russian records that the Near Island Aleut polity moved from Agattu to Attu in the 18th century not only to participate in the Russian fur trade but also to avoid 'continuous raids from the Central and Eastern Aleuts, which continued well into the Russian period.' Possibly raiding or other interactions occurred from west to east as well. Corbett and others (1997) reported that in the occidental part of the archipelago 'Available artifactual evidence, and the island's location, suggests both Rat and Near Islanders used Buldir, perhaps concurrently.' The dated organics from Shemya Island indicate that Aleuts were participating in dangerous sea voyages through rough ocean passages prior to 3000 years ago (Corbett *et al.* 2010; West *et al.* 1999). In short, our research and the ethnographic record show that eastern and central Aleuts were well aware of each other, that their interactions might be congenial or hostile and that obsidian might, and probably did, change hands through any number of personal interactions. Regardless of the societal or cultural driver behind the movement of obsidian, it persists in Adak sites over 6000 years of occupation. It seems unlikely that all obsidian on Adak was procured and transported in a single instance.

Why does obsidian abundance decrease with time in Adak sites?

The small size and number of obsidian flakes and lack of large tools recovered from ADK-171 and ADK-011 is tantalizingly suggestive. It is likely that use of obsidian decreased as contact with Russian explorers and traders provided metal tools to islanders after AD 1600 (e.g., McCartney 1971). The paucity of obsidian artifacts in Component 2 of ADK-011 relative to the circa 6000 years BP site of ADK-171 may testify to this change in technology. Obsidian technology would have been gradually abandoned. However, obsidian found in Component 1 of ADK-011 (circa 2500 years BP) is similarly rare and implies that obsidian supplies on Adak were not abundant nor routinely replenished. We suggest that obsidian was a distant, difficult to obtain resource; once procured, tools made of this exotic material were used to exhaustion prior to discard. The older site contains a much higher proportion of obsidian debitage than the younger site. This trend of decreasing obsidian use on Adak possibly mirrors a decrease in its use throughout Aleutian archaeological sites (Hatfield 2010; Wilmerding and Hatfield Chapter 12 this volume). Regardless of how Adak Aleuts were able to acquire the initial supply of obsidian, its paucity suggests that replenishment of the resource did not frequently occur. If trade along the island chain regularly occurred, surely we might expect to find more tools, cores, or larger debitage fragments in Adak archaeological sites. Perhaps trade networks became disrupted by interisland clashes as resource competition increased in intensity as populations expanded across the Aleutian Islands in the pre-contact period.

Summary

In conclusion, we documented the compositional similarity of four samples of obsidian debitage from two distinct sites spanning an archaeological record representing 6000 years of occupation on Adak Island; this material is compositionally indistinguishable and likely came from a single source. We reviewed the geological and archaeological literature and noted the absence of any geologic source for obsidian in the central or western Aleutian Islands. Instead, we found that the Okmok source of obsidian on Umnak Island was valued in ethnographic accounts and that the Okmok source matched the geochemical composition of Adak debitage. We presented a geologic description and geochemical characterization of five samples from the Akutan obsidian source, a source also mentioned in the ethnographic accounts of Aleuts at the time Russian contact. This alternate source proved geochemically distinct both from the Okmok source and from the Adak debitage. We propose that Aleuts transported Okmok obsidian to the central Aleutians as a valued resource, though the material itself may have exchanged hands as gift items, in raids, or in trade. Possibly the Aleuts gathered and transported obsidian during their migration westward along the island arc. In the absence of provenance information for obsidian recovered on Amchitka, this study, then, documented the furthest western transport of obsidian by humans from the Okmok source. Finally, this study presented an example of the use of ICP-MS analysis to perform a rapid and effective method for archaeological provenance investigations.

Acknowledgements

We thank Dr. Dominique Weis for making the PCIGR lab available for our analyses. In particular, the dissolution and analysis was completed by Wilma Praetorius with the aid of co-author Taylor Johson. We thank Ms. Pretorius for her careful attention and thorough explanations, which enabled the acquisition of accurate and precise results. Discussion of this project with Natalia Slobodina brought the Okmok source locality to our attention and she generously shared AAOD unpublished trace element data for that sample with us. We thank Drs. R. J. Speakman and Jeff Rasic whose comments improved this chapter. We further thank the University of Kansas and Whitman College for financial support for the participation of Nicolaysen and Johnson. Further funding was provided by National Science Foundation grant (OPP-0353065).

References

Aigner, Jean S. 1978. Activity Zonation in a 4000 Year Old Aleut House, Chaluka Village, Umnak Island, Alaska. *Anthropological Papers of the University of Alaska* 19(1), 11-25.

Aigner, Jean S., Bruce Fullem, Douglas Veltre and Mary Veltre. 1976. Preliminary Reports on Remains from Sandy Beach Bay, a 4300-5600 BP Aleut Village. *Arctic Anthropology* 13(2), 83-90.

Bank, Theodore. 1950. *The University of Michigan Expedition to the Aleutian Islands, 1948-1949*. Ann Arbor, University of Michigan Botanical Gardens and Museum of Anthropology.

Begét, J. E., J. F. Larsen, C. A. Neal, C. J. Nye, and J. R. Schaefer. 2005. Preliminary Volcano-hazard Assessment for Okmok Volcano, Umnak Island, Alaska, *Alaska Division of Geological and Geophysical Surveys Report of Investigation* 2004-3, 32 p., 1 sheet, scale 1:150,000.

Binford, Lewis. 1979. Organization and Formation Processes: Looking at Curated Technology. *Journal of Anthropological Research* 35, 255-273.

Black, Lydia. 1984. *Atka: An Ethnohistory of the Western Aleutians*. R. A. Pierce (ed.). Kingston, Ontario, Canada, The Limestone Press.

Boyd, Francis R. 1961. Welded Tuffs and Flows in the Rhyolite Plateau of Yellowstone Park, Wyoming, *Geological Society of America Bulletin* 72 (3), 387-426.

Byers, Jr., F. M. 1959. Geology of Umnak and Bogoslof Islands, Aleutian Island, Alaska. *US Geological Survey Bulletin* 1028-L, 107–367.

Castro, Jonathan M. and Celestine Mercer. 2004. Microlite Textures and Volatile Contents of Obsidian from the Inyo Volcanic Chain, California. *Geophysical Research Letters* 31 (18), doi: 10.1029/2004GL020489.

Coats, Robert R. 1956. *Geology of Northern Adak Island, Alaska*, US Geological Survey Bulletin 1028-C, 47-67.

Cook, J. P. 1995. Characterization and Distribution of Obsidian in Alaska. *Arctic Anthropology* 32(1), 92-100.

Corbett, Debra. 1990. Archaeological Survey and Testing on Shemya Island, Western Aleutians, Alaska in June 1990. Final Report for the Geist Fund, Fairbanks, Alaska State Museum.

Corbett, Debra, Christine Lefèvre, Thomas J. Corbett, Dixie West, and Douglas Siegel-Causey. 1997. Excavations at KIS-008, Buldir Island: Evaluation and Potential. *Arctic Anthropology* 34(2), 100-117.

Corbett, Debra, Christine Lefèvre, and Douglas Siegel-Causey. 1997. The Western Aleutians: Cultural Isolation and Environmental Change. *Human Ecology* 25(3), 459-479.

Corbett, Debra, Dixie West and Christine Lefèvre (eds.) 2010. *The People at the End of the World: The Western Aleutians Project and the Archaeology of Shemya Island.* Anchorage, Aurora, Alaska Anthropological Association Monograph Series VIII.

Denniston, Glenda Boyd. 1972. Ashishik Point: An Economic Analysis of a Prehistoric Aleutian Community. Unpublished PhD. Dissertation, Department of Anthropology, University of Wisconsin, Madison.

Desautels, Roger J., Albert J. McCardy, James D. Flyn, and Robert R. Ellis. 1971. *Archaeological Report, Amchitka Island, 19691970.* Archaeological Research, Inc., US Atomic Energy Commission Report TID-25481.

Dillian, Carolyn D. and Carolyn L. White. 2010a. Introduction: Perspectives on Trade and Exchange. In C. D. Dillian and C. L. White (eds.), *Trade and Exchange: Archaeological Studies from History and Prehistory,* 3-14. DOI 10.1007/978-1-4419-1072-1_1.

Dumond, Don and Richard Knecht. 2001. An Early Blade Site in the Eastern Aleutians. In Don Dumond (ed.), *Archaeology in the Aleut Zone of Alaska: Some Recent Research,* 9-34. University of Oregon Anthropological Papers No. 58. Eugene, University of Oregon Press.

Earle, Timothy. 2010. Exchange Systems in Prehistory. In C. D. Dillian and C. L. White (eds.), *Trade and Exchange: Archaeological Studies from History and Prehistory,* 205-217. DOI 10.1007/978-1-4419-1072-1_1.

Frohlich, Bruno and David Kopjanski. 1975. Aleutian Site Survey, 1975: Preliminary Report. Laboratory of Biological Anthropology, Storrs, University of Connecticut.

George, Rhiannon, Simon Turner, Chris Hawkesworth, Charles R. Bacon, Chris Nye, Pete Stelling and Scott Dreher. 2004. Chemical Versus Temporal Controls on the Evolution of Tholeiitic and Calc-alkaline Magmas at Two Volcanoes in the Alaska-Aleutian Arc. *Journal of Petrology* 45 (1), 203-219.

Glascock, M. D. 2002. Obsidian Provenance Research in the Americas. *Accounts of Chemical Research* 35, 611-617.

Grebennikov, A. V., V. K. Popov, M. D. Glascock, R. J. Speakman, Y. V. Kuzmin, A. V. Ptashinsky. 2010. Obsidian Provenance Studies on Kamchatka Peninsula (Far Eastern Russia): 2003-2009 Results. In Y.V. Kuzmin and M.D. Glascock (eds.), *Crossing the Straits: Prehistoric Obsidian Source Exploitation in the North Pacific Rim,* 89-120. Oxford, British Archaeological Reports International Series 2152.

Gualtieri, Lyn, Brenn Sarata, Mitsuru Okuno, and Dixie West. 2011. Did Holocene Paleoenvironmental Factors Affect Occupation and Settlement in the Central Aleutians? Chapter 3 in D. West, V. Hatfield, E. Wilmerding, C. Lefèvre, and L. Gualtieri (eds), *The People Before: The Geology, Paleoecology and Archaeology of Adak Island, Alaska.* Oxford, British Archaeological Reports.

Hatfield, Virginia L. 2010. Material Culture Across the Aleutian Archipelago. *Human Biology* 82 (5-6), 525-556.

Henry, Christopher D., and J. A. Wolff. 1992. Distinguishing Strongly Rheomorphic Tuffs from Extensive Silicic Lavas. *Bulletin of Volcanology* 54 (3), 171-186.

Hoffman, Brian W. 2002. The Organization of Complexity: A Study of Late Prehistoric Village Organization in the Eastern Aleutian Region. Unpublished PhD. Dissertation, Department of Anthropology, University of Wisconsin, Madison.

Hrdlička, Aleš. 1945. *The Aleutian and Commander Islands and Their Inhabitants.* Philadelphia, Wistar Institute of Anatomy and Biology.

Jicha, Brian, David Scholl, Brad Singer, Gene Yogodzinski, and S. M. Kay. 2006. Revised Age of Aleutian Island Arc Formation Implies High Rate of Magma Production. *Geology* 34, 661-664.

Kay, Suzanne M., Robert W. Kay, H. K. Brueckner, James L. Rubenstone. 1983. Tholeiitic Aleutian Arc Plutonism; the Finger Bay Pluton, Adak, Alaska, *Contributions to Mineralogy and Petrology 82(1), 99-116.*

Kelemen, Peter B., Gene M. Yogodzinski, and David W. Scholl. 2003. Along-strike Variation in the Aleutian Island Arc; Genesis of High Mg# Andesite and Implications for Continental Crust. In John M. Eiler (ed.), *Inside the Subduction Factory. Geophysical Monograph,* 138, 223-276. Washington, DC, American Geophysical Union.

Knecht, Richard, Richard Davis, and Gary A. Carver. 2001. The Margaret Bay Site and Eastern Aleutian Prehistory. In Don Dumond (ed.), *Archaeology in the Aleut Zone of Alaska: Some Recent Research. University of Oregon Anthropological Papers 58,* 35-69. Eugene, University of Oregon Press.

Knecht, Richard and Richard Davis. 2001. A Prehistoric Sequence for the Eastern Aleutians. In Don Dumond (ed.), *Archaeology in the Aleut Zone of Alaska, University of Oregon Anthropological Papers 58,* 269-288. Eugene, University of Oregon Press.

Knecht, Richard and Richard Davis. 2003. Archaeological Evaluation of Tanaxtaxak, the Amaknak Spit Site (UNL-055) Final Report. On file. Museum of the Aleutians, Unalaska.

Knecht, Richard and Richard Davis. 2005. South Channel Bridge Project No. MGSSTPBR 0310(S)/52930, Amaknak Bridge Site Data Recovery Project Final Report. On File: Alaska Department of Transportation and Public Facilities, Anchorage.

Knecht, Richard A. and Richard S. Davis. 2008. The Amaknak Bridge Site: Cultural Change and the Neoglacial in the Eastern Aleutians. *Arctic Anthropology* 45, 61-78.

Kuzmin, Y. V. 2010. Prehistoric Obsidian Exchange in Northeast Asia. In Y.V. Kuzmin and M.D. Glascock (eds.), *Crossing the Straits: Prehistoric Obsidian Source Exploitation in the North Pacific Rim,* 137-153. Oxford, British Archaeological Research International Series 2152.

Kuzmin Y.V., and M. D. Glascock (eds.) 2010. *Crossing the Straits: Prehistoric Obsidian Source Exploitation in the North Pacific Rim.* Oxford, British Archaeological Research International Series 2152.

Larsen, Jessica F., Christina Neal, Janet Schaefer, Jim Begét, and Chris Nye. 2007. Late Pleistocene and Holocene Caldera-forming Eruptions of Okmok Caldera, Aleutian Islands, Alaska. In J. Eichelberger, E. Gordeev, P. Izbekov, M. Kasahara, and J. Lees (eds.), *Volcanism and Subduction: The Kamchatka Region: Geophysical Monograph* 172, 343-364. Washington, DC, American Geophysical Union.

Laughlin, William S. and Gordon H. Marsh. 1954. The lamellar Flake Manufacturing Site on Anangula Island in the Aleutians. *American Antiquity* 20, 27-30.

Liapunova, Roza G. 1996. *Essays on the Ethnography of the Aleuts.* [J. Shelest, trans.], Rasmusson Library, Historical Translation Series 9. Fairbanks: University of Alaska Press [Original Russian edition, Leningrad, 1975].

Luttrell, Mark and Debra Corbett. 2000. Archaeological Investigation on Northeast Adak Island, July 1999, Alaska. Report. Report on File, US Fish and Wildlife Service, Anchorage, Alaska.

MacDonald, R., Smith, R. L., Thomas, J. E. 1992. *Chemistry of the Subalkaltc Silicic Obsidians.* Washington, DC, US Government Printing Office.

McCartney, Allen P. 1971. A Proposed Western Aleutian Phase in the Near Islands, Alaska. *Arctic Anthropology* VIII (2), 92-142.

McCartney, Allen P. 1972. An Archaeological Site Survey and Inventory for the Aleutian Islands, National Wildlife Refuge, Alaska, 1972. Report submitted to US Fish and Wildlife Service, Wilderness Studies Branch, Anchorage.

McCartney, Allen P. 1974. Prehistoric Cultural Integration Along the Alaska Peninsula. *Anthropological Papers of the University of Alaska* 16 (1), 59-84.

McCartney, Allen P. 1977. Prehistoric Human Occupation of the Rat Islands. In M.L. Merrit and R.G. Fuller (eds.), *The Environment of Amchitka Island, Alaska,* 59-113. Technical Information Center, Energy Research and Development Administration, TID-26712.

McCartney, Allen P. 1984. Prehistory of the Aleutian Region. In D. Damas (ed.), *Arctic, Handbook of North American Indians* 5, 119-135. Washington, DC, Smithsonian Institution.

McCartney, Allen P. and Douglas W. Veltre. 2002. Longhouses of the Eastern Aleutian Islands, Alaska. In B. Frohlich, A.B. Harper, and R. Gilberg (eds.), *To the Aleutians and Beyond: the Anthropology of William S. Laughlin*, 249-266. Publications of the National Museum, Ethnographical Series 20. Copenhagen, The National Museum of Denmark Department of Ethnography.

Miller, T. P., R. G. McGimsey, D. H. Richter, J. R. Riehle, C. J. Nye, M. E. Yount, and J. A. Dumoulin 1998. *Catalog of the Historically Active Volcanoes of Alaska.* US Geological Survey Open-File Report 98-582, 104 p.

Myers, James D. and Bruce D. Marsh. 1987. Aleutian Lead Isotopic Data: Additional Evidence for the Evolution of Lithospheric Plumbing Systems. *Geochimica et Cosmochimica Acta* 51(7), 1833-1842.

Okuno, Mitsuru, Keiji Wada, Toshio Nakamura, Lyn Gualtieri, Brenn Sarata, Dixie West, Masayuki Torii. 2011. Holocene Tephra Layers on the Northern Half of Adak Island in the west-central Aleutians, Alaska. Chapter 4 in D. West, V. Hatfield, E. Wilmerding, C. Lefèvre and L. Gualtieri (eds.), *The People Before: The Geology, Paleoecology and Archaeology of Adak Island, Alaska.* Oxford, British Archaeological Reports.

O'Leary, Matthew. 1998. New Site No. 2 ('Tutiakoff Site'). Unpublished Report on File at US Department of the Interior, Bureau of Indian Affairs, ANCSA Projects Office, Anchorage.

O'Leary, Matthew. 2001. Volcanic Ash Stratigraphy for Adak Island, Central Aleutian Archipelago. In Don Dumond (ed.), *Archaeology in the Aleut Zone of Alaska: Some Recent Research, University of Oregon Anthropological Papers No. 58*, 215-233 Eugene, University of Oregon Press.

Phillips, S. Colby. 2010. Bridging the Gap Between Two Obsidian Source Areas in Northeast Asia: LA-ICP-MS Analysis of Obsidian Artefacts from the Kurile Islands of the Russian Far East. In Y.V. Kuzmin and M. D. Glascock (eds.), *Crossing the Straits: Prehistoric Obsidian Source Exploitation in the North Pacific Rim*, 121-136. Oxford, British Archaeological Research International Series 2152.

Phillips, S. Colby and Robert J. Speakman. 2009. Initial Source Evaluation of Archaeological Obsidian from the Kuril Islands of the Russian Far East Using Portable XRF. *Journal of Archaeological Science* 36, 1256-1263.

Pretorius, Wilma, Dominique Weis, Gwen Williams, Diane Hanano, Bruno Kieffer, and James Scoates. 2006. Complete Trace Elemental Characterisation of Granitoid (USGS G-2, GSP-2) Reference Materials by High Resolution Inductively Coupled Plasma-mass Spectrometry. *Geostandards and Geoanalytical Research* 30 (1), 39-54.

Reuther, J. D., Natalia S. Slobodina, J. Rasic, J. P. Cook, and R. J. Speakman. 2011. Gaining Momentum: Late Pleistocene and Early Holocene Archaeological Obsidian Source Studies in Interior and Northeastern Beringia, In T. Goebel and I. Buvit (eds.), *Explaining Lithic Assemblage Variability Across Beringia*, 270-286. College Station, Texas A & M University Press.

Richter, D. H., C. F. Waythomas, R. G. McGimsey, and P.L.

Stelling. 1998. Geological Map of Akutan Island, Alaska. US Geological Survey Open-File Report 98, 135p.

Rogers, J. S., M. R. Yarbrough, and C. L. Pendleton. 2009. An Anangula Period Core-and-blade Site on Amaknak Island, Eastern Aleutians. *Alaska Journal of Anthropology* 7 (1) 153-165).

Savinetsky, Arkady B., Dixie L. West, Zhanna A. Antipushina, Bulat F. Khassanov, Nina K. Kiseleva, Olga A. Krylovich, Andrei M. Pereladov. 2011. The Reconstruction of Ecosystems History of Adak Island (Aleutian Islands) during the Holocene, Chapter 5 In D. West, V. Hatfield, E. Wilmerding, C. Lefèvre, and L. Gualtieri (eds.) *The People Before: The Geology, Paleoecology and Archaeology of Adak Island, Alaska.* Oxford, British Archaeological Reports.

Scutter, C. R. 1998. Facies Architecture and Origin of a Submarine Rhyolitic Lava Flow-dome Complex, Ponza, Italy, *Journal of Geophysical Research* 103(B11) 27, 551-566.

Shackley, M. S. 2005. *Obsidian: Geology and Archaeology in the North American Southwest.* Tucson, University of Arizona Press.

Siegel-Causey, Douglas, Debra Corbett, Christine Lefèvre, and Stephen Loring. 1995. Report of the Western Aleutian Archaeological Project. Shemya Island Excavations, August 1994. Report on File. Washington, DC, National Geographic Society.

Spaulding, Albert. 1962. *Archaeological Investigation on Agattu, Aleutian Islands.* Anthropological Papers of the Museum of Anthropology No. 18. Ann Arbor, University of Michigan.

Tuffen, Hugh and Jonathan M. Castro. 2009. The Emplacement of an Obsidian Dyke Through Thin Ice: Hrafntinnuhryggur, Krafla Iceland. *Journal of Volcanology and Geothermal Research* doi:10.1016/j.jvolgeores.2008.10.021

Turner, Christy G. II and Jacqueline A. Turner. 1974. Progress Report on Evolutionary Anthropological Study of Akun Strait District, Eastern Aleutians, Alaska, 1970-1971. *Anthropological Papers of the University of Alaska* 16 (1), 27-57.

Veltre, Douglas. 1997. An Archaeological Survey of Selected Sites on Northern Adak Island, Aleutian Islands, Alaska. Report submitted to the Aleut Corporation, Anchorage.

Veltre, Douglas. 2011. One Hundred Forty Years of Archaeology in the Central Aleutian Islands. Chapter 2 In D. West, V. Hatfield, E. Wilmerding, C. Lefèvre, and L. Gualtieri (eds.) *The People Before: The Geology, Paleoecology and Archaeology of Adak Island, Alaska.* Oxford, British Archaeological Reports.

Veniaminov, Ivan. 1984. *Notes on the Islands of the Unalashka District.* [Lydia Black and R. H. Geoghegan, trans.]. Kingston, Ontario, The Limestone Press.

Waythomas, Christopher. 1995. Surficial Geologic Map of Northern Adak Island, Alaska. US Geological Survey, OFR 95-128.

West, Dixie, Christine Lefèvre, Debra Corbett, Arkady Savinetsky. 1999. Radiocarbon Dates for the Near Islands, Aleutian Islands, Alaska. *Current Research in the Pleistocene* 16, 83-85.

West, Dixie, Michael Crawford, Arkady Savinetsky. 2007. Genetics, Prehistory and the Colonisation of the Aleutian Islands. *Earth and Environmental Science Transactions of the Royal Society of Edinburgh* 98, 47-57.

West, Dixie, Christine Lefèvre, Elizabeth Wilmerding, Virginia Hatfield and Lyn Gualtieri. 2011. The Central Aleutians Archaeological and Paleobiological Project. Chapter 1 In D. West, V. Hatfield, E. Wilmerding, C. Lefèvre, and L. Gualtieri (eds.), *The People Before: The Geology, Paleoecology and Archaeology of Adak Island, Alaska.* Oxford, British Archaeological Reports.

Wilmerding, Elizabeth G. 2005. The Culture Sequence at the Nunik Site, Chernabura Island, Alaska. Unpublished PhD. Dissertation, Department of Anthropology, Washington State University, Pullman.

Wilmerding, Elizabeth and Virginia Hatfield. 2011. Six Thousand Years of Lithic Technology on Adak in a Broader Aleutian Context. Chapter 12 in D. West, V. Hatfield, E. Wilmerding, C. Lefèvre and L. Gualtieri (eds), *The People Before: The Geology, Paleoecology and Archaeology of Adak Island, Alaska.* Oxford, British Archaeological Reports.

Chapter 12
Six Thousand Years of Lithic Technology on Adak in a Broader Aleutian Context

Elizabeth G. Wilmerding and Virginia Hatfield

Abstract: *This chapter describes lithic artifacts from three prehistoric sites around Clam Lagoon and discusses: (1) chipped stone reduction and manufacture techniques, (2) the use of chipped stone, groundstone, pecked stone, and abraded stone tools and, (3) the overall technological and material choices of Central Aleutian Unangan hunter gatherers. The Tutiakoff site (ADK-171), the Dozered site (ADK-012), and the Zeto Point Village site (ADK-011) span approximately six millennia from 6500-170 years before present. The evidence from these sites documents the use of bifacial and irregular flake technology and a greater reliance on chert and obsidian for formal tools in the earlier occupations compared to a shift, by later populations, to basalts and andesites associated with more expedient technology. Abraders are present in all time periods, but pecked stone tools appear after 2000 BP, and ulus and beveled edged tools appear only after AD1000. These sites also document changes in the availability of, and/or access to, lithic materials and possibly a reorganization in social and subsistence strategies of the central Aleutian Unangan people.*

Keywords: *Unangan hunter gatherers, chipped stone tools, groundstone tools, core and flake technology, blade, microblade, debitage analysis*

Introduction

This chapter presents detailed descriptions of stone tools and debitage recovered from the Tutiakoff site (ADK-171), the Dozered site (ADK-012), and the Zeto Point Village site (ADK-011). Together these sites span some 6000 years of central Aleutians prehistory with a 3500-year hiatus occurring between ADK-171 and ADK-011, Component 1.

Stone tools recovered from the three Clam Lagoon sites were analyzed in terms of their morphology and associated function. Kay (Chapter 13 this volume) evaluates actual use of a selection of tools via use-wear analyses. Our research reveals that informal flake tools increase relative to more formal unifacial and bifacial tools over time. Abraders, although present at all sites, also substantially increase in number over time, perhaps indicating by proxy an increase in bone and wood tools. Groundstone ulu knives appear only in the youngest site (Component 2 of ADK-011). Debitage analysis documents a change from chert and obsidian to basalt and andesites that correlates with the shift to informal tools. In addition, the debitage illustrates that Adak Unangan people selected a much wider range of lithic materials for tools than the finished stone tools themselves represent.

Based on previous investigations along the Aleutian archipelago, the artifact assemblages from Adak can be compared to the sequence of Aleutian stone tool manufacture through time. In a pan-Aleutian artifact study, McCartney (1984,135) concluded that some local variation existed in tools across the archipelago, but overall, lithic assemblages exhibited a tremendous amount of continuity over time, regardless of distance. More recently, archaeologists (Corbett, *et al.* 2010; Hatfield 2006; Knecht, Davis, and Carver 2001; Knecht and Davis 2001, 2008; Wilmerding 2006a, 2006b, 2008) have suggested that regional and local variations might be due to lithic resource availability, local innovations, or retention of older ideas. Our research goal is to determine if and why Adak artifact types changed over time and how these types fit into a larger framework of Aleutian prehistory.

Methodology

All lithic artifacts were recovered from 1m square units excavated using trowels. Matrix was screened through 1/4th inch or 1/8th inch mesh. Wilmerding analyzed all material recovered from 1/8th inch mesh, which included very small debitage and tool fragments. Hatfield analyzed all material recovered from 1/4th inch mesh and artifacts hand collected during excavation, which included all the larger tools, tool fragments, and debitage. Analytical categories of the lithic artifacts, based on technological categories, included chipped stone, groundstone, pecked or battered stone, griddle stone, and pigment minerals including ocher. We defined chipped stone artifact categories based on morphological and technological attributes. These categories included projectile point/knives (tools used as either projectiles or as hafted knives or both), bifaces (including preforms and tools), celts, core tools, flake tools, blades and blade tools. Given their significance, blades with no retouch are grouped with blade tools to distinguish them from other debitage.

Debitage analyses in the Aleutians are relatively rare,

occurring, for example, at the Anangula Blade site (Aigner *et al.* 1976), the Hog Island sites (Dumond 2001), the Amaknak Bridge site (Knecht and Davis 2008), the Nunik site (Wilmerding 2005), and in one pan-Aleutian comparison (Hatfield 2006). Our research demonstrates that debitage analysis, although time consuming, provides valuable information on tool production and maintenance as well as lithic material choices through time. Cores, a source of debitage, were divided into bipolar, irregular (or multiple platform), tabular, and single platform types. In addition to cores, specific debitage categories were identified by visual inspection. These categories have been experimentally defined and tested by Root (Root and Ferguson 2003) through replicative experiments. These categories include abraded flakes, alternate flakes, biface thinning flakes, bipolar flakes, cobble flakes, complex flakes, notching flakes, pressure flakes, primary decortication flakes, platform rejuvenation flakes, simple flakes, shatter, and small undiagnostic flakes. Notching flakes were produced when a shouldered projectile point was made and the stem was clearly delineated from the blade edges. Pressure flakes were produced when the edges of an almost-finished biface were trimmed or when a broken or dulled biface was resharpened. When a groundstone tool was resharpened, flakes abraded on the platform or ventral surface were created. Biface thinning flakes resulted when a roughed out flake blank was further shaped into a biface like a projectile point or knife (Root and Ferguson 2003).

Groundstone tool categories include abraders and abrasive materials (both generally made of pumice or scoria), manos, net sinkers, girdled cobbles, lamps, manos, ulus, and miscellaneous groundstone fragments. Pecked or battered stone tools include hammerstones, anvils, and pitted stones. Pigments refer to ocher or other minerals used as color sources. We also included griddle stones as a lithic category. These are flat tabular pieces of rock, usually an andesite or related material, believed to have been used for cooking.

We categorized chipped stone materials into groups of chert, basalt, andesite, and obsidian. Variations within each group were further defined by unaided visual inspection to identify differences in color and inclusions. Some of these correlate with materials identified by other researchers; for example, Kay (Chapter 13 this volume) identified a banded serpentine that is described as a chert type in our analysis. No analysis of the variations within chert, andesite, or basalt is attempted in this chapter. Instead, comparisons are limited to the use of chert, obsidian, basalt, and andesite at each archaeological site.

Clam Lagoon Lithic Assemblages

The Clam Lagoon archeological sites considered here include (1) ADK-171—the Tutiakoff site (the *Clinocardium* midden, Pits 1 and 2 and Test Pit 55-1, (2) ADK-011—the Zeto Point Village site (Components 1 and 2), and (3) ADK-012—the Dozered site. Detailed descriptions and ages of these archaeological sites are presented elsewhere in this volume (West *et al.* Chapter 1 this volume). Here we provide a brief overview of the archaeological units, cultural layers, and radiocarbon dates.

ADK-171, dating circa 6000 BP, represents the oldest identified human settlement in the central Aleutians (Luttrell and Corbett 2000; O'Leary 2001; Savinetsky *et al.* Chapter 5 this volume). CAAPP excavated three areas: (1) the *Clinocardium* midden, (2) Pits 1 and 2, and (3) Test pit 55-1. The *Clinocardium* midden unit measured 0.5m X 1.3m. Midden deposits sit immediately above Intermediate volcanic ash that was deposited approximately 7000 BP (Okuno *et al.* Chapter 4 this volume). Pits 1 and 2, measuring 1m X 2m and excavated over 1m deep, were culturally sterile except for one isolated artifact. Pit 55-1, a 1.3m X 1m unit, was excavated to intercept a shallow depression believed to be a house pit. Cultural lenses, near the bottom of the 140cm deep Pit 55-1 dated to 6040±30 BP (NUTA2-10967), a time corresponding to the *Clinocardium* midden. Because the *Clinocardium* midden and Pit 55-1 excavations represent midden and occupation areas, artifacts from each are herein considered separately.

Ten 1 X 1m units at ADK-011 revealed two distinct occupations separated by a 10cm thick, nearly sterile dark brown sand. The oldest component (1) dated between 2590 and 2390 BP; the youngest component (2) dated between 160 and 415 BP. In Component 1, organic preservation was poor, and cultural materials chiefly comprised chipped stone tools and debitage. Two distinct areas, a midden (units B-F) and a house floor (units G-K), comprise the ADK-011, Component 2 deposits. The midden was dominated by a dense shell and bone deposit containing many stone artifacts. The house had a less dense scatter of material and contained two shallow depressions each clearly demarcated with a 1-2cm thick, blackened ash layer interpreted as house floors. Because the ADK-011 house pit contained two components (1 and 2) separated by a 2000-year hiatus, we considered lithics from these components separately.

ADK-012 (the Dozered Site) sits on the north end of barrier beach separating Sitkin Sound from Clam Lagoon. World War II bulldozing and road construction pushed cultural materials into a large berm, essentially destroying the site. An excavated 110cm thick profile in a knoll revealed a 70cm thick deposit of cultural and natural lenses bulldozed out of context covering a 30cm thick undisturbed stratum. Dated organics range between 2470 and 1515 BP. Dates on terrestrial plants and birds suggest a date 1865-1515 BP is probably an accurate age estimate for the single component ADK-012 deposits.

Table 12.1 shows the types and counts of lithic tools, blades, and cores recovered from the three sites. Table 12.2 illustrates the range and kinds of lithic debitage from each site and component.

Table 12.1. Stone tools from three Clam Lagoon sites.

Artifact Description	***ADK-171 Clinocardium Midden***	***ADK-171 Test Pit 55-1***	***ADK-011 Component 1***	***ADK-012 Dozered Site***	***ADK-011 Component 2***
	6041-5735 BP	***6041-5735 BP***	***2590-2390 BP***	***1865-1515 BP***	***400-170 BP***
Abraders	3		19		28
Abrasive material	3	2	2	2	49
Anvil					2
Biface/Biface fragments	3	1	3	6	25
Biface preform					1
Blades/blade tools			4		1
Bifaces, recycled					4
Celts					1
Cores, bipolar	3			2	9
Cores, irregular	2		2	1	7
Cores, tabular	2				
Cores, single platform	1				5
Core tools					2
Flake tools	1		11	7	24
Girdled cobbles				1	
Griddle stones			3		70
Griddle stone fragments		1	5	32	80
Groundstone fragments					7
Hammerstones			1	2	8
Lamps					2
Manos					8
Manuport cobbles			1	5	40
Pigment, ocher				2	
Pitted stones					2
Projectile point/ knives	1	1	6	3	11
Net Sinker				1	5
Ulus & ulu fragments					12
Total	19	5	57	64	403

The Tutiakoff Site, ADK-171: 6041-5735 BP

ADK-171-The Clinocardium Midden

O'Leary (2001, 222) described three formal tools collected from an eroding talus below the ADK-171 midden exposure. These tools included two unifacially retouched small stemmed points made on flakes and a small unifacial scraper made on a fine-grained basalt or obsidian flake. Adding to this lithic inventory, CAAPP archaeologists recovered ten stone tools and eight cores (Table 12.1), as well as 249 pieces of debitage (Table 12.2). The tools included a point and three bifaces (Figure 12.1:A-D). The asymmetrical point is unstemmed and made of beige-tan chert (Figure 12.1:A). It is relatively thin in cross section and finely, but not regularly, flaked on both sides. Use-wear on this specimen suggests that it was once symmetrical and became irregular through use and retouch sequences (Kay Chapter 13 this volume). Two biface midsection fragments, made of gray chert, were reworked or recycled using a bipolar technique (Figure 12.1:B-C) and show evidence of use as wedges (Kay Chapter 13 this volume). The recycling of these broken bifacial artifacts was probably in response to scarcity of, or restricted access to, high quality chert (Bamforth 1986). The third biface is also broken and irregularly shaped (Figure 12.1:D). This is an isolated find from ADK-171, Test Pits 1 and 2. The deposits from these two test pits were culturally sterile except for this biface. Because the test pits were located within two meters of the midden and the tool was found at a depth corresponding to the midden, the biface is combined with the *Clinocardium* midden assemblage. Use-wear analysis showed that this piece had been hafted and used as a cutting tool (Kay Chapter 13 this volume). A bifacially retouched or utilized chert flake was also identified and may have been used for scraping or engraving (Kay Chapter 13 this volume).

Three abraders, one made from scoria and two from pumice, had grooves on one or more faces indicating they were used as smoothers and/or shapers. The scoria abrader, a rounded half circle in cross section, possessed a flat face with several parallel, u-shaped troughs (Figure 12.1:H). This smoothing tool was probably used to shape bone awls, needles, or something similar. The pumice abraders had both u-shaped and v-shaped grooves for shaping rounded or beveled bone

Table 12.2. Flake types from three Clam Lagoon sites.

Flake Type	ADK-171 *Clinocardium* Midden	*ADK-171 Test Pit 55-1*	*ADK-011 Component 1*	*ADK-012 Dozered Site*	*ADK-011 Component 2*
	6041-5735 BP	*6041-5735 BP*	*2590-2390 BP*	*1865-1515 BP*	*400-170 BP*
Abraded					2
Alternate	1	1		3	13
Biface Thinning	1	2	19	12	49
Bipolar	21			14	70
Blade					7
Cobble Flake					6
Complex	15	13	50	159	808
Notching					5
Pressure	5	11		43	77
Primary Decortication	19		1	53	110
Platform Rejuvenation	1		2		16
Shatter	16		5	46	242
Simple	124	26	32	389	2683
Small Undiagnostic	46	75	1	697	985
Total	249	128	110	1416	5073

or wooden tools. Three additional pumice pieces showed no signs of use and may have been spares.

The core and flake industry was not strictly governed by one method of manufacture, and the flakes were irregularly shaped. Of the seven identified cores, five were small, irregular, and possessed more than one platform for flake removal. The cores include three made of andesite, one of basalt, two of chert, and one of a metamorphosed sedimentary rock. One very small, regular core, made on banded tan and black chert, was conical, had a single platform, and was probably expended (Figure 12.1:G). Another core, made of a dark brown-black chert, had platforms typical of a bipolar (hammer and anvil) reduction technique. A single basalt core possessed several tabular platforms, indicating hard hammer percussion flake removal. Both of the chert cores were small, indicating either smaller starting nodules or a long use-life.

The ADK-171 *Clinocardium* midden debitage (Table 12.2) suggests that Unangan people relied on a wide range of materials for tools. The debitage (N=249) was comprised of 56% chert, 28% andesite, 12% basalt, and 4% other lithic material. Like the tools, this distribution suggests a preference for fine-grained, rather than coarse-grained, materials for chipped stone tools. The debitage was largely undiagnostic, and many of the flakes could have been produced at different manufacturing stages or during the production of more than one kind of tool. Identified diagnostic flake types included bipolar flakes, several pressure flakes (evidence for the final stages of reduction or sharpening), one biface thinning flake, and one alternate flake (evidence of tabular basalt being reduced from its original slab shape). These flake types suggest that the site occupants utilized both bifacial and irregular flake/core technologies.

Test Pit 55-1

Test pit 55-1 contained five stone tools (Table 12.1) and 128 debitage pieces (Table 12.2). A biface midsection, thin in cross section and finely worked on both sides, had both tip and base removed by bend fractures. This projectile point fragment probably broke during use and was discarded (Figure 12.1 E). Use-wear on this specimen indicates it was used as a projectile point (Kay Chapter 13 this volume). A roughly triangular, basalt biface fragment, thick in cross-section, was bifacially worked on one edge, unifacially worked on another, and exhibited a bend fracture on a third (Figure 12.1:F). This piece probably broke during manufacture and was subsequently discarded. Other tools included two scoria fragments probably used for shaping and smoothing wood, bone, or leather (Figure 12.1:J). Both were oblong, rounded on all sides, and bore faint scratches but not deep grooves or definite flattened sides. A tabular slab fragment of porphyritic basalt thought to represent a griddle stone was also recovered.

The debitage was dominated by obsidian (56%), followed by basalt (23%), chert (16%), and andesite (5%). Most of the debitage was undiagnostic regarding reduction stage or manufacture type. However, eleven pressure flakes (chert=7, basalt=2, obsidian=2) indicate late stage manufacture or repair to bifacial tools. Two basalt biface thinning flakes indicate early biface manufacture and one basalt alternate flake suggests initial tabular core reduction. Again, debitage supports both bifacial and irregular flake technologies similar to the *Clinocardium* midden materials.

Given that the *Clinocardium* midden and Test Pit 55-1, layer 6 were simultaneously used, the absence of obsidian in the midden is puzzling. Obsidian tools were re-sharpened within the area of Test Pit 55-1, but no such activity

Figure 12.1 Chipped stone and groundstone artifacts from ADK-171: A) Projectile point/knife, Clinocardium midden; B) Biface fragment, Clinocardium midden; C) Biface fragment, Clinocardium midden; D) hafted bifacial knife, Clinocardium midden-Test pit 1,2; E) projectile point/knife fragment, Test Pit 55-1, F) biface fragment, Test Pit 55-1, G) core, Clinocardium midden; H) pumice abrader, Clinocardium midden; I-J) pumice abraders, Test Pit 55-1.

apparently occurred in the midden. This difference might be related to the small size of the assemblages or differences in activities represented in the dumped materials within the midden versus materials deposited at Test Pit 55-1. Regardless, obsidian was clearly part of the Adak toolkit 6000 years ago, and the greater reliance on chert relative to other lithic materials during the earliest occupation, sharply contrasts with coarser grained stones used at later sites on north Adak.

The Zeto Point Village Site, ADK-011 (Component 1): (2590-2390 BP)

ADK-011, Component 1 tools (Table 12.1) included six projectile point/knives, a bifacial drill, two bifacial knives, 11 flake tools, one blade tool, 19 abraders, and a hammerstone. In addition, several griddle stones were documented (see also Jeannotte *et al.* Chapter 14 this volume). Of the projectile point/knives, four were bifacially

Figure 12.2. Chipped stone artifacts from ADK-011, Component 1: A-F) projectile point/knives and fragments; G-H) biface fragments; I) bifacial drill; J) flake-tool knife; K) flake-tool scraper; L) flake-tool knife; M) possible blade tool retouched bifacially, used as a knife; N) microblade.

retouched (Figure 12.2:A-D) and two were unifacially retouched (Figure 12.2:E-F). The only complete projectile point was bifacially worked and had slightly rounded shoulders and a long contracting stem (Figure 12.2:A). This specimen possessed a straight base and was probably once hafted. Another projectile point/knife was bifacially worked and was missing its base (Figure 12.2: B). These points were extensively reworked, suggesting multiple episodes of use and resharpening. Point fragments included bases and tips broken during manufacture or use (Figure 12.2:C-F). A basalt bifacial projectile point preform, which appears to have been discarded after a manufacturing failure, was also recovered from this component (Figure 12.2: D). Two bifacially retouched basalt fragments made on flake blanks that broke during manufacture were also recovered (Figure 12.2: G-H). Another biface had a drill bit on one end (Figure 12.2:I).

Eleven flake tools were recovered from ADK-011 (Figure 12.2:J-L). Four flake tools (three basalt, one chert) exhibited acutely angled edges suggesting use as knives. Three flake tools (one basalt, two chert) had thicker, more steeply angled edges, suggesting they were used as scrapers. One basalt flake was probably used as a wedge. One chert flake tool had a flake removed at an angle to its long axis, creating a burin with a thick cross section that could have been used for engraving wood or bone. The Component 1 assemblage also included two general-purpose basalt flakes with retouched edges and a flake with an isolated protrusion possibly used as a drill or graver.

Three possible blades/microblades and a blade tool were also recovered. These few specimens may either be the by-products of irregular flake tool manufacture or the result of blade technology as was employed in eastern Aleutian sites. One of the blades, made of dacite, is bifacially retouched for use as a cutting and/or scraping tool (Figure 12.2:M). The bifacial retouch obscures the ventral surface, and the dorsal surface has retouch perpendicular to the dorsal blade scars. It has a flat to triangular cross section and is broken. It is questionable whether this is actually a blade given the bifacial retouch, which has obscured the dorsal flake scars; however, its morphology suggests that it is a blade. The remaining specimens are unmodified and include two microblades (Figure 12.2:N) and another possible blade. Because the sample is small, it is difficult to determine whether these were purposely created as part of a blade core technology, but it is unlikely. All of the ADK-011 cores represent irregular flake technology, suggesting the four possible blades/microblades recovered from Component 1 were incidental by-products of chipped stone reduction. Perhaps these few specimens reflect that a blade/microblade technology comprised a small proportion of the technology at this time, but such a conclusion cannot currently be confirmed.

Other tools recovered include 19 abraders (18 pumice and one scoria). Of these, two pumice abraders and the scoria piece possessed one or more distinctly flattened surfaces from smoothing wood or bone. Fifteen abraders showed distinct grooves suggesting they perhaps were used as smoothing tools for weapon shafts, baidarka (kayak) parts, or needles. The remaining two pieces of pumice were unmodified. Only five griddle stones were recovered.

Component 1 contained 110 pieces of debitage (Table 12.2). A total of 88 undiagnostic flakes (80%) were simple or complex. Nineteen biface thinning flakes represent the middle stages of biface manufacture. A single primary decortication flake suggests that very little early shaping of tools from cores occurred on site. Two platform rejuvenation flakes were produced when the flint knapper needed to refresh the platform edge and remove mistakes. A paucity of very small, late stage debris and resharpening flakes may well be a result of 1/4th inch mesh used during screening Component 1 matrix. Basalt and chert were the most common lithic materials at ADK-011, Component 1. Obsidian flakes were present, but rare. Andesite flakes were also present but were not a significant percentage of the assemblage.

Lithic Artifacts at ADK-011, Sand Deposit

The nearly sterile 10cm sand deposit separating Components 1 and 2 contained ten artifacts (Table 12.1): three abraders, three flake tools, one pecked cobble resembling a net sinker, and three groundstone ulu fragments. The pecked cobble and ulu fragments suggest an affiliation with Component 2 or an occupation after AD 1000; however, this deposit is not directly dated.

The Dozered Site, ADK-012: (1865-1515 BP)

CAAPP archaeologists recovered 56 tools, three cores, and five manuports from ADK-012 (Table 12.1). Chipped stone tools included three projectile point/knives, six bifaces including projectile point/knife fragments, and seven flake tools (Figure 12.3:B). Other tools included 32 tabular slabs (griddle stone fragments), two pieces of abrasive material (Figure 12.3:D), two hammerstones (Figure 12.3:E), one girdled cobble (Figure 12.3:F), one net sinker (Figure 12.3:G), and two pieces of red ochre. Five of the six biface fragments were made from basalt and one was made from chert. These included fragments of the proximal, lateral, and distal portions of projectile point/knives. The largest distal fragment is made of basalt and represents a small, narrow, almost parallel-sided projectile point (Figure 12.3:A). One asymmetrical basalt knife is missing its base and had a blunt tip made on a thick flake. Two basalt knife fragments, made on large, thick flakes, show use wear. The projectile point fragments, girdled cobbles, and net sinkers indicate that both hunting and net fishing occurred at the site.

The ADK-012 assemblage included a small core and flake industry in addition to the bifacial industry. Three cores, two made of basalt and one from chert, were recovered. Of these, the chert core shows bipolar percussion (Figure 12.3:C) and the remaining cores are irregular showing

Figure 12.3. Chipped and groundstone artifacts from ADK-012. A) tip of a projectile point; B) flake-tool knife; C) core; D) abrasive material; E) hammerstone fragment; F) girdled cobble; G) net sinker.

evidence of free hand percussion. Debitage included 1416 pieces (Table 12.2); of the total 1289 (91%) were undiagnostic. It is remotely possible that military bulldozing could have created some of these undiagnostic flakes making them a byproduct of the disturbance and not a result of prehistoric flint knapping. However, because most of the undiagnostic flakes did not appear crushed or freshly broken, they are more likely to be prehistoric artifacts. Several manufacturing stages are reflected in the remaining 127 pieces. Primary decortication, alternate, and bipolar flakes represent early stages of manufacture. Of the 70 flakes, 42 were andesites, 25 were basalt, and three were chert. The dominance of andesite indicates that cobble reduction for expedient tools was an important part of the technology. Early stage basaltic flakes, with 100% cortex, could represent small nodules brought from their source for testing or reduction into finer-grained flake tools. However, bulldozing may be the most parsimonious explanation for these cortical flakes. Twelve biface thinning flakes represent the middle stages of biface manufacture and 43 pressure flakes indicate that final stages of manufacture and repair.

Basalts (85.3%) dominated the lithic assemblage, followed by andesites (11.4%), cherts (3.1%), and two obsidian flakes. Fifteen of the finished tools were made of basalt, one was made of chert, and both girdled cobbles were made of andesite. Chert and obsidian were rare in these deposits.

The Zeto Point Village Site, ADK-011 (Component 2): (400-170 BP)

The younger occupation at the Zeto Point Village site provided the largest and richest lithic assemblage of any location excavated by CAAPP. This is due, at least in part, to the size of the excavated area (see West *et al.* Chapter 1 this volume). The lithic tools recovered (N=400) From ADK-011, Component 2 represent 24 tool types (Table12.1).

Chipped stone tools included eleven projectile point/knives (ten bifacial and one unifacial), 25 biface fragments, one bifacial preform, four recycled bifaces, and 24 flake tools (Table 12.1). All projectile point/knife forms were made from local basalt. Three very small projectile points were recovered (Figure 12.4:A-C). One of these points (Figure 12.4:A) had a short, wide blade and a relatively long stem, suggesting that it had been extensively resharpened while in the haft. The larger bifacial projectile point/knives included one laurel leaf shaped bipoint (Figure 12.4:H), as well as five points with discernable stems (Figure 12.4:D-G, I). All of the larger specimens were possible dart or harpoon heads. Other projectile point/knife fragments, bifaces, and biface fragments were also recovered (Figure 12.4:K-R).

Other tools included a handful of flake tools (Figure 12.4:S-U). These were irregularly shaped flakes struck from andesite or basalt cores and were probably used to cut meat, fish, and/or hides to make clothing and other articles. An obsidian flake tool fragment (Figure 12.4:U) had previously been broken and was reused. A core with retouch on one end indicating it was used as a tool was also recovered (Figure 12.4:V). In addition, there was one unmodified, blade-like flake (Figure 12.4:W), which might suggest the presence of microblade technology, however a single possible microblade is insufficient to confirm this technology at this site at this time.

Twelve ulus and ulu fragments were recovered and mark the appearance of these tools in the Adak tool kit (Figure 12.5). Several ulus have beveled edges. Ethnographically, Unangan people used ulus (knives with ground and beveled edges) to cut fish or separate meat and fat from hides. Use-wear analyses on several of the ulus, referred to as 'food cutting knives' (Kay Chapter 13 this volume) indicate they were used as general utility knives, most likely for cutting food. One groundstone celt fragment (Figure 12.6:A) retains only a portion of the originally prepared edge. Use-wear analysis indicates it had been reworked and used as a wedge (Kay Chapter 13 this volume). Abraders are also represented at the Zeto Point Village during this time (Figure 12.6: B-C). Pecked stone tools comprised a third type of groundstone at ADK-011, Component 2. Andesite cobbles were used for lamps (Figure 12.6:D), manos (Figure 12.6:E), anvils (Figure 12.6:F) and hammerstones (Figure 12.6:G), as well as plummets, and net sinkers. Unmodified cobbles (manuports) were also documented. These could have been used as weights or as 'feet' to prop tabular stones above the fire, for other utilitarian tasks, or have been collected for future use.

Seventy griddle stones and fragments of tabular slabs, both large and small, were scattered throughout the midden and possibly reflect intensified cooking activities at ADK-011. Naturally occurring slabs of andesite are found in quarries within walking/paddling distance of the site and may have served as the source materials for these griddle stones (Jeannotte *et al.* Chapter 14 this volume).

ADK-011, Component 2 contained 5073 flakes (Table 12.2); 93% (N=4718) were undiagnostic. The remaining 7% (N=355) provide information about tool making techniques in late prehistory at this site. This assemblage represents all manufacturing stages. The alternate flakes illustrate that naturally occurring tabular slabs of basalt were roughed out for further shaping into bifaces. The bipolar, cobble, and primary decortication flakes of andesite show the manufacture of flakes on local, coarse-grained material to perhaps make expedient tools. Chert and basalt biface thinning flakes represent middle stages of tool manufacture. Late stages of tool manufacture or resharpening of a broken or dulled tool were also present and include notching, pressure, and abraded flakes. A few blade-like flakes were also recovered. These are rare and probably represent the accidental by-product of irregular flake/core reduction.

Lithics from ADK-011, Component 2, Layer 3E, House Floor

Layer 3E, a house floor from Component 2 in units G, H,

Figure 12.4. Chipped stone artifacts from ADK-011, Component 2. A-C) small projectile points; D-I) stemmed projectile points; J-L) projectile point fragments; M-P) bifaces and biface fragments; Q-U) flake tools; V) core tool: W) microblade.

J and K, consisted of a shallow bowl-shaped depression below 40 Year Ash and recognizable by a layer of blackened earth. The tools and debitage from this provenience were found either on the house floor or within a couple of centimeters of the surface. Table 12.3 lists artifacts and debitage from this feature. The artifacts and pieces of debitage, included in the previous Components 2 analysis, are here considered within the specific context of the ADK-011 house floor.

The lithic materials (50% basalt, 34% chert, 16% andesite) constitute the same primary material types seen in other parts of the assemblage. Quartz is the only other material represented. The house floor contained less material variety overall and, in particular, less andesite than the midden area or other non-floor areas. The small flakes represented on the floor feature included 94% undiagnostic types (complex, simple, shatter, and small undiagnostic combined). Biface thinning, bipolar, and primary decortication flakes were rare and variable in size, compared with the midden area. Clearly some tool maintenance and/or late stage production occurred inside the 'house,' but it left a very different

Figure 12.5. ADK-011 ulus.

distribution than in the midden area. Probably most tool production took place outside the house area, due to better visibility. Or perhaps tool production occurred inside, took place on mats or other removable surfaces, and debris was removed outside during clean up episodes (Binford 1978; Hayden and Cannon 1983).

The 17 artifacts found on or very near the floor feature are listed in Table 12.3. These include broken bifacial pieces, abrasive material fragments, griddle stones, a single projectile point/knife, and three ulus, as well as five manuport cobbles. These artifacts were embedded with the overall discussion of the Component 2 assemblage and are illustrated in Figures 12.4-12.6. The projectile point/knife (Figure 12.4:H) and the ulus represent activities such as food and/or hide cutting. Again this suggests that (1) the

Table 12.3. Zeto Point Village house floor artifacts and flakes.

Artifact Description	***Quantity***
Abrasive Material	1
Biface edge	2
Biface tip	1
Cobbles	5
Flakes	1
Platform Reduction Flake	1
Projectile point/knife form	1
Tabular Slab fragment	2
Ulus	3
Total	17

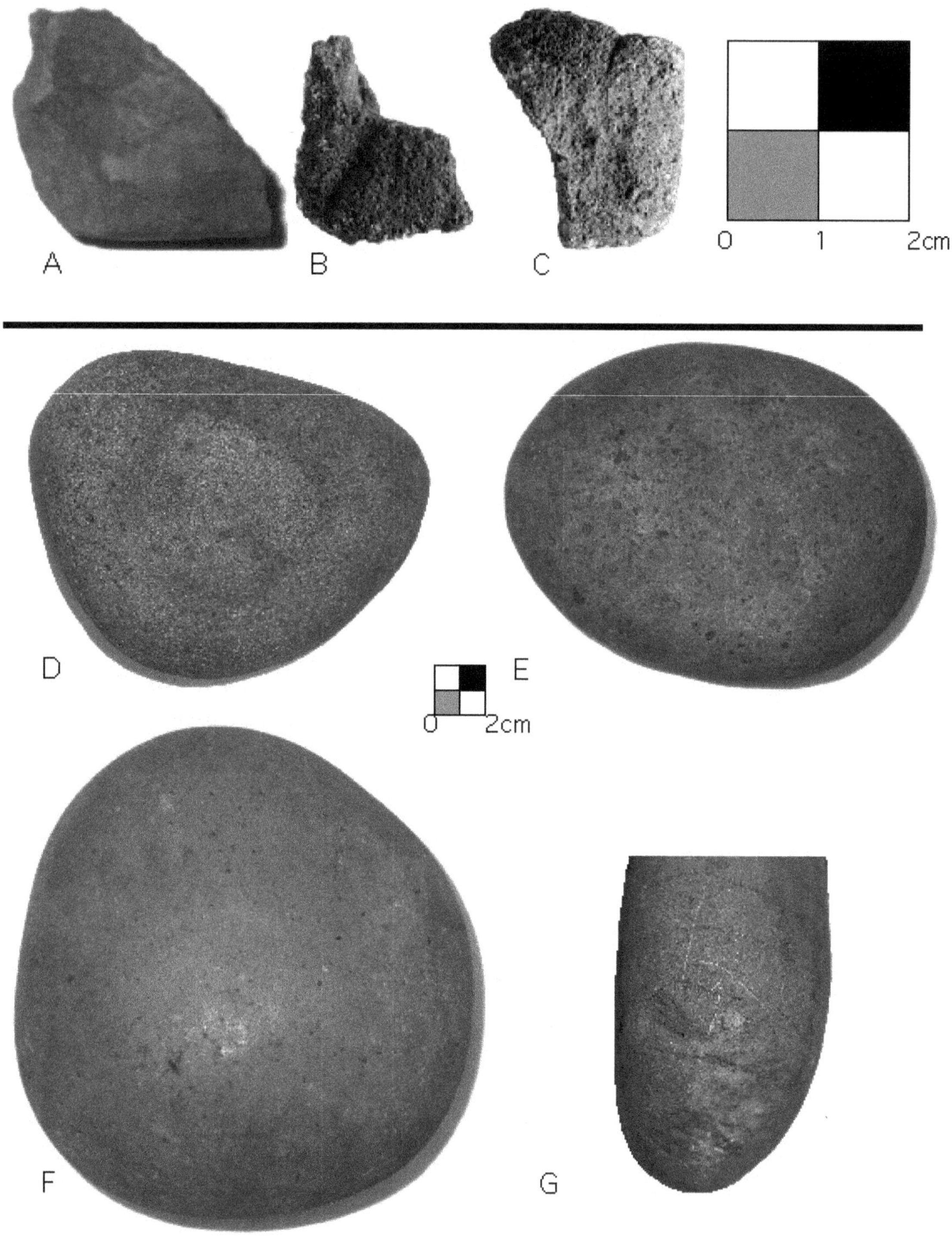

Figure 12.6. Groundstone tools from ADK-011, Component 2. A) celt, B) scoria abrader, C) pumice abrader, D) lamp, E) mano, F) anvil, G) hammerstone.

floor was cleaned prior to abandonment, (2) the remaining tools were lost or deliberately left behind (Stevenson 1982), or (3) most activities took place outside). Although artifacts and debitage in the 'house' area were similar to those of the midden, the deposition patterns differed. Less variety in lithic material and fewer artifact types occurred in the 'house.' Apparently, larger pieces had been removed from the house, leaving behind smaller fragments among the debris.

Discussion in a Broader Aleutian Context

The lithic assemblages from these north Adak sites conform fairly well to the technologies documented in the eastern Aleutians by Knecht and Davis (2001), with the exception of the lack of the blade/microblade technology that characterizes the earliest technologies in the eastern Aleutians. The limited evidence for blade/microblade technology on Adak may be a product of sampling error.

The assemblages documenting the blade technology in the eastern Aleutians are significantly larger samples than any of the Adak assemblages. In addition, in later phases, blade technology represents a relatively small proportion of these assemblages compared to the irregular flake technology of the eastern Aleutian chipped stone assemblages and would be encountered less often. Other differences between the eastern and central Aleutians include access to obsidian and the timing in the appearance of ulu style tools.

The most recent work in the Aleutians has delineated possible relationships to other Alaskan populations, e.g., Arctic Small Tool tradition (Dumond 2001b, 289), and defined settlement patterning and behavioral changes through time for eastern, central, and western Aleutian prehistory (Corbett *et al.* 1997a, 2001, 2010; Dumond and Knecht 2001; Johnson and Wilmerding 2001; Knecht *et al.* 2001; Knecht and Davis 2001; Lefèvre *et al.* 1999, 2001; Maschner and Jordan 2001; O'Leary 2001; Veltre 2001).

A sequence of occupation 'phases' has been established for the eastern Aleutians by Knecht and Davis (2001), based in part on technological attributes, and documents technological shifts in stone and bone technology. In discussing Adak Island technology, we will apply the phases defined by Knecht and Davis (2001), thought to pertain across the entire Aleutian chain. However, due to small sample sizes, the applicability of these phases to the central and western Aleutians has yet to be fully demonstrated by the material evidence.

9000-7000 BP: The Early Anangula Phase

Only three Aleutian sites thus far represent the Early Anangula phase (Table 12.4) An archaeological signature based on so few sites may over- or under-emphasize aspects of a population; however, this bias is currently unavoidable and can only be remedied with discovery and study of more sites dating to this time. The earliest sites in the eastern Aleutians date between 8500 and 7000 years BP and are all located on small islands within protective bays of larger islands. All three sites are located on terraces 17 to 35m above modern sea level (Knecht and Davis 2001, 272). They were revealed by erosion as they occur below volcanic tephras, one-to-two meters below modern ground surface. Thus far, no sites of this time period have been documented in the central or western Aleutians, possibly because investigations have been limited to large village sites located on first and second terraces.

The Early Anangula phase sites documented in the eastern Aleutians are the Anangula Blade site on Anangula Island, near Umnak Island, and the Russian Spruce and Oiled-Blade sites on Hog Island, near Unalaska Island (Table 12.4) (Aigner 1978; Dumond and Knecht 2001; Knecht and Davis 2001; Laughlin and Aigner 1966). These sites are characterized by large prismatic blade core/blade technology and microblade technology, but apparently lack bifacial technology (Dixon 1975; Knecht and Davis 2001). Knecht and Davis (2001) characterize the Early Anangula phase as blade and microblade technologies with irregular, rotated, and extensively recycled cores (Knecht and Davis 2001, 272). Tools are primarily made on blades using unifacial retouch, and include knives, end scrapers, and transverse burins. Groundstone tools also occur and include lamps, pumice abraders, net sinkers, bowls, and ochre grinders.

7000-4000 BP: The Late Anangula Phase

The demise of the Early Anangulans has been attributed to adverse climate changes that may have 'extinguished Anangula Blade and related communities on Unalaska' (Mason 2001, 118). This hypothesis in part derives from the lack of sites dating between 7000 and 6000 years BP, with the exception of two sites—the Amaknak Quarry site in the eastern Aleutians and ADK-171 on Adak (Figure 12.7). The proposed hiatus in occupations is attributed to catastrophic volcanism or climatic factors (Dumond 2001, 290; Mason 2001, 117-118). Research herein regarding ADK-171 supports an extremely cold climate at this time. The discovery of saffron cod, a cold-loving species at ADK-171, correlated with the ratio of boreal and north-alpine diatoms recovered at Haven Lake peat deposits leads to the inference that the Unangan people settled Adak during an extremely cold, but highly bioproductive, period circa 6000 years BP (Savinetsky *et al.* Chapter 5 this volume). However, a lack of sites may also reflect sampling error due to site location strategies or geomorphology, especially given the small sample of seven sites in the eastern Aleutians, the two sites in the central Aleutians, and the four sites in the western Aleutians that date before 4000 years BP.

Little is known about the Amaknak Quarry site, which is only mentioned briefly in a discussion about the Amaknak Bridge (Table 12.5) site as dating to around 7069 calibrated BP and having artifacts with bifacial technology among its assemblage (Knecht and Davis 2008). Other than this site, the Late Anangula phase sites in the eastern Aleutians date to after 6000 years BP (Table 12.5) and are all located eight to 29m above modern sea level. According to Knecht and Davis (2001, 271, 274), these sites are characterized by a continuance of blade, microblade, and irregular flake technology, with tools made on blades and flakes, including stemmed projectile points, bell-shaped scrapers, gravers, and transverse, polished, and mitten burins (Knecht and Davis 2001, 274). Bifacial technology and bone artifacts first appear during this phase (Knecht and Davis 2001, 274), although the late appearance of bone artifacts may be due to factors of preservation.

The Margaret Bay site (Figure 12.7) on Unalaska has two occupation components dating to this phase (Table 12.5). The Margaret Bay, Level 5 (5470±140 years BP) deposits contain discrete features, including charcoal and red ochre concentrations. The site's stone tool assemblage includes microblade, blade, and irregular flake technology, and some bone and groundstone tools (Knecht *et al.* 2001, 51). The

Table 12.4. Early Anangula Phase sites and radiocarbon dates.

Island, Site	***Years BP± sigma***	***Source***
Anangula, Anangula Blade	8480±350	Mason 2001
Hog, Russian Spruce	8320±220	Dumond and Knecht 2001
Anangula, Anangula Blade	8170±240	Mason 2001
Hog, Russian Spruce	8110±50	Dumond and Knecht 2001
Hog, Russian Spruce	8090±50	Dumond and Knecht 2001
Anangula, Anangula Blade	8060±240	Mason 2001
Hog, Oiled Blade	8050±60	Knecht and Davis 2001
Hog, Russian Spruce	8050±80	Dumond and Knecht 2001
Umnak, Sandy Beach Bay	8045±390	Mason 2001
Hog, Oiled Blade	8040±50	Knecht and Davis 2001
Hog, Russian Spruce	7950±90	Dumond and Knecht 2001
Hog, Russian Spruce	7930±60	Dumond and Knecht 2001
Anangula, Anangula Blade	7870±260	Mason 2001
Anangula, Anangula Blade	7793±116	Mason 2001
Anangula, Anangula Blade	7180±250	Mason 2001
Anangula, Anangula Blade	7175±240	**Mason 2001**

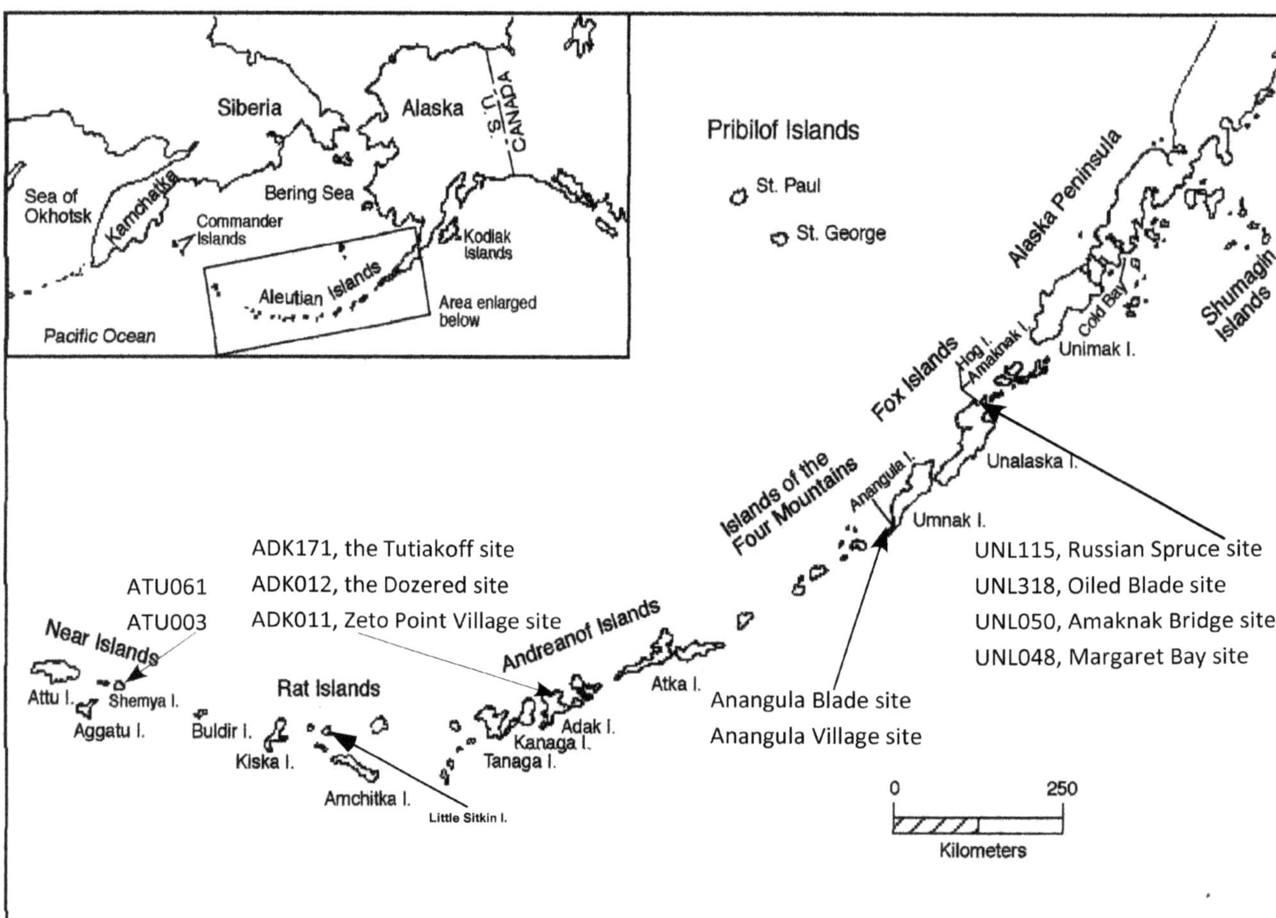

Figure 12.7. Map of the Aleutian Islands with the location of selected sites discussed in the text (illustration by R. Rubicz, with modifications).

Table 12.5. Late Anangula Phase sites and radiocarbon dates.

Island, Site	***Years BP+sigma***	***Source***
Eastern Aleutians		
Amaknak, Amaknak Quarry	7060	Knecht and Davis 2008
Anangula, Anangula Village	5920±80	Laughlin 1975:512; Mason 2001:109
Anangula, Anangula Village	5750±65	Laughlin 1975:512; Mason 2001:109
Amaknak, Margaret Bay	5470±140	Knecht, *et al.* 2001
Umnak, Sandy Beach Bay	5370±240	Mason 2001:110
Anangula, Anangula Village	5340±80	Laughlin 1975:512; Mason 2001:109
Amaknak, Margaret Bay	5250±70	Knecht, *et al.* 2001
Anangula, Anangula Village	5180±100	Laughlin 1975:512; Mason 2001:109
Unalaska, Agnes Beach	5120±120	Knecht, *et al.* 2001
Amaknak, Margaret Bay	4700±40	Knecht, *et al.* 2001:42
Amaknak, Margaret Bay	4660±80	Knecht, *et al.* 2001:42
Umnak, Sandy Beach Bay	4655±160	Mason 2001:110
Amaknak, Margaret Bay	4520±60	Knecht, *et al.* 2001:42
Anangula, Anangula Village	4510±125	Laughlin 1975:512; Mason 2001:109
Umnak, Sandy Beach Bay	4385±200	Mason 2001:110
Umnak, Sandy Beach Bay	4295±200	Mason 2001:110
Umnak, Sandy Beach Bay	4205±180	Mason 2001:110
Umnak, Idaliuk Bay	4165±170	Mason 2001:110
Amaknak, Margaret Bay	4130±40	Knecht, *et al.* 2001:42
Central Aleutians		
Adak, ADK171, Tutiakoff	6525±94	this volume
Adak, ADK171, Tutiakoff	6410±60 BP	O'Leary 2001:221
Adak, ADK171, Tutiakoff	6180±60 BP	O'Leary 2001:221
Adak, ADK171, Tutiakoff	6141±123	this volume
Adak, ADK171, Tutiakoff	6172±192	this volume
Adak, ADK171, Tutiakoff	6005±20	this volume
Adak, ADK171, Tutiakoff	5960±60	this volume
Adak, ADK171, Tutiakoff	5750±60	O'Leary 2001:221
Adak, ADK171, Tutiakoff	5735±30	this volume
Adak, ADK171, Tutiakoff	5405±15	this volume
Adak, ADK012/013	4620±100	O'Leary 2001:221
Adak, ADK012/013	4530±150	O'Leary 2001:221
Western Aleutians		
Amchitka, RAT-070	4780±270	BIA ANCSA n.d.
Amchitka, RAT-068	4610±110	BIA ANCSA n.d.
Amchitka, RAT-068	4510±230	BIA ANCSA n.d.
Amchitka, RAT-024	4440±90	**BIA ANCSA n.d.**

second component, Margaret Bay, Level 4 (4700±40 years BP), is a well preserved midden yielding, along with other fauna, walrus, ringed seal, and polar bear remains, all of which are ice-edge adapted animals and uncommon in the region today. Elements of blade, microblade, irregular flake, and bifacial technology were recovered from these deposits, although cores are absent (Knecht *et al.* 2001). The bifacial tools include large bifaces (also identified in Sandy Beach Bay), bi-pointed knives, and stemmed projectile point/knives (Knecht *et al.* 2001, 51). Many bone tools were recovered, including needles, harpoon points, throwing board pins, and composite fish hook elements. This assemblage provides excellent evidence for bone technology dating to this time (Knecht *et al.* 2001, 53). Regional variation within the eastern islands may occur because there are no reports of microblades at the Anangula Village site on Anangula Island; however, bifacially flaked tools are documented (Laughlin and Aigner 1975; Mason 2001).

In the western Aleutians, Amchitka Island's oldest sites (RAT-070, 068, and 024) are lithic scatters and midden deposits identified during BIA surveys and occur at elevations 30m or more above modern sea level. Elements of bifacial and irregular flake core technology were recovered from these sites. Flake tools include burins, burin spalls, projectile points, gravers, scrapers, abraders, and choppers (Young n.d.). No blade, microblade, or bone technology was recovered; however, limited excavations again impede evaluation of the overall technology of these peoples. Further investigation of these sites would enhance better understanding of their technological relationships with the eastern islands.

On Adak Island, the Tutiakoff site (ADK-171) dates to this phase, 6041 to 5735 years BP (Figure 12.7). This site is located on a second terrace, 20m above modern sea level, and today overlooks a large lagoon (Luttrell and Corbett 2000) that was a sheltered rocky coastline 6000 years ago. As mentioned above, current research indicates this was a very cold time on Adak (Savinetsky *et al.* Chapter 5 this volume). Herein we have documented stemmed projectile points, biface fragments, wedges, a unifacial scraper, a flake tool, as well as abraders and a tabular slab griddle stone fragment. The ADK-171 assemblage also includes bipolar and irregular flake cores and debitage including bifacial thinning flakes. Also from this site, a bone point, several awls, a needle preform, two worked ribs, and a worked bone scrap were recovered (West and Hatfield Chapter 15 this volume). The technologies represented at ADK-171 are consistent with the Late Anangula phase based on the presence of (1) bifacial and irregular flake technologies, (2) stemmed projectiles, (3) abraders and (4) the bone tools. The Tutiakoff site assemblage indicates that the earliest identified occupants of Clam Lagoon preferred chert, in marked contrast to the emphasis on basalt material for Adak tools in later times. Like the assemblages recovered from sites on Amchitka Island, no blade or microblade technology was recovered. This, again, may be a product of sampling or due to the fact that this technology was not part of the toolkits of the first migrants who settled the central and western Aleutians. Further research of sites across the archipelago dating between 7000 and 4000 years BP will be necessary to evaluate this variability.

4000-3000 BP: The Margaret Bay Phase

According to Mason (2001, 113), the Aleutian landscape changed between 4000 and 3500 years BP due to sea level rise, which forced the abandonment of some sites and allowed settlement of others. The sites dating to this period (Table 12.6) document a significant shift in technology, including a notable decrease in blade and microblade technology, which are absent by 3000 years BP (Knecht and Davis 2001). An irregular flake core industry becomes the dominant chipped stone industry following 3000 years BP. Introduced during this time are well-flaked, small tools (e.g., *Qaxaq* points) characteristic of the Arctic Small Tool tradition (ASTt), suggesting communication with or movement of people from the Alaska Peninsula and coastal mainland Alaska (Dumond 2001, 292; Knecht and Davis 2001, 271). This contact appears to begin sometime around 3500 years BP; however, given the absence of tools characteristic of the ASTt at the Amaknak Bridge site (Figure 12.7), around 3000 years BP, this contact or exchange appears to have ended by 3000 years BP (Knecht and Davis 2008, 71).

Tools that continue at this time include polished and transverse burins, end and bell-shaped scrapers made on blades, and assorted flake tools. Less common are the stemmed points from the Late Anangula phase. More common are groundstone polished adzes, slate ulu blades and lances, and ground jet ornaments (Knecht and Davis 2001, 276). A great many bone artifacts are recovered during this time representing a well-developed technology comprised of unilaterally barbed harpoons, end-blade lance slots, socket components, fish hook barb components, wedges with drill sockets, bi-points, and adze-blade holders (Knecht *et al.* 2001, 44-47). Margaret Bay, Level 3, and the Amaknak Bridge site (Figure 12.7) provide evidence for substantial semi-subterranean stone-lined houses with elaborately constructed hearths and other linear features. Tools from these assemblages include flake scrapers, large bifaces, and stemmed projectile points; however, the small *Qaxaq* points were only recovered from the Margaret Bay, Level 3 component. Groundstone and bone technology are present as well as blade and microblade technology (Knecht *et al.* 2001, 51). On Umnak, Chaluka's lowermost cultural horizon does not contain any blade elements, which may be due to incorrect dating or regional variation (Dumond 2001, 291).

In the central and western Aleutians, sites dating to this time have been documented on Adak, Amchitka, and Shemya (Table 12.5). On Adak Island, Margaret Bay phase sites are located on terraces ranging from ten to 40m above sea level (O'Leary 2001, 222), but little is known about these sites. The sites on Amchitka and Shemya document irregular flake, bifacial, and bone technology, not inconsistent with the eastern Aleutian Margaret Bay phase descriptions. However, the western sites lack definitive evidence for blade/ microblade technology and, thus far, no ASTt technology has been identified in these assemblages. On Shemya, ATU-061 (Figure 12.7), a small site along the southwestern coast, documents stone and bone artifacts dating between 3500 and 3000 years BP and represents the earliest known occupation in the Near Island group (Corbett *et al.* 2010). The lithic artifacts included several bifaces, unifacial scrapers, several flake tools, abundant hammerstones, choppers, a few griddle stones, several abraders, a few anvils, one groundstone ulu, a groundstone lamp, a chipped stone chisel, two chipped stone wedges and debitage representing irregular flake core technology (Corbett *et al.* 2010). Several macroblade and microblade elements were identified in the debitage analysis of this site (Hatfield 2006). Bone technology included wedges, bird bone awls, bird bone-splinter points, a flaker, and several pieces of bone scrap from mammals and birds (Corbett *et al.* 2010). Another possible blade tool was recovered from Shemya Island in the Near Islands (Corbett, personal communication). However, other than this one specimen and the pieces from ATU-061, the evidence for blade and microblade technology in the Andreanof, Rat, or Near Island groups is very limited.

3000-1000 BP: The Amaknak Phase

The sites dating between 3000 and 1000 years BP in the

Table 12.6. Margaret Bay Phase sites and radiocarbon dates.

Island, Site	***years BP ± sigma***	***Source***
Eastern Aleutians		
Umnak, Chaluka	4000	Laughlin 1975
Umnak, Chaluka	3674 (1724±55 BC)	Corbett 1991:48
Amaknak, Margaret Bay	3630±70	Knecht, *et al.* 2001:42
Amaknak, Margaret Bay	3610±110	Knecht, *et al.* 2001:42
Umnak, Chaluka	3603 (1643±60 BC)	Corbett 1991:48
Amaknak, Amaknak Bridge	3470±70	Knecht and Davis 2005
Umnak, Chaluka	3460 (1510±150 BC)	Corbett 1991:48
Amaknak, Amaknak Bridge	3470±70 BP	Knecht and Davis 2008
Unalaska, Agnes Beach	3450±60 BP	Knecht and Davis 2001
Amaknak, Amaknak Bridge	3370 ±60	Knecht and Davis 2008
Amaknak, Amaknak Bridge	3360±95 BP	Davis and Knecht 2001
Amaknak, Amaknak Bridge	3310±1100	Knecht and Davis 2008
Amaknak, Margaret Bay	3280±70 BP	Knecht, *et al.* 2001:42
Amaknak, Margaret Bay	3270±70 BP	Knecht, *et al.* 2001:42
Amaknak, Amaknak Bridge	3240±90	Knecht and Davis 2008
Umnak, Chaluka	3148 (1198±60 BC)	Corbett 1991:48
Umnak, Chaluka	3113 (1163±60 BC)	Corbett 1991:48
Amaknak, Margaret Bay	3110±60	Knecht, *et al.* 2001:42
Umnak, Chaluka	3106 (1156±60 BC)	Corbett 1991:48
Umnak, Chaluka	3070 (1120±60 BC)	Corbett 1991:48
Umnak, Chaluka	3031 (1081±60 BC)	Corbett 1991:48
Amaknak, Amaknak Bridge	3000±70	Knecht and Davis 2008
Central Aleutians		
Adak, ADK-187	3600±50	O'Leary 2001:225
Adak, ADK-187	3590±50	Luttrell and Corbett 2000
Amchitka, AA-12090	3300±70	O'Leary 2001:225
Adak, ADK-012/013	3300±200	O'Leary 2001:225
Adak, ADK-181, Grassy Knoll	3300±50	Luttrell and Corbett 2000
Adak, ADK-171, Tutiakoff	3210±20	this volume
Adak, ADK-194	3050±70	O'Leary 2001:225
Adak, ADK-194	3050±70 BP	Luttrell and Corbett 2000
Western Aleutians		
Amchitka, RAT-017	3640±90BP	Young n.d.; BIA ANCSA
Amchitka, RAT-024	3520±130 BP	Young n.d.; BIA ANCSA
Shemya, ATU-061	3540±60	Corbett, *et al.* 2010
Shemya, ATU-061	3255±60 BP	Lefèvre, *et al.* 2001:239
Shemya, ATU-061	3120±80 BP	Corbett, *et al.* 2010
Shemya, ATU-061	3096±155 BP	Corbett, *et al.* 2010
Shemya, ATU-061	3080±110	Corbett, *et al.* 2010

eastern Aleutians (e.g., Chaluka and Summer Bay) occur eight to ten meters above modern sea level (Knecht and Davis 2001, 271). Substantially more sites of this time have been discovered (Table 12.7). This may be due to limited landscape alteration (sites occur in places we recognize as good locations today) and/or may be due to a substantial increase in prehistoric population in the archipelago and more and larger village locations creating more visible archaeological signatures. In addition to the increase in number of sites recorded, *umqans* (burials on mountain slopes marked by 'A' shaped trenches) appear, suggesting a shift in social organization or belief system (Knecht and Davis 2001, 277).

Absent from all of the sites following 3000 years BP is blade and microblade technology (Knecht and Davis 2001, 278). Ground slate tools (ulus) become abundant. Stemmed projectile points, asymmetrical flake tool knives and scrapers, symmetrical bell shaped knives, and cobble spall scrapers are also common. The stone tools are larger than during the previous phase. Bone technology is more diverse and elaborate with multiple harpoon types with variable barbing including bilateral barbs; harpoons also now have line holes, toggles, sockets, and/or wedge shaped bases (Knecht and Davis 2001, 271, 277-278).

Several sites dating to this time period in the central and

western Aleutians have been recorded and document irregular flake, bifacial, and bone technology, with little change from the previous period (Corbett *et al.* 2001; Lefèvre *et al.* 2001; O'Leary 2001, 226). On the southwestern shore of Shemya, ATU-003 (Figure 12.7; Table 12.7) dates to this period and consists of midden deposits (Corbett *et al.* 2010). Two occupations were noted. The older dates to 2500 years BP and consists of hearth features and sea urchin lenses. The younger dates between 2100 and 1500 years BP with much more complicated occupation activity. Stone artifacts include stemmed points and parallel-sided points, as well as net sinkers. Bone artifacts included harpoon foreshafts, unilaterally barbed bone points, toggles, fishhook shanks and fishhook points. Sites appear to vary compared to the eastern Aleutians especially given the absence of any *Umqan* burials in the central or western Aleutians, suggesting there are significant sociopolitical differences. However, the increasing complexity in bone tool technology suggests some similarities with eastern Aleutian sites.

From the current Adak research, the Dozered site (ADK-012) dates to this time period (Figure 12.7). The significant damage from bulldozing at the Dozered site impacts some of the interpretations of this site. However, the chipped and pecked stone tools represent definitive evidence of prehistoric human occupation at this site. These tools include projectile point/knives and fragments, flake tools, as well as net sinkers, hammerstones, abraders (including abrasive materials) and griddle stones. However, no ulus were recovered. Bone tools recovered from ADK-012 include awls, needles, a fish hook shank, cut bird bone, wedges, worked mammal rib fragments, and worked mammal and bird bone (West and Hatfield, Chapter 15, this volume).

Component 1 at the Zeto Point Village site (ADK-011) (Figure 12.7) also dates to this phase and lithic tools included bifaces, flake tools, abraders, a possible blade tool, and a hammerstone as well as several griddlestones fragments. The cores and flakes predominantly reflect irregular flake technology and some bifacial technology. Four specimens may reflect blade/microblade technology, but this sparse sample limits confidence in such an assessment.

1000-200 BP: The Late Aleutian Phase

The Late Aleutian phase (Table 12.8) in the eastern Aleutians documents massive midden deposits and the appearance of longhouses. These sites occur on terraces close to the modern sea level. Technologically, this period is characterized more by groundstone than by chipped stone technologies. In the eastern Aleutians, ground slate straight and semi-lunar ulu blades and lance heads are common (Knecht and Davis 2001, 279). Irregular core and flake industries occur but less frequently than before, and chipped stone tools are poorly formed (Knecht and Davis 2001, 279). The chipped stone industries reflect expedient technologies. The bone technology, however, reflects elaborate harpoon morphologies (with and without line holes) and cylindrical socket pieces (Knecht and Davis 2001, 279).

Similarly, for the central and western Aleutians, Corbett (1991, 46) describes the continuance of irregular flake, bifacial, and groundstone technologies, and increasingly complex bone technology with distinctive fore shaft and socket pieces and elaborately barbed harpoon and lance points (Corbett *et al.* 2010). On Amchitka, around 1000 years ago, ground slate tools, bone awls, needles, symmetrically barbed harpoon heads, conically tanged projectile tips, and conical socket pieces occur (Desautels *et al.* 1970, 249; Corbett 1991, 47). On Adak, the Zeto Point Village site's Component 2 assemblage included bifacial projectile point/knives (one laurel leaf shaped point and five stemmed points), a unifacial point, and flake tools. Ground and pecked stone tools include ulus, abraders, lamps, plummets, net sinkers, anvils, hammerstones, and griddle stones. From the debitage there was one possible blade-like flake, but mostly irregular flake and bifacial technology are represented. Bone technology included bone pendants and other decorative bone artifacts, fish hooks, a bilaterally carved harpoon point, asymmetrically barbed bone points, a flat bone point, awls, a needle preform, a gorge, bone wedges, bird bone with scoring, and other worked bone fragments, as well as bones that show use as abraders (Luttrell and Corbett 2000; West and Hatfield, Chapter 15, this volume).

Overall, there are some differences in the Adak assemblages compared to the eastern Aleutian sequence. These differences primarily center on the presence/absence of microblade and blade technologies. A few potential artifacts may reflect blade and microblade technologies in the Adak assemblages; however, they are easily regarded as accidental by-products of irregular flake/core technologies. Possibly, the deficit of blade and microblade technologies in late Adak sites is due to the relatively small sample of sites and assemblage sizes recovered thus far. The microblade and blade technologies in the eastern Aleutians are reflected within assemblages that are comparatively large. Also, many more sites have been documented in the eastern Aleutians. The few blade and microblade artifacts from the western Aleutians (from Shemya, ATU-061) suggest this technology may occur across the chain, but perhaps was far less common than in the eastern Aleutians.

There also seems to be some variation in the timing of the appearance of ulus, which are found dating to the Margaret Bay Phase (4000-3000 years BP) in the eastern Aleutians (at the Margaret Bay site, Level 3) and on Shemya (ATU-061) and are abundant during the Amaknak Phase (3000-1000 years BP) at several sites in the eastern and western Aleutians. On Adak, ulus are found, so far, only during the Late Aleutian Phase (1000-200 years BP). The occurrence of other groundstone tools are documented at all of the Adak sites reported herein and include abraders recovered from

Table 12.7. Amaknak Phase sites and radiocarbon dates.

Island, Site	***years BP ± sigma***	***Source***
Eastern Aleutians		
Amaknak, Amaknak Bridge	2970±60	Knecht and Davis 2008
Umnak, Chaluka	2908 (958±60 BC)	Corbett 1991:48
Umnak, Chaluka	2902 (952±60 BC)	Corbett 1991:48
Umnak, Chaluka	2896 (946±60 BC	Corbett 1991:48
Umnak, Chaluka	2853 (903±60 BC)	Corbett 1991:48
Umnak, Chaluka	2844 (894±60 BC)	Corbett 1991:48
Amaknak, Amaknak Bridge	2840±90	Knecht and Davis 2008
Amaknak, Amaknak Bridge	2780±70	Knecht and Davis 2008
Amaknak, Amaknak Bridge	2670±70	Knecht and Davis 2008
Amaknak, Amaknak Bridge	2590±90	Knecht and Davis 2008
Amaknak, Amaknak Bridge	2540±60	Knecht and Davis 2008
Unalaska, Summer Bay	2470±190	Knecht and Davis 2001
Unalaska, Summer Bay	1860±60	Knecht and Davis 2001
Umnak, Chaluka	1444 (AD 506±46)	Corbett 1991:48
Central Aleutians		
Adak, ADK193, Ocean Bluff	2900±40	Luttrell and Corbett 2000
Adak, ADK013, Ocean View	2900±40	this volume
Adak, ADK011, Zeto Point	2590±20	this volume
Adak, ADK012, Dozered site	2510±50	this volume
Adak, ADK011, Zeto Point	2490±50	Luttrell and Corbett 2000
Adak, ADK012, Dozered Site	2470±15	this volume
Adak, ADK012, Dozered Site	2455±20	this volume
Adak, ADK011, Zeto Point	2440±20	this volume
Adak, ADK011, Zeto Point	2420±15	this volume
Adak, ADK011, Zeto Point	2395±15	this volume
Adak, ADK011, Zeto Point	2390±15	this volume
Adak, ADK102	2310±50	Luttrell and Corbett 2000
Adak, Came Cove, house	2250±130	O'Leary 2001:225
Adak, ADK-011, Zeto Point	2160±40	West, *et al.* 1999
Adak, ADK-012, Dozered Site	2085±15	this volume
Adak, Clam Lagoon	2060±40	Luttrell and Corbett 2000
Adak, Clam Lagoon Peninsula	2020±40	this volume
Adak, ADK181, Grassy Knoll	1950±40	Luttrell and Corbett 2000
Atka, Korovinski	1930±100	Veltre 2001:207
Adak, Lake (ADK-182)	1920±40	Luttrell and Corbett 2000
Atka, Korovinski, Unit 39	1910±90	Veltre 2001:207
Adak, ADK182, Lake Site	1900±40	this volume
Adak, ADK-009, Sweeper Cove	2080±55	Savinetsky, *et al.* 2010
Adak, ADK-012, Dozered Site	1865±15	this volume
Adak, ADK-013 Ocean View	1850±60	this volume
Atka, Korovinski, Unit 16	1830±175	Veltre 2001:207
Adak, ADK009, Sweeper Cove	1710±70	Luttrell and Corbett 2000
Adak, 3 Arm Bay	1530±50	O'Leary 2001:225
Adak, ADK012, Dozered Site	1515±15	this volume
Adak, ADK009, Sweeper Cove	1700±75	this volume
Adak, ADK009, Sweeper Cove	1560±80	this volume
Adak, ADK178, Sweeper Cove 3	1360±40	this volume
Adak, ADK011, Zeto Point	1325±14	this volume
Adak, ADK009, Sweeper Cove	1240±70	this volume
Adak, ADK011, Zeto Point	1240±70	this volume
Adak, ADK011, Zeto Point	1235±15	this volume
Adak, Staten Island	1190±60	O'Leary 2001:225
Adak, ADK009, Sweeper Cove	1180±80	this volume
Adak, ADK009, Sweeper Cove	1180±90	this volume
Adak, ADK011, Zeto Point	1180±20	this volume

Island, Site	***years BP ± sigma***	***Source***
Adak, ADK009, Sweeper Cove	1180±40	Luttrell and Corbett 2000
Adak, ADK009, Sweeper Cove	1360±110	this volume
Adak, ADK171, Tutiakoff	1062±16	this volume
Adak, ADK009, Sweeper Cove	1040±70	this volume
Western Aleutians		
Shemya ATU-023, Pit 2	2905±80	Lefèvre, *et al.* 2001
Shemya ATU-061, Pit 1	2880±155	Lefèvre, *et al.* 2001
Shemya ATU-061, house	2865±110	Lefèvre, *et al.* 2001
Agattu, AG-3	2680±70	BIA-ANCSA
Shemya ATU-023	2680 (730±70 BC)	Corbett 1991:48
Shemya AA-11925	2680±70	BIA-ANCSA
Agattu, Krugloi Point	2630±300	Spaulding 1962
Shemya, ATU-061	2630±60	Corbett, *et al.* 2010
Shemya, ATU-061	2570±140	Corbett, *et al.* 2010
Shemya, ATU-003	2555±126	Corbett, *et al.* 2010
Agattu AG-2	2550 (500±300 BC)	Corbett 1991:48
Amchitka Rat-31	2550 (600±95 BC)	Corbett 1991:48
Agattu Krugloi Point	2500±300	Spaulding 1962
Amchitka, RAT-17	2430±190	Corbett 1991:48
Shemya ATU-613	2380±40	Corbett, *et al.* 2010
Shemya ATU-061,House	2355±140	Lefèvre, *et al.* 2001
Buldir, KIS-008, pit 7	2350±85	Savinetsky, *et al.* 2010
Buldir, KIS-008, pit 7	2347±84	Young n.d.
Shemya ATU-003	2340±125	Lefèvre, *et al.* 2001
Amchitka RAT-031	2250±95	Desautels, *et al.* 1970
Amchitka, Rat 036, 2N	2245±95	Desautels, *et al.* 1970
Shemya, ATU-066	2244±182	Savinetsky, *et al.* 2010
Attu, ATU-193, F44	2210±60	Lefèvre, *et al.* 2001
Amchitka Rat 036, 2N,1E	2190±95	Desautels, *et al.* 1970
Shemya, ATU-003	2148±70	Corbett, *et al.* 2010
Shemya, ATU-613	2110±40	Corbett, *et al.* 2010
Shemya ATU-062	2110 (110±90BC)	Corbett 1991:48
Shemya ATU-062	2060±90	BIA ANCSA 1992
Amchitka RAT-036, 3N/1E	2055±90	Desautels, *et al.* 1970
Shemya ATU-022, Pit 2	2047±82	Savinetsky, *et al.* 2010
Shemya ATU-033, Pit 3	2030±70	Lefèvre, *et al.* 2001
Shemya ATU-066, Pit 1	2030±180	Savinetsky, *et al.* 2010
Shemya. ATU-022	2020±110	Lefèvre, *et al.* 2001
Shemya ATU-021.Pit1,L2	1980±60	Lefèvre, *et al.* 2001
Shemya, ATU-003,Pit4,L2	1935±70	Lefèvre, *et al.* 2001
Amchitka, RAT-031	1890±95	Corbett 1991:48
Amchitka, RAT-032	1865±135	Cook, *et al.* 1972
Attu, Cairn Creek, F43	1860±70	Lefèvre, *et al.* 2001
Attu, Murder Point	1860±70	Lefèvre, *et al.* 2001
Shemya, ATU-003	1860±90	Lefèvre, *et al.* 2001
Shemya, ATU-022	1856±69	Corbett, *et al.* 2010
Shemya, ATU-022	1830±80	Lefèvre, *et al.* 2001
Shemya, ATU-003, Pit3, L3	1810±60	Lefèvre, *et al.* 2001
Shemya, ATU-003	1790±60	Corbett 1991:48
Shemya, ATU-003,Pit1,L4	1790±70	Lefèvre, *et al.* 2001
Shemya, ATU-003,Pit 2,L1	1770±120	Lefèvre, *et al.* 2001
Shemya ATU-003	1720±70	Lefèvre, *et al.* 2001
Shemya ATU-021	1700±70	Corbett, *et al.* 2010
Shemya, ATU-003,Pit1,L4	1700±70	Lefèvre, *et al.* 2001
Shemya, ATU-022,Pit1,L9	1640±70 BP	Lefèvre, *et al.* 2001
Attu, Rocky Point, F6	1630±60 BP	Lefèvre, *et al.* 2001
Shemya ATU-062, Pit 1	1625±105 BP	Lefèvre, *et al.* 2001
Attu, Ballentine Creek, F73	1590±70 BP	Lefèvre, *et al.* 2001

Island, Site	***years BP ± sigma***	***Source***
Shemya, ATU-022	1589±60 BP	Corbett, *et al.* 2010
Shemya, ATU-613	1570±40 BP	Corbett, *et al.* 2010
Shemya, ATU-022,Pit1,L10	1375±60 BP	Lefèvre, *et al.* 2001
Attu, Waterfall, F24	1320±70 BP	Lefèvre, *et al.* 2001
Attu, Ballentine Creek,F194	1310±100 BP	Lefèvre, *et al.* 2001
Agattu, Krugloi Point	1300±150 BP	Corbett 1991:48
Attu, Ballentine Creek, F73	1190±70 BP	Lefèvre, *et al.* 2001
Attu, ATU-198 site, Burial 1	1160±60 BP	Lefèvre, *et al.* 2001
Attu, Rocky Point, F36	1160±50 BP	Lefèvre, *et al.* 2001
Buldir, KIS-008, Pit2, L7	1160±50 BP	Corbett, *et al.* 1997b
Attu, Waterfall, F89	1030±80 BP	Lefèvre, *et al.* 2001
Attu, Loran site, house	1030±60 BP	Lefèvre, *et al.* 2001

Table 12.8. Late Aleutian Phase sites and radiocarbon dates.

Site	***years BP ± sigma***	***Source***
Eastern Aleutians		
Amaknak, UNL-055	560±50	Knecht and Davis 2001
Umnak, Chaluka	424 (AD 1526±46)	Corbett 1991:48
Unalaska, Reese Bay	380±50	Veltre and McCartney 2001
Amaknak, UNL-055	360±70	Knecht and Davis 2001
Unalaska, Reese Bay	360±50	Veltre and McCartney 2001
Unalaska, Reese Bay	340±50	Veltre and McCartney 2001
Umnak, Chaluka	318 (AD 1632±46)	Corbett 1991:48
Unalaska, Reese Bay	250±60	Veltre and McCartney 2001
Unalaska, Reese Bay	240±50	Veltre and McCartney 2001
Unalaska, Reese Bay	210±50	Veltre and McCartney 2001
Unalaska, Reese Bay	190±60	Veltre and McCartney 2001
Unalaska, Reese Bay	170±70	Veltre and McCartney 2001
Central Aleutians		
Adak, ADK-011, Zeto Point	840±40	West, *et al.* 1999
Adak, ADK177, Sweeper Cove 2	840±40	this volume
Adak, ADK 185	810±40	Luttrell and Corbett 2000
Adak, ADK185, Ow-ee Site	790±40	this volume
Atka, Korovinski, Unit 15	785±65	Veltre 2001:207
Adak, ADK-014, Lake Andrew	760±50	Luttrell and Corbett 2000
Adak, ADK-009, Sweeper Cove	970±80	this volume
Adak, ADK193,	2485±20	this volume
Adak, Shagak Bay, base	660±100	O'Leary 2001:225
Adak, ADK-011, Zeto Point	615±15	this volume
Atka, Korovinski, Unit 39	535±80	Veltre 2001:207
Adak, Andrew Bay	520±40	Luttrell and Corbett 2000
Adak, Bay of Waterfalls	500±70	O'Leary 2001:225
Adak, ADK-084 Andrew Lake	500±40	this volume
Adak, Clam Lagoon	460±40	Luttrell and Corbett 2000
Adak, ADK-011, Zeto Point	440±40	West, *et al.* 1999
Adak, ADK183, Clam Lagoon Overlook	450±40	this volume
Adak, ADK-011, Zeto Point	435±20	this volume
Adak, ADK-011, Zeto Point	415±20	this volume
Adak, ADK-011, Zeto Point	410±20	this volume
Adak, ADK-011, Zeto Point	400±20	this volume
Adak, ADK-011, Zeto Point	390±20	this volume
Adak, ADK-011, Zeto Point	385±25	this volume
Adak, ADK-011, Zeto Point	375±20	this volume
Adak, ADK-011, Zeto Point	370±20	this volume
Adak, ADK-011, Zeto Point	365±20	this volume

Site	***years BP ± sigma***	***Source***
Adak, Lake Andrew	340±50	Luttrell and Corbett 2000
Adak, ADK-011, Zeto Point	340±15	this volume
Adak, ADK-011, Zeto Point	340±20	this volume
Adak, ADK-014, Andrew Lake	340±50	this volume
Adak, ADK-011, Zeto Point	335±15	this volume
Adak, ADK-011, Zeto Point	335±20	this volume
Adak, ADK-084, Andrew Bay	335±20	this volume
Adak, ADK-011, Zeto Point	325±20	this volume
Adak, Campers Point	320±40	O'Leary 2001:225
Adak, ADK-011, Zeto Point	320±20	this volume
Adak, ADK-011, Zeto Point	320±25	this volume
Adak, ADK-011, Zeto Point	315±20	this volume
Adak, ADK-084, Andrew Bay	290±40	Luttrell and Corbett 2000
Adak, ADK-011, Zeto Point	285±15	this volume
Adak, ADK-011, Zeto Point	285±20	this volume
Adak, ADK-084, Andrew Bay	280±20	this volume
Adak, ADK-011, Zeto Point	265±20	this volume
Adak, Boot Bay	230-60	O'Leary 2001:225
Adak, Shagak Bay	230±40	O'Leary 2001:225
Adak, ADK-011, Zeto Point	220±50	Luttrell and Corbett 2000
Adak, Shagak Bay	210±40	O'Leary 2001:225
Adak, ADK-011, Zeto Point	180±60	Luttrell and Corbett 2000
Adak, ADK-011, Zeto Point	170±25	this volume
Adak, Airport site	160±30	O'Leary 2001:225
Adak, ADK-011, Zeto Point	160±20	this volume
Adak, ADK-010, Palisades Overlook	150±60	Luttrell and Corbett 2000
Adak, Eddy Island	102±1.1	O'Leary 2001:225
Adak, ADK-011, Zeto Point	75±20	this volume
Western Aleutians		
Shemya, ATU-023, surface	980±70 BP	Lefèvre, *et al.* 2001
Attu, Goltsov Point, F20	960±60 BP	Lefèvre, *et al.* 2001:247
Attu, Ballentine Creek	950±100 BP	Lefèvre, *et al.* 2001:247
Amchitka, RAT-031	890±90 BP	Desautels, *et al.* 1971
Attu, ATU-198, burial 1	790±90 BP	Lefèvre, *et al.* 2001:247
Agattu, ATU-036, F49	760±780 BP	Corbett, *et al.* 2001:263
Agattu. ATU-035	760±70 BP	BIA ANCSA 1996
Buldir. KIS-008	760±60 BP	Young n.d.
Buldir, KIS-008	630±60 BP	Young n.d.
Attu, NE Bluff	540±80 BP	Lefèvre, *et al.* 2001:247
Buldir, KIS-008	530±60 BP	Corbett, *et al.* 1997b
Buldir, KIS-008	530±60 BP	Corbett, *et al.* 1997b
Attu, Ballentine Creek	490±60 BP	Lefèvre, *et al.* 2001
Buldir, KIS-008	460±50 BP	Corbett, *et al.* 1997b
Attu, Ballentine Creek	430±60 BP	Lefèvre, *et al.* 2001:247
Buldir, KIS-008	420±60 BP	Corbett, *et al.* 1997b
Attu, Coastal Plain	390±50 BP	Lefèvre, *et al.* 2001:247
Attu, ATU-198	390±40 BP	Lefèvre, *et al.* 2001:247
Buldir, KIS-008	380±190 BP	Young n.d.
Little Sitka, RAT-085	360±60 BP	Corbett, *et al.* 2001:263
Agattu, ATU-035	360±60 BP	BIA ANCSA 1996
Agattu, ATU-036	360±60 BP	Corbett, *et al.* 2001:263
Attu, NE Bluff	360±60 BP	Lefèvre, *et al.* 2001:247
Rat, RAT-079	360±90 BP	Corbett, *et al.* 2001:263
Buldir KIS-008	350±80 BP	Corbett, *et al.* 1997b
Buldir, KIS-008	320±60 BP	Corbett, *et al.* 1997b
Buldir, KIS-008	320±50 BP	BIA ANCSA 1990
Amchitka, RAT-035	310±60 BP	Young n.d
Kiska, KIS-010	310±60 BP	Corbett, *et al.* 2001:263

Site	*years BP ± sigma*	*Source*
Attu, Ballentine Creek	280±70 BP	Lefèvre, *et al.* 2001:247
Buldir, KIS-008	280±50 BP	Corbett, *et al.* 1997b
Buldir, KIS-008	250±70 BP	Young n.d.
Buldir, KIS-008	240±60 BP	Corbett, *et al.* 1997b
Attu, Golstov Point	230±60 BP	Lefèvre, *et al.* 2001:247
Rat, RAT-079	230±90 BP	Corbett, *et al.* 2001:263
Buldir, KIS-008	220±70 BP	Young n.d.
Rat, RAT-079	180±50 BP	Corbett, *et al.* 2001:263
Attu, ATU-198	150±60 BP	Lefèvre, *et al.* 2001:247
Attu, ATU-198	100±70 BP	Lefèvre, *et al.* 2001:247
Attu, ATU-198, entrance	100±40 BP	Lefèvre, *et al.* 2001:247

ADK-171 (within the *Clinocardium* midden deposits and test pit 55-1), from ADK-011, Component 1 (2400-2500 BP), and from ADK-011, Component 2. These abraders were made of scoria, pumice, or fine-grained sandstone and were used for different smoothing and shaping tasks. Other groundstone tools include girdled cobbles and net sinkers from ADK-012, pecked stone tools, including hammerstones and anvils, as well as groundstone tools such as plummets, lamps, and net sinkers from ADK-011, Component 2. But only within the ADK-011, Component 2, 200-400 year old layers were ulus recovered, and they mark the appearance of these tools in the Adak toolkit. No ulus were recovered from ADK-012 or ADK-171. Possibly there is a sampling error, due to smaller assemblages or fewer known and investigated central Aleutian sites dating to the time period in question, compared to the eastern and western Aleutians. Further research is needed to evaluate the central and western Aleutian sites with respect to their similarities and differences with each other and the eastern Aleutians sites.

The use of griddle stones (flat tabular rocks presumably used in cooking activities) is documented at all three sites (and both components at ADK-011) on Adak, and appears to increase in use through time. Jochelson (1925) ethnographically documented the use of griddle stones to heat food. How this compares to other Aleutian sites is not well understood and may be more variable than hunting and processing technologies, given the wide variety of ways of cooking food and/or heating houses.

Unangan Lithic Technological and Material Choices

Through time, across the archipelago, the Aleutian sites document a shift from formal chipped stone tools requiring some time investment for manufacture to more expedient chipped stone tools. This is seen in the decrease in bifacial technology and the increase in flake tools through time. Bifaces are flaked on both sides, require planning, and involve several steps towards shaping the final tool. In contrast, most flake tools capitalize on the natural shape and cutting angles of the stone with little or no additional shaping and require less time to be invested in manufacture. This transition from formal to expedient tools co-occurs with an increased investment in bone technology, which becomes more complex through time. In addition, ground and pecked stone tools also increase and perhaps replace certain chipped stone tools. The changes from formal to expedient tools, as well as the shift toward groundstone, pecked stone, and bone technology may reflect the availability or access to fine-grained materials used for chipped stone tools and/or increased sedentism (Parry and Kelly 1987). This assumes, of course, that the ground and pecked stone tools are made from locally available materials.

The shift in proportion from fine-grained to coarse-grained materials used for chipped stone tools on Adak corresponds with similar shifts in the eastern Aleutians. The fine-grained materials comprise a higher proportion of certain tool types earlier in time and decrease in frequency through time as formal tools give way to informal, more expedient tools. The shift toward coarse-grained materials could also reflect population density increases or restricted access to materials that are rare. Access to lithic materials is the most likely reason for the shift in importance from formal to informal/expedient lithic technologies and from lithic to bone technologies. However, the shift to coarse-grained materials may have been driven by an increase in the use of bone and ground/pecked stone technologies or, alternatively, an increase in the use of coarse-grained material may have influenced the shift to bone and wood tools and more pecked stone objects.

Lithic material comparisons are uncommon in the Aleutians. Hatfield (2006) documented the changes in lithic materials at five Aleutian sites [the Russian Spruce site (UNL-115), Oiled Blade Site (UNL-318), the Margaret Bay site (UNL-048), the Amaknak Bridge site (UNL-050) and Shemya 6 (ATU-061)]. These sites date between 8000 and 3000 RCYBP and document a change in material use through time corresponding to the shifts documented in the Adak assemblages. Throughout the Aleutians, chert and obsidian are geologically far less common than basalt and andesite. Through time, the use of obsidian and chert decreases, indicating either that the access to these materials is more difficult or changes in technology were such that the importance of these materials decreased. This was noted in

Unalaska Bay where obsidian and chert decrease through time.

The use of different materials by industry indicates that blade and microblade technology rely primarily on chert (defined here to include chert, jasper, chalcedony and other cryptocrystallines that are of moderate to high quality and durability) and basalt. Obsidian is rare in bifacial technology due to the fragile nature of this glass-like material with less durable edges than basalts, chert, or almost any other material and is most common in blade technology (Hatfield 2006).

Both bifacial and flake technology are dominated by basalt, with chert well represented. On Shemya, at ATU-061, very little chert and no obsidian are represented and most of the materials are made of basalts and andesites reflecting locally available materials (Hatfield 2006; Corbett *et al.* 2010). The lithic materials represented in bifacial technology suggest that this technology required materials that are durable and at least of moderate quality. The fine-grained chert tools would have made longer-lasting tools for cutting or scraping. Bifacial technology did not reflect change in lithic material use through time in the comparison of the five sites by Hatfield (2006). Bifaces occasionally are made from obsidian or andesite; however, the preferred or default materials are basalt and, to a lesser degree, chert. This again suggests that bifacial technology is dependent on certain lithic material attributes. Because basalt provided many of those attributes, and was relatively ubiquitous, it was the preferred choice for bifaces.

At ADK-171 lithic material used for chipped stone artifacts included only one obsidian tool and several chert tools. ADK-171 cores were made of andesite, basalt, chert, and metamorphosed sedimentary rocks. The debitage similarly represents a preference for fine-grained materials over the coarse grained materials for chipped stone tools. Andesitic cores could produce flakes for cutting, scraping, or chopping, but these flakes would have quickly become dulled and be discarded. Thus andesites are typically associated with expedient tools. The high number of chert flakes and tools compared with those made of andesite and basalt suggests that early Unangan settlers on Adak, as represented by ADK-171, had a comparatively broad territorial range. Chert is available only in small seams between basalt flows in the Clam Lagoon environs (Lyn Gualtieri, Seattle Univ., personal communication) and would have required Unangan people to travel to specific locales and farther afield to collect suitable nodules. Large territories provided access to higher quality stone that Unangan people used until each core was exhausted. The locally available, but poorer quality, andesite and basalt were more easily acquired and dependence on these materials later in time perhaps reflects a smaller region of travel or trade. Andesite and basalt are ubiquitous at Clam Lagoon, occurring in outcrops and as cobbles on beaches.

There are no known obsidian sources farther west than Umnak and Akutan islands in the eastern Aleutians (see Nicolaysen *et al.* Chapter 11 this volume). If these were the only obsidian sources in the archipelago, Unangan people at ADK-171 carried the highly valued material with them when they migrated to Adak, traded with eastern Unangan groups who had access to this material, or made special long-distance voyages to the east to acquire it first hand. Like their ADK-171 ancestors, Unangan people at ADK-011 either faced long sea voyages to collect obsidian or traded/raided with eastern Unangan groups who had primary access to this prized resource. When obsidian projectile points or knife forms failed, Adak Unangan hunter gatherers certainly would have recycled broken tools into other types of tools in order to capitalize the last bit of usefulness.

Minimal andesite flakes indicate that lower-quality, expedient tools from local cobbles were a less insignificant part of the tool kit. At the Dozered site (ADK-012), the lithic material is dominated by andesite and basalt in debitage and tools and indicates that cobble reduction for expedient tools was an important part of the technology (Parry and Kelly 1987). The shift from pronounced chert use at the ADK-171 site to minimal use at ADK-012 suggests that this material was hard to come by or that access was restricted. Similarly at ADK-011, Component 2, the many flakes made of andesite indicate an increased use of local coarse-grained material and a corresponding dependence on the use of expedient tools.

Acknowledgements

The initial research for collecting and analyzing the lithic material was accomplished with a grant from the National Science Foundation, Office of Polar Programs (#OPP0353065). The fieldwork could not have been done without the help of the community of Adak. Kansas University students Chris Hare and Morgan Shirley, and Kansas State University student Sarah Meitl participated in fieldwork (2006). Vassar students Hannah Scranton and Will Thannhauser, through the Undergraduate Research Summer Institute of Vassar College, helped both with fieldwork and with sorting screen samples (2006). Matthew J. Root of Rainshadow Research in Pullman, WA made invaluable observations on the manufacture and breakage of the lithic tools as well as provided guidance on manufacturing stages and lithic debitage. Lyn Gualtieri and Brenn Sarata, CAAPP geologists, helped archaeologists establish an inventory of stone material types used for categorizing Adak artifacts. Kirsten Nicolaysen and Taylor Johnson (Whitman College, WA) verified and sourced obsidian fragments recovered from Adak excavations (see Nicolaysen *et al* Chapter 11 this volume). Nicolaysen sourced flat andesite slabs (see Jeannotte *et al.* Chapter 14 this volume). Lucy Johnson (Vassar College, NY) reviewed the chapter and provided valuable suggestions.

References

Aigner, Jean S. 1978. The Lithic Remains from Anangula, an 8500 year old Aleut Coastal Village. *Urgeschichtiliche Materialhefte 3*. Germany, Verlag Archaeologica Ventoria Institut für Urgeschlichte der Univesität Tübingen.

Aigner, Jean S. and Alan M. Beiber, Jr. 1976. Preliminary Analysis of Stone Tool Distributions and Activity Zonation at Anangula, an 8500B.P. Coastal Village in the Aleutian Islands, Alaska. *Arctic Anthropology* 13(2), 46-59.

Bamforth, Douglas. 1986. Technological Efficiency and Tool Curation. *American Antiquity* 51(1), 38-50.

Binford, Lewis R. 1978. Dimensional Analysis of Behavior and Site Structure: Learning from an Eskimo Hunting Stand. *American Antiquity* 43(3), 330-361.

Cook, John P., James Dixon, and C. E. Holmes. 1972. *Archaeological Report, Site 49RAT32, Amchitka Island.* United States Atomic Energy Commission Report.

Corbett, Debra G. 1991. *Aleut Settlement Patterns in the Western Aleutians, Alaska*. Unpublished MA thesis, Department of Anthropology, University of Alaska, Fairbanks.

Corbett, Debra G., Christine Lefévre, and Doug Siegel-Causey. 1997a The Western Aleutians: Cultural Isolation and Environmental Change. *Human Ecology* 35(3), 459-478.

Corbett, Debra G., Christine Lefévre, Thomas J. Corbett, Dixie L. West, Douglas Siegel-Causey. 1997b. Excavations at KIS-008, Buldir Island: Evaluation and Potential. *Arctic Anthropology* 34(2), 100-117.

Corbett, Debra G., Dixie L. West, and Christine Lefévre. 2001 Prehistoric Village Organization in the Western Aleutians, in D. E. Dumond (ed.), *Archaeology in the Aleut Zone of Alaska, Some Recent Research,* 251-266. University of Oregon Anthropological Papers No. 58. Eugene, University of Oregon Press.

Corbett , Debra G., Dixie West, and Christine Lefévre (eds.) 2010. *The People at the End of the World: The Western Aleutians Project and the Archaeology of Shemya Island.* Alaska Anthropological Association Monograph Series—VII. Fairbanks, Alaska, Aurora.

DeSautels, Roger J., Albert J. McCurdy, James D. Flynn and Robert R. Ellis. 1971. *Archaeological report. Amchitka Island, Alaska 1969-1970*. Los Angeles, United States Atomic Energy Commission, TID-25481.

Dixon, E. James. 1975. The Gallagher Flint Station, An Early Man Site on the North Slope, Arctic Alaska, and its Role in Relation to the Bering Land Bridge. *Arctic Anthropology* 12(1), 68-75.

Dumond, Don E. and Richard A. Knecht. 2001. An Early Blade Site in the Eastern Aleutians, in Don E. Dumond (ed.) *Archaeology in the Aleut Zone of Alaska, Some Recent Research*, 9-34. University of Oregon Anthropological Papers No. 58. Eugene, Oregon, University of Oregon Press.

Dumond, Don E. 2001. Toward a (Yet) Newer View of the (Pre)History of the Aleutians, in D. E. Dumond (ed.), *Archaeology in the Aleut Zone of Alaska, Some Recent Research*, 289-309. University of Oregon Anthropological Papers No. 58. Eugene, University of Oregon Press.

Hatfield, Virginia L. 2006. *Historical Continuity from Shemya to Dutch Harbor: An Evolutionary Analysis of Chipped Stone Technology in the Aleutian Islands*. Unpublished PhD dissertation, Department of Anthropology, Lawrence, University of Kansas.

Hayden, Brian and Aubrey Cannon. 1983. Where the Garbage Goes: Refuse Disposal in the Maya Highlands. *Journal of Anthropological Archaeology* 2, 117-163.

Jeannotte, Richard, Kirsten Nicolaysen, Libin Yao, Floyd E. Dowell, Elizabeth Maghirang, Taylor Johnson, and Dixie West. 2011. Griddlestones: Their Provenance and Associated Organic Molecules, Adak Island, Alaska Chapter 14 in D. West, V. Hatfield, E. Wilmerding, C. Lefèvre, L. Gualtieri (eds.), *The People Before: The Geology, Paleoecology and Archaeology of Adak Island, Alaska*. Oxford, British Archaeological Reports.

Jochelson, Waldemar. 1925. *Archaeological Investigations in the Aleutian Islands*, Publication No. 367. Washington, D. C., Carnegie Institution of Washington.

Johnson, Lucille Lewis and Elizabeth G. Wilmerding. 2001. Bringing the House Down: Modeling Construction and Deconstruction of Aleut Semisubterranean Houses, in D. E. Dumond (ed.) *Archaeology in the Aleut Zone of Alaska, Some Recent Research*, 127-149. University of Oregon Anthropological Papers No. 58. Eugene, University of Oregon Press.

Kay, Marvin. 2010. Clam Lagoon Archaeology and Technology. Chapter 13 in D. West, V. Hatfield, E. Wilmerding, C. Lefèvre, L. Gualtieri (eds.), *The People Before: The Geology, Paleoecology and Archaeology of Adak Island, Alaska*. Oxford, British Archaeological Reports.

Knecht, Richard A., Richard S. Davis and Gary A. Carver. 2001. The Margaret Bay Site and Eastern Aleutian Prehistory, in D. E. Dumond (ed.) *Archaeology in the Aleut Zone of Alaska, Some Recent Research*, 35-70. University of Oregon Anthropological Papers No. 58. Eugene, Oregon, University of Oregon Press.

Knecht, Richard A. and Richard S. Davis. 2001. Prehistoric Sequence for the Eastern Aleutians, in D. E. Dumond (ed.) *Archaeology in the Aleut Zone of Alaska, Some Recent Research*, 269-288. University of Oregon Anthropological Papers No. 58. Eugene, Oregon, University of Oregon Press.

Knecht, Richard A. and Richard S. Davis. 2008. The Amaknak Bridge Site: Cultural Change and the Neoglacial in the Eastern Aleutians. *Arctic Anthropology* 45(1), 61-78.

Laughlin, William S. 1975. Aleuts: Ecosystems, Holocene History, and Siberian Origin. *Science* 189(4202), 507-515.

Laughlin, William S. and Jean S. Aigner. 1966. Preliminary Analysis of the Anangula Unifacial Core and Blade Industry. *Arctic Anthropology* 3(2), 41-56.

Laughlin, William S. and Jean S. Aigner. 1975. Aleut

Adaptation and Evolution, in W. Fitzhugh (ed.), *Prehistoric Maritime Adaptations of the Circumpolar Zone*, 181-202. Mouton, The Hague.

Lefévre, Christine, Dixie West and Debra G. Corbett. 2001. Archaeological Surveys in the Near Islands: Attu Island and Shemya Island, in D. E. Dumond (ed.) *Archaeology in the Aleut Zone of Alaska, Some Recent Research*, 235-250. University of Oregon Anthropological Papers No. 58. Eugene, Oregon, University of Oregon Press.

Luttrell, Mark and Debra G. Corbett. 2000. *Archaeological investigations on Northeast Adak Island, Alaska, July, 1999*. United States Fish and Wildlife Service report.

Maschner, Herbert D. G. and James W. Jordan. 2001. The Russell Creek Manifestation of the Arctic Small Tool Tradition on the Western Alaska Peninsula, in D. E. Dumond (ed.), *Archaeology in the Aleut Zone of Alaska, Some Recent Research*, 151-171. University of Oregon Anthropological Papers No. 58. Eugene, University of Oregon Press.

Mason, Owen K. 2001. Catastrophic Environmental Change and the Middle Holocene Transition in the Aleutian Islands, in D. E. Dumond (ed.), *Archaeology in the Aleut Zone of Alaska, Some Recent Research*, 105-121. University of Oregon Anthropological Papers No. 58. Eugene, University of Oregon Press.

McCartney, Allen P. 1984. Prehistory of the Aleutian Region, in D. Damas (ed.) *Handbook of North American Indians, Arctic*, 5, 119-135. Washington D. C., Smithsonian Institution.

Nicolaysen, Kirsten, Taylor Johnson, Elizabeth Wilmerding, Virginia Hatfield, Dixie West, and Robert G. McGimsey. 2011. Provenance of Obsidian Fragments Recovered from Adak Island, Central Aleutian Islands: Evidence for Long-distance Transport of Raw Lithic Material. Chapter 11 in D. West, V. Hatfield, E. Wilmerding, C. Lefèvre, L. Gualtieri (eds.), *The People Before: The Geology, Paleoecology and Archaeology of Adak Island, Alaska*. Oxford, British Archaeological Reports.

Okuno, Mitsuru, Keiji Wada, Toshio Nakamura, Lyn Gualtieri, Brenn Sarata, Dixie West, Masayuki Torii. 2011. Holocene Tephra Layers on the Northern Half of Adak Island in the west-central Aleutians, Alaska. Chapter 3 in D. West, V. Hatfield, E. Wilmerding, C. Lefèvre, L. Gualtieri (eds.), *The People Before: The Geology, Paleoecology and Archaeology of Adak Island, Alaska*. Oxford, British Archaeological Reports.

O'Leary, Matthew. 2001. Volcanic Ash Stratigraphy for Adak Island, Central Aleutian Archipelago, in D. E. Dumond (ed.) *Archaeology in the Aleut Zone of Alaska, Some Recent Research*, 215-233. University of Oregon Anthropological Papers No. 58. Eugene, University of Oregon Press.

Parry, William J. and Robert L. Kelly. 1987. Expedient Core Technology and Sedentism, in J. K. Johnson and C. A. Morrow (eds.), *The Organization of Core Technology*, 285-304. Boulder, CO, Westview Press.

Root, Matthew J. and Daryl E. Ferguson. 2003. Protocols for Inventory and Analysis of Ground Stone and Chipped-Stone Artifacts. Project Report No. 58. Rainshadow Research, Pullman, Washington.

Savinetsky, Arkady B., Nina Kiseleva and Bulat Khassanov. 2010. The Natural Environment of Shemya Island, in D. West, D. Corbett and C. Lefèvre (eds.), *The People at the End of the World: The Western Aleutians Project and the Archaeology of Shemya Island.* Alaska Anthropological Association Monograph Series—VII. Fairbanks, Alaska, Aurora.

Savinetsky, Arkady B., Dixie West, Zhanna A. Antipushina, Bulat F. Khassanov, Nina K. Kiseleva, Olga A. Krylovich, and Andrei M. Perelodov. 2011. The Reconstruction of Ecosystems History of Adak Island (Aleutian Islands) During the Holocene. Chapter 5 in D. West, V. Hatfield, E. Wilmerding, C. Lefèvre, L. Gualtieri (eds.), *The People Before: The Geology, Paleoecology and Archaeology of Adak Island, Alaska.* Oxford, British Archaeological Reports.

Siegel-Causey, Doug, Christine Lefévre and Arkady Savinetsky. 1991. Historical Biodiversity of Cormorants and Shags from Amchitka Island, Alaska. *The Condor* 93, 840-852.

Spaulding, Albert C. 1962. Archaeological Investigations on Agattu, Aleutian Islands. *University of Michigan Museum of Anthropological Papers* 18.

Stevenson, Marc G. 1982. Toward an Understanding of Site Abandonment Behavior: Evidence from Historic Mining Camps in the Southwest Yukon. *Journal of Anthropological Archaeology* 1, 237-265.

United States Bureau of Indian Affairs. 1985. *Report of Investigations for BLM AA-12080. The Aleut Corporation.* By William L Sheppard and Debra Corbett. On File at the US Bureau of Indian Affairs, ANCSA Office.

United States Bureau of Indian Affairs. 1991. *Compilation of Investigation Reports, 1983 Sites.* By William L Sheppard and Gary Navarre. On File at the US Bureau of Indian Affairs, ANCSA Office.

United States Bureau of Indian Affairs. 1992a. *Report of Investigation for Rat Islands overview.* The Aleut Corporation, BLM AA-11927. On File at the US Bureau of Indian Affairs, ANCSA Office.

United States Bureau of Indian Affairs. 1992b. *Report of Investigations for Adak Island, BLM AA-12088, The Aleut Corporation (a.k.a. Adak Overview)*. By Thomas J. Turck. On File at the US Bureau of Indian Affairs, ANCSA Office.

United States Bureau of Indian Affairs. n.d. *Radiocarbon dates for sites AA-11960, 11967, 12011, 12013.* On File at the US Bureau of Indian Affairs, ANCSA Office.

Veltre, Douglas W. 2001. Koroviniski: Archaeological and Ethnohistorical Investigations of a Pre- and Post-Contact Aleut and Russian Settlement on Atka Island, in D. E. Dumond (ed.) *Archaeology in the Aleut Zone of Alaska, Some Recent Research*, 187-213. University of Oregon Anthropological Papers No. 58. Eugene, Oregon, University of Oregon Press.

Veltre, Douglas W. and Alan P. McCartney. 2001. Ethnohistorical Archaeology at the Reese Bay Site,

Unalaska Island, in D. E. Dumond (ed.) *Archaeology in the Aleut Zone of Alaska, Some Recent Research*, 87-104. University of Oregon Anthropological Papers No. 58. Eugene, Oregon, University of Oregon Press.

West, Dixie L, Christine LeFévre, Debra G. Corbett and Arkady B. Savinetsky. 1999. Radiocarbon dates for the Near Islands, Alaska. *Current Research in the Pleistocene* 16:83-85.

West, Dixie L., Christine Lefèvre, Elizabeth Wilmerding, Virginia Hatfield, and Lyn Gualtieri. 2011. The Central Aleutians Archaeological and Paleobiological Project. Chapter 1 in D. West, V. Hatfield, E. Wilmerding, C. Lefèvre, L. Gualtieri (eds.), *The People Before: The Geology, Paleoecology and Archaeology of Adak Island, Alaska*. Oxford, British Archaeological Reports.

West, Dixie L. and Virginia Hatfield. 2011. Bone Artifacts and Debris from Three Archaeological sites on Adak Island, Alaska. Chapter 15 in D. West, V. Hatfield, E. Wilmerding, C. Lefèvre, L. Gualtieri (eds.), *The People Before: The Geology, Paleoecology and Archaeology of Adak Island, Alaska*. Oxford, British Archaeological Reports.

Wilmerding, Elizabeth G. 2005. The Cultural Sequence at the Nunik Site, Chernabura Island, Alaska. Unpublished PhD dissertation, Department of Anthropology, Washington State University.

Wilmerding, Elizabeth G. 2006a. ADK-171 Unit 3 and 3 annex, *Clinocardium* Clam Midden, Lithic Analysis. Unpublished report for Central Aleutians Archaeological and Paleobiological Project, National Science Foundation, Office of Polar Programs award #0353065. Manuscript in possession of the author.

Wilmerding, Elizabeth G. 2006b. ADK-171 Unit 55-1: Lithic Analysis. Unpublished report for Central Aleutians Archaeological and Paleobiological Project, National Science Foundation, Office of Polar Programs award #0353065. Manuscript in possession of the author.

Wilmerding, Elizabeth G. 2008. Report on Lithic Analysis for ADK-011, Summer 2006 Field Work. Unpublished report for Central Aleutians Archaeological and Paleobiological Project, National Science Foundation, Office of Polar Programs award #0353065. Manuscript in possession of the author.

Young, Allison. n.d. Aleut Technological Organization, Amchitka Island, Alaska. Unpublished manuscript in possession of the author.

Chapter 13
Clam Lagoon Archaeology and Technology

Marvin Kay

'Weather forecast for tonight: dark.' -- George Carlin

Abstract: *Although exotic lithics occur in an archaeological context on Adak Island, Alaska, the predominant pattern was procurement of local stone for ground and chipped stone tool production over several thousand years. Igneous rocks of various shapes, sizes and textures used for chipped and ground stone production were deliberately selected apparently from outcrops and shoreline beach deposits on north Adak. Distinctive production chains exist for ground and chipped stone artifacts. Microscopic use-wear analysis indicates a wide array of chipped stone tool functions, long use-lives, and recycling. This patterning suggests careful selection and conservation of even clearly local stone resources.*

Keywords: *Aleutian Islands, stone tools, production chains, microscopic use-wear analysis*

Introduction

The object of this study is Aleut stone technology inclusive of byproducts of manufacture or repair and tools on Adak Island, Alaska. Overall, the research is exploratory and is a first glimpse of microscopic use-wear methods discussed more broadly elsewhere (Hardy *et al.* 2001; Kay 1996, 1998; Kay and Martens 2004; Kay and Solecki 2000; Root *et al.* 1999) applied to collections from the Aleutian Islands. The microscopic use-wear assessment follows procedures that bridge traceological (Semenov 1964) and polish formation (Keeley 1980) approaches. Most important are striations and associated traceological wear that often crosscut what are likely silica gel residues, or microplates, on tool surfaces and edges. The microscopy uses a differential-interference binocular microscope with polarized light and Nomarski optics (Hoffman and Gross 1970) at magnifications of 100, 200, and 400 diameters. This microscope affords three-dimensional views with exceptional clarity of microtopography that enhances microscopic details, and improves resolution as magnification increases; all are essential to a use-wear evaluation. As previously described (Kay 2006), the analysis addresses (1) differentiation between taphonomic alterations and evidence of tool use or hafting, (2) determination of production stages of manufacture-use-recycling, and (3) tool function and likely contact material for those artifacts employed as tools. The use-wear analyses are not intended to be independent, stand-alone assessments, but are a sampling of illustrated results of Adak Island artifact use-wear. Use-wear analyses are balanced against other lines of information about the artifacts and the archaeological sites from which they came.

Descriptions of Aleut life and technology stem from Jochelson's (1925) and later scholars' (see McCartney 1984 and for a still reasonable overview) research. Stone tools are the most common and easily preserved archaeological material remains. By employing use-wear analysis founded on realistic experiments in the manufacture and use of stone tools, new insight is offered in regards to the engineering design and use of Aleut technology. Samples used in this analysis were recovered from the Andreanof Islands, specifically, the area around Clam Lagoon on northeast Adak. Artifacts from sites on the lagoon's east side are the prime focus of this study. Studies specifically targeted an egg-shaped, small island in Clam Lagoon and the larger landmass incorporating Zeto Point, its freshwater ponds and embayment.

In addressing why stone technology is vital and appropriate, this chapter expands upon a global analytical model for the Aleutian Islands. Simply put, stone technology of the Aleuts, or any other group, is analytically robust in that stone tools are produced, used, repaired and recycled in a readily understood, stepwise fashion and modeled as a production chain (Bleed 2001). William Henry Holmes (1890, 1892, 1894, 1897 and as compiled in 1919) developed its key features well over a century ago. Johnson (1993, 151-152) offers an astute summary of Holmes' research and why it is vital: '[Holmes' research was] based on a detailed examination of thousands of artifacts combined with experimentation, refitting, a keen insight into the strategy of bifacial reduction, and a commitment to systematic generalization. These factors make Holmes' publications relevant in a way that many more recent studies are not.'

Aleut Technology Production Chain Modeling Concepts

Holmes' studies roughly correspond to when the Aleuts themselves were a prime source of technological information for ethnologists who cited him (Holmes 1919; see Jochelson 1925, 57, 69) and also recognized successive stages in artifact manufacture, use, and repair. The degree to which Holmes' seminal studies contributed to Jochelson's work is uncertain. But the parallels in logic might well indicate a direct connection. Often providing Aleut terminology first developed in 1846 by Veniaminov (Collins 1984, 13), Jochelson offers a detailed, well illustrated critique of Aleut stone and bone technology upon which I build.

Jochelson's excavations of settlement middens were often deep and, over the course of 47 working days, spanned the archipelago. Moreover, Jochelson's team also included Aleuts who provided him with the Aleut names and the function of the artifacts that he recovered. This would have to rank his as among the most amazing and largely productive excavation programs in the annals of Aleutian archaeology, even given significant improvement in artifact recovery since. Included are western (Attu Island on the west and speaking the Attuan subdialect, Atka Island at the east end of the Andreanof Islands and speaking the Atkan subdialect) and eastern dialect (Umnak Island, and Amaknax and Ukundax islands near Unalaska Island) groups (see Woodbury 1984). The dialectical differences, however, are fairly minor and according to Woodbury (1984, 49) 'overall differences are less considerable than those found among the Inuit-Inupiaq dialects.' Woodbury estimates the split between Aleut and Eskimo languages about 6000 years or so ago. And that Aleut 'must be viewed as a complex of dialects where innovations and retentions overlap in different ways' (Woodbury 1984, 51). Undoubtedly time differences exist among the items and terminology that Jochelson (1925) describes.

Although Jochelson is given credit for exemplary descriptions (Bank 1953), later archaeological research does not address nor take advantage of Jochelson's review of the uses of the tools and how the Aleuts described them. Rather, as McCartney (1984, 122) noted, 'Artifact functions and their use contexts cannot always be identified, but the major deficiency is the lack of even cursory typological studies prerequisite to establishing major time-space divisions for the [Aleutian] chain. Within the middens, slow changes in artifact styles occurred, but few horizon markers are known….' Jochelson and most others since (McCartney 1984; Woodbury 1984) have not found major chronological differences in Aleut language or technology although geographical ones for the archipelago seem more evident. Others have failed to take advantage of Jochelson's functional technological descriptions, implicit employment of a production chain approach, and the gift he gave us of Aleut terminology for the tools themselves. The latter allows us in many cases to make intelligible gender and finer-grained technological distinctions.

The general principles of the model proposed here guide the research and apply to virtually any domestic craft where the maker is most likely the user, although not necessarily the sole one (Braun 1983). A subtractive technology model is outlined where individual artifacts or parts of compound tools are produced by the removal of mass. The model, developed some time ago (Kay 1984, 1998), draws from earlier efforts (Bradley 1975; Collins 1975; Schiffer 1972) and incorporates ideas from Bleed (1986).

Ecological and cultural contexts are the two domains under which any technology operates (Figure 13.1). Systemic, subtractive technology processes are conceived as mutually exclusive for either domain. Initial resource procurement pertains solely to the ecological context while all subsequent technological manipulation depends on the cultural context. A technological analysis, in the case of material procurement, specifies the ecological context of a resource and how it was procured. Processes leading to a technological product are strictly within the cultural domain. Material sorting followed by initial and subsequent reduction lead to technology products, or artifacts. Tool use may follow any step or linear combination of steps. Optional modifications or maintenance procedures may occur during artifact production and during the artifact use life. Therefore, optional modifications, maintenance procedures, and recycling are viewed as secondary, or attendant, processes. These may coincide with any, or a linear combination of, primary procedures that result in a technological product, tool use, discard, or specialized disposal. Discard or specialized disposal are the final cultural processes that remove artifacts from a cultural system. The two distinguish significant variation in the removal process. Discard behaviors encompass planned or unplanned loss of technology products. Specialized disposal entails an exclusive, planned removal of technological products in which their ultimate context is a mortuary context or a cache. Discard or specialized disposal can occur after any of the preceding primary or secondary subtractive technology processes.

Bleed (1986) differentiates daily tool use and repair by maker-users (maintainable systems) from occasional tool use and repair by specialists (reliable systems). He offers other insights about a fundamental dichotomy in engineering design of technological systems or a continuum (maintainable and reliable systems) that might bring oppositions into a common sense harmony (see also Ahler and Geib 2000; Hayden *et al.* 1996). Bleed has explicit expectations about these different technological systems. The technological system illustrated in Figure 13.1 relates to task performance, or cycles of use (+), repair (-), and downtime (horizontal arrow). Task performance describes the anticipated and actual use history of tools and the conceptual differences in the way technological systems are designed to satisfy fundamental needs. Inferences about the production chain and technological system stem from the archaeological context, writ large. In most cases, archaeologists work back from the artifacts and any supporting documentation to explore and reconstruct the responsible ecological context and systemic cultural context.

Key organizing concepts of extractive (i.e., resource

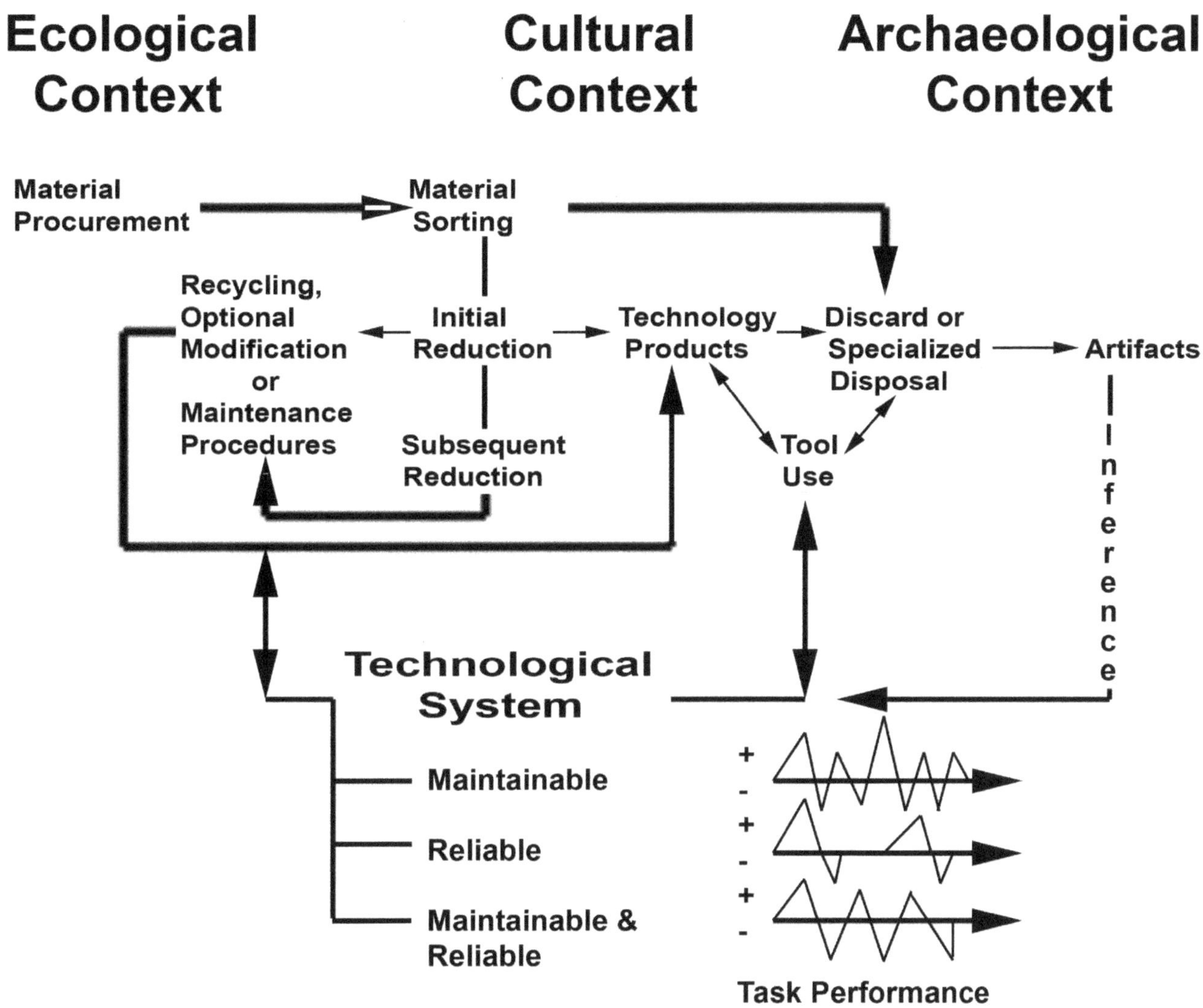

Figure 13.1. Subtractive technology model adapted from Kay (1984) and Bleed (1986).

procurement and exploitation) and maintenance (i.e., fabrication and repair) tools are derived from Binford and Binford (1966); sociotechnic (i.e., status markers) and ideotechnic (i.e., ritually symbolic and cosmological) are from Binford (1962). Production chain terminology basically divides tools (patterned and ad hoc or expedient) from non-tool elements; and, within the non-tool category, refers to preforms (items intended for further modification as patterned formal tools) of any one of several production stages and debitage (discarded manufacture or optional maintenance debris plus cores). From Jochelson's (1925) commentary, gender distinctions are added whenever practical that, ultimately, are important in discussing Clam Lagoon artifacts.

The basic approach to the use-wear analysis is to first assess taphonomic factors such as trampling damage or soil movement. These instances of pseudo-wear are distinct from but can be confused with actual wear traces; they also provide information about the abandonment of artifacts or their post-depositional history. Next, use-wear and organic microremains provide direct evidence of stone tool usage, materials worked (i.e., contact materials) and decisions made about tool maintenance, recycling or discarding. The use-wear approach addresses facets of a technological production chain not as readily or unambiguously available by other means; included are the delineation of functional types actually employed by the makers and users of the artifacts plus tool use history and potential utility (Figure 13.2).

Adak Island artifact wear traces are largely striated residues, or additive 'microplating' features, that develop progressively with tool use (Kay 1996, 1998). Microplating residues are impervious to ultrasonic cleaning with concentrated strong bases (KOH) and acid (HCL), occur on siliceous artifacts from varied depositional environments, and are found on tools in excess of at least 100,000 years. Experimentation demonstrates that microplating residues develop and harden coincident with tool use, are a biochemical byproduct of moisture and direct contact with a material worked by a stone tool or adhering to it, and in an elegant way express tool motion kinematics (or friction-related features of surface-to-surface contact). They exhibit

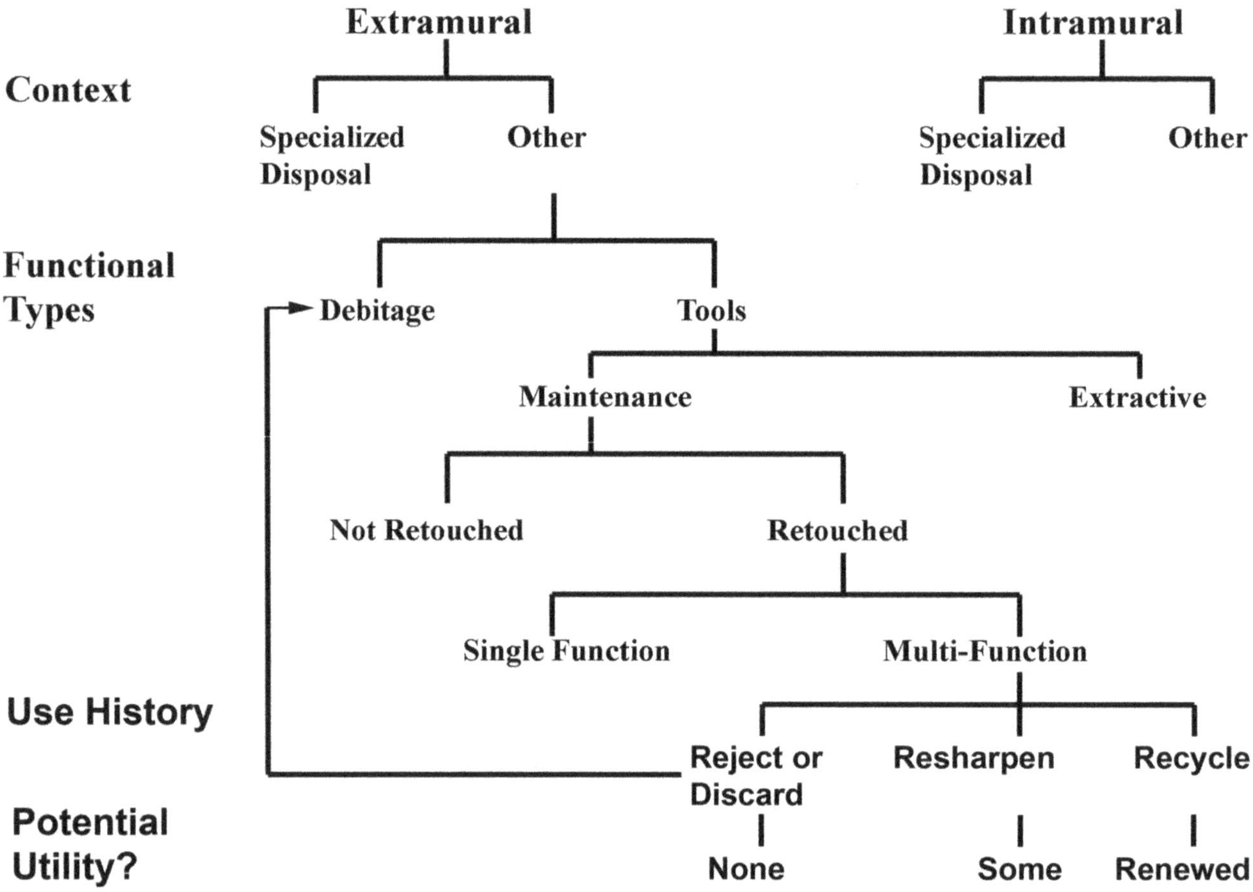

Figure 13.2. Use-wear supported evaluations of lithic technology. Note for reasons of space the hierarchical relations illustrated only for extramural contexts would apply equally for intramural contexts. See Bleed 2001 for a general discussion of production chains; Binford and Binford 1966 for differences in tool function.

flow characteristics of a viscous liquid, and desiccation cracks as they harden. Microplating in-fill striations becomes striated whenever abrasive particles strike, and crystallizes as brilliant white translucent filaments on the trailing (i.e., opposite from the) border of contact with a worked or manipulated material, and thus opposite the direction(s) of movement of a tool stroke. They are also instructive of hand-holding the tool or complementary movement of the tool in its handle. Microplating features are ubiquitous on the artifacts and overprint other tool use-related abrasion and abrasive wear traces.

Microplates are among the more valuable wear traces, because of their long-term stability, sensitivity to motion and affects on the microgeometry of a tool edge. Microplates bond to stone surfaces and edges as use continues. Experimental controls indicate hardening occurs coincident with use, or shortly thereafter. Unless deliberately removed, microplating so dulls an edge that it can no longer function. So among the advantages of this analysis is to identify when a tool no longer had a usable edge plus the steps taken just before discard of spent tools.

Jochelson's Aleut Tool Taxonomy

In referencing individualized and highly specific terminology, the Aleut folk taxonomy, informally presented by Jochelson (1925), exhibits a classic pattern of splitting and a predisposition to identifying rocks and minerals and their engineering qualities vital to understanding tool function (Table 13.1) The general term for stone is *nux̑*. 'Andesite is called *umgi'dax̑*; slate is called *i'gax̑*, the hard schists (hornstone and quartz schists) are called *a'n·kix*; green hornstone schist is called *nu'gim-cidgayū'*, i.e., the green stone; tuff of andesite, *uya'ñix̑*; vesicular andesitic lava, *amdi'xtix̑*; microgrannular andesitic lava, *cignu'nix̑*' (Jochelson 1925, 57). Aleuts perceived relative differences in hardness or abrasive quality of stone implements, especially of maintenance tools used in the fabrication of other objects.

With the above qualifiers in mind, Figure 13.3 diagrams Jochelson's (1925) Aleut stone tool and other finished artifact terminology. Extractive tools clearly dominate Jochelson's descriptions, are differentiated more than any other artifact class, and have gender distinctions. For both

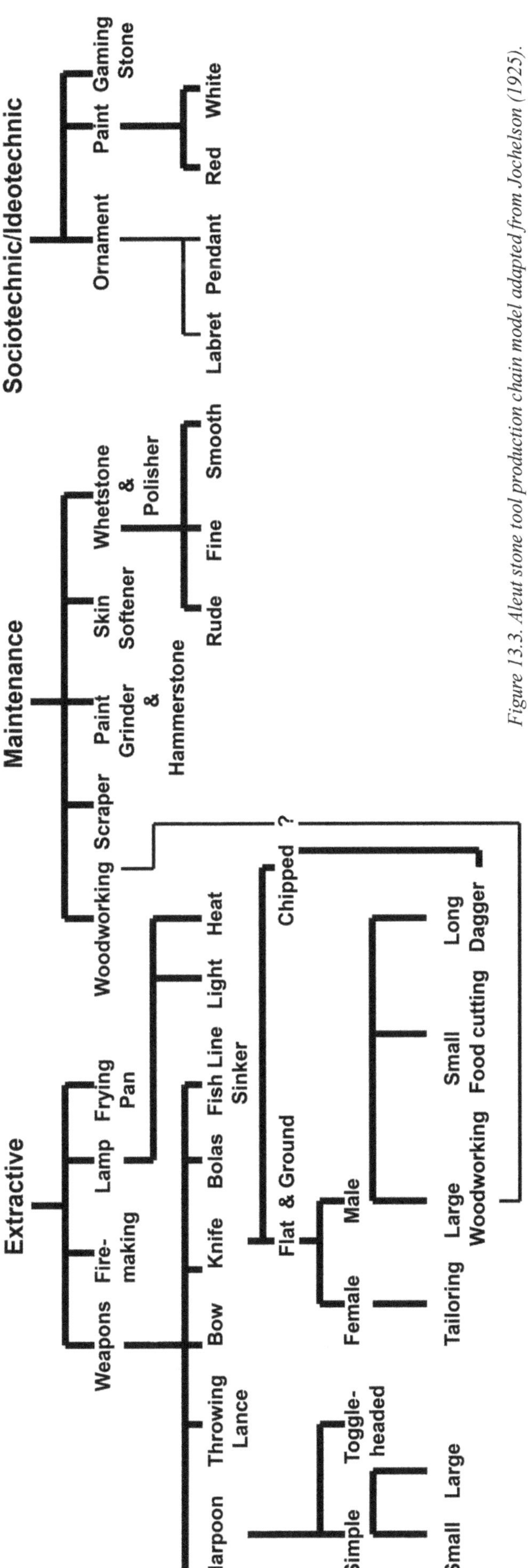

Figure 13.3. Aleut stone tool production chain model adapted from Jochelson (1925).

the toggle-headed harpoon (*akagu'six́*) and the throwing lance (*qulga'yax́*) there are single bifacial chipped stone point end blades identified by one term (*ka'dax́*), but the point shapes differ. The toggle-headed harpoon point fits within a split socket and is ovate shaped; the shouldered lance (or arrow) point with a contracting stem fits into a spoon-shaped socket. Simple harpoons are not similarly tipped and have bone points. Knives are either bifacially chipped or ground. Jochelson refers to the ground knives as flat and polished. He does not use the now conventional *ulu* terminology for the Aleuts (see McCartney 1984 as one example) and distinguishes gender- and use-related flat knives, with women's tailoring (*iga'dax́*) and men's woodworking (*utu'qnux́*) and weaponry--long daggers (*cunu'six́*). It seems likely the small food cutting knife, or *ma'gyix́* in the Attu dialect (Jochelson 1925, 64, Figure 3), is a man's knife too, as discussed later. This presents a minor problem for classification, although clearly not for the Aleuts, if only in conflating extractive and maintenance tools when describing women's and men's flat knives. But it has the greater advantage of showing how the Aleuts regarded their knives. As shown later, the Aleut approach to flat knives is easy to follow and largely coincides with technological descriptions made independently of Jochelson. The long dagger, a man's weapon, is ground and double bladed with shoulders. It superficially resembles bifacially chipped tools of similar form. Aleuts referred to both the ground and chipped daggers using a single term (*cunu'six́*). Fish line sinkers and bolas stones are small, round to oval forms with engraved lines, notches or grooves to attach an encircling line; they are referred to by a single term (*kayu' xtax́*). These differ from similarly sized, but not lined or grooved, stones of similar rock types used for games (*a'gix́*). Stone lamps are divided into types for either heat and warmth (*a'ñux́*) or types for light and illumination (*i'xtax́*). Generally smaller, the illumination lamps also provided warmth and were carried by individual hunters away from settlements. Frying pans (*cu'ñlux́*) (or griddle stones) are stone slabs that were used as is or with minor modification and set above a fire, often by placing small stones beneath (For further details on griddle stones see Jeannotte *et al.* Chapter 14 this volume).

Excluding weaponry, Jochelson (1925) differentiates roughly to the same degree other extractive and maintenance tools and sociotechnic/ideotechnic items. Driftwood was worked by several tools: split by bone wedges struck by stone hammers; planed by adzes or celts; and further shaped or cut by a man's large woodworking knife. Each had its own individual term.

Paint grinders and hammerstones were used to pound and grind mineral pigments (red ochre of either red felsite or hematite, and white volcanic tuff) and stone objects. The paint grinder (*ikica'six́*) was used against an anvil stone (*hinkulu'gix́*) and was also referred to as '*hinkulu'gim-cā*, i.e., the arm of *hinkulu'gix́*' (Jochelson 1925, 66, Figure 22; see also Jochelson 1925, 74-75, Figure 4 for a variation of this terminology and for descriptions of a paint anvil and pestle).

Jochelson (1925, 68) divides hammerstones into two types: 'an oblong rounded boulder found on the shore' (*tuga'six̂*) and 'used in accordance with the hardness of the material to be beaten' such as grooving a fishhook sinker or 'rude pecking out' a stone lamp; and pecking implements made of the hardest materials (*itmagu'six̂*). The latter were 'egg-shaped, round, or discoidal in form, and served for the final chipping of stone implements and for hollowing out stone lamps.'

Other expediently used pieces include skin softening stones (*cigu'nix̂*) and whetstones to polish stone implements of unshaped 'micro-vesicular andesitic lava' (Jochelson 1925, 64-65, Figure 22 and Figure 23); these two categories would be almost impossible to separate without Aleut guidance. Jochelson (1925, 72) continues this emphasis of Aleut engineering design characteristics in describing polishing implements for final preparation of flat knives, adzes, and some lamps plus bone implements. Those of andesitic lava were either 'of fine and rude vesicularity...The polishing implement of coarse vesicular andesitic lava, called by the Aleut *sña'gix̂* or *ska'gix̂*, was used in the first phase of the process. For the second step, after the coarse unevenness of the stone was removed by the *sña'gix̂*, an implement of fine vesicular andesitic lava, called by the Aleut *cuktu'six̌*, was used. The third stage consisted in grinding with a smooth whetstone, called *cumli'six̌*, chiefly made of volcanic tuff... With this implement the edges were also sharpened. In order to polish the surface the Aleut used the horsetail, *Equisetum nyemale*, which they called *q'i'lim alucaya'ñin*, i.e., the needles of the mythical monster, *Qi'lix̌*. Chiefly, however, bone objects received their final polish with this plant.'

Turning now to other sociotechnic or ideotechnic items, Jochelson (1925, 59) describes potential gender differences and ritual connotations of red ochre. And it seems no wonder he dwelt on this detail: '...pieces of red felsite... occurred, but nowhere were implements of this material found. We were told that felsite, when powdered and mixed with oil, furnished an excellent and lasting paint. More frequently than felsite were found pieces of hematite..., which evidently had some religious significance. It was called *a'six̂*, a term applied to all objects supposed to have miraculous power. The present-day Aleut call hematite, *ca'gax̂*. Pieces of hematite were worn by every hunter, for they were said to have the power of attracting sea mammals, and particularly whales.' Of other ornaments Jochelson (1925, 57) mentions labrets of marble 'obtained from the Alaskan Peninsula through barter.' He (1925, 67) also illustrates a possible stone disk pendant of volcanic andesitic tuff and stone bow drills.

Clam Lagoon Archaeology

During and since World War II, Clam Lagoon lost its more-or-less pristine character. Roads and rock quarries that outline the lagoon adversely impact prehistoric Aleut settlements, including several summarized here. The lagoon itself served during World War II as an American training location for pilots of amphibious aircraft and Quonset huts once dotted Zeto Point. The island landscape is now a honeycomb of subterranean burrows due to introduction of rats and foxes. This burrowing degrades habitat and adversely affects Clam Lagoon archaeology.

The following brief review of Clam Lagoon archaeological sites (ADK-171, ADK-011, ADK-012, ADK-187, and the Swetzoff Site) supplements the more detailed descriptions found elsewhere (West *et al.* Chapter 1 this volume). The excavations, overall, covered small areas, effectively no more than one or two 1-meter squares to substantially less than a dozen units of this size. Excavation was in arbitrary levels no more than 10cm thick within natural stratigraphic layers. Excavated matrix was processed through 1/4-inch dry screens and 1/8-inch mesh sizes for water screening. These procedures resulted in sufficiently large samples to characterize lithic debitage and the general approaches of stone tool manufacture covered elsewhere in this volume (see Wilmerding and Hatfield Chapter 12 this volume). Less confidence, however, can be placed in the range in variability of stone tool elements due to the recovery of comparatively small samples.

Site ADK-171 faces the lagoon from the east and is on the west flank of Zeto Point. This site includes three excavation areas: a *Clinocardium* midden located above a lens of Intermediate Ash, Pits 1 and 2, a 1m X 2m test pit just upslope from the midden site; and Test pit 55-1, a 1m X 1m excavation. ADK-171 dates to approximately 6000 BP (See Chapter 1 Appendix 1.B) and represents the oldest archaeological component in the Andreanof Islands; roughly, the time when the Eskimo and Aleut language split. All artifacts available for analysis from this site were microscopically analyzed for use-wear; included are four chipped stone tools from the *Clinocardium* midden, two from pit 55-1, and one from Test pits 1 and 2.

Site ADK-012, the Dozered site, is located on a barrier beach on the northeast side of Clam Lagoon. Its radiocarbon assays range from about 2300 to 1500 BP (See also Appendix 1.B West *et al.* Chapter 1 this volume). All 15 artifacts from this site were microscopically analyzed for use-wear.

Site ADK-011, on Zeto Point east of Clam Lagoon and facing Sitkin Sound, has an early component and a late house village settlement. It has the largest collection of lithic artifacts I considered, numbering just shy of 100 chipped and ground stone items. For purposes of this discussion only, the artifacts are divided into early (i.e., ca. 2390-2590 B.) and late (i.e., ca. 170-415 BP) components. Artifacts from these components were selectively analyzed for microscopic use-wear.

The Swetzoff site (ADK-283) is located on a small island

in Clam Lagoon and is informally named after Michael Swetzoff, an Aleut, who suggested CAAPP should survey the island in 2005 and provided a boat to get there. In many ways the Swetzoff site is an anomaly. It has no fresh water, is at an inconvenient distance from terrestrial resources, and offers scant protection from the elements. Island activity could be fairly well observed from the shore of the lagoon or its surrounding uplands and it could be easily reached by boat or by wading. It takes little effort to climb a few meters up a gentle incline from its beach to the nearly level top. The island would have offered little in the way of natural defensive features. It is the only island within the lagoon and quite literally offers a bird's-eye view.

A house depression test hole in 2005 produced a stone seal oil lamp from which charred residues were radiocarbon dated to approximately 3835±20 years ago. During the 2007 CAAPP excavations, a magnetic north-oriented trench of five 1-meter squares of which the two northernmost were deepest and most important, incorporated the original test hole and produced charcoal from levels below the lamp that radiocarbon dated to approximately 1700 years ago (see Chapter 1 Appendix 1.B). While the dating is subject to further review, the Swetzoff site appears to be a contemporary of the Dozered site settlement and fits in the Zeto Point sequence, but the Swetzoff Site dates approximately 4,000 years after the earliest documented Zeto Point habitation at ADK-171. The Swetzoff site has poor organic preservation, and animal burrows riddle the house depressions on its southwest side. This burrowing mixed the archaeological remains to an uncertain degree and blurred stratigraphy. Nonetheless, one house pit not only truncates another but also was nearly destroyed by mechanical excavations at the edge of the island (see West *et al.* Chapter 1 this volume). Despite these problems, the Swetzoff site is worthy of study. A surface collection of its encircling beach showed that stone artifacts ring the entire island.

An opportunistic surface collection of stone artifacts from blowouts at a fifth site, ADK-187, was done in 2007 and these artifacts are also considered in this chapter. This site is above the outlet to the lagoon on its west side. A single radiocarbon sample dates ADK-187 to 3590±50 years BP (Luttrell and Corbett 2000).

Technological Assessments

Raw Materials: Igneous and igneous-based metamorphic stone sources around Clam Lagoon and in the nearby vicinity fall into three major categories, and account for most of the lithic material found in Adak archaeological sites. Gravel-to boulder-sized, rounded cobbles typically mantle the beaches of the lagoon, Kuluk Bay and Sitkin Sound. These vary in rock type, color, shape and size; basalt dominates and occurs as bedrock as well as in cobble form. Relatively soft and granular, round-to-ovoid, basalt cobbles, ranging anywhere from grapefruit to basketball size, were primarily exploited as hammers, abraders, anvils or lamps. Smooth, hard cobbles of fine-grained andesite are smaller and are either rounded or oblong. When broken, these exhibit conchoidal fracture and are among the more common sources of chipped stone artifacts. Other stone suitable for both chipped and ground tool manufacture come from volcanic sills and dikes in the hills fronting the lagoon. Among these is a buff colored, banded, serpentine-like, tabular stone with good conchoidal fracture. The Aleuts at Clam Lagoon almost invariably used this 'banded serpentine' to exhaustion. Another, a coarse- to finer-grained andesite was used for ground knives. An outcrop of Mount Adagdak and fronting on Sitkin Sound northeast of the lagoon is the likely source of griddle stones, as discussed in Jeanotte *et al.* (Chapter 14 this volume) and are not technologically assessed here. The only well-established exotic stone material found at several of these sites is obsidian; its sourcing and implications for archipelago exchange systems are covered in Nicolaysen *et al.* (Chapter 11 this volume). Slate also occurs and apparently originated from elsewhere and not the immediate lagoon area.

Use-wear analysis: Kay (2006) microscopically analyzed artifacts from two Clam Lagoon archaeological sites, ADK-171 and ADK-012. This study did not similarly evaluate every item from Zeto Point, Swetzoff and ADK-187, in part because the coarse nature of basalt artifacts could not be evaluated consistently for use-wear. Use-wear observations illustrated in this chapter are photomicrographs properly oriented to their placement on an artifact edge or surface. The photomicrographs are identified by a notation system of a solid symbol (a square or circle) and, for multiple photomicrographs, an alphabetically ordered letter that references their placement on the artifact and the sequence of analytical observations. (So a photomicrograph labeled 'c' would refer to the third microscopic image on an artifact but need not appear on the same side as observations 'a' and 'b'.) For this study, macroscopic technological attributes, combined with Jochelson's (1925) illustrated descriptions, were used to extend the information from the microscopic use-wear analysis. In a deliberately selective way, chipped stone debitage and tool elements from the three archaeological sites were subjected to a microscopic spot check but not a thorough analysis. This often resulted in item reclassification as debitage, a tool or tool element (see Figure 13.3).

For macroscopic photography, some artifacts were 'smoked' with water-soluble ammonium chloride to form a uniform, thin, white coating. Under oblique light, this coating shows flake scar patterns and edge damage. Simple immersion in water dissolves the ammonium chloride.

Use-wear Observations

The overall scope of the wear traces, in comparison with experimental analogues, are presented using two basalt tools recovered from ADK-171 and ADK-012. Although the two artifacts differ in appearance and likely represent different time periods, they are both general utility tools that

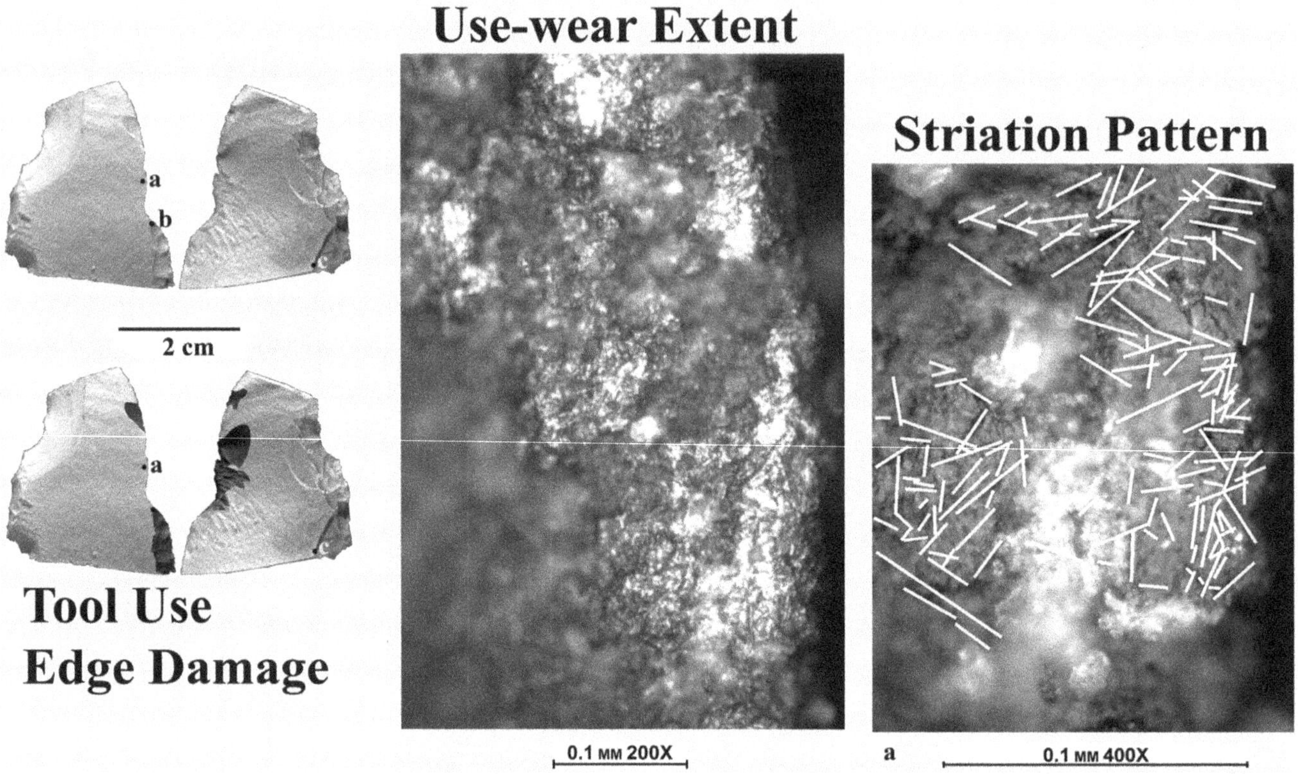

Figure 13.4. Macroscopic edge damage and microscopic wear traces (at a) of ADK-012 general utility tool. Note the edge rounding and crushing, and crosscutting striations along the edge (From Kay 2006).

Figure 13.5. Microplating domes (at b) on ADK-012 general utility tool compared with experimental replications. (From Kay 2006).

would have been used for cutting and scraping. They likely functioned in much the same way as many of the men's ground flat knives described by Jochelson (1925). They share general modes of manufacture, and haft preparation, but neither the working edge nor the haft element is readily apparent from a macroscopic inspection.

Figure 13.4 depicts the critical use-wear observations of the working edge for the ADK-012 tool fashioned from a tabular flake fragment. The use-wear is dominated by striated microplating, and crystallization filaments.

Damage, restricted to the *unretouched* side, consists of small flake scars and crushing of the edge. This, in fact, is the tool edge, as clearly shown microscopically by the nominally invasive extent of use-wear, the rounding and crushing of the edge, and the crosscutting striations along the edge. The latter show the tool stroke was both parallel and oblique to the edge, and involved repetitive motions in either general direction. These striations are typical evidence for both cutting strokes parallel to the edge and scraping strokes in motions oblique to nearly perpendicular to the edge. The rounding and crushing of the edge, along with the macroscopic edge damage, show the working edge contacted a relatively hard material and failed, and the tool was not repaired.

The tool edge wear traces included microplating domes, or spots of residue buildup containing desiccation cracks and crystallization filaments. In this case, the crystallization filaments are predominately on the interior facing side of the tool edge. Domes often, but not exclusively, form when tools are used against hard contact materials, including bone or ivory. A dome on the ADK-012 tool was compared to domes produced experimentally by rubbing a tool edge against antler and cutting fat and meat (Figure 13.6, respectively, experiments 1 and 2).

The opposite side of this tool was prepared by simple, or crude, bifacial retouch into a v-shape. Use wear along the interior margin of the flake scars and onto the adjacent surface (Figure 13.6c) is indicative of dynamic, invasive edge and surface contact. Both the high density and consistent orientation of striations are commonly observed in the slotting (Figure 13.6a) or grooving (Figure 13.6b) of bone or antler, or other similarly hard material. These wear traces also have features (bidirectional crystallization filaments and abrasive particles at opposite ends of striations) consistent with a back-and-forth movement within a groove or slot. The extent and invasiveness of these wear traces opposite the tool edge are diagnostic of haft wear. Most likely the handle was made from sea mammal bone that had been grooved into a v-shape to facilitate hafting of the tool.

Similar problems of functional tool zone definition are presented by the ADK-171 artifact (Figure 13.7) from test pits 1 and 2. The sides of this large flake asymmetrically converge to a rounded point. The larger, and more curved, of the two edges has bifacial marginal retouch; the straighter but shorter side has unifacial retouch on the ventral surface that forms an occasional large serration along its length. Large, negative flake scars on the dorsal and ventral surfaces are adjacent to, and originate from, the striking platform end; their removal thinned and flattened the proximal end and adjacent surfaces.

These differences in edge and surface preparation correspond with functional prerequisites of hafting and tool edge performance. The hafted part of the tool is defined by wear traces (Figure 13.7) of general, but highly oriented, abrasion of surfaces adjacent to the striking platform and extending invasively above the bifacial retouched, curved edge. Along the dorsal flake arris (e.g., flake scar ridge line) these haft wear traces are roughly perpendicular to the bifacially retouched edge. They also are densely striated with a consistent parallel orientation, much like the haft wear traces on the ADK-012 tool. Considering the extent and overall placement of the haft wear traces, it is likely the hafting groove of the handle encased most of the bifacially retouched edge, the proximal end of the flake, and extended beyond the original dorsal flake arris that goes to the rounded tip. This left the serrated and nominally straight edge exposed.

Wear traces are found along this edge and just inward from it, especially along the broadly toothed, or serrated, surfaces on the dorsal face. This is the flatter surface and would have been the leading side in tool use. This edge was repeatedly sharpened and re-serrated; remaining serrations represent damaged remnants beyond the point of further edge repair. The use-wear striations are dense, fine, and mostly u-shaped (Figure 13.8). They are parallel, transverse, and oblique to the blade edge. Spot crystallization occurs and has a 'cloud-like' appearance along the blade edge. There are also abrasive particles, but these are better represented in the haft element (Figure 13.7c). The tool edge is damaged but not rounded and has final edge angles measuring between 80 and 85 degrees. This would have resulted in a stronger edge than the ADK-012 tool, whose tool edge angle measures about 35 degrees. Considering the non-invasive edge wear and damage, the tool edge failure was likely due to contact with relatively hard contact material. Functional zones, developed from the use-wear assessments, are reconstructed in Figure 13.9 for these two tools.

Individual Sites and Components

The following discussion reviews technological evidence chronologically arranged from earliest to youngest and splits Zeto Point into two separate components based on their age estimates.

ADK-171

The seven artifacts from ADK-171 are of basalt (Figure 13.10a,b,g) and banded serpentine (Figure 13.10c-f). The banded serpentine artifacts are from the *Clinocardium*

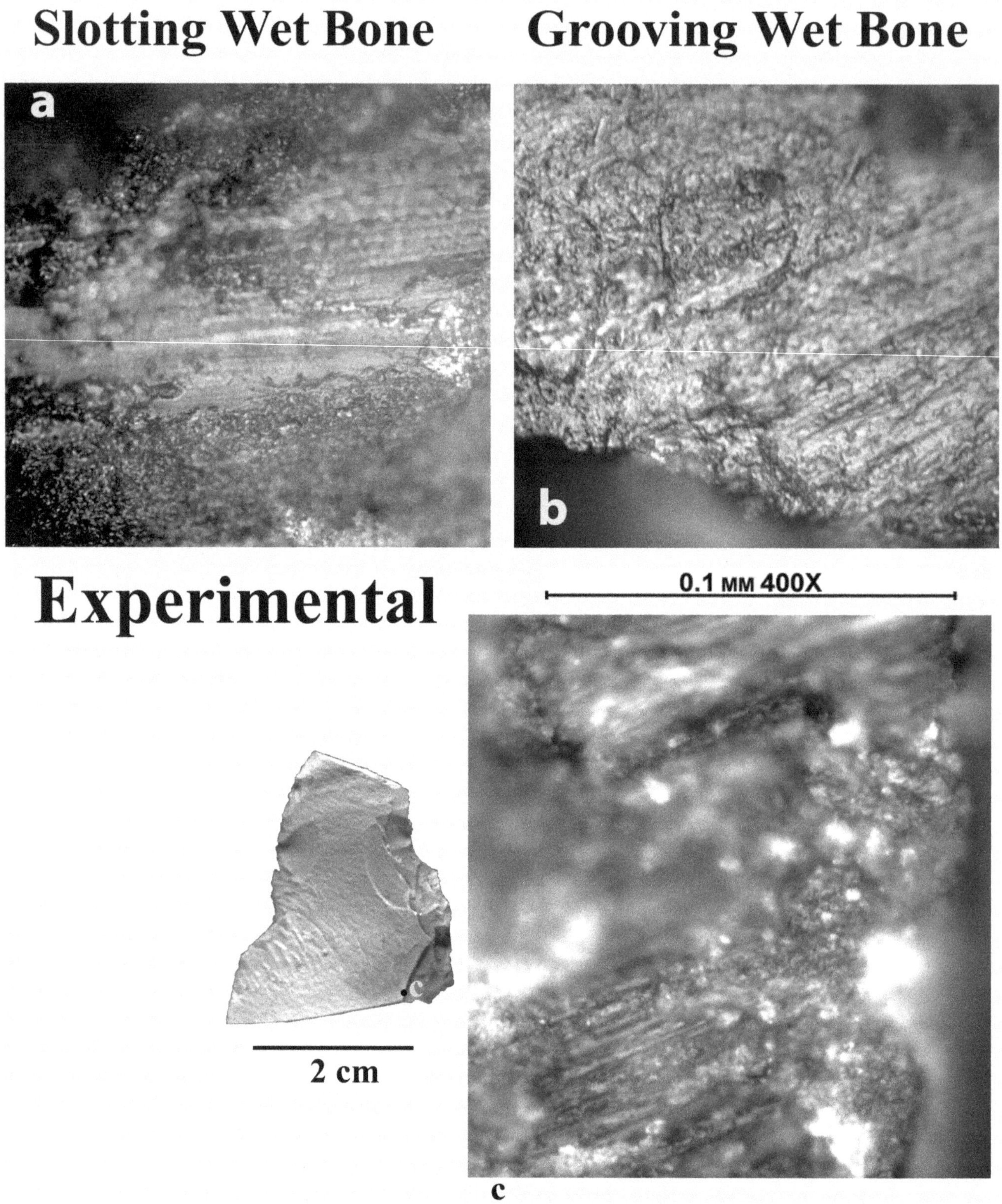

Figure 13.6. Haft wear (at c) on ADK-012 general utility tool compared with experimental replications. (From Kay 2006).

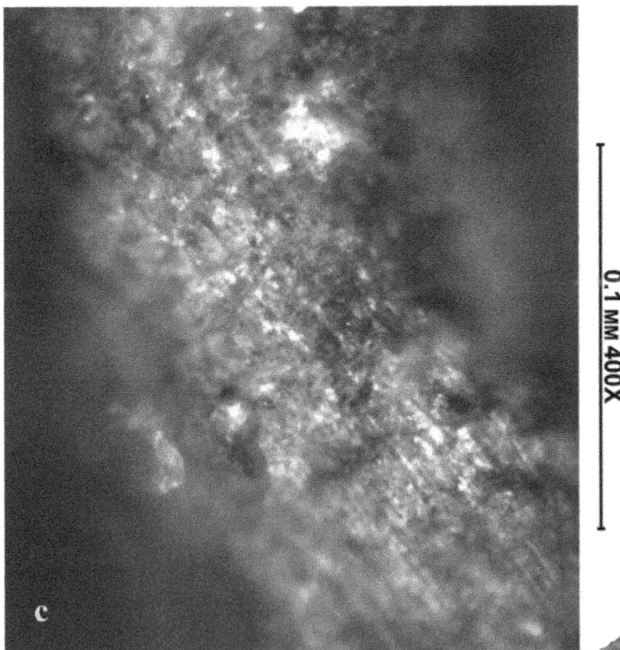

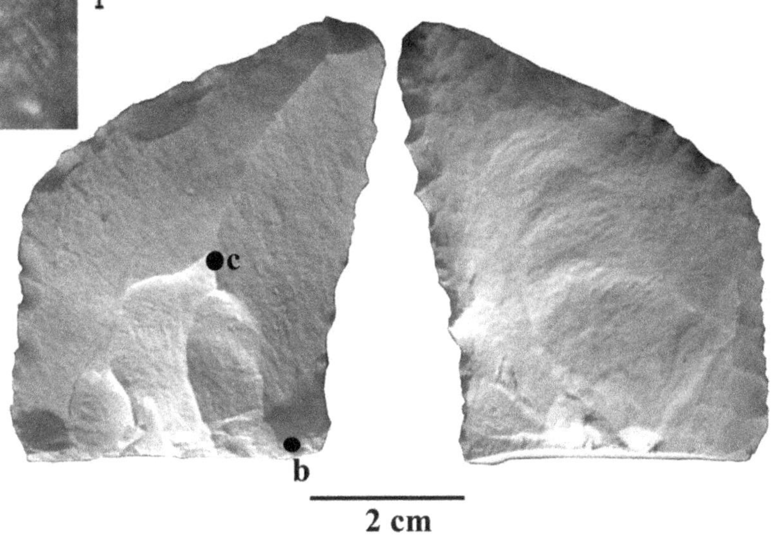

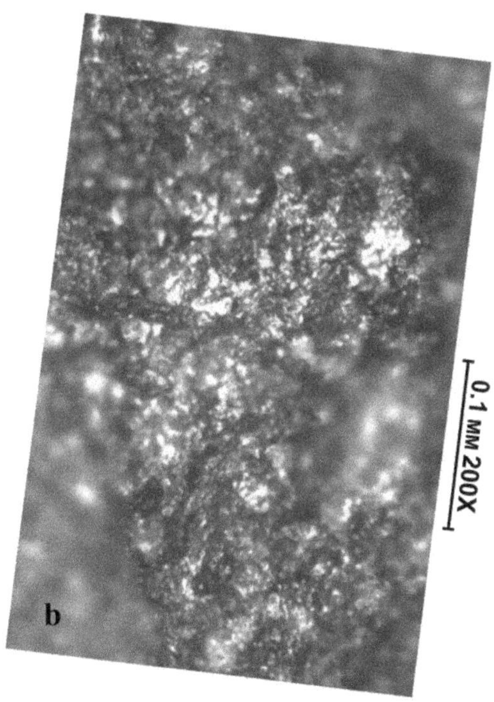

Figure 13.7. Haft wear traces for ADK-171 general utility tool from the Clinocardium midden. (From Kay 2006).

midden. The basalt artifacts include two fragmentary ones (Figure 13.10b,g) from test pit 55-1 and the large, direct percussion flake with a bifacially trimmed lateral edge (Figure 13.10a) from test pits 1 and 2. The latter's use-wear assessment is described above. The basalt is fine grained with macroscopic white phenocrysts; the banded serpentine is about equally fine grained, buff to olive color and opaque. One banded serpentine specimen, a teardrop shaped bifacial point (Figure 13.10c), is clearly fashioned from a flake blank; the others (Figure 13.10d-f) are all small fragments—among them, a biface segment (Figure 13.10f) and bipolar percussion debris (or *piecès escaillées*, Figure 13.10d,e).

Six ADK-171 artifacts exhibit wear traces indicative of tool use. Functional categories include two basalt general utility tools (Figure 13.10a,b), one each from test pit 55-1 and test pits 1 and 2; from the *Clinocardium* midden, a banded serpentine projectile point (Figure 13.10c) and three banded serpentine wedges. One wedge was recycled from a broken biface (Figure 13.10f) and two (Figure 13.10d,e) were produced by bipolar percussion. Figure 13.11 depicts use-wear zones for the wedges. Two wedges (Figure 13.11a,c) each had two tool edges, or bit edges, opposite hammer impact scars and sequentially rotated 90 degrees from each other. The general utility tool from test pits 1 and 2 (Figure 13.10a) exhibits wear traces indicative of hafting,

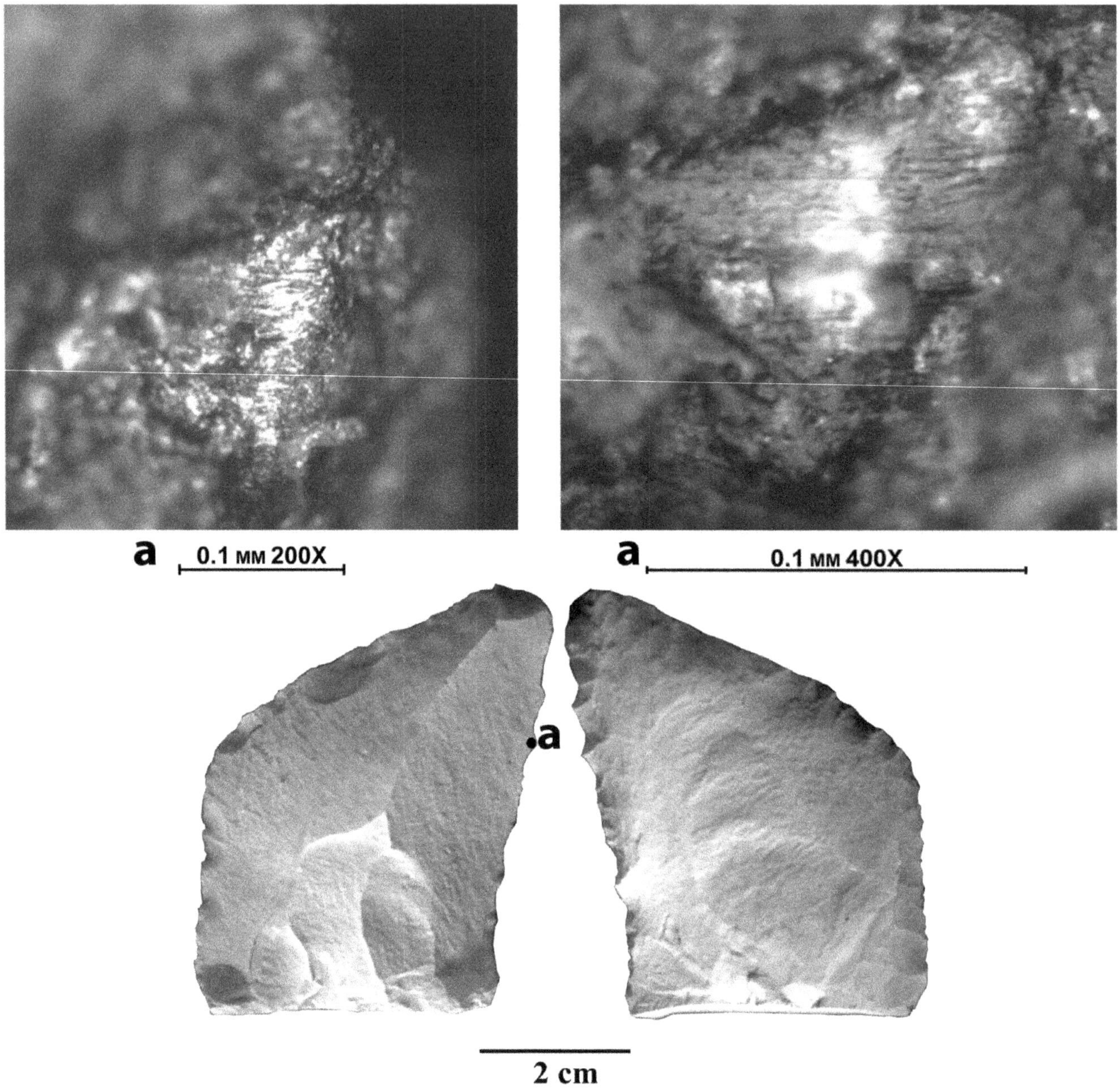

Figure 13.8. Tool edge wear traces for ADK-171 general utility tool from the Clinocardium midden. (From Kay 2006). The white, cloud-like blotches overlying the broad, u-shaped striations in the center of the use-wear and best seen at 400x are spot crystallization features.

described previously (see Figure 13.7). The overall form of the projectile point is reminiscent of toggle-headed harpoon end blades (Jochelson 1925:55, Figure 8). All tools appear to have reached a point beyond which they could be used.

The two basalt artifacts from above the Intermediate Ash (Figure 13.10b,g) have roughly the same petrographic characteristics as the test units 1 and 2 artifact (Figure 13.10a). One, a general utility tool, is a distal segment with bifacial, marginal retouch on the flake blank (Figure 13.10b). This tool is too fragmentary to be evaluated for possible hafting and differs in optional maintenance from the complete general utility tool from test units 1 and 2. It would be classed as debitage, but it had a longer use history. This tool is distinctive because the right edge was resharpened in an alternating unifacial fashion (Figure 13.12) in contrast to the bifacial sharpening of the general utility tool from test units 1 and 2. The other artifact (Figure 13.10g) is a lenticular biface midsection that lacks wear traces. Its overall flaking, size and thickness are indicative of a bifacial point. It may have broken in manufacture, and was apparently discarded; as such, it is classed, too, as debitage.

ADK-011 Early Component

Twenty-four items are from the basal unit, or Layer 4 (2390-2590 B.P.). The sample is a small subset of Layer 4

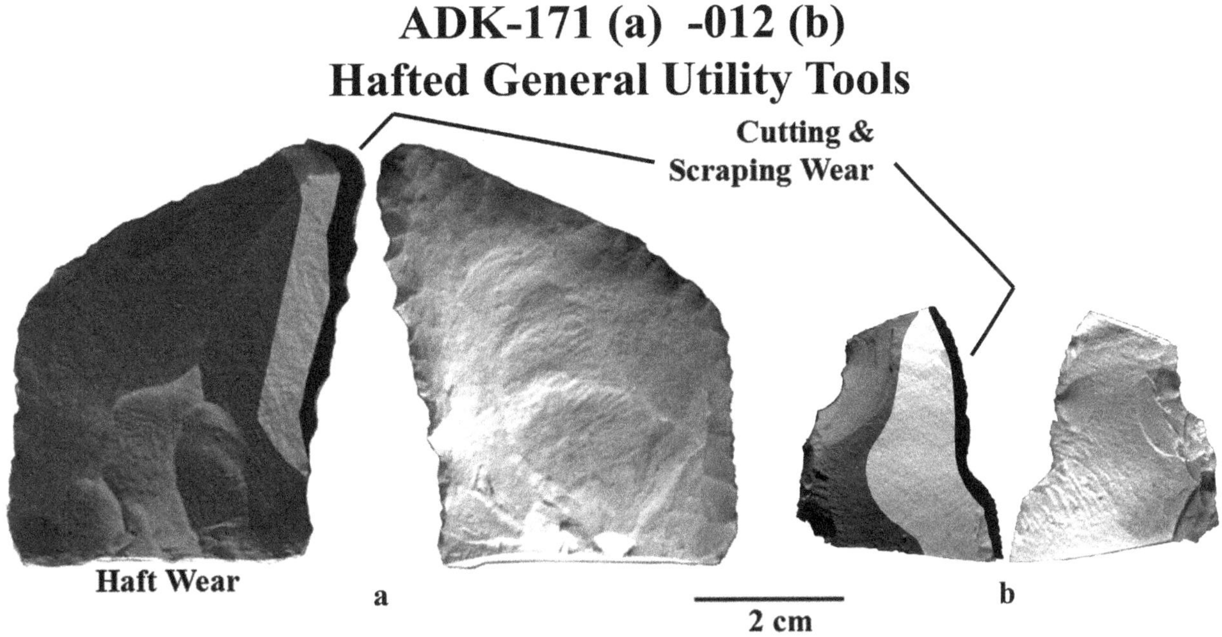

Figure 13.9. Reconstruction of functional zones from use-wear evidence. (From Kay 2006).

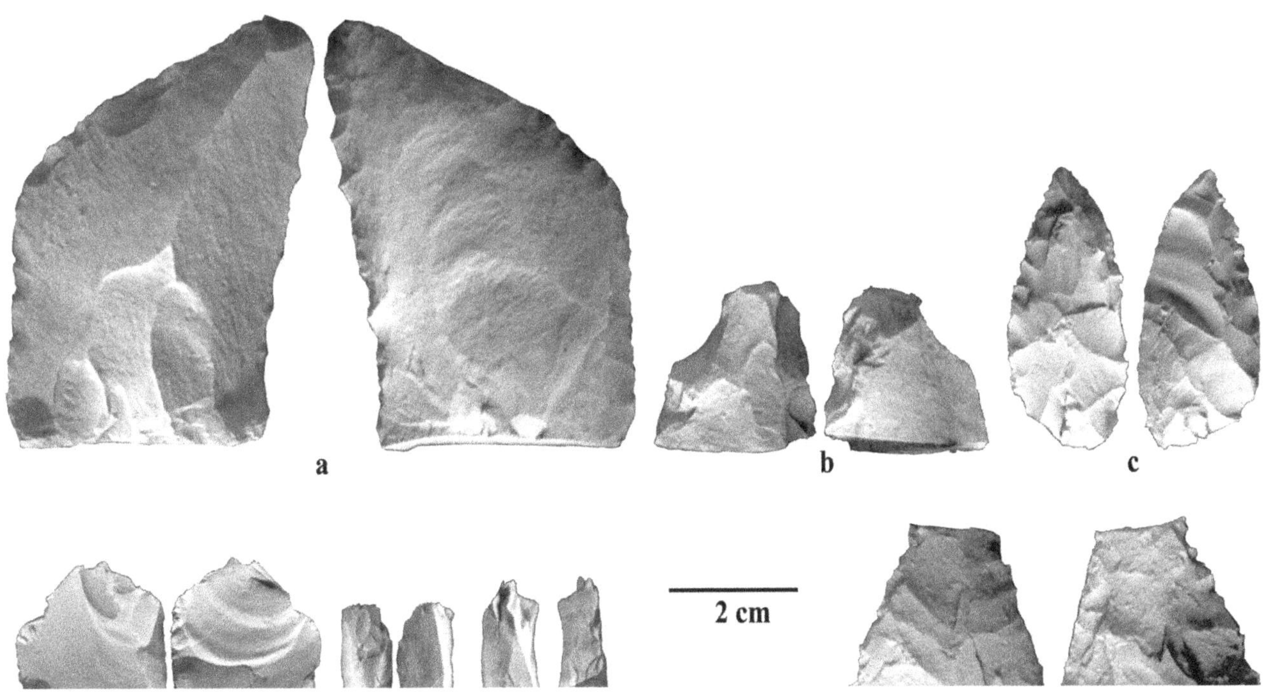

Figure 13.10. Adak Island artifacts from ADK-171. (a: from test pits 1 and 2; b, g: from lithic scatter above Intermediate Ash in test pit 55-1; c-f: from the Clinocardium midden). Note all items have been coated (or "smoked") with ammonium chloride for photography. (From Kay 2006).

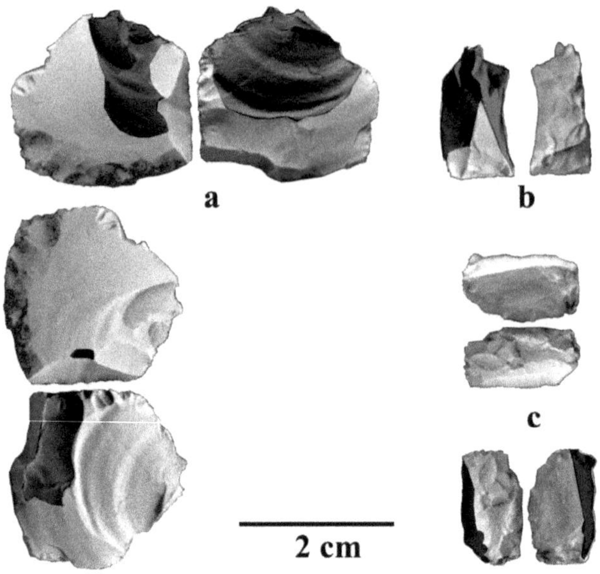

Figure 13.11. Wedges with highlighted hammer-impact negative flake scars from ADK-171 Clinocardium Midden. The wedge edge is at the bottom of each illustration. Note the lower illustrations of wedges a and c are shown as having been rotated to create a new wedge edge opposite a final hammer platform; b is recycled from a broken bifacial edge segment.

lithic debris but includes virtually all potentially patterned chipped stone artifacts and 12 likely tool elements previously identified (Wilmerding and Hatfield 2008; Wilmerding and Hatfield Chapter 12 this volume). No artifacts were analyzed from the overlying Layer 3, which lacks organics for radiocarbon dating but whose age is bracketed by radiocarbon assays for the underlying Layer 4 and overlying Layers 2 and 1. The other Zeto Point artifact samples are from Layers 3 east (a subset of Layer 2), 2 and 1 that date after about 400 years ago and represent a small part of excavated lithic debris.

Seven Layer 4 samples are vitrified ash with distinct layering and air or gas holes. Their masses vary from approximately 1g to almost 37g. Relative to their masses, the individual pieces have low bulk densities and probably would float. For the most part the vitrified ash surfaces are rounded lumps and appear to have been wind tumbled. But two small fragments have cut, flat surfaces. These two artifacts are likely whetstones or skin smoothers as described by Jochelson (1925). Two other items are thermal fractures of basalt and, at best, are simply debitage or rejects.

Piecès esquillées, or bipolar fragments, include three artifacts, all likely wedges. These have been rotated 90

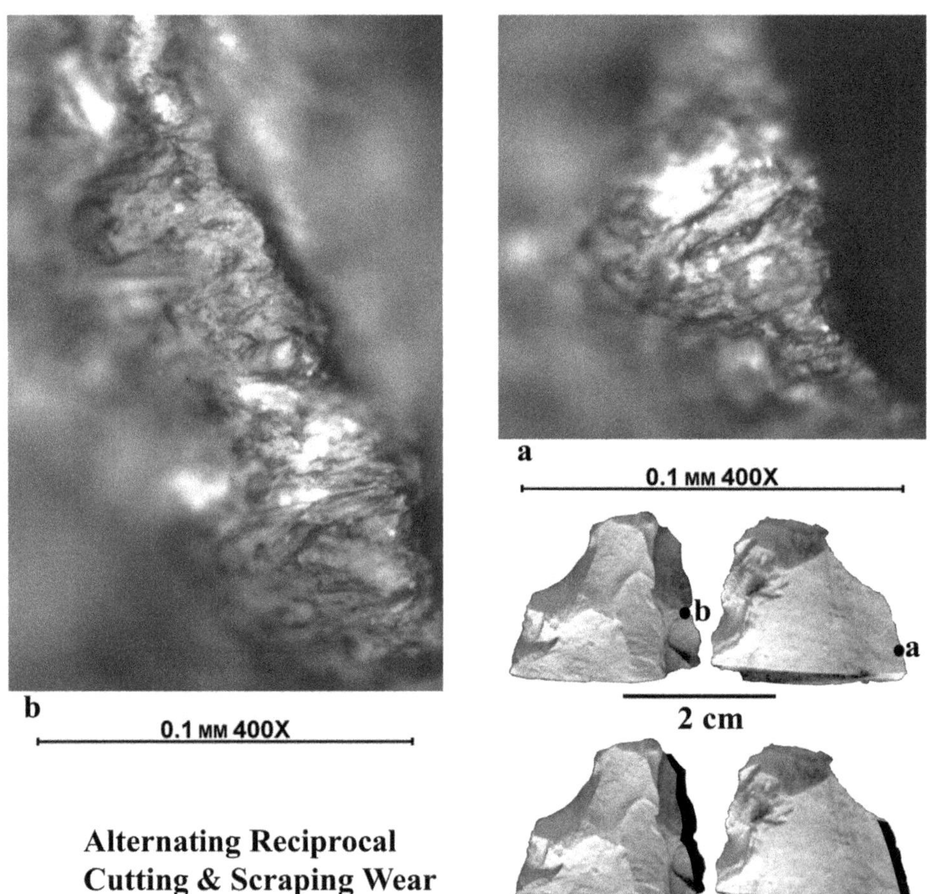

Figure 13.12. General utility tool wear traces and functional tool edge zones for basalt distal fragment of a flake with marginal bifacial retouch, ADK-171 test pit 55-1 above Intermediate Ash. See also Figure 9 (From Kay 2006).

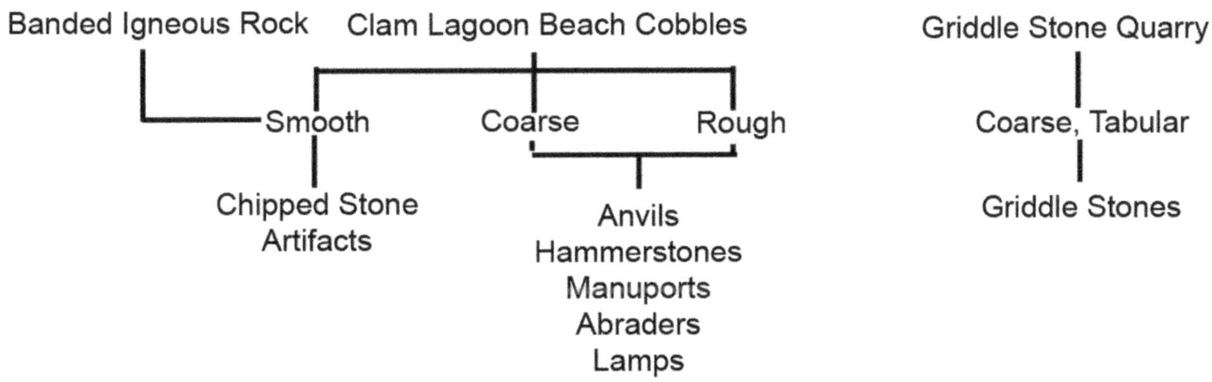

Figure 13.13. Stone resource and artifact production for the Swetzoff site, ADK-283.

degrees from the initial wedge edge and re-struck, creating the final break that led to rejection. In form and use, they are almost identical to the wedges from ADK-171 (see Figure 13.11) but generally are larger. One was recycled from a ground fragment of possibly a flat knife, one from a biface segment, and one achieved its final form strictly through the process of bipolar breakage and may not have been recycled. A tiny basalt flake that technically would be a microblade has a pointed tip that shows microscopic evidence of graver use. The tip is somewhat rounded in micro cross-section, which suggests its use against a fairly yielding and soft contact material such as skin or hide. Jochelson (1925, 60-61) describes bifacial items as lancets for bleeding that, I think, are similar in function but not in form or preparation.

A radially broken flake from a biface showed microscopic use-wear diagnostic of having been used as a gouging burin at the area of the break. This distal edge has a greater than 90° angle and is fairly straight. There is a final cleaning stroke along the edge that would have removed detritus as the burin was used; this typically occurs just prior to tool rejection and discard. This artifact and the three wedges, represent hard material fabricating tools, or what likely were used to split or gouge wood or bone. Other examples of burins on radially broken flakes from the Swetzoff site are described below.

A flake with a straight, backed denticulate edge and a flake scraper with a rounded edge were indefinite for use. No microscopic evidence of use could be judged for the backed flake because it is difficult to see wear traces on coarse-grained basalt material, which is not unusual for Adak Island artifacts. But the artifact had edge rounding that might indicate use against relative soft and yielding contact materials. The scraper of a harder andesite exhibited no evidence of use.

The eight remaining artifacts are bifacial elements, all of local basalt or hard andesite other than one segment likely of obsidian. Lacking use-wear are the likely obsidian item, an arrowhead or lancet base, a thick plano-convex biface tip without secondary flaking, and a potential lance or toggle-headed harpoon point tip. All but the likely obsidian item were fashioned from flake blanks. They all appear to be manufacture failures. Two arrowhead tips show microscopic and macroscopic impact fractures, and possibly were returned in an animal carcass. The same might well be true for a resharpened whole point with impact use-wear and whose form is that of a toggle-headed harpoon. The final item is a broken drill recycled as a wedge.

ADK-283

Figure 13.13 is a guide to the stone tool production chain for the Swetzoff site. Although exploitation of basalt from its beach undoubtedly occurred, the collection shows efficient use of stone from sources at some distance from the island. Procurement would have been most practical during warmer months, when stone cobbles could be dislodged more easily from unfrozen sediments. The predominant source is likely to have been the nearby Clam Lagoon beach where rounded cobbles are readily available to make ground and chipped stone tools. These cobbles occur in a variety of sizes, shapes, hardness and surface textures or roughness. The ones transported to the Swetzoff site tend to be among the larger but not the largest beach cobbles, depending on surface textures or roughness ranging from coarse to smooth (Figures 13.14,13.15, 13.16).

Moderate to very coarse textured cobbles are among the larger ones transported to the island; four excavated cobbles are clearly modified in apparently an ad hoc fashion. These expedient tools include a hand-held hammerstone or coarse whetstone (Figure 13.17) and three stationary tools—an abrader and two anvils (Figures 13.18, 13.19). One anvil is rudely cupped out on both faces (Figure 13.19) as though it were in the initial stage of production. A red residue is concentrated on both faces, however, and clearly the tool was used in pigment production. Interestingly, a similar colored red residue occurs on both sides of a broken, basalt biface, which microscopic evaluation shows to have been mixed in a gel of fat or other liquid that congealed and cracked on drying (Figure 13.20). These two artifacts are

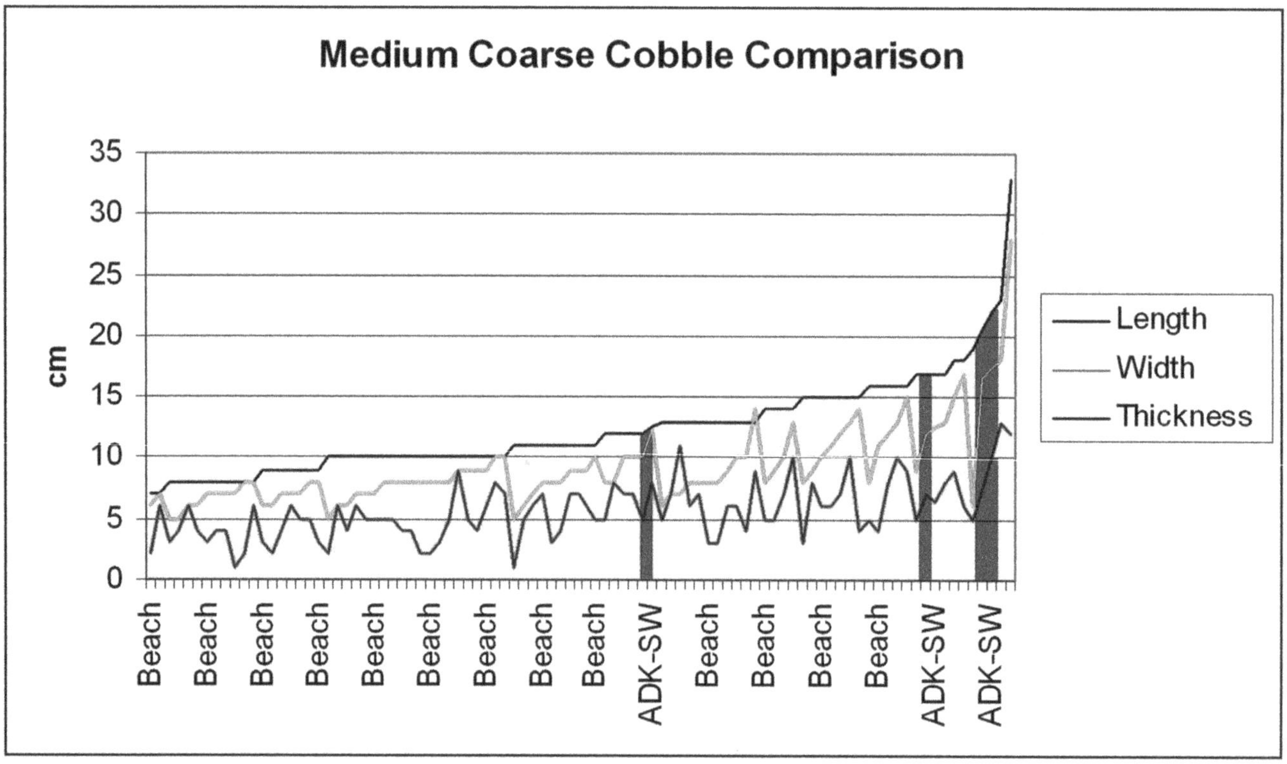

Figure 13.14. Size comparisons for medium coarse, rounded cobbles from the Swetzoff site, ADK-283, and the beach of Clam Lagoon. Beach data for May 2007 came from three randomly chosen 1-meter square areas for this and the next figure.

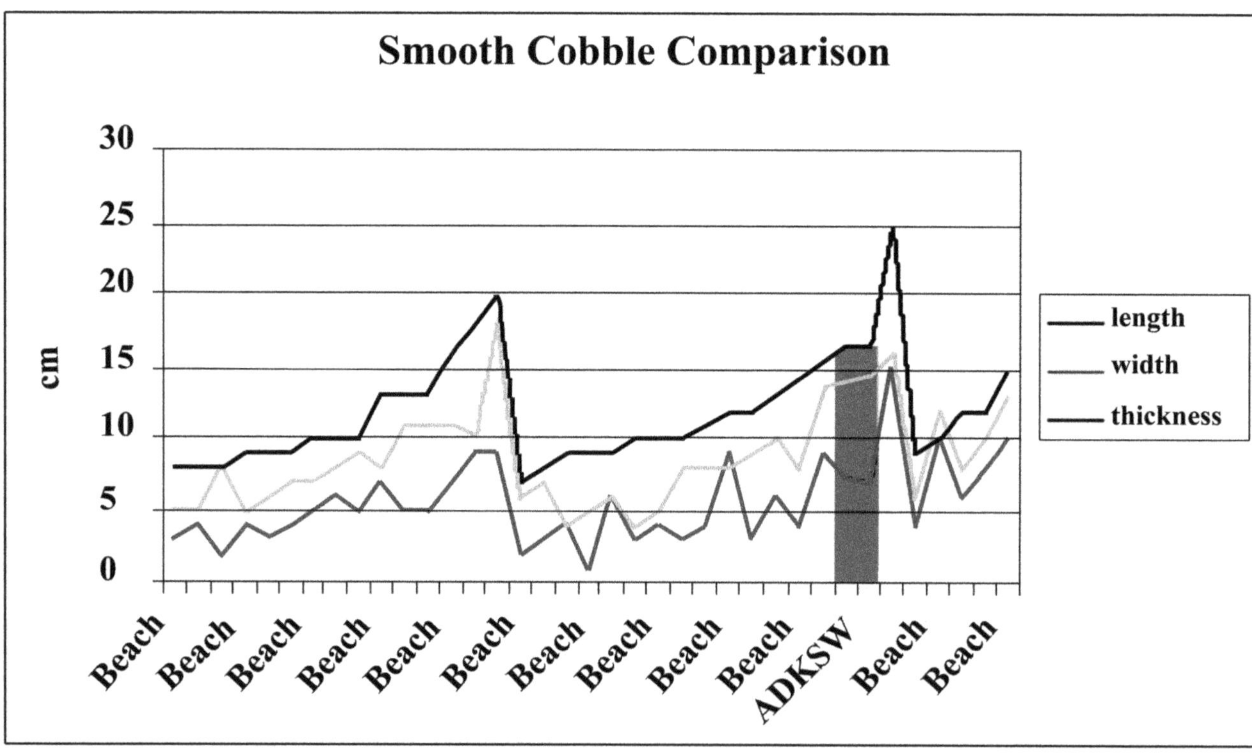

Figure 13.15. Size comparison for smooth, rounded cobbles from the Swetzoff site, ADK-283, and the beach of Clam Lagoon.

Figure 13.16. Examples of cobbles from the Clam Lagoon beach.

Figure 13.17. Hammerstone or coarse whetstone from the 2007 excavation of the Swetzoff site, ADK-283.

Figure 13.18. An anvil stone (top) and abrader (bottom) from the 2007 excavation of the Swetzoff site, ADK-283.

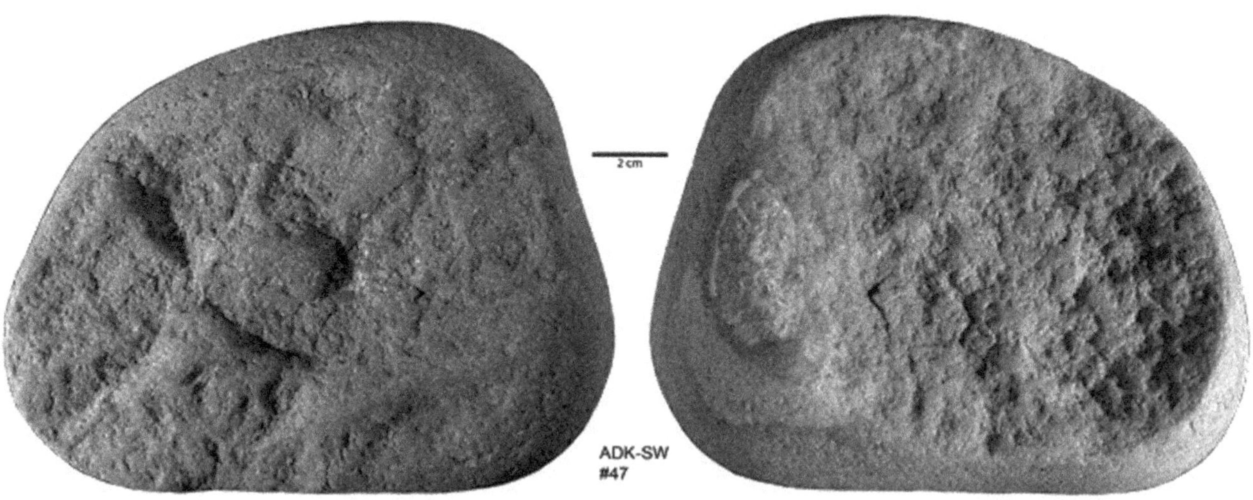

Figure 13.19. An anvil stone with red pigment residue on both faces from the 2007 excavation of the Swetzoff site, ADK-283, compare also with the anvil stone in Figure 13.17.

our best evidence of red paint production. Placement on the surfaces of a stone tool remains a mystery.

Also excavated were two small, rough cobbles in the size range of bola stones or fish line sinkers. The two lack grooves to attach a line; possibly they served as gaming stones or as unused hammerstones.

The smaller, smooth cobbles of fine-grained andesite together with tabular pieces of banded serpentine also came from extra-local sources and were used for chipped stone tool production (Figure 13.21). After transport to the island, these were split by bipolar percussion, an easy way to break rounded cobbles. This resulted in an occasional large bipolar flake with the smooth, rounded, outer surface still intact and opposing compression fractures on the interior surface (Figure 13.22a), and a tabular chunk or bipolar core.

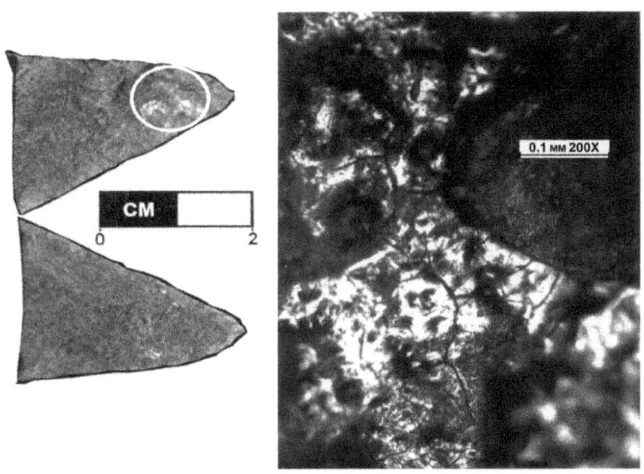

Figure 13.20. Biface tip of basalt with desiccated red pigment residue on both faces from the 2007 investigation of the Swetzoff site, ADK-283; center of area circled is the photomicrograph location.

Following this step, direct and indirect percussion were used to detach other flakes from split smooth cobbles or, when reducing already tabular basalt and banded serpentine pieces, to initiate bifaces and flakes (Figure 13.22b,d). The one complete biface (Figure 13.22c) is of basalt, not the best of materials, and came from the island's beach collection. Seemingly it was a practice piece, or what a novice knapper would make and likely throw away.

Other bifacial tool elements came from the Swetzoff site excavation. Included are the previously described biface tip smeared with a red pigment and three other points. These represent either toggle-headed harpoon or lance heads. The three points are of basalt; two possess rounded bases and one has an impact scarred tip. They are too short (bases vary from >12mm – 18mm, the tip is 24.8mm) for effective repair, and none appears to have been used after breakage.

Flakes were unifacially retouched and used as cutting tools against relatively soft (Figures 13.23-13.24) and harder (Figures 13.25-13.26) contact materials. These contact materials would include driftwood and animal byproducts—everything from skins and meat to bone or ivory.

A further embellishment was the selective, deliberate breaking of large flakes of basalt and andesite by radial fracture (see Root *et al.* 1999), which created tough, obtuse angled tool edges. Although these radially fractured flake tools represent expedient technology, their manufacture indicates a well executed, but simple, pattern of burin manufacture. These were attached to handles and used to slot or groove hard contact material (Figures 13.27-13.28). This pattern was, in fact, complemented by ad hoc selection of snapped flakes having similar tough obtuse angled edges (Figures 13.29-13.30). Of the latter, the flakes often stem from a bifacial edge (Figure 13.22d). These bifacial thinning flakes are prime evidence of a more expansive

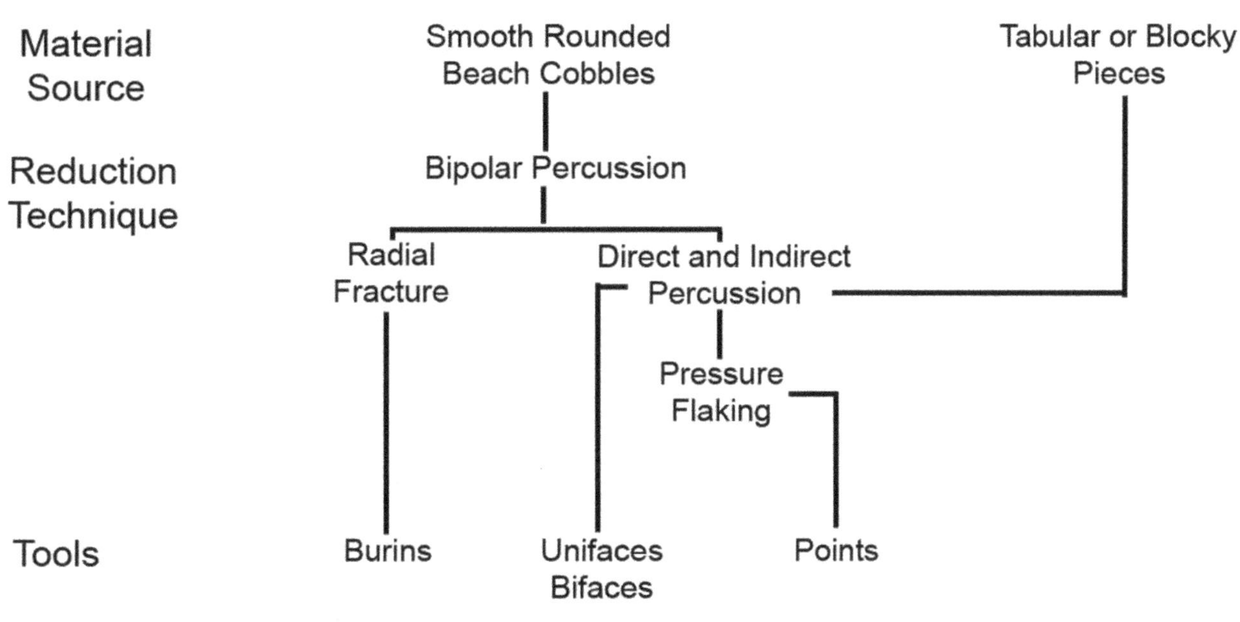

Figure 13.21. Flow chart model of chipped stone production chain for the Swetzoff site, ADK-283.

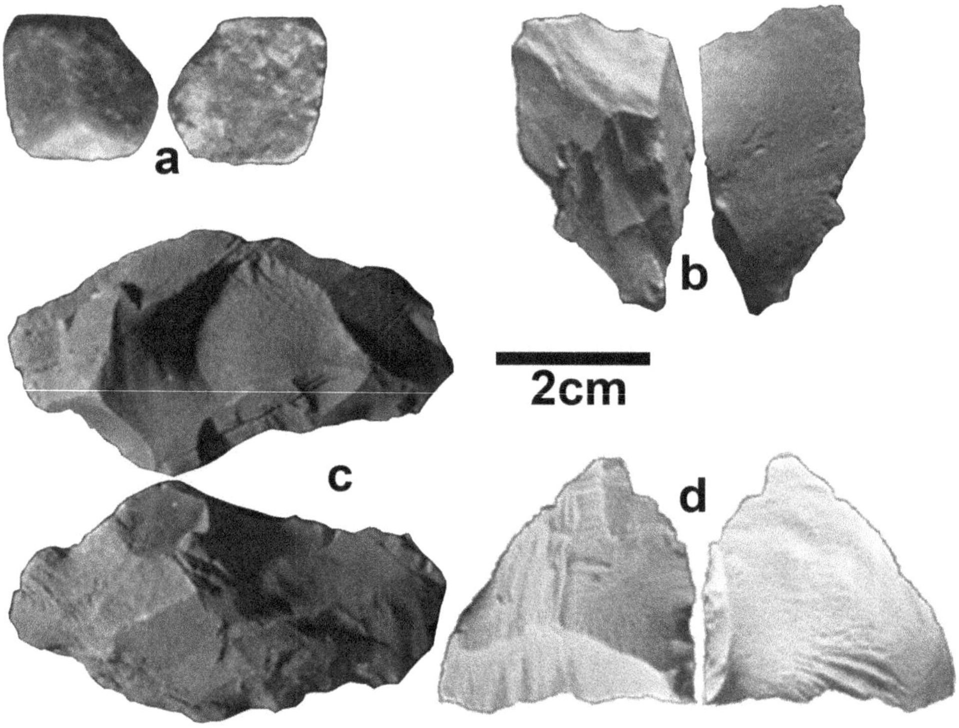

Figure 13.22. Examples of bipolar and other percussion artifacts from the 2007 investigation of the Swetzoff site, ADK-283.

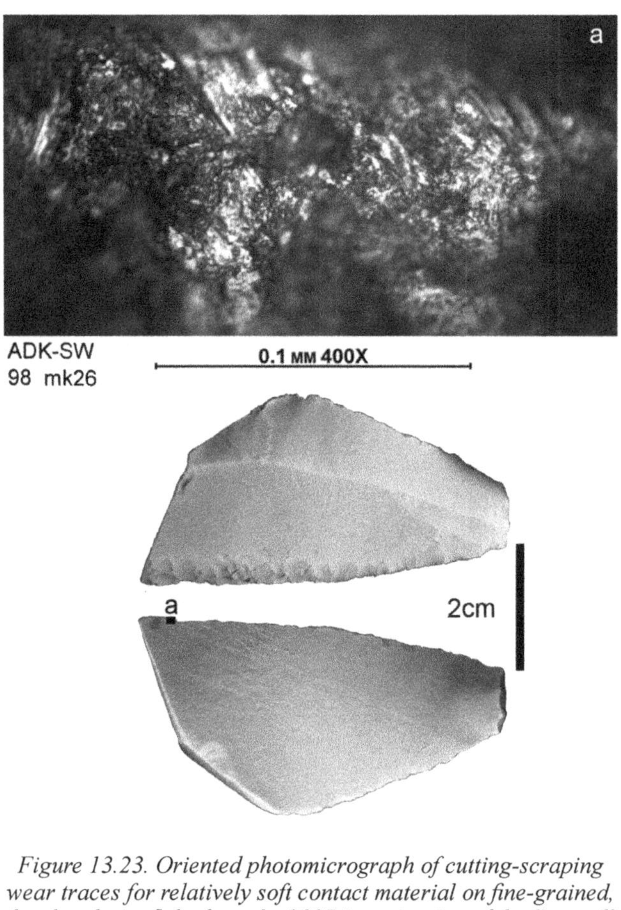

Figure 13.23. Oriented photomicrograph of cutting-scraping wear traces for relatively soft contact material on fine-grained, hard andesite flake from the 2007 investigation of the Swetzoff site, ADK-283.

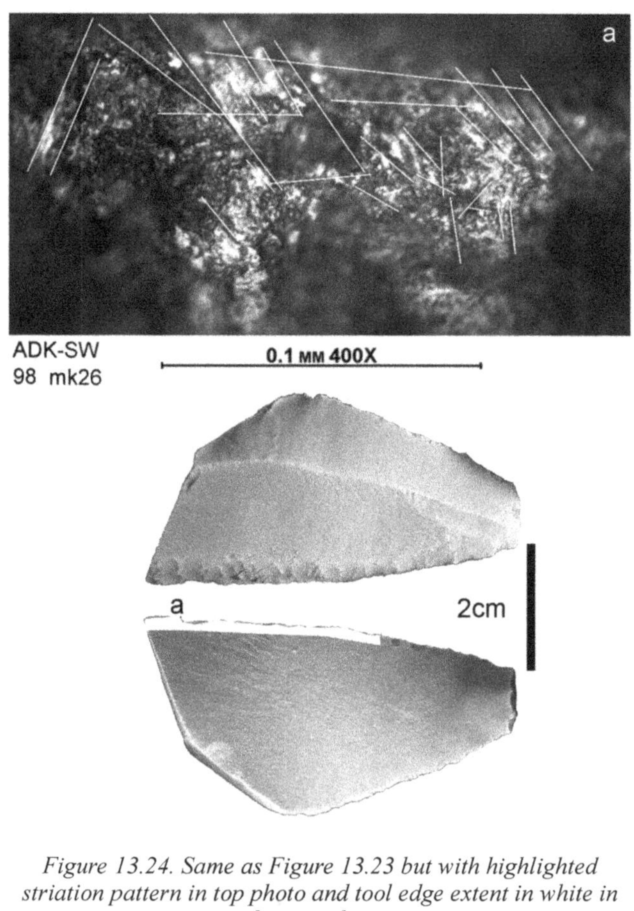

Figure 13.24. Same as Figure 13.23 but with highlighted striation pattern in top photo and tool edge extent in white in bottom photo.

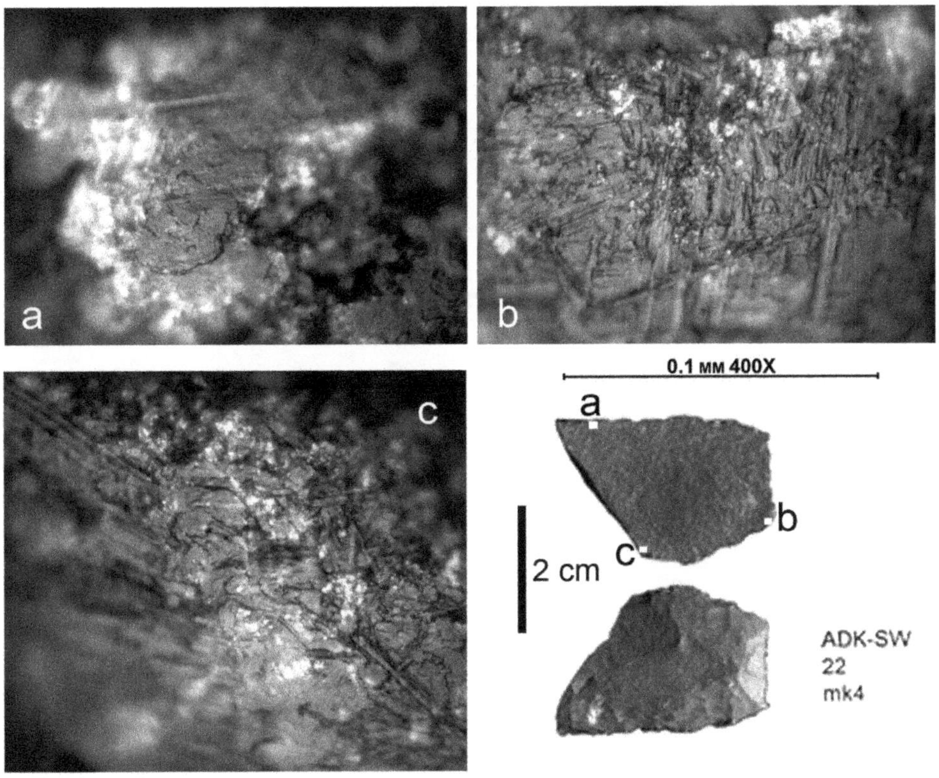

Figure 13.25. Oriented photomicrograph of burin (c) wear traces for relatively hard contact material and haft wear (a, b) on fine-grained, hard basalt flake from the 2007 investigation of the Swetzoff site, ADK-283. Note tool edge is snapped, haft element is deliberately backed.

Figure 13.26. Same as Figure 13.25 but with highlighted striation pattern in photos a,b, and c; in bottom right photo the snapped burin tool edge is located between points a and c; the haft element is located between points a and b.

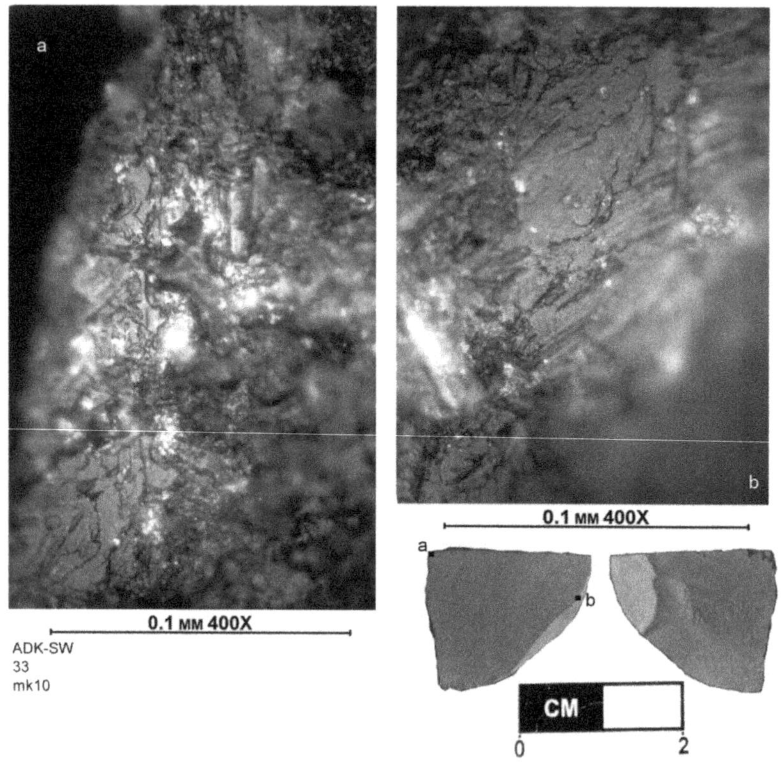

Figure 13.27. Oriented photomicrograph of burin (a) wear traces for relatively hard contact material and haft wear (b) on fine-grained, hard basalt radial fractured flake from the 2007 investigation of the Swetzoff site, ADK-283.

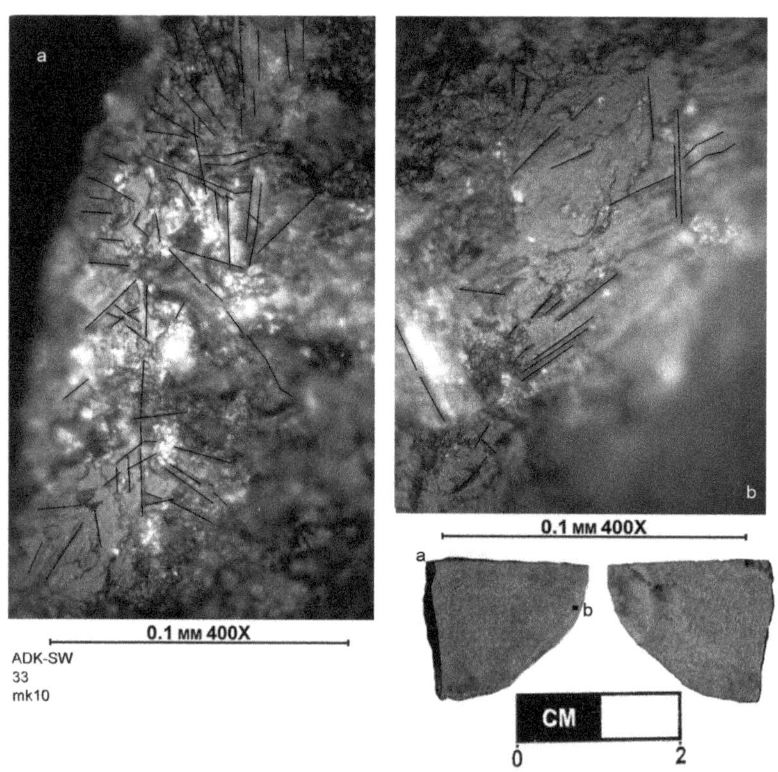

Figure 13.28. Same as Figure 13.27 but with lowlighted striation pattern in photos a and b; in bottom right photo the radial fractured burin tool edge (in black) extends on the tool edge below "a."

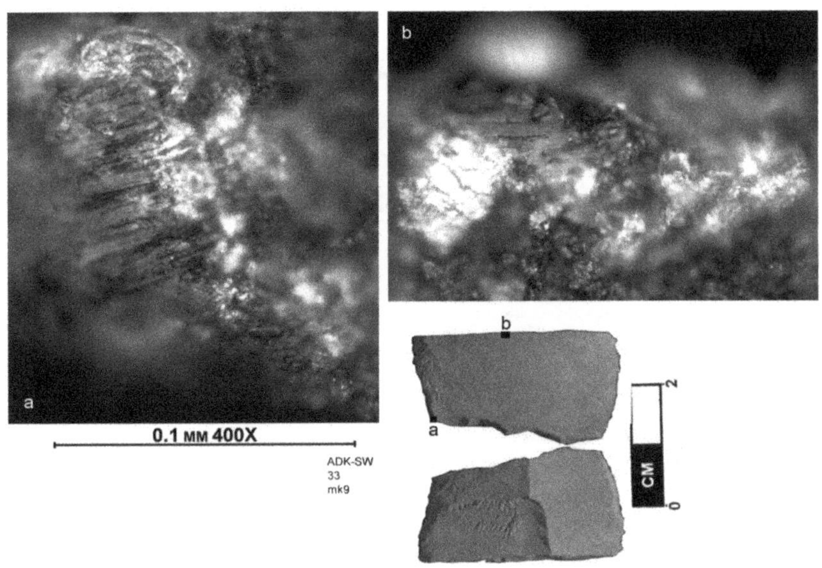

Figure 13.29. Oriented photomicrograph of burin (b) wear traces for relatively hard contact material and haft wear (a) on fine-grained, hard andesite flake from the 2007 investigation of the Swetzoff site, ADK-283.

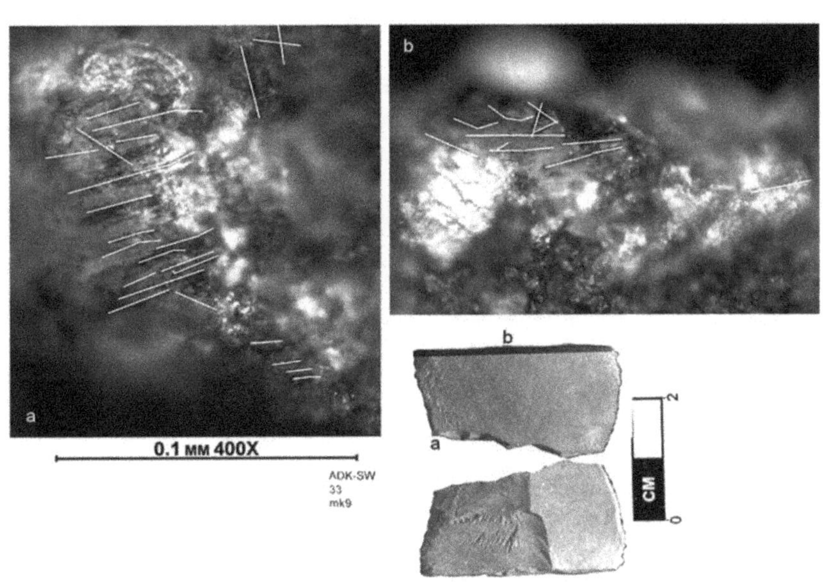

Figure 13.30. Same as Figure 13.29 but with highlighted striation pattern in photos a and b; in lower right photo the snapped burin tool edge extent (black) runs just below the "b.".

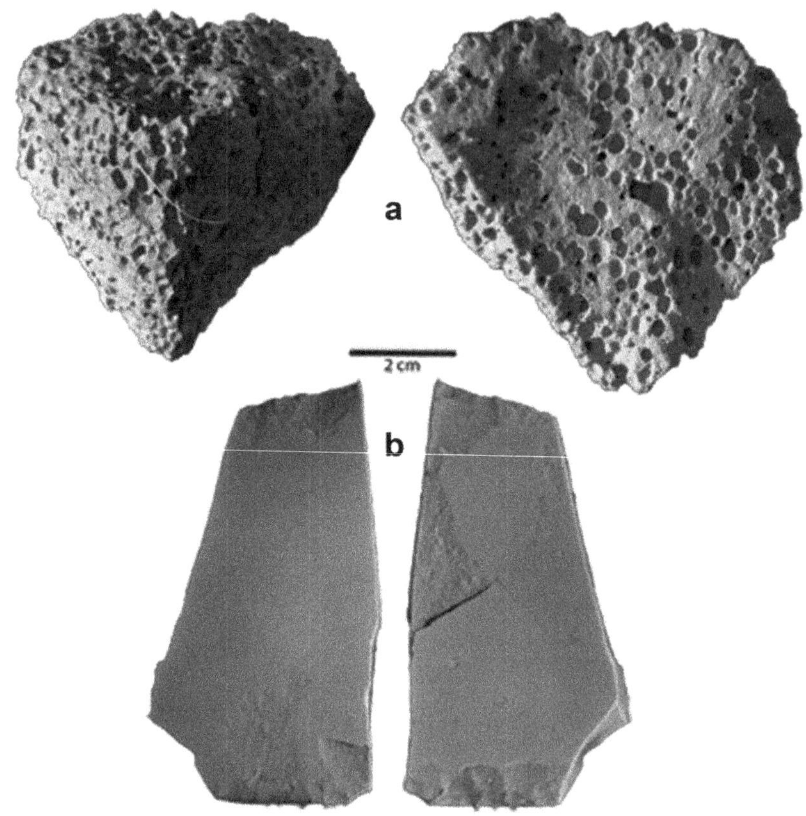

Figure 13.31. Stone artifacts from 2007 investigations of blowouts at ADK-187.

program of biface manufacture and, likely, subsequent transport of manufactured bifaces from the island.

ADK-187

Two artifacts from ADK-187 include a chunk of vesicular andesite (Figure 13.31a) and a slate, double-bladed knife fragment (Figure 13.31b). The former could well be a rough abrading stone and is similar to other examples from Zeto Point. The slate knife fragment has marginal bifacial retouch on both blade edges. In this respect, it differs significantly from the flat knives from Zeto Point, discussed later. It is biconvex in cross-section and has ground smooth surfaces. It measures 82mm between the two blade edges and 6.4mm in thickness. Unfortunately, wind blasting obscures any potential use-wear. It is impossible to associate this tool with any of Jochelson's flat knife categories.

ADK-012

The ADK-012 artifacts were made from basalt (N=11) and banded serpentine (N=4); they have been previously described and illustrated (Kay 2006). Although the banded serpentine is roughly similar to that from ADK-171, the basalt is more variable in quality. It ranges from a majority of relatively fine grained basalt with white phenocrysts to four others more heavily laden with phenocrysts, and accordingly more granular in texture. All of the artifacts are coated with a mineral residue that is more pronounced than for ADK-171. Artifact forms include three biface edge segments (of which two are banded serpentine), two basalt biface tips, a small basalt biface base, six edge damaged flakes (of which two are banded serpentine), and three basalt, possibly retouched, flake blanks.

Five items from ADK-012 are inconclusive for use-wear and include a banded serpentine biface edge segment, and four basalt artifacts; among these, a biface tip, a bifacially retouched flake, a possible knife preform, and what might have been the haft element of a bifacial knife. Flake debitage includes two of basalt and one of banded serpentine with extensive edge damage. Seven artifacts exhibit wear traces indicative of tool use or of patterned tool manufacture. Tools include a basalt biface edge segment used in whittling or butchering, the distal portion of a basalt broken arrowhead with impact striations and damage at the tip, and two general utility tools used to cut and scrape medium hard to hard material. Evidence of patterned tool manufacture includes a bifacial edge--either platform preparation or ground for hafting; a possible arrowhead base with lateral edges rounded and crudely ground; and a blank or simple bifacial perform with trampling damaged surfaces.

ADK-011 Late Component

Using Wilmerding and Hatfield's (2008) summary, I grouped together 72 examined artifacts dating to the 170

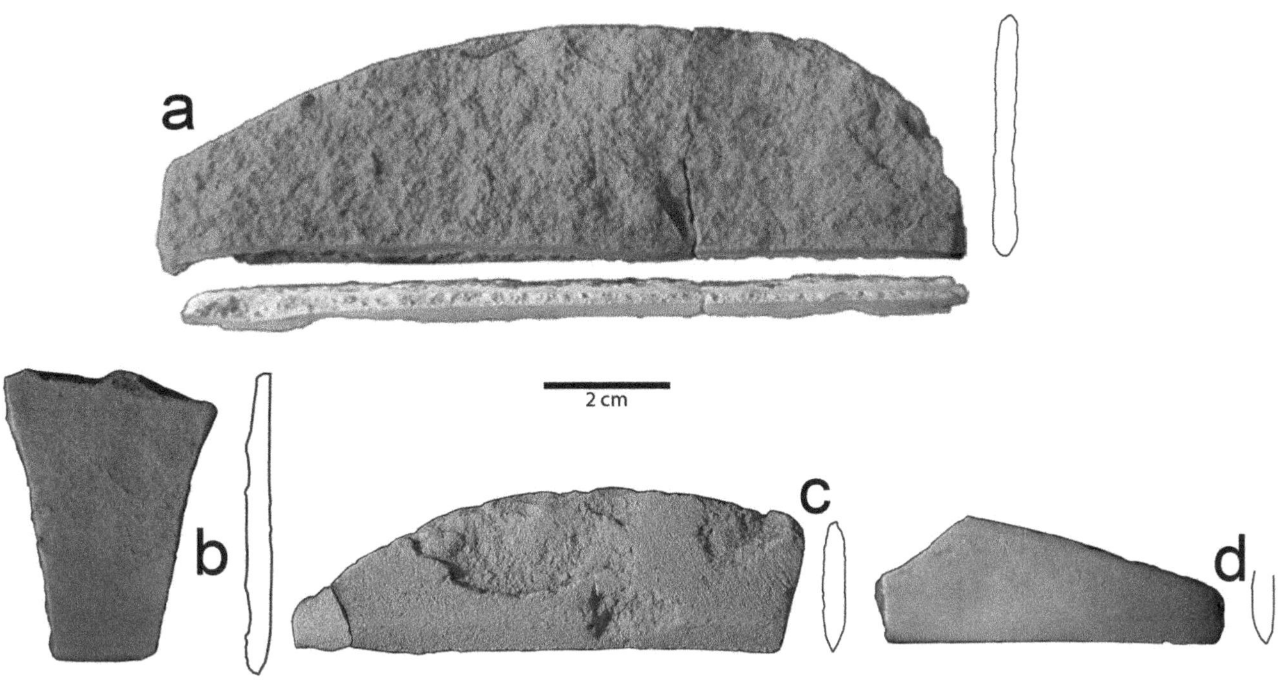

Figure 13.32. Men's large woodworking knives from 2006 excavations of late component of Zeto Point, ADK-011.

to 400 BP layers. Wilmerding and Hatfield (2008) initially regarded these items as potential tool elements and illustrate many.

Abrasive Raw Materials: These vary from micro- to coarse-vesicular andesite chunks to pieces of weakly silicified volcanic ash that measure from less than a gram to 105g. Each could easily fit in the palm of one's hand. None has obvious evidence of modification either intentional or through use. All 18 items are potentially abrasive sources of local origin.

Abraders: All five items of local origin are clearly modified and appear to have been used in an ad hoc fashion to grind or smooth. Thus, they are expedient maintenance tools with masses varying from slightly more than 7g to 45g. Relatively small, they would have been hand-held as opposed to stationary. Ground substances are likely to have included edges and surfaces of chipped or other ground stone, bone, or skin items. The two samples from Layer 3 east are fine-grained and dense silicified ash; one is four-sided with sharpening grooves, the other a plano-convex whetstone or skin softener on the convex side. Layer 2 has three whetstones or skin softeners. Of equally dense, fine-grained, silicified ash to those from Layer 3 east, the largest weighs 31.3g, has broad sharpening furrows on one surface opposite a ground flat side and with rounded corners, creating a squashed loaf shape. Its flat side has a dark brown organic or metallic residue (hematite?) filling in voids in the microtopography. Not much smaller but substantially lighter (15.9g) is a second item, a wedge- or pie- shaped whetstone of finely vesicular andesite with converging concave ground surfaces and broad linear grooves. From either Layer 1 or 2 but tabulated with the latter, the third tool weighed 14.4g, but is significantly denser and compact silicified ash. It is a rounded, bar-shape with a ground, flat side and two adjacent, ground surfaces that converge into an overall inverted 'U' appearance in cross-section. The two items from Layer 1 include a whetstone fragment of dense, compact volcanic ash (38.2g) with linear sharpening grooves on one surface and a significantly less dense whetstone or skin softener (7.1g) of finely vesicular andesite with one convex grinding surface.

Other Ground Stone: With one obvious exception, the 16 artifacts are all patterned tool elements divided into two general groups, flat knives and celts. The exception is a potential griddle stone seemingly split by bipolar percussion from a smooth beach cobble; it has a plano-convex cross section and oval shape and weighs 211g. It does not appear to have been in a fire. Examined microscopically, its interior surface exhibits movement across a hard substrate, or what normally would be termed trampling damage.

Flat knives include 14 items of a fine grained andesite and of slate. These fall into straight and rounded blade specimens, or what Jochelson (1925, 64-65) identifies as men's large woodworking and small food cutting knives. Jochelson differentiates a man's from a woman's knife by the presence on a woman's knife of drilled perforations through which a cord would secure the blade to its handle. From Jochelson's (1925, 64-65) illustrations, there can be no doubt his classification would apply to ADK-011.

The four men's large woodworking knives have a ground straight blade opposite a percussion-prepared chipped and then rudely ground edge of variable shape but lunate in form. The largest (Figure 13.32a) is from Layer 1 and represents a preliminary manufacture stage: it was first roughed out from a tabular piece of andesite. Its straight blade edge has girdling grooves cut parallel to and along the length of the blade from both faces. These do not meet and the blade edge was created by snapping across the girdling grooves. Experimentation with stone saws indicates use of a similar tool to create the girdling grooves. The blade edge measures 152mm. The specimen exhibits no surface grinding. Grinding would have been a subsequent preparatory step of both the tool edge and adjacent surfaces. This specimen measured slightly more than 45 mm wide and just over 6mm thick; these two dimensions express the upward boundaries for further tool shaping and optional maintenance or tool edge resharpening evident on the remaining three tools from Layer 2.

The three fragments exhibit carefully ground surfaces and still-sharp blade edges. Microscopic examination shows that blade sharpening consisted of grinding strokes parallel to the edge, or what superficially looks like initial blade preparation; whereas cutting strokes were parallel and oblique to the edge and resulted in crisscross striations and streaks extending from the blade edge onto the adjacent surfaces. The three tools appear to have been repeatedly sharpened. The most obvious result is reduction in overall width and less so in thickness without sacrificing blade length to an appreciable degree. This appears to have been an intended engineering design attribute that allowed for repeated and prolonged maintenance. Blade breakage seems most responsible for discontinued use; even the smallest fragments had carefully maintained, sharp tool edges.

One artifact of fine-grained andesite (Figure 13.32b) is measured slightly wider (48.4mm) but significantly thinner (4mm) than the unfinished specimen. Residues on the cutting edge and an adjacent ground surface indicate final use to cut or saw hematite. Its blade edge angle is 65°. A second andesite specimen (Figure 13.32c) is substantially less wide (26.7mm) but of comparable maximum thickness (6.3mm) to the unfinished specimen, and has a blade edge angle of 55°. The last tool (Figure 13.32d) is carefully ground slate and roughly similar in width (21mm) but less thick (5mm) and with a blade edge angle of 50°.

Two each from Layer 3 east and Layer 2 are too fragmented to identify based on overall form. An andesite specimen from Layer 2 is a spall from a ground surface, likely a flat knife. The three of slate (two Layer 3 east specimens [see Figure 13.33f for one with a tool edge angle of 55°] and one from Layer 2) are most similar in tool edge preparation and use to the straight blade specimens. Based on their use-wear, these three would be classified as discarded fragments of men's large woodworking knives. Their thicknesses of about 4mm or less suggest they were repeatedly sharpened prior to breakage.

One can tentatively conclude that these represent men's large woodworking knives and that the preferred material was slate. Slate seems to have held a better edge, or one that could be more easily sharpened, or both. The slate knives reflect longer use lives in being generally both narrower in width and thinner than those of andesite.

Six small food-cutting knives have a ground blade rounded or curved in outline opposite a ground flat edge. Overall, they seem more carefully prepared than the woodworking knives. Each has extensive grinding of surfaces, the intentional flat backed haft edge, and the opposite blade edge. More generally, however, their blades edges are not as sharp as the woodcutting knives and appear more damaged. Microscopic examination confirms the greater edge damage; the use-wear is quite similar to that of general utility tools discussed earlier. If only in width and thickness, the largest are actually bigger than the woodworking knives; calling them 'food cutting knives' might be more appropriate. They represent both andesite and slate. Final grinding preparation of the tool is represented by one of slate from Layer 3 east (Figure 13.33a) that broke during manufacture. Its blade is hinted at by a clear narrowing of the edge opposite a flat ground edge 9mm thick; the specimen's maximum width is 44.3mm.

Three other finished tools from Layer 2 measure between 51.25mm to 57.6mm and range in thickness between 7.65 mm to 10.2mm. Two of these are andesite (Figure 13.33b-c), the third (Figure 13.33d) slate. The largest of this group (Figure 13.33b) is the only whole or complete tool in the sample and has a blade length of over 90mm, and a tool edge angle of 70°. The two ends are not deliberately shaped but are coated with a residue (probably a mastic remnant and possibly with hematite) that extends across the flat backed edge and adjacent surfaces up to the point of blade edge sharpening. This is the clearest macroscopic evidence for tool hafting among the sample. The other two specimens have blade edge angles of 55° and 50°; one (Figure 13.33c) is likely to have cut hematite as this residue is on the tool edge and one adjacent ground surface. The remaining two items are carefully ground and extensively sharpened knife fragments. The largest of andesite (Figure 13.33e) is from Layer 1 and is 26.2mm in maximum width, 3.7mm thickness and with a tool edge angle of 30°. The other of slate is from Layer 2 (Figure 13.33g) and slightly smaller in width, about the same thickness, and with an edge angle of 65°.

Celts are represented by a single fragment of andesite (Layer 1) weighing 6.4g. It is heavily ground on opposing surfaces and an adjacent rounded, broad edge; and is 8.25mm thick. Exactly how wide the blade would have been is unclear. The celt measures 27mm from the ground edge to the opposite side: the original width would have been somewhat larger. Likely, the celt first broke and the bit, or working, end was retained. Fractures of the bit and adjacent edge are bipolar percussion in origin. Bipolar blows began first opposite the bit edge on its original break. Later the tool was rotated 90° or so and hammered from the remaining rounded ground edge, which is hammer impact -scarred. The broken bit and adjacent ground surfaces have adhering hematite residues. From these details, it is likely

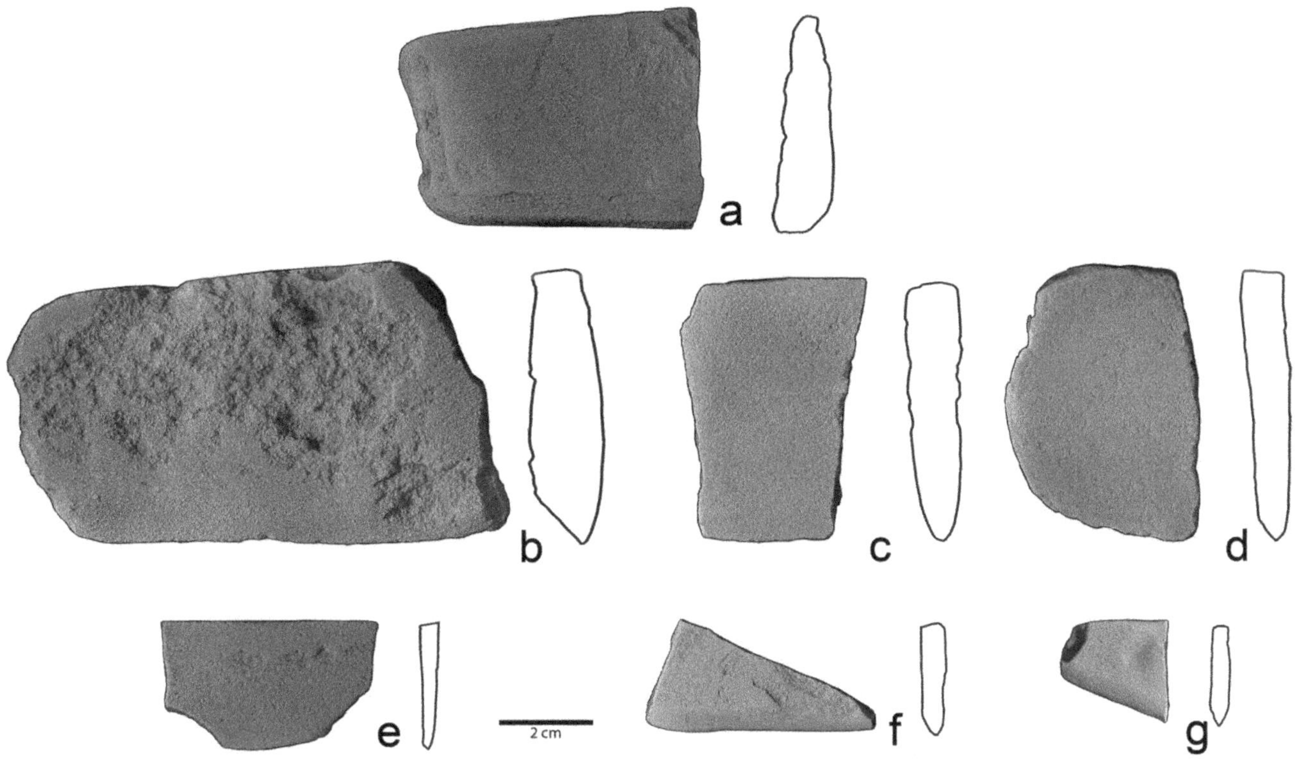

Figure 13.33. Small food cutting knives and one fragment (f) of a man's large woodworking knife from 2006 excavations of late component of Zeto Point, ADK-011.

the celt broke during use and, in an ad hoc fashion, was recycled as a wedge used, last, to split hematite.

Pieces Esquillées and Cores: For all practical purposes and consistent with the recycled celt, six of the eight artifacts had the same function, and were bipolar wedges of either basalt or andesite. They share the same ad hoc attributes of wedges previously described, including the rotation of about 90° after the first wedge edge failed and continued hammering until the second edge broke. Away from the bits the opposing, more-or-less flat surfaces also show microscopic abrasion, streaks and striations that crosscut one another at roughly a right angle. These details are fully consistent with a wedge's deep penetration into a groove. Given the fractures on the bit edges, it seems likely that none could have been further used successfully as wedges. The wedges include one from Layer 3 east, four from Layer 2, and one of basalt with negative burin facets whose layer is unrecorded. Two from Layer 2 were identified initially as cores and vary in mass from over 21g to more than 64g. At the other extreme are the two small ones also from Layer 2 that weigh less than a gram to about 3g; and one initially described as an aborted biface. The two remaining *pieces esquillées* from Layer 3 east are small, smashed chunks of 2g or less and lack use-wear. One, light red in color, is possibly felsite.

Flakes: Of the 10 flakes selected as potential tools, three (all from Layer 2) appear to have been used as tools. One of basalt has invasive cutting or hafting wear on a retouched edge. One is an andesite flake detached from a flat knife, celt or adz. One of basalt has a rounded prepared scraper edge and use striations oriented oblique to the edge. A large, ovate basalt flake from Layer 1 lacks use-wear but has bifacial marginal retouch. It possibly is an early production stage for a lance point, and is consider later with them.

Three broken bifaces are included. From Layer 3 east is a radially fractured segment that lacks use-wear. Two of basalt (one each from layers 2 and 1) represent early production stage failures but are inconclusive for use-wear due to material difficulties.

Points: These are further subdivided into likely toggle-headed harpoon (ovate form, N=1, from Layer 2) or lance heads (shouldered, N=5, 4 from Layer 2, 1 from Layer 3 east) and substantially smaller and thinner arrowheads (N=4, from all but Layer 3 east) (Table 13.2). All but the toggle-headed harpoon point are of basalt. The basalt artifacts, with one exception, could not be judged for use-wear based on difficulties with the raw material; however, they all show macroscopic evidence of use, damage in use, and repair to blade elements. The lance heads are divided into two subgroups on the basis of differing blade-to-haft element proportions; all are contracting stemmed tools.

The single ovate point of andesite from Layer 2 has extensive haft abrasion, impact striations and damage to the tip, and easily falls into the toggle-headed harpoon point category. Although longer than the Layer 4 point, the two are similar in haft lengths (>24mm) and significantly greater than those of the lance points (Table 13.2).

It is likely that only a single lance point would have been serviceable. It is one of three with shorter haft elements, has sharp blade edges and a needle-like tip. Its blade edge angles are 45° and 40°. These edge angles compare favorably with the arrowheads and from the same subgroup. Had it not been for an impact fractured tip, another lance point would have been as long (>37mm). The bottom line on the blade edge angles is that the fragmentary arrow points, including a detached point tip, are probably better indicators of the initial engineering design of these weapon heads because they either broke in manufacture or early in use. Their blade edge angles are similar to the two sharpest lance points but not ones further damaged. Undoubtedly, overall steeper blade edge angles are a function of use-related damage and repair.

The likely still serviceable lance point is one of the two largest; the other is from the same subgroup. The other slightly larger point has a narrower and dulled blade with much steeper edge angles and was not subjected to further repair. Assuming the points were not lost, the implication of these observations is that when this subgroup's points became shorter than 37mm and/or less than 21mm in width they were discarded intentionally, not recycled.

The other lance point subgroup has two specimens, both with far shorter and highly damaged and broken blades but longer haft elements, and one each from Layers 3 east and 2. The Layer 3 east specimen is a nearly complete shouldered lance point whose tip impact fracture is consistent with microscopic use-wear. These two artifacts are too dulled and damaged to have been further repaired. They, too, appear to have been discarded intentionally, not recycled.

A basalt flake with bifacial marginal retouch and previously noted as not having use-wear is of interest too. Its oval form with a pointed tip and overall size make it if only an abandoned early production stage of a lance point.

Three arrowhead fragments of basalt (a base from Layer 2, a tip from Layer 1, a base from a not recorded layer) could not be evaluated for use-wear, as noted earlier. The bases are all rounded not notched. From an unrecorded layer is a final basalt specimen, a flake with marginal retouch and a rounded base; this is likely an unfinished early production stage arrowhead failure.

Implications and Reflections

The Aleuts used compound tools, or tools of separate parts; however, archaeologists study the parts and only rarely the whole. The parts generally are insufficient to fully characterize the technology that is prone to differential breakage, repair and replacement. When poor preservation of organic materials is factored in, archaeologists are often left with an unenviable task. Guesswork is common when considering the durability and viability of this technology, even given exquisite diagrams and Aleut accounts of the technology itself (Jochelson 1925).

It comes as no surprise that archaeologists resort to a 'fuzzy' logic, as Adams (2002, 13-15) describes it, in which our classification and understanding are less than perfect. No matter that our intentions are beyond reproach. Our analyses are highly partial. Successes are few. But we can improve the odds of making sense of the past by a shift in focus from objects as art to the telltale signs of a technology's production, use, repair, and abandonment. Doing so loosens the bonds of static typology and squares with the reality of what technology is: its object is task performance. To assess performance, we need an engineering design perspective along with a use-wear approach. The latter is a systematic way to document how a tool is used, contacts worked material, is held or attached to a handle, experiences failure, and the nature if not the identity of the contact material.

The basic question of a compound tool becomes, what is the weakest link in the chain? Or stated differently, what part or parts require if not daily attention then regular upkeep and, if needed, replacement? To do otherwise assures failure not just of the part but of the whole, and would occur well before the potential utility (i.e., theoretical use life) of the tool is exhausted. Although conceptually different, single piece tools can be addressed similarly; their potential utility would be reached at the point when the drop in tool efficiency is exceeded by maintenance costs.

Aleut usage of basalt for chipped stone tools is curious as better quality material seemingly was as readily available about Clam Lagoon. This preference is most peculiar for lance points and arrowheads, the one part of compound weapons that had to work as expected and to do so every time without fail. The Aleut approach is inexplicable because it struck a poor balance between tool use efficiency and maintenance costs. Indeed, reduced tool efficiency would have increased maintenance at progressively greater cost. But that seems to be what the Aleut preferred, even while it led to greater replacement rates for spent basalt tools. The tools found were mostly left over or discarded after further recycling as simple wedges and obtuse angled burins, and would have been at the end of their use lives. They would be less desirable for stylistic comparisons. The ideal would be tools in near mint condition that incorporate as many aspects of the entire production chain as possible. Items cached in Aleut burial caves might be a likely source of this information. Absent these observations, the little progress made in major time-space divisions for the Aleutian archipelago is not all that surprising.

Adak Island's Clam Lagoon archaeology represents the fully developed Aleutian cultural tradition. Without diminishing the contribution of simple beach cobble ground stone tools, its technology mainstays are a bipolar core-flake and biface industry that varies little over 6000 years. This conservativeness is almost as remarkable as the Aleut's finely parsed distinctions of this technology. It may reflect a larger reality: where overall stability in adaptation is the norm and sources of innovation are highly directed and isolated or remote, tinkering with technology lessens. One might expect that were there a major shift in adaptation, there would be a corresponding technological response. Perhaps the seemingly late switch to ground flat knives preferentially of slate is one example, while the alternate case of long-term technological and symbolic continuity is usage of hematite, ground red pigments and paint.

When the Aleut technological record is parceled out as stone or bone tools, for instance, rather than stone *and* bone tools, it reduces the chance to recognize a technological response to environmental change. Small scale excavation or other recovery programs well suited to highlight environmental conditions and subsistence trends are not always as adept at measuring technological response. It should be clear too that systematic technological analyses require, as a minimum, an adequate sampling of the settlement system. Rarely, if ever, does that system fit conveniently in a 1-meter square or even several of them, no matter how nice it might be if it did.

Clam Lagoon and its environs represent a significant slice of the variability in an Aleut marine centered settlement system. Especially conspicuous is littoral exploitation witnessed early on by the *Clinocardium* midden of ADK-171 and much later at the Zeto Point, Swetzoff and Dozered sites. If nothing else, this analysis and related ones in this volume point out the need for artifact samples and sampling programs to more fully reflect settlement variability in an unforgiving Arctic environment.

Acknowledgments

Clam Lagoon archaeology is the subject of an interdisciplinary team of researchers, of whom I am the latest addition. After working with stone artifact collections from two Clam Lagoon Aleut settlements (Kay 2006) and one nearby, I accompanied Dixie West to Adak Island in May, 2007. While there I determined likely areas of stone raw material procurement relative to the Clam Lagoon archaeological sites, got a better understanding of its archaeology, and most important met Aleuts to the extent time and circumstance allowed.

Other than helping my faculty colleague and friend, the late Allen P. McCartney, with graduate student research on the Reese Bay project of Unalaska Island and talking with Allen about Anangula, the most ancient site of the archipelago, I had had no prior Aleutian Islands experience nor had I ever visited. And now especially, I make no claim to Aleutian Islands expertise but to a deeply abiding affection and fascination.

My heartfelt appreciation goes to the people of Adak Island, to Dixie West, and to Jim Prince and David Fraser. This research was supported, in part, by the Office of Polar Programs, Arctic Social Sciences Division, National Science Foundation (OPP Award # 0353065).

References

Adams, Jenny L. 2002. *Ground Stone Analysis: A Technological Approach*. Salt Lake City, Utah, University of Utah Press.

Ahler, Stanley A. and Phil R. Geib. 2000. Why Flute? Folsom Point Design and Adaptation. *Journal of Archaeological Science* 27(9), 799-820.

Bank, Theodore P., II. 1953. Cultural Succession in the Aleutians. *American Antiquity* 19(1), 40-49.

Binford, Lewis R. 1962. Archaeology as Anthropology. *American Antiquity* 28, 217-225.

Binford, Lewis R. and Sally R. Binford. 1966. A Preliminary Analysis of Functional Variability in the Mousterian of Levallois Faces. In J. D. Clark and F. C. Howell (eds.), *Recent Studies in Paleoanthropology,* American Anthropologist Special Publication No. 68(2-2), 238-295.

Bleed, Peter. 1986. The Optimal Design of Hunting Weapons: Maintainability or Reliability. *American Antiquity* 51, 737-747.

Bleed, Peter. 2001 Trees or Chains, Links or Branches: Conceptual Alternatives for Consideration of Stone Tool Production and Other Sequential Activities. *Journal of Archaeological Method and Theory* 8(1), 101-127. (available online).

Bradley, Bruce A. 1975. Lithic Reduction Sequences: A Glossary and Discussion. In (Earl Swanson, ed.), *Lithic Technology Making and Using Stone Tools,* 5-13. The Hague, Mouton Publishers.

Braun, David P. 1983. Pots as Tools. In James A. Moore and Arthur S. Keene, (eds.), *Archaeological Hammers and Theories*, 108-134. New York, Academic Press.

Collins, Henry B. 1984. History of Research Before 1945. In (D. Damas, ed.), *Handbook of North American Indian, Volume* 5, Arctic, 8-16. Washington, DC, Smithsonian Institution.

Collins, Michael B. 1975. Lithic Technology as a Means of Processual Inference. In Earl Swanson (ed.), *Lithic Technology Making and Using Stone Tools,* 15-34. The Hague, Mouton Publishers.

Hardy, Bruce L., Marvin Kay, Anthony E. Marks, and Katherine Monigal. 2001. *Stone Tool Function at the Paleolithic Sites of Starosele and Buran Kaya III, Crimea: Behavioral Implications.* Proceedings of the National Academy of Sciences of the United States of America 98(19), 10972-10977.

Hayden, Brian, Nora Franco, and Jim Spafford. 1996. Evaluating Lithic Strategies and Design Criteria. In George H. Odell (ed.), *Stone Tools: Theoretical Insights into Human Prehistory*, 9-45, New York, Plenum Press.

Hoffman, R. and L. Gross. 1970. Reflected-light Differential-interference Microscopy: Principles, Use, and Image Interpretation. *Journal of Microscopy* 91(3), 149-172.

Holmes, William Henry. 1890. A Quarry Workshop of

the Flaked-stone Implement Makers in the District of Columbia. *American Anthropologist* 3, 1-26.

Holmes, William Henry. 1892. Modern Quarry Refuse and the Paleolithic Theory. *Science* 20, 295-297.

Holmes, William Henry. 1894. Natural History of Flaked Stone Implements. In *Memoirs of the International Congress of Anthropology*, 120-139. Chicago, Illinois.

Holmes, William Henry. 1897. Stone Implements of the Potomac-Chesapeake Tidewater Province. *Bureau of American Ethnology Annual Report* 1893-94, 13-157. Washington, DC, Washington Government Printing Office.

Holmes, William Henry. 1919. *Handbook of Aboriginal Antiquities. Part I. Introductory. The Lithic Industries.* Bureau of American Ethnology, Bulletin 60. Washington, DC, Washington Government Printing Office.

Jochelson, Waldemar. 1925. *Archaeological Investigations in the Aleutian Islands*. Publication 367. Washington, Carnegie Institution of Washington.

Johnson, Jay K. 1993. North American Biface Production Trajectory Modeling in Historic Perspective. *Plains Anthropologist* 38-143, 151-162.

Kay, Marvin. 1984. Late Caddo Subtractive Technology in the Red River Basin. In N. Trubowitz, (ed.), *Cedar Grove: An Interdisciplinary Investigation of a Late Caddo Farmstead in the Red River Valley*, 174-206. Research Series No. 23. Fayetteville, Arkansas Archeological Survey.

Kay, Marvin. 1996. Microwear Analysis of Some Clovis and Experimental Chipped Stone Tools. In George H. Odell (ed.), *Stone Tools: Theoretical Insights into Human Prehistory*, 315-344. New York, Plenum Press.

Kay, Marvin. 1998. Scratchin' the Surface: Stone Artifact Microwear Evaluation. In Wilson-Leonard (ed.), *An 11,000-year Archeological Record of Hunter-Gatherers in Central Texas. Volume III: Artifacts and Special Artifact Studies, Assembled and Edited by M. B. Collins*, 743-794. Studies in Archeology 31, Austin, University of Texas Archeological Research Laboratory.

Kay, Marvin. 2006. Adak Island Use-wear Assessments. Report to Dixie West—Central Aleutians Archaeological and Paleobiological Project, Kansas University.

Kay, Marvin and Richard E. Martens. 2004. Clovis Scrapers from the Martens Site. *The Missouri Archaeologist* 65, 44-67.

Kay, Marvin and Ralph Solecki. 2000. Pilot Study of Burin Use-wear from Shanidar Cave, Iraq. *Lithic Technology* 25(1), 30-41.

Keeley, Lawrence H. 1980. *Experimental Determination of Stone Tool Uses: A Microwear Analysis.* Chicago, University of Chicago Press.

McCartney, Allen P. 1984. Prehistory of the Aleutian Region. In D. Damas (ed.), *Arctic, Handbook of North American Indians*, vol. 5, 119-135. Washington, D. C., Smithsonian Institution.

Root, Matthew J., J. D. Williams, Marvin Kay, and L. K. Shifrin. 1999. Folsom Ultrathin Biface and Radial Break Tools in the Knife River Flint Quarry Area. In D. S. Amick (ed.), Folsom Lithic Technology: Explorations in Structure and Variation, 144-168. *International Monographs in Prehistory Archaeological Series* 12, 144-168. Ann Arbor, Michigan.

Schiffer, Michael B. 1972. Archaeological Context and Systemic Context. *American Antiquity* 37, 156-165.

Semenov, Sergei A. 1964. *Prehistoric Technology*. London, Cory, Adams, and MacKay.

West, Dixie, Christine Lefèvre, Elizabeth Wilmerding, Virginia Hatfield, and Lyn Gualtieri. 2011. The Central Aleutians Archaeological and Paleobiological Project. Chapter 1 in D. West, V. Hatfield, E. Wilmerding, C. Lefèvre and L. Gualtieri (eds), *The People Before: The Geology, Paleoecology and Archaeology of Adak Island, Alaska*. Oxford, British Archaeological Reports.

Wilmerding, Elizabeth G. and Virginia L. Hatfield. 2008. Zeto Point (ADK-011) Lithic Analysis from the Central Aleutians Archaeological and Paleobiological Project 2006 Investigations. Presentation at Symposium on the Central Aleutian Islands Archaeological and Paleobiological Project, Alaskan Anthropological Association, Anchorage, Alaska.

Wilmerding, Elizabeth and Virginia Hatfield, 2011. Six Thousand Years of Lithic Technology on Adak in a Broader Aleutian Context. Chapter 12 in D. West, V. Hatfield, E. Wilmerding, C. Lefèvre and L. Gualtieri (eds), *The People Before: The Geology, Paleoecology and Archaeology of Adak Island, Alaska*. Oxford, British Archaeological Reports.

Woodbury, Anthony C. 1984. Eskimo and Aleut Languages. In D. Damas (ed.), *Arctic, Handbook of North American Indians, vol.* 5, Arctic, 49-63. Washington, DC, Smithsonian Institution.

Chapter 14
Griddlestones from Adak Island, Alaska: Their Provenance and the Biological Origins of Organic Residues from Cooking

Richard Jeannotte, Kirsten Nicolaysen, Floyd E. Dowell, Taylor Johnson and Dixie West

Abstract: *Burned stone slabs, historically called griddlestones, were recovered from Components 1 (2390-2590 RCYPB) and 2 (170-415 RCYBP) at archaeological site ADK-011 on Adak Island, Aleutian Islands, Alaska. The griddlestones show evidence of fire exposure and have a dark, often greasy, matrix of decomposed organic residues. The griddlestones consist of hornblende andesite that cleaves into flat fragments suitable for a cooking platform. It is unclear how prehistoric Aleuts actually used these flat slabs. Petrographic analysis reveals that Aleuts collected most, if not all, of the andesite slabs at a natural outcrop 5km from ADK-011. Stable isotope signatures recovered from the stones (-24.89 ± 0.23) were within in the same range as signatures found on the stones collected from the natural quarry (-25.90), plants (-26.23 ± 0.38), soils (-25.83 ± 0.12) and tephras (-27.20 ± 0.13). The $\delta^{13}C$ signatures found on the griddlestones could be attributed to soils, decaying plants, and/or tephras associated with the archaeological site along with contributions of 'seafood' $\delta^{13}C$ signatures (-4.03 ± 1.27). This could explain the slight enrichment in ^{13}C observed in griddlestones compared to soils, plants and tephras. The analysis of the molecular signatures of the stones by VIS/NIR spectroscopy confirmed the stable isotope data. The sources of the chemical fingerprints on the griddlestones could be multiple. A more detailed analysis using biomarkers and their compound-specific isotopic ratios will be required to test the hypotheses raised in this chapter.*

Keywords: *Aleutians, cooking stones, organic residues, stable isotope analysis, visible/near-infrared spectroscopy, hornblende andesite*

Introduction

Parasitological evidence suggests that Aleuts ate some, if not most, of their food raw (Bouchet *et al.* 2001; West *et al.* Chapter 16 this volume). However, their cooking, or at least heating, of food is recorded in Russian ethnographic accounts translated by Lydia Black (1984), the journal of Father Veniaminov (1984), and an archaeological publication of Waldemar Jochelson (1925). Sometime before 1761, Prokopii Lisenkov visited the western and central Aleutians, writing a report ultimately filed with the Governor of Irkutsk (Black 1984, 5, 43). Lisenkov reported, 'Mostly, they eat [food] raw, but also they cook [their food] sometimes. Instead of kettles, they cook on stone slabs smeared round the edges with clay' (Black 1984, 53). Veniaminov (1984, 278) noted that 'The Aleuts eat almost everything raw except codfish which, raw—and not completed [sic] cooked—is very harmful. The meat of sea mammals, though cooked or rather stone boiled, may be said only to be warmed up.' Jochelson (1925, 73) wrote, 'When the Aleut made a driftwood fire it was usually in the open air or in the special semi-underground kitchen huts. There they fried meat of sea animals, birds and fish on flat stone frying-pans called *cu'nlux*. Many such stone frying pans were found on all the islands....In use, the edges of the frying pan (sic) were supported by stones, so wood could be burned under it.' In the Aleutians, driftwood was a coveted resource because of the lack of woody trees and shrubs on the islands. Even a small stand of Sitka spruces, transplanted to Unalaska Bay in the 1940's, barely survives the windy conditions (Alden and Bruce 1989, 1-2). Fire making would have required tools including a drill, bow, head for the drill, dry grass or wood, and sulphur (Jochelson 1925, 74). Aleuts primarily used seal and whale oil burned in stone lamps; they supplemented this with dry grass, cancellous whale bone, and spoiled seal flesh (Hrdlička 1945, 55).

The Aleutian ethnographic and archaeological records support the comparatively unique status of the cooking hearth (Table 14.1). Hrdlička (1945, 54) reported: 'Neither the Aleuts nor their predecessors in the islands had any regular fireplaces. That this was so from pre-Aleut times, was shown by a complete want of fireplaces in the deposits, and by the absence of any accumulations of ashes or charcoal. The only marks of fire in all the islands where excavations were made by us were occasional moderate-sized, undressed, more or less fire blackened pieces of stone on which they had probably burned oil-soaked grass or moss, rather than wood, for which most of the slabs were too small.' Jochelson (1925, 109) reported: 'That the Aleut used fire in the earliest period of the accumulation of kitchen refuse is shown by the discovery of stone 'frying-pans' and traces of charcoal and ashes of fire-places in the lowest layers of remains.'

In the central and western Aleutians archaeologists rarely find hearths. An exception is a semi circular arrangement

Table 14.1. Aleutian hearth provenance and dates.

Site	*Location*	*Type*	*Date*	*Reference*
Russell Creek	Alaska Peninsula	Box shaped hearth	3110-3335 BP	Mashner and Jordan 2001
Amaknak Bridge	Unalaska	Chimneys and sub floor channels	2590±90 BP to 3470 ±70 BP	Knecht and Davis 2004
Margaret Bay	Amaknak	Hearths with slab stone storage features	3110±60 BP to 3280±70 BP	Knecht *et al.* 2001
Reese Bay		Ash concentrations; possible stone features	170±70 BP to 380±50 BP	McCartney and Veltre 2001
ATU-194	Attu	Circular ash concentration	nd	West *et al.* (2002)
Pit 4	Buldir	Amorphous ash concentration near whale skull	220±70 BP ; 250±70 BP; 390±80 BP	West *et al.* (1997, 2001); Corbett *et al.* 2008
Swetzoff Site	Adak	Fire cracked stone concentration	1755±20 BP	West (Chapter 1 this volume); Kay (Chapter 13 this volume)

of fire cracked stones and charcoal dated between 1735 and 1755 BP at the Swetzoff Site (ADK-283) on a small island in Clam Lagoon, Adak Island (West *et al.* Chapter 1 this volume). The Western Aleutians Archaeological Project reported a dense, ashy deposit believed to be a simple hearth in a whalebone structure on Buldir Island (Corbett *et al.* 2008; West *et al.* 1997). A concentration of ash in a house feature on Attu suggests some type of fireplace (West *et al.* 2002). Generally, central and western Aleutian research supports the idea that occidental Aleuts did not use complex hearth structures. However, the recovery rate of cultural charcoal at many archaeological sites across the archipelago attests that driftwood was burned on a regular basis.

Hearth types range from simple to complex in eastern Aleutian prehistoric sites. Hearths at Reese Bay were not especially elaborate or particularly well defined (Veltre and McCartney 2001). Of three that were most obvious in Longhouse 2, two were near the northeast end in the central part of the floor. The third was located at the west end wall and may actually have been outside the house itself. All three were associated with midden refuse. Three other areas in the central floor of the longhouse may also have represented hearths. Veltre and McCartney found tabular, charred stones, believed to be griddlestones, randomly distributed throughout the excavations (Douglas Veltre, U. of Alaska, pers. comm.). The Russell Creek site on the Alaska Peninsula, near Cold Bay provides several examples of simple, yet effective, cooking areas. Excavation by Maschner and Jordan (2001, 157-159) revealed two complete houses and parts of other houses where four different hearths were found. Each hearth consisted of four stones, two that were elongate, thin slabs resting on their long sides and two smaller stones that were more irregular or semi-rounded in shape. These stones were embedded in the house floor so the top of the hearth was flush with the surrounding floor surface. Maschner and Jordan suggest that small trenches in the sand next to the hearth provided ventilation to the fire. Knecht and Davis (2004, 195) reported that the Amaknak Bridge Site houses, including a multi-roomed type, featured 'exterior chimney[s] and stone lined sub floor channels that may have served as heating systems.'

Early Aleutian observers reported that the Aleuts used flat stones for cooking or warming food. Because these stones are common in many Aleutian sites, archaeologists assume that these were employed for some type of culinary activity. However, historical interpretations are subjective, and biases can influence the interpretation of the observer. The objectives of this work were to: (1) identify the source of flat, andesite stones recovered from archaeological site ADK-011, (2) study the degraded organic residues found at the surfaces of the stones, many of which are charred, and (3) determine if they were actually used to heat or cook food.

Sourcing the ADK-011 Griddlestones

On-site Selection of Stone Samples

Archaeologists recovered 85 tabular slab fragments (See Wilmerding and Hatfield Chapter 12 this volume), and nearly all were similar in mineral type and texture. These were recovered from two occupations, Components 1 and 2, at archaeological site ADK-011. Component 1, dated at 2390-2590 RCYBP, represents a series of thin ashy 'floors' separated by thin sand lenses. Component 2, relatively recent (170-415 RCYPB) and well preserved, represents part of a prehistoric house depression adjacent to a shell and bone midden. The Component 2 house depression and kitchen midden are roughly contemporaneous. Medium gray, volcanic slabs, horizontally oriented, were randomly distributed on the Component 1 occupation surfaces, the Component 2 house feature floor, and within its associated refuse pile. Stone size, shape, and orientation suggest that these are not remnants of actual hearths, nor do they constitute a pavement.

Seventy-three of the 85 encountered slabs were inspected, weighed and field identified. Maximum length of the rock slabs was 5.5 to 41.5cm and weight varied from 30 to 4200g though the median maximum length was 13.5cm and the median weight was 359g. Fifty-nine percent of slabs had a minimum thickness of 4-10cm. Geologically, most of these slabs would qualify as angular to sub-angular cobbles that have a tabular or platy shape.

Natural processes can scatter stones episodically through an inhabited site, but none of the four candidate geologic processes appear responsible for the primary deposition of these stones. Though Herculaneum and Pompeii provide characteristic examples of volcanic activity that blankets a habitation site with airfall tephra or violent pyroclastic flows, the shape and density of the slabs and the lack of adjacent pumice or scoria of similar or greater size is completely inconsistent with the slabs being volcanic fragments ejected and deposited by an explosive eruption. Our rejection of this hypothesis for the presence of the slabs does not preclude the ashy and sandy layers of Component 1 deriving wholly or in part from distant eruptions. Two processes involving water might distribute large rock within a site. Sea-level rise to the level of the site could strand large rock fragments, such as during the creation of a cobble beach or shingle from an adjacent rocky headland. No such headland exists adjacent to the site –except that which underlies the site – and the wave action would simultaneously round the corners and edges of the slabs while washing away any fine-grained material within which they lay. This cannot explain the presence of the rock fragments. A river or stream could deposit cobbles of lava, yet topography of the site on a headland height and the primary angularity of the rocks speak against this. Likeliest of the geologic processes that could entrain subangular to angular lava fragments within a fine-grained matrix is the past occurrence of a debris flow prior to the Component 1 occupation. A debris flow is like a runny milkshake with raisins in it. The thicker the milkshake, the larger the raisins (e.g., lava fragments) it can carry as it slurries down slope. When volcanic, these often start from an eruption melting snow or mixing with heavy rain. Waythomas (1995) maps lahar or volcanic debris flows on the lower slopes of Mount Adagdak far short of the archaeological site. Only a debris avalanche might be sufficient to carry lava fragments the distance to Zeto Point.

A debris avalanche is a landslide from the collapse of the entire side of the volcano. The spectacular 1980 eruption of Mt. St. Helens began because the combination of gravity and the ballooning pressure from magma inside the volcano caused the north flank to slough off. Because debris avalanche deposits can have a mixed character with rocks as large as boulders (>64cm) embedded in a matrix of soil and sediment, this hypothesis is worth considering further. ADK-011 is approximately 10km from the summit of Mount Adagdak. Ui *et al.* (2000, 621) defined a relationship between the maximum runout length of a debris avalanche and the height at which it initiates. At the ~10km distance, the required initiation height of an avalanche depositing cobble-sized fragments at ADK-011 would have been 900-1600m high, yet Mount Adagdak is only 650m high. This is not conclusive, because the past summit of Adagdak may have been higher, but we find this argument persuasive that the flat slabs at ADK-011 were not deposited in an avalanche debris flow. Furthermore Waythomas (1995) maps a solitary landslide deposit (Qcl) on the lower margins of Adagdak, but this is yet ~5km short of site ADK-011. We conclude that humans transported these flat slabs to ADK-011.

Description of Slabs Recovered at ADK-011

Many of the flat stones are blackened with what appears to be charred, greasy material covering the surface. Only 11% of described clasts do not have a blackened, charred or dark surface. Of the remaining stones, 72% are darkened, black or charred on both flat surfaces. Centuries of soil microbe activity have not destroyed the charred, greasy residue.

We visually inspected the 73 characterized samples and placed them into one of three categories of lava: porphyritic hornblende andesite (84%), coarse-grained plagioclase porphyritic andesite (~10%) and plagioclase phyric andesite (~6%). The slabs of porphyritic, hornblende-bearing andesite were generally without vesicles, the small holes in volcanic rock created by gases exhaled from a cooling magma. The medium gray matrix contained plagioclase, a feldspar rich in calcium, aluminum and silica, that was tabular or equant as exposed in the broken surface and appeared to make up about 20% of the area of a surface. An elongate black mineral, approximately twice the thickness of a pencil lead and about 1cm long, is an amphibole, probably hornblende and makes up about 5% of the samples that contain it. Amphibole, an iron-magnesium silicate, has a significant amount of water bound in its mineral structure; its presence is diagnostic of the rock type as andesite. The hornblende andesites of Adak are relatively rich in magnesium and are the type example of the eponymous *adakite*. Andesite is relatively common lava of the great chains of volcanoes along continental margins above a subduction zone: the Cascades, the Andes, the Japanese and Russian volcanoes of Kamchatka and the Kuriles. The andesite variety adakite has erupted, though infrequently, at many subduction zone volcanoes; in the Aleutians, it is most abundant on Adak and on islands located to the west of Adak (Kelemen *et al.* 2003).

Potential Stone Sources

The human accumulation of stones with flat, platy features suggests function. But the importance and prevalence of these stones may be revealed also by how far they were transported from source to place of use. The shape and density suggest the slabs formed in the interior or at the base of a lava flow that was viscous enough, when a moving molten mass, that shear stresses associated with differences in flow speed over topography overcame the tensile strength of the melt (e.g., Cordonnier *et al.* 2009; Webb and Dingwell 1990). Andesitic and rhyolitic lavas, while still molten, have relatively high viscosity because their high silica content promotes the polymerization or linking of silica tetrahedra. Tetrahedra form when silicon coordinates and bonds with four oxygen atoms, and the oxygen atoms may be shared by adjacent silicon atoms. Because one, two, three or all four oxygen atoms may be shared, the melt matrix has nascent silicate chains and networks at a molecular level even without these being incorporated into the structure of minerals growing in the melt. These chains and networks in the matrix impede the flow of the lava, raising its viscosity by several orders of magnitude compared to fluid basalt. Andesitic flows would act rather more like a ball of silly putty slowly flattening on a desk due to gravity than like ketchup or molasses poured from a jar. When liquids flow like water (i.e., Newtonian conditions), the atoms of the melt readily accommodate the influence of gravity and hillslope. Andesites instead act like non-Newtonian fluids or even fail brittlely – essentially tearing the silica polymers at a microscopic scale - where stress may be concentrated in specific areas of the flow (Cordonnier *et al.* 2009; Smith 1996; Webb and Dingwell 1990). These breakages of the silicate networks and chains within the lava matrix manifest themselves years later as joints or planes of weakness within the rock. Joints propagate through the matrix that has devitrified and grown clays as the rock begins the slow process of attaining equilibrium with a wet, low temperature environment (e.g., Fuller 1938). Ultimately the orientation of these planes of weakness reflects the past movement of the lava flow. The advantage to humans is that these processes create rock slabs the size and shape of a rectangular cutting board.

Though the ADK-011 site is built upon remnants of the extrusive igneous bedrock (i.e., lava) and eroded domes (Figure 14.1), the surface deposit is Quaternary eolian sand (Waythomas 1995). Inspection of the jointing patterns of the rock making up the headlands closest to the site shows that this material breaks preferentially into blocks that are considerably thicker than the platy lava fragments recovered at the site. No clear candidate for the lava slabs outcrops adjacent to ADK-011.

During 2006 and 2007 the Central Aleutians archaeological team surveyed north Adak in vicinity of Clam Lagoon for convenient outcrops of lava that seemed likely to break into platy slabs (Figure 14.2). Two potential sources were found. One possible quarry site is located approximately 1km north of the most northeastern part of Clam Lagoon and is approximately 5km from ADK-011. As a candidate for the original source of the griddlestones, this site is superior in offering an abundance of platy, blue-grey andesite that is relatively fresh (i.e., little alteration of the matrix to clay minerals due to chemical weathering in the presence of water). The ~8m high outcrop spalls suitably flat, thin slabs from its interior (Figure 14.2) directly on to the beach berm and would have been accessible by kayak even had sea-level been significantly higher in the past.

The second possible quarry site is on a point near Candlestick Bridge at the south end of Clam Lagoon roughly 1.5km from site ADK-011. This exposure is approximately 50m long and 2-3.5m high at the water's edge. During high water stands, this outcrop would likely have been covered by water. The lava breaks into blocky fragments unlike most of the slabs recovered at the archaeological site, though there are a few places where the potential source breaks into flat slabs (Figure 14.2). Comparative sample 06ADK-03 is from this locality (UTM1, N5751967, E529392).

Analysis of Minerals and Textures of Site Slabs and Potential Source Samples

Thin sections (0.03mm thick) of two comparative samples and five griddlestones were examined using a microscope with polarizing filters. This is an extremely effective tool for identifying minerals and the rock textures formed by minerals. Minerals are visible in this cross-polarized light because their ordered structure impedes and splits the light transmitted through the rock; glass matrix appears black. In the following discussion, we provide idealized mineral compositions to orient the reader to the differences between commonly occurring minerals in lavas.

The two possible source locations differ distinctly in mineral composition and abundance (Table 14.2). Comparative sample ADK-1 has large (1-2mm) clots of plagioclase ($CaAl_2Si_2O_8$), augite ($Ca[Mg,Fe][Al,Si]_2O_6$) and an opaque mineral that is likely an iron-titanium oxide ($[Fe,Ti]_3O_4$). Plagioclase is also abundant throughout the rock at a variety of grain sizes (0.4-1.5mm; seriate texture). Furthermore this rock, in hand sample, has elongate black minerals typical of amphibole, a hydrous iron-magnesium silicate. Microscopic inspection of these black minerals reveals a complicated form that consists of extremely fine opaque minerals and clinopyroxene where the individual grain size is essentially <0.01mm (Figure 14.3). The shape formed by this cloud-like, multimineralic aggregate is diamond-shaped in cross section. Within the interior of some of these diffuse rims, an anhedral sliver of the amphibole hornblende remains. The textural relationship between the diffuse opaque-clinopyroxene cloud that shows the shape of an intact hornblende and bits of the relict mineral in the interior indicates that this entire multimineralic assemblage was previously one well-formed hornblende amphibole. Amphibole reaction textures may form by two mechanisms. One possibility is that the parental magma moves through the crust rapidly prior to eruption, losing significant amounts of water. Although the hornblende was stable under the high pressure-high water conditions early in the magma's history, it began reacting to a low-pressure, anhydrous mineral assemblage (e.g., Rutherford and Hill 1993). Browne *et al.* (2003) find that decompression rims are thin and typically consist of calcium-poor orthopyroxene, plagioclase and opaque iron-titanium oxides. In contrast, some reaction rims grow when an amphibole is plucked from country rock and incorporated into a magma as that magma traverses through a crustal conduit. These rims are thick and consist of Ca-rich clinopyroxene and Fe-Ti oxides with small amounts of orthopyroxene and plagioclase (Browne *et al.* 2003). For either process, eruption and quenching of the magma interrupted the breakdown of the amphibole. The reaction rims of ADK-1 are primarily augite+ Fe-Ti oxides with

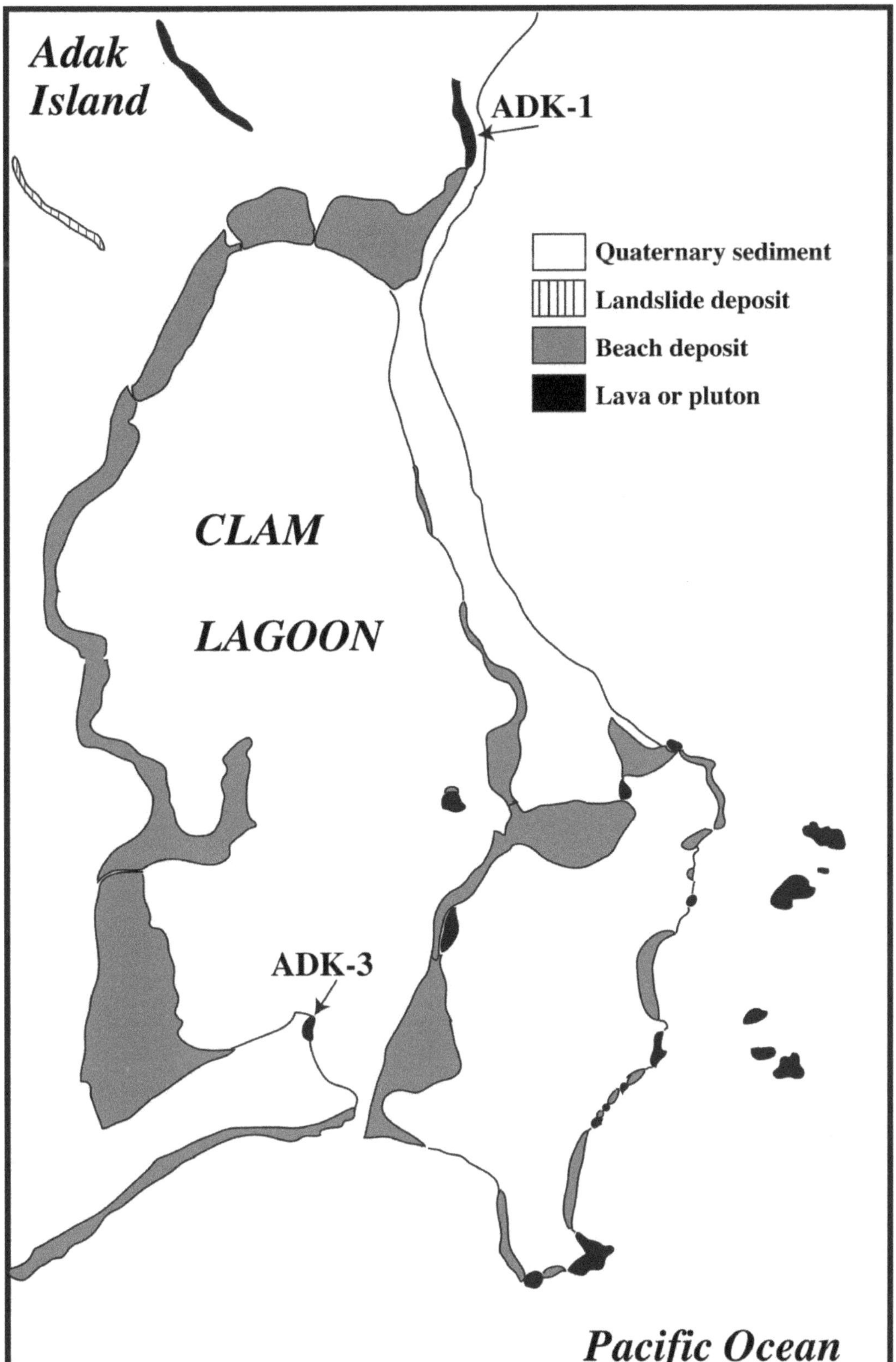

Figure 14.1. Map of Clam Lagoon and surrounding areas. Bedrock outcrops are outlined in black and are largely covered by Quaternary sediment (after Waythomas 1995). Locations of two potential source localities for slabby lava fragments are indicated.

Figure 14.2. Photographs of the two possible source locations for the lava slabs recovered from site ADK-011. (Left) Although this andesitic lava flow is considerably further from the archaeological site, the platy jointing of its flow interior spalls abundant flat slabs into the beach berm a few meters from the ocean shore. Comparative sample ADK-1 is from this location. Co-author T. Johnson is approximately 2m tall for scale. (Right) Near the south exit of Clam Lagoon, this lava breaks primarily into equant blocks though a few areas with platy jointing could provide rock slabs. Comparative sample ADK-3 is from this location.

Table 14.2. Petrographic characteristics of comparative samples from possible quarry sites.

Characteristic	***ADK-1***	***ADK-3***
Plagioclase phenocrysts $(Ca,Na)Al_{1-2}Si_{2-3}O_8$	euhedral to subhedral 0.4-1.5 mm maximum length ~25% abundance deformation twinning internal compositional zoning thin compositional rim seriate texture	euhedral to subhedral 1-5 mm maximum length 1-3% abundance complex internal zoning poikilitic texture
Hornblende phenocrysts $(Na,K)_{0-1}Ca_2(Mg, Fe^{2+}, Fe^{3+}, Al)_5$ $(Si,Al)_8O_{22}(OH)_2$	pseudomorph of hornblende to pyroxene-opaque aggregate 1.5-2 mm maximum length rare relict hornblende in core	euhedral to subhedral pyroxene-opaque aggregate in core or part of outer mantle 0.25-4 mm maximum length twinned, undulatory extinction seriate texture ~10% abundance
Augite phenocrysts $(Ca, Mg, Fe^{2+}, Fe^{3+}, Al)_2$ $(Si,Al)_2O_6$	subhedral to anhedral in 1-2 mm dia. Crystal clots some twinned some enclose opaques (poikilitic) occasionally low Ca pyroxene as a reaction rim	euhedral to subhedral small (0.2 mm dia) in matrix
Opaque mineral probable Titanomagnetite $(Fe,Ti)_3O_4$	two populations by size 1st: subhedral, equant ~0.25 mm dia. 2nd: euhedral, 0.02 mm in matrix	
Quartz SiO_2	none	Rare subhedral phenocryst ~3 mm dia.

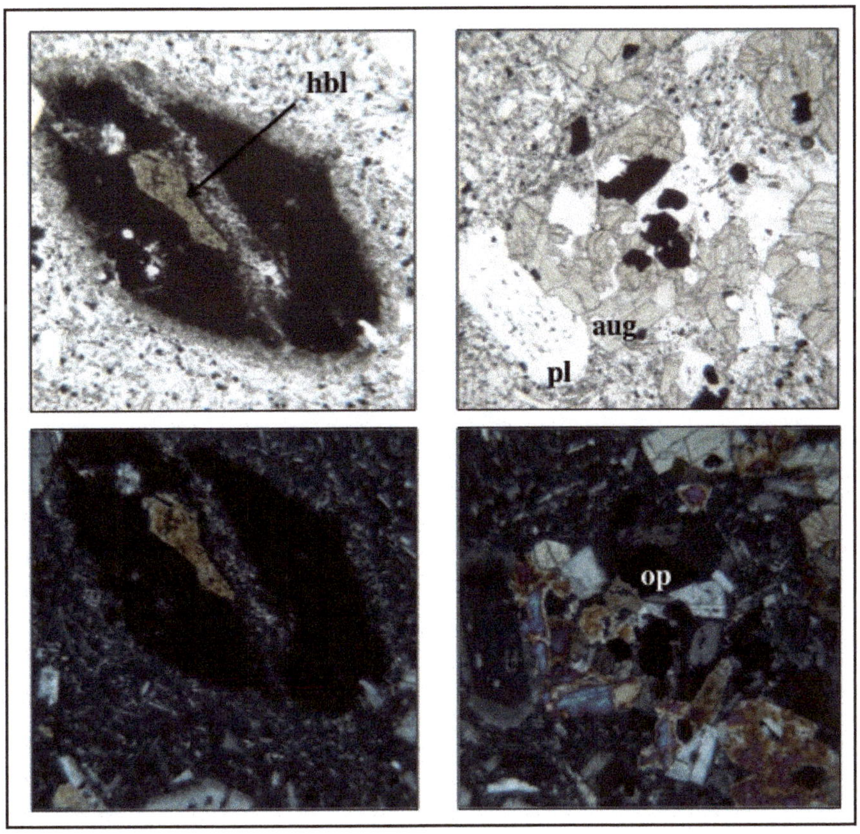

Figure 14.3. Microphotographs of sample ADK-1. (Top left) Plane polarized light (PPL) view of relict hornblende (hbl) in the center of a pseudomorph aggregate of opaque oxides and tiny clinopyroxenes. (Bottom left) Cross-polarized (XPL) view of same pseudomorph displays the inclined extinction and interference colors typical of hornblende. Plagioclase microlites are the white-gray laths surrounding the reaction rim. (Top right) Plane polarized light view of the diagnostic augite+opaques+plagioclase (aug+op+pl) crystal clots. The opaque mineral is probably a titanium-rich iron oxide. (Bottom right) The cross-polarized light view of the crystal clots shows that the augite are 1-2 mm and generally do not have good crystal form (anhedral). These are sometimes twinned and have a lower calcium rim. Field of view is 5mm in all panels.

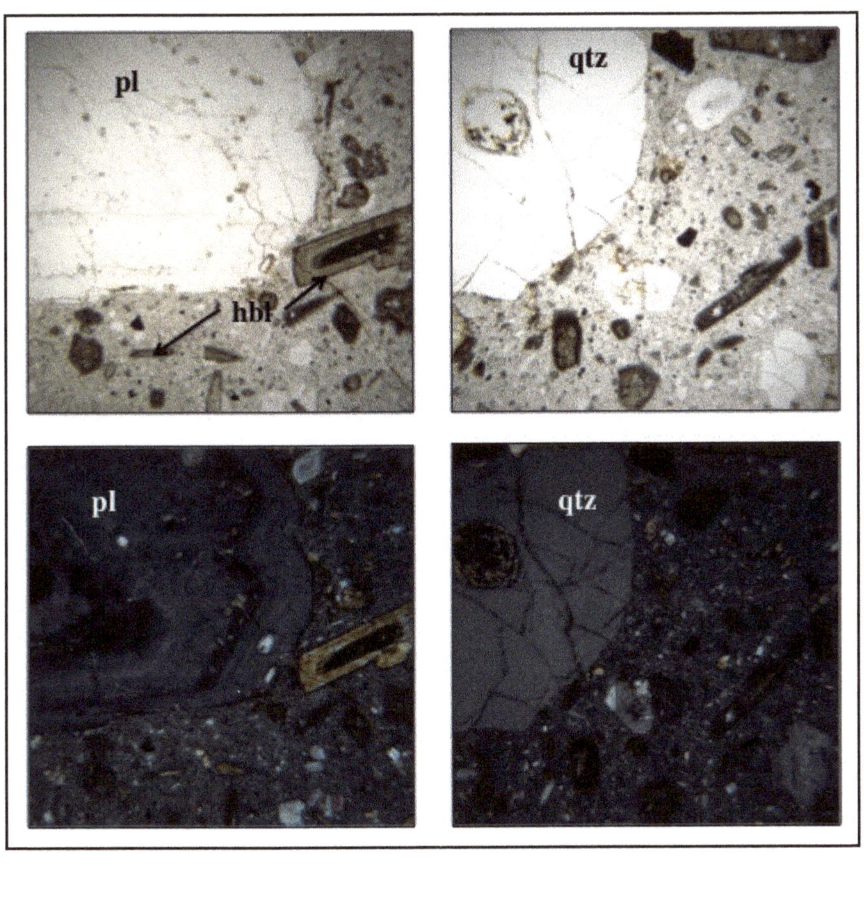

Figure 14.4. Photomicrographs of comparative sample ADK-3 (Top left) PPL view of a large plagioclase (pl) phenocryst surrounded by abundant hornblende (hbl) phenocrysts that vary in size. Although cores and mantles of some hornblende phenocrysts have a dusty opaque region, suggestive of dewatering and disequilibrium, these crystals are far more intact than in ADK-1. (Bottom left) The XPL view shows the concentric compositional zoning characteristic of plagioclase. (Top right) PPL view of a subhedral quartz (qtz) crystal with central inclusion. The lava matrix contains scattered, numerous small hornblende phenocrysts. (Bottom right) Unlike plagioclase, quartz does not have compositional zoning in XPL. This identification was confirmed by the uniaxial positive interference figure of the quartz. Field of view is 2 mm for all panels.

minor plagioclase, suggesting the hornblende are xenocrysts reheated by the host magma. These pseudomorphed and relict hornblendes and the augite+plagioclase+oxide crystal clots are extremely diagnostic features of ADK-1. Based on the macroscopic and microscopic characteristics of ADK-1, it should be called a porphyritic hornblende-bearing andesite.

Comparative sample ADK-3 is similar to ADK-1 in that it contains plagioclase and hornblende, but that is the limit of the resemblance. Notably, there is an absence of the pyroxene-plagioclase clots seen in ADK-1. Furthermore the plagioclase groups into two size populations: tiny microlites in the matrix and large phenocrysts (1-5mm). Furthermore the plagioclase phenocrysts are only 1-3% of the rock. Even more unusual, this lava has large, rare crystals of quartz though these show signs of being in disequilibrium with the host lava, which perhaps assimilated them en route to the surface. Finally the hornblende in this sample is abundant (10-15%), ranges in size (0.25-4mm length), and is far more intact than the hornblende in ADK-1 (Figure 14.4). The presence of quartz argues that this rock should be named a hornblende rhyolite, though if the quartz is xenocrystic (foreign to the magma), hornblende dacite would be a more appropriate name.

The Source of ADK-011 Slabby Griddlestones

Thin sections of four of the griddlestones (ADK-011: 1075, -1076, -1077, 1078) from the site were selected from the group identified, based on macroscopic features, as porphyritic hornblende andesite because 84% of the griddlestone slabs belong to this group (Table 14.3). About 10% of the griddlestones consisted of coarse-grained plagioclase porphyritic andesite and one of these was chosen for petrographic analysis. This sample (ADK-011:403) had a distinct, greasy charred surface. Petrographically, samples ADK-011: 1075, -1076, -1077, 1078 all have crystal clots of augite and plagioclase and have dusty pseudomorphs of amphibole created by fine-grained dusty opaques and clinopyroxene crystals. The mineralogical assemblages, size ranges, proportions, and overall texture all suggest that these slabs originated from the lava exposed north of Clam Lagoon (comparative sample ADK-1).

Sample ADK-403 is a tougher case for provenance identification. This thin section indicates pervasive alteration to clay and cracking consistent with heating in a wet environment. The original mineralogy is disturbed enough that it is unclear whether this is from ADK-3 outcrop or from some as yet unidentified site. Most interestingly, the blackened organic residue on this sample is clearly seen in thin section as coating one surface and penetrating cracks that extend from that surface into the interior. All of these features are consistent with sustained use of this rock as a griddlestone or for cooking.

In summary, we describe the myriad of flat slabs recovered from the excavation of ADK-011 as griddlestones, though not all of these show signs of charring or organic residue. Manifestly, the Aleut people who inhabited Adak-011 during 2500 years and between 400 and 170 years ago placed enough value on the function of these lava fragments, that they carried them to the site. They especially prized stones that were thin, flat and relatively fresh such that they sought the materials from a location approximately 5km distance from the site over poorer quality slabs a mere 1km away. We examined and photographed closer sites exposed in the headlands south and east of Clam Lagoon, but the jointing of these rocks was blocky, and platy slabs were not formed. We did not examine isolated outcrops offshore to the east, though these would have been accessible to the kayak based culture of the Aleut. The ADK-1 site is a beautiful source for highly functional stone. That Aleuts selected thin, platy stones from ADK-1 is probably not surprising. Thinner stones would warm and transfer heat more quickly, and would require less fuel. This might be the reason that Aleuts preferred using andesite slabs.

Organic Residues on the ADK-011 Griddlestones

Analysis of Organic Residues in Archeology

The integrity of archaeological samples, hundreds or thousands years old, can be compromised by environmental perturbations and organic degradation. Acidic soils, aerobic microbial activity, water leaching, and other natural processes can influence or even destroy archaeological samples. Consequently, macroscopic organics introduced to a prehistoric site may or may not leave a visible signal for typical analytical methods. Archaeological layers and amorphous organic residues contained on artifacts such as

Table 14.3. Characteristics and source affinity for analyzed griddlestones.

Characteristic	*ADK11-403*	*ADK11-1075*	*ADK11-1076*	*ADK11-1077*	*ADK11-1078*
Unit	E	G	F	C-D	H
Layer	2	4	2	2	4
Level	6	9b	5	1	7
Maximum length (cm)	12	23.5	27	27.5	13
Maximum width (cm)	8.5	18	19.5	15.5	9
Maximum thickness (mm)	9.8-11.2	17.3	11.5	30.5	18-21
Mass (g)	154	836	966	1695	353
Source affinity	ADK3?	ADK1	ADK1	ADK1	ADK1

stones, however impoverished of visual data, potentially hold valuable information that can be explored at the molecular level (Buonasera 2005, 2007; Evershed 1993, 2008). Using analytical chemistry applied to organic residue analysis, it is possible to recover chemical signatures from degraded archaeological samples; these can be used to identify past human activity (Barnard *et al.* 2007; Evershed, 2008).

Organic residue analysis in archeology is based on the selective preservation of biochemical molecules under a variety of environmental conditions at archeological sites. Lipid molecules are relatively resistant to decay compared to carbohydrates, proteins, and nucleotides. Because of their hydrophobic nature, lipids often remain at their original site of deposition, making them good candidates for biomarkers in archaeological research. This 'entrapment' has been shown to enhance preservation of lipids absorbed on carbonized surfaces of archeological ceramics and charred surface residues (Buonasera 2007; Evershed 2008).

A suite of analytical chemical techniques has been employed to determine the composition of these unseen 'molecular artifacts' (Barnard *et al.* 2007; Evershed 2008). Stable isotope analysis, including $\delta^{13}C$ and $\delta^{15}N$, has been widely used in diet reconstruction (Ambrose 1993). However, stable isotope analysis of specific biomolecules is more challenging. The procedure, from the extraction to separation and analysis of the biomolecules, using chromatography and mass spectrometry, is very time-consuming and impractical when analyzing large numbers of samples. Fingerprinting methods including infrared, Raman and nuclear magnetic resonance spectroscopy can provide information about the overall composition of the residues (Evershed 2008) and are more appropriate with large sample sets.

VIS/NIR Spectroscopy Applied to Archeology

Visible (VIS, 400-750 nm wavelengths) and near-infrared (NIR, 750-2500 nm) absorbance (or reflectance) spectra are the product of complex patterns of scattering and absorption by biochemical molecules. NIR compound signatures reflect the stretching and vibration by the C-H, N-H, and O-H groups composing biochemical molecules. The spectral characteristics of different compounds are unique, but they are also broad and frequently overlap, making it difficult to interpret the VIS/NIR absorbance spectra (Ciurczack 2001). However, the VIS/NIR spectrum of a sample provides a summary of its overall biochemical composition (Foley *et al.* 1998).

Assuming a direct connection between the spectral pattern and the sample biochemical composition, samples that differ biochemically also vary in their VIS/NIR absorbance characteristics. Thus, the spectral fingerprints of samples can be used to determine their biochemical similarity or dissimilarity. Multivariate analysis techniques are then required to examine overall similarity and dissimilarity among the sample biochemical compositions (Richardson *et al.* 2003). Compared with traditional methods used for organic residue analysis (Barnard *et al.* 2007; Evershed 2008), VIS/NIR spectroscopy is rapid, non-destructive, reproducible and cost-effective (Cecillon *et al.* 2009). This method has been used to predict soil properties, monitor soil quality (Cecillon *et al.* 2009), and fingerprint soil organic matter (Ertlen *et al.* 2010). It is important to remember that when VIS/NIR is used to analyze organics on stones, the stones themselves might contribute to the VIS/NIR signals (Bowitz and Ehling 2008). This can complicate the identification of the organic residues covering the stones.

Using stable isotope analysis of ^{13}C content and visible/near-infrared spectroscopic analysis, we analyzed organic residues on andesite slabs. We performed the same analyses on samples (plants, seafoods, soils, tephras) collected at ADK-011 and elsewhere on Adak Island. The results were compared to determine the biological origins of the griddlestone residues.

Collection of Samples, Their Storage and Preparation for Analysis

The griddlestones and other samples (plants, soils, seashells, fish bones, etc.) were collected at the archeological site ADK-011 and elsewhere on Adak Island (Appendix 14A). Samples were bagged and stored at room temperature until analysis. The organic residues on griddlestone surfaces were difficult to quantitatively recover. Because our analysis requires representative and homogeneous samples, we crushed and milled the stones, knowing we would 'dilute' any organic residue chemical signal. Griddlestones were coarsely crushed using a Rocklabs hydraulic crusher with a tungsten carbide crushing surface. The stones were milled in a Spex 8500 shatterbox with the tungsten carbide disk mill. All other samples were manually crushed and finely milled using an 8000D Mixer/Mill, made by SPEX SamplePrep LLC. No heat treatment was applied on any sample prior to stable isotope and VIS/NIR spectroscopy analysis in order to simulate the cooking process using the griddlestones.

Stable Isotope Analysis

The $\delta^{13}C$ of all the samples listed in Appendix 14A was performed at the Stable Isotope Mass Spectrometry Laboratory (SIMSL) located in the Division of Biology at Kansas State University. Between 1 to 10mg of finely ground material was weighed and analyzed with a ThermoFinnigan Delta Plus mass spectrometer with a CE 1110 elemental analyzer and Conflo II interface in continuous flow mode (CF-IRMS). $\delta^{13}C$ stable isotope ratios were calculated as follows:

$$\delta^{13}C_{sample}\ (‰) = [(R_{sample} / R_{PDB})-1]\times 1000$$

Table 14.4. Summary of the $\delta^{13}C$ and % C for the samples used in this study.

Type of Material	*Number of samples per type*	$\delta^{13}C$	*%C*
		Average ± Confidence Interval (95%)	
Griddlestone from quarry	1	-25.90	0.07
Grddlestones with carbonized residues	8	-24.89 ± 0.23	0.25 ± 0.03
Seafoods (modern and archeological)	8	-4.03 ± 1.27	2.34 ± 0.06
Tephras	8	-27.20 ± 0.13	1.48 ± 0.13
Plants	41	-26.23 ± 0.38	39.87 ± 0.81
Soils	83	-25.83 ± 0.12	5.60 ± 0.36

where PDB stands for Pee Dee Belemnite (PDB) standard (see Savinetsky *et al.* Chapter 5 this volume).

Description and Discussion of the $\delta^{13}C$ Stable Isotope Data

The $\delta^{13}C$ signatures of the residues of organic materials that were heated, cooked or burned should be similar to the signatures of the original materials (Ambrose, 1993). Burning is not a mechanism that will selectively enrich or deplete ^{13}C in a given material. Therefore, we would expect that the ^{13}C signature of the griddlestone organic residues should be similar to whatever came into contact with the stone (Ambrose 1993). However, the temperature of the fire during the cooking/heating may determine the nature of the compounds that are deposited on the stones; some compounds may burn more or less easily than others and, thus, have been deposited on the stone.

More than 140 samples were analyzed to determine their $\delta^{13}C$ ratios (Appendix 14A). Aleuts consumed fish and invertebrates (Crockford Chapter 6 this volume; Savinetsky *et al.* Chapter 5 this volume; West *et al.* Chapter 10 this volume); therefore, we included ancient and modern examples of these for our analysis. We analyzed Adak soils, tephras, and plants that might have come into contact with the stones. We also included an andesite stone from the natural outcrop ADK-1. The dataset of $\delta^{13}C$ and %C values presented in this paper should be a useful reference for future archeological work in the Aleutian Islands.

We regrouped the samples by type (Appendix 14A and Table14.4; Figures 14.5, 14.6, 14.7). Figure 14.5 shows a biplot of $\delta^{13}C$ and %C values detailed in Appendix 14A. Figure 14.6 is an enlargement of the region from -30 to -20‰ for $\delta^{13}C$ and from 0 to 5 %C. The group 'seafood' is the most heterogeneous ranging from archeological cod to modern clam. Seafood $\delta^{13}C$ ratios widely vary from -16.64 to -0.34 ‰ and %C data range from 1.90 to 2.70. Griddlestones with organic residues $\delta^{13}C$ ratios range from -25.78 to -23.04 ‰ and from 0.10 to 0.50 for %C. The platy andesite recovered at the quarry has the lowest %C (0.07) and a $\delta^{13}C$ ratio of -25.90 ‰. Terrestrial plant $\delta^{13}C$ signatures vary from -31.40 to -14.94 ‰ and from 14.63 to 50.80 for %C. Soil samples recovered at archaeological site ADK-011 and nearby show a range of $\delta^{13}C$ from -27.62 to -13.63 and a range of %C from 0.13 to 36.49. Finally, the values for the tephras vary from -28.13 to -24.64 ‰ for $\delta^{13}C$ and from 0.52 to 2.20 %C.

The analysis of the confidence interval, at 95%, (Figure 14.7 and Table 14.4) for the $\delta^{13}C$ data obtained from each group demonstrated clearly that the values from plants, soils and platy andesite overlap, but the ones from the griddlestones with organic residues do not. $\delta^{13}C$ data from plants and trephas do not overlap with any other group and are found at the extremes of the data range (toward more positive $\delta^{13}C$ for plants and toward more negative values for tephras).

The andesite slab collected from the ADK-1 outcrop has nearly the same $\delta^{13}C$ signature as the stones covered with organic residues. A larger quarry stone sample would have been useful to consider the $\delta^{13}C$ variability in these natural stones. In relation to the organic residues, important precautions (gloves were used to collect and sample the slab) were taken to prevent any exogenous carbon from contacting the samples. No apparent organic residue could be seen on the quarry stone, and volcanic rock does not contain carbonates. However, the observed ^{13}C must come from the quarry itself. Because the slab is from an open air quarry, we cannot rule out that the natural andesite was exposed to: (1) leaching organic matter from the surrounding soils and vegetation, or (2) carbon carried in by ocean sprays. Yasuhara and Kazahaya (1999, 156) note that, 'The partial pressure of CO_2 in the soil atmosphere… is generally 2-3 orders of magnitude greater than that of the atmosphere, presumably as a result of plant root respiration and decay in the soil by bacterial activities.' Therefore, it might not be surprising that the noncultural andesite slab contains carbon signatures. Although the $\delta^{13}C$ signatures of all andesite samples are very similar, the % C in the quarry stone is to 3 to 20 times lower than the values from the griddlestones containing residues and the tephras (Appendix 14A and Table 14.4). The more positive $\delta^{13}C$ values of griddlestone samples falls on a mixing line between the more negative value of the quarry stone and the less negative values of average seafood, suggesting that both contribute to the analyzed samples from the ADK-011.

The $\delta^{13}C$ signatures found on the griddlestones could be attributed to soils, decaying plants, and/or tephras associated with the archaeological site along with a contribution of seafood $\delta^{13}C$ signatures, through natural processes of carbon distribution including leaching and deposition. Rather than

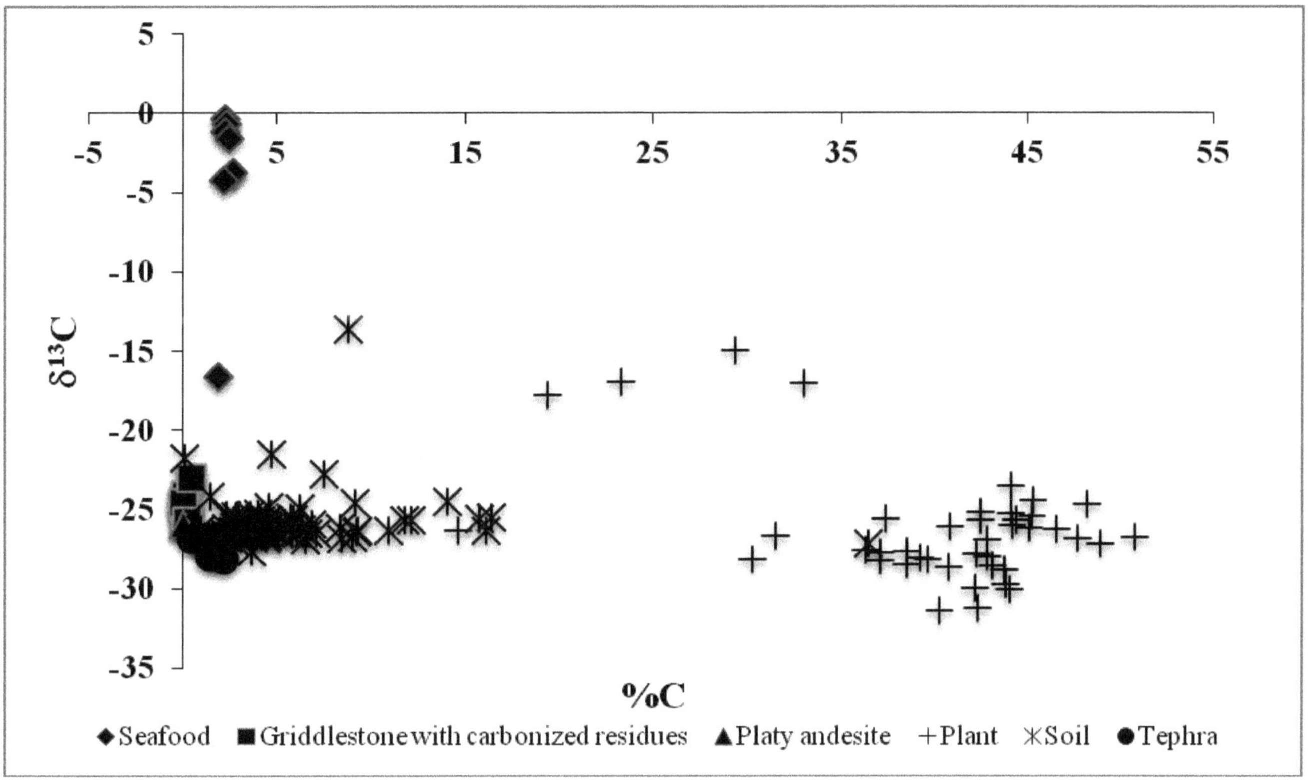

Figure 14.5. Biplot of $\delta^{13}C$ and %C results for the griddlestones with carbonized residues compared with other site samples (described in Appendix 14A).

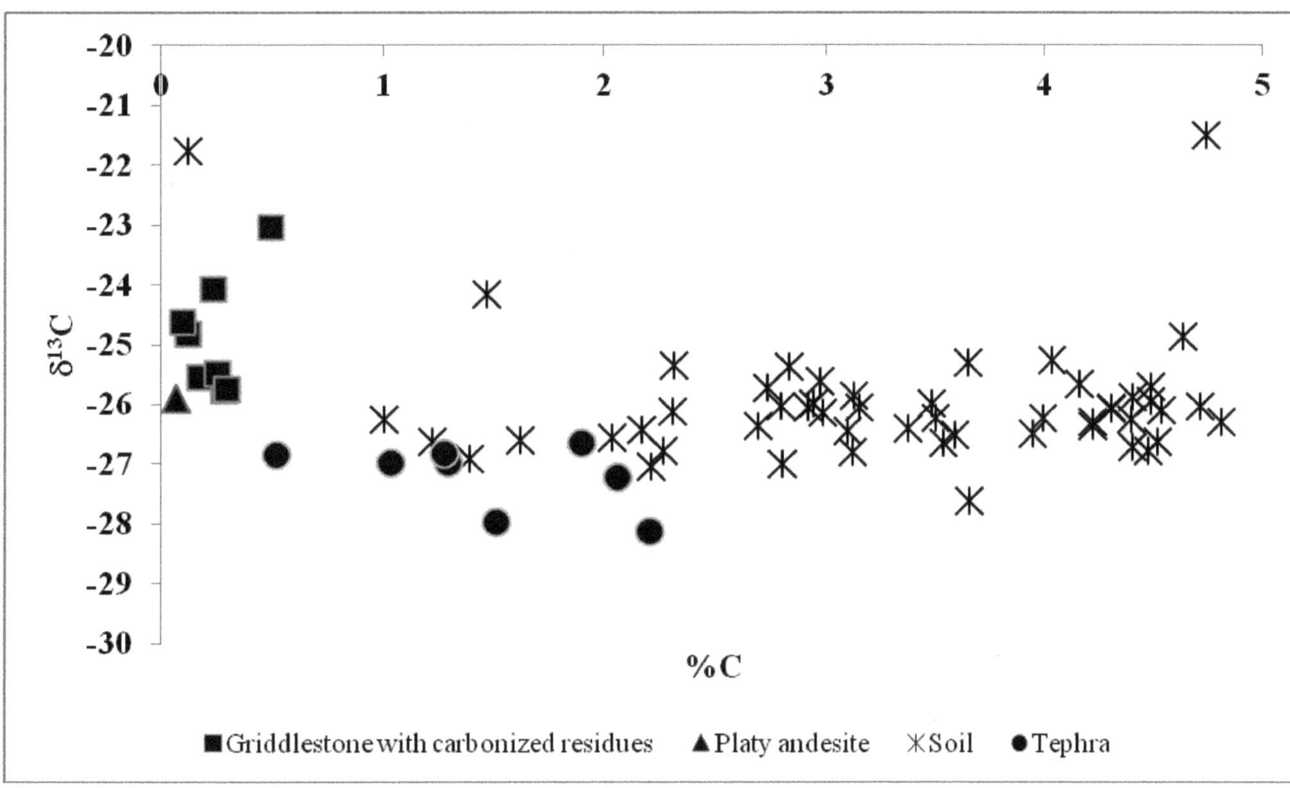

Figure 14.6. Enlarged version of Figure 14.5 in the area corresponding to -30 to -20 ‰ of $\delta^{13}C$ and from 0 to 5 %C.

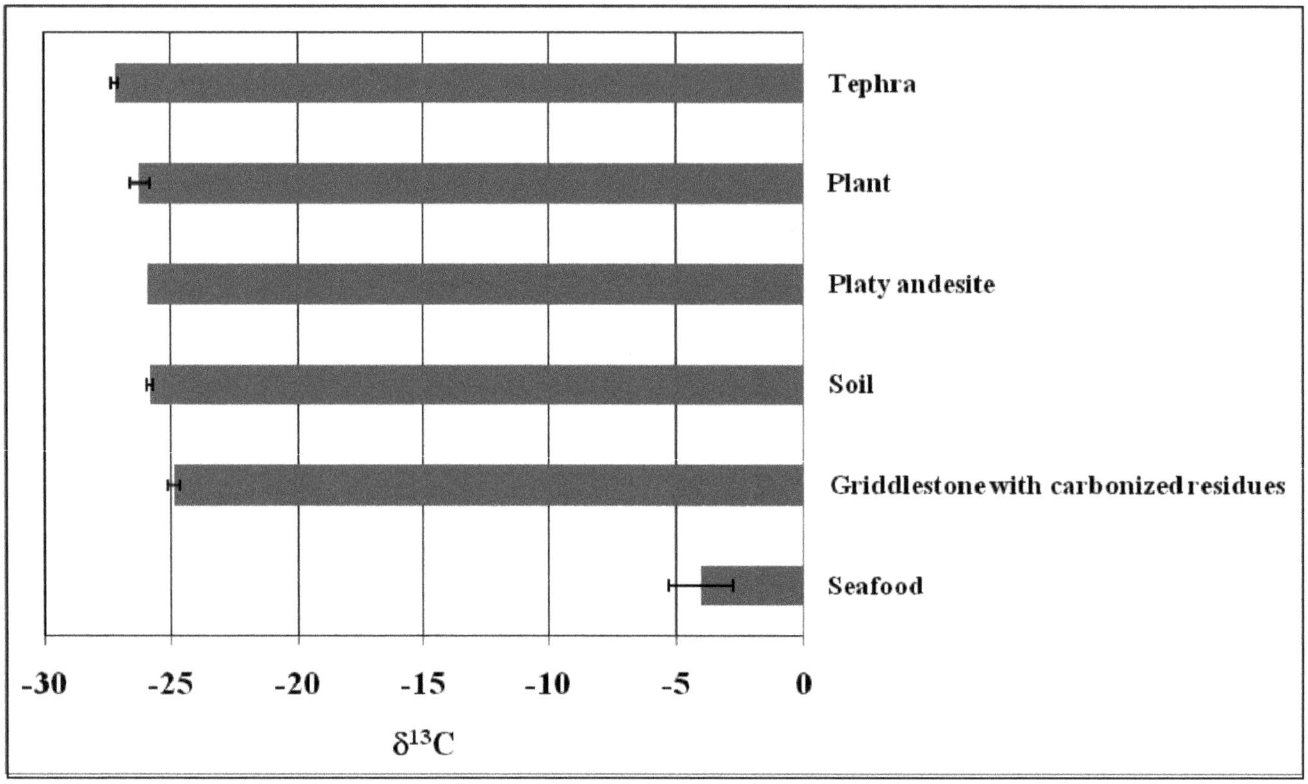

Figure 14.7. Summary of the $\delta^{13}C$ isotopic results showing 95% confidence interval of the mean.

representing organic residues from food processing, the slab residues could represent the combustion of the material used as a fuel source. Arkady Savinetsky and colleagues, in ongoing research, have determined that some of the driftwood in the central and western Aleutians is spruce (*Picea* sp.). The comparatively large amount of ADK-011 charcoal could belong to this species. We did not test how the burned residues found on the ADK-011 andesite slabs biochemically compares with spruce or charcoal recovered at ADK-011. This study will be performed and published in the future.

In conclusion, the $\delta^{13}C$ signatures of the griddlestone residues originate from four possible sources: (1) the driftwood fuel used to heat the stones, (2) roots in the archaeological site, (3) other decayed archaeological site matrix, and/or (4) burned organic residues associated with the 40 Year ash.

Organic Residue Analysis by VIS/NIR Spectroscopy

The chemical fingerprints of the griddlestone organic residues were also investigated using VIS/NIR spectroscopy. VIS/NIR spectroscopy allowed us to obtain a fingerprint related to the chemical composition of each sample. We used this technique to demonstrate its potential in organic residue analysis applied to paleodiet reconstruction.

NIR Spectra were collected for all the samples listed in Appendix 14A using a QualitySpec Pro spectrometer (ASD Inc, Boulder, CO). The spectrometer measures absorbance from 350nm to 2500nm using silicon and indium-gallium-arsenide sensors. An HL-2000 halogen light source (Mikropack, Ostfildern, Germany) was used for illumination. For each sample, a 1.7mL microcentrifuge tube was used to measure samples for scanning. In cases of small sample size, the sample was placed in the center of the 3.2cm diameter container to maintain a comparable depth with larger samples. The distance from the probe (8mm) was maintained for all samples. The instrument was set to collect 20 absorbance spectra that were stored as an average spectrum. ASD software RS3 version 3.1 was used to collect all spectra. RS3 output was converted to an Excel (.xls) file using ASD ViewSpecPro software. No pre-processing was applied to the spectra prior to statistical analysis. Multivariate statistical analysis was performed with Systat software (Version 9.1, Systat Software Inc., Richmond, CA, USA). VIS/NIR spectra of the samples were used as inputs in the principal component analysis (PCA). PCA was used to explore the dataset and assess the spectral differences among the various samples.

An average spectrum per group was calculated, and these average spectra are compared in Figure 14.8. Clearly, each group has distinct spectra. The average spectrum from the griddlestone with organic residues is only slightly different from the spectrum of the andesite from the quarry. The main difference is in the visible zone of the spectrum. VIS/NIR spectra reveal that the signatures of the organic residues on the stones were different from VIS/NIR signatures of all other analyzed materials. These spectral differences

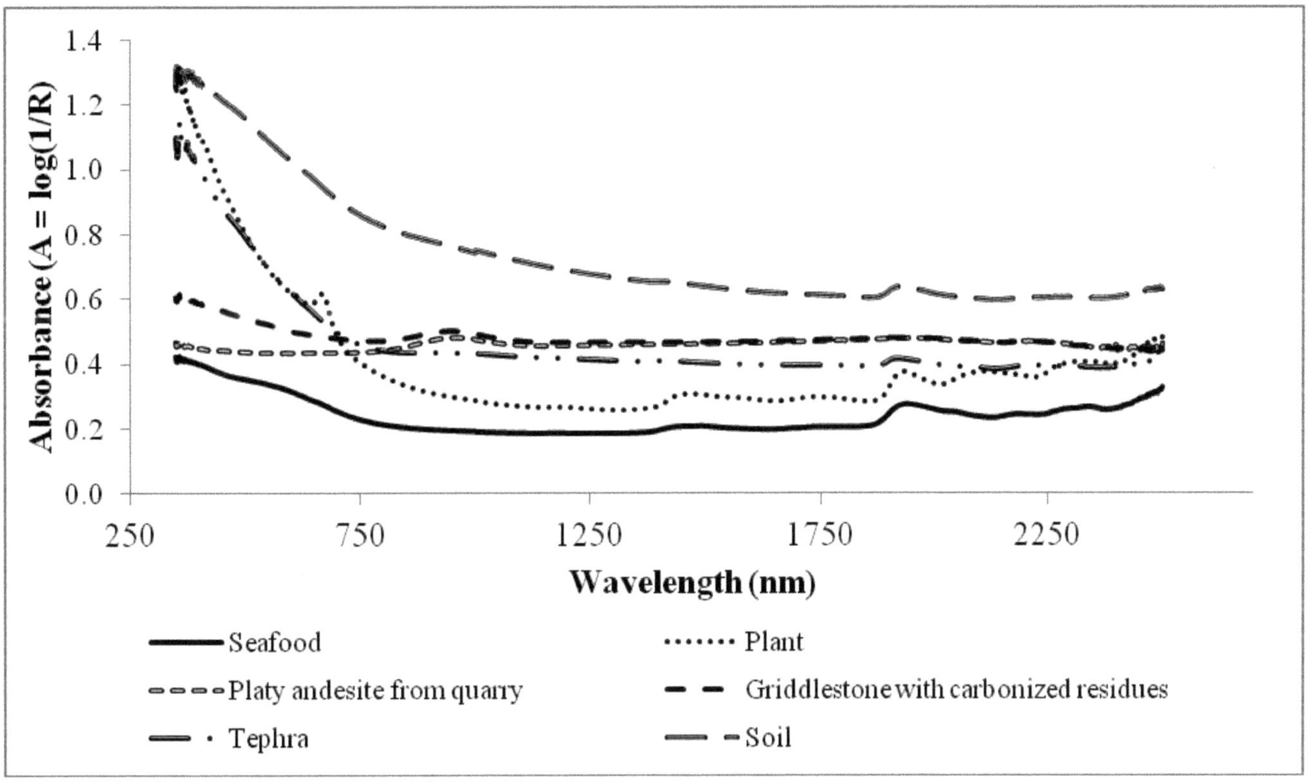

Figure 14.8. Visible/near infrared (VIS/NIR) absorbance spectra, from 350 to 2500 nm, of dried and ground materials. For each sample described in Appendix 14A, a VIS/NIR spectrum was acquired. Spectra represent the mean for each of the categories defined in Appendix 14A.

might be due to differences in chemical composition and/or abundance of compounds.

Principal components analysis reduces the number of variables in complex datasets by calculating mathematical variables, the principal components, which are a combination of the original variables (Richardson *et al.* 2003). Five factors explain more than 99% of the variation of the data. The first two components, displayed in Figure 14.9, explain more than 98%. PC1 (88.4%) separates the samples in four main groups along its axis: soil, griddlestones (with/without organic residues)-tephras-plants, and seafoods. PC2 (9.7%) separates the group griddlestones-tephras-plants along its axis. Griddlestones, tephras, and plants share spectral similarities. However, griddlestones are not similar to either seafoods or soils. Figure 14.8 shows that the raw spectra of the griddlestones with residues supported similar features, but also differences, with the spectrum of the quarry stone. PCA confirms this observation (Figure 14.9), but more spectra from the stones without organic residues would allow us to clearly demonstrate the differences.

In conclusion: (1) chemical fingerprints were recovered, using VIS/NIR spectroscopy, from degraded samples collected in Aleutian archaeological sites; (2) VIS/NIR spectroscopy, coupled with principal components analysis, enabled differentiation of the various stones and biological samples obtained; (3) the sources of the chemical fingerprints on the griddlestones could be multiple, as hypothesized by the $\delta^{13}C$ stable isotope data, and its complexity is clearly demonstrated by the principal components analysis. Stable isotope signatures are not altered by cooking/burning; however, chemical composition is. The chemical fingerprints from the candidate Aleut foods were taken from uncooked materials. We may need to simulate the heating of the quarry stones in the presence of the hypothesized potential sources of food and fuel; (4) The chemical fingerprints on the stones are definitively different from fingerprints of the soils; and (5) a more detailed analysis using lipid biomarkers and their compound-specific isotopic ratios can be used to test the hypotheses raised in this chapter. Finally, VIS/NIR spectroscopy is a potentially useful tool in archaeology, both in field and laboratory analysis; it is relatively inexpensive, fast, nondestructive, easily portable, and can collect spectra on a wide variety of materials.

Conclusions

Griddlestones are nearly ubiquitous in Aleutian archaeological sites. As stone artifacts, their transport, use and discard provide valuable information about past Aleut activities. Here we attempted to source and perform biochemical analyses on griddlestones recovered at archaeological site ADK-011. We wanted to see how far prehistoric Aleuts were willing to venture to collect a suitable "frying pan". Our results show that Aleuts were as

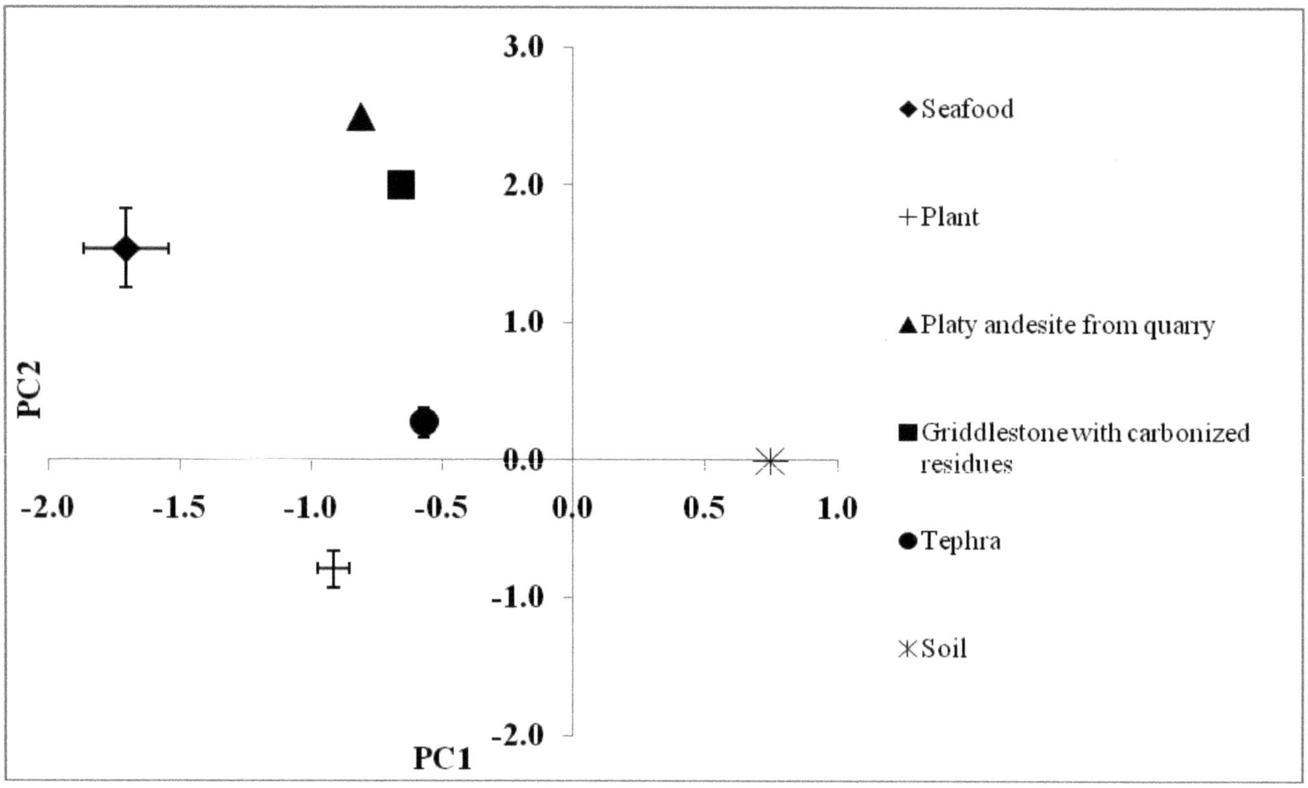

Figure 14.9. Biplot of the first two principal components derived from the VIS/NIR absorbance spectra of the samples analyzed. The principal component analysis was computed using the original and untransformed data. The two first components explain more than 98% of the total variance of the dataset. The samples used in the analysis are described in Appendix 14A.

somewhat selective in their cooking utensils as they were in choosing stones for other hunting/processing tasks. Aleuts at ADK-011 opted to select thin, andesite slabs from a large outcrop 5km from Clam Lagoon although flat stones of lesser quality were located less than 1km from their village. The superior quality of the andesite slabs at ADK-1 prevailed over stones of lesser quality at an outcrop near Candlestick Bridge.

The function of the ADK-011 griddlestones remains elusive. Although we are convinced by the ethnographic literature that these flat stones were used to heat or even cook something, what that 'something' was remains unclear. Stable isotope signatures of greasy residues on the stones apparently represent either the driftwood or grasses used to heat the stones, leaching of ^{13}C rich groundwater, or the decaying plant/soil matrix in which the stones were embedded at the archaeological site as well as possible contribution from seafood signatures. These humble, but functional, stones were probably used in multiple ways—to radiate heat and/or cook foods. More analyses are necessary to begin to tease out the chemical signatures on their surfaces.

Acknowledgements

This research was made possible by a Research Education for Undergraduates (REU) grant awarded as a supplement to the University of Kansas by the Office of Polar Programs, Arctic Social Sciences Division, National Science Foundation (Award # 0353065). Nicolaysen thanks Andrea Seymour for assistance in preparing the map figure. Jeannotte thanks Libin Yao (Division of Biology, KSU) and Elizabeth Maghirang (USDA) for their help in the preparation of samples and acquisition of VIS/NIR spectra. Jeannotte thanks also Dr. John M. Blair (Division of Biology, KSU) and Dr. Matthew E. Brueseke (Department of Geology, KSU) for the use of equipment for milling and grinding samples in their laboratories. Jeannotte was supported by the Functional Genomics Consortium initiative of Kansas State University's Targeted Excellence program. Kansas Lipidomics Research Center was supported by National Science Foundation (EPS 0236913, MCB 0455318, DBI 0521587), Kansas Technology Enterprise Corporation, K-IDeA Networks of Biomedical Research Excellence (INBRE) of National Institute of Health (P20RR16475), and Kansas State University.

References

Alden, J., and D. Bruce. 1989. *Growth of Historical Sitka Spruce Plantations at Unalaska Bay, Alaska.* US Department of Agriculture, Forest Service General Technical Report PNW-GTR-236, 19 pages.

Ambrose, Stanley H. 1993. Isotopic Analysis of Paleodiets. Methodological and Interpretive Considerations. In Mary K. Sandford (ed.), *Investigations of Ancient Human*

Tissue. Chemical Analyses in Anthropology, 60-130. Langhorne, Gordon and Breach Science Publishers.

Barnard, H., S. H. Ambrose, D. E. Beehr, M. D. Forster, R. E. Lanehart, M. E. Malainey, R. E. Parr, M. Rider, C. Solazzo, and R. M. Yohe II. 2007. Mixed Results of Seven Methods for Organic Residue Analysis Applied to one Vessel with the Residue of a Known Foodstuff. *Journal of Archaeological Science* 34(1), 28-37.

Black, Lydia T. 1984. *Atka: An Ethnohistory of the Western Aleutians*. Kingston, Limestone Press.

Bouchet, François, Dixie West, Christine Lefèvre, and Debra G. Corbett. 2001. Identification of Parasitoses in a Child Burial from Adak Island (Central Aleutian Islands, Alaska). *Comptes Rendus Academy of Sciences* 324, 123-127.

Bowitz, Jörg, and Angela Ehling. 2008. Non-destructive Infrared Analyses: A Method for Provenance Analyses of Sandstones. *Environmental Geology* 56 (3-4), 623–630.

Browne, B. L., Gardner, J. E., and J. Larsen. 2003. Amphibole Reaction Rims in Response to Decompression Compared to Heating: An Experimental Approach. *Eos Transactions of the American Geophysical Union* 84, V#12A-551.

Buonasera, Tammy. 2005. Fatty Acid Analysis of Prehistoric Burned Rocks: A Case Study from Central California. *Journal of Archaeological Science* 32(6), 957-965.

Buonasera, Tammy. 2007. Investigating the Presence of Ancient Absorbed Organic Residues in Groundstone Using GC-MS and Other Analytical Techniques: A Residue Study of Several Prehistoric Milling Tools from Central California. *Journal of Archaeological Science* 34(9), 1379-1390.

Cecillon, L., B. G. Barthes, C. Gomez, D. Ertlen, V. Genot, M. Hedde, A. Stevens, and J. J. Brun. 2009. Assessment and Monitoring of Soil Quality Using Near-infrared Reflectance Spectroscopy (NIRS). *European Journal of Soil Science* 60(5), 770-784.

Ciurczack, Emil W. 2001. Principles of Near Infrared Spectroscopy, in Donald A. Burns and Emil W. Ciurczack (eds), *Handbook of Near-Infrared Analysis*, 7-18. New York, Marcel Dekker.

Corbett, Debra G., Douglas Causey, Mark Clementz, Paul L. Koch, Angela Doroff, Christine Lefèvre, and Dixie West. 2008. Aleut Hunters, Sea Otters, and Sea Cows. In C. Rick Torben and Jon M. Erlandson (eds), *Human Impacts on Ancient Marine Ecosystems: A Global Perspective*, 43-75. Berkeley, University of California Press.

Cordonnier, B., K. U. Hess, Y. Lavallee, and D.B. Dingwell. 2009. Rheological Properties of Dôme Lavas: Case Study of Unzen Volcano. *Earth and Planetary Science Letters* 279(3-4), 263-272.

Crockford, Susan. 2011. Archaeozoology of Adak Island: 6000 Years of Subsistence History in the Central Aleutians. Chapter 6 in D. West, V. Hatfield, E. Wilmerding, C. Lefèvre and L. Gualtieri (eds.), *The People Before: The Geology, Paleoecology and Archaeology of Adak Island*, Alaska. Oxford, British Archaeological Reports.

Dumond, Don E. 2001. *Archaeology in the Aleut Zone of Alaska: Some Recent Research.* Eugene, University of Oregon Press.

Ertlen, Damien, Dominique Schwartz, Martine Trautmann, Richard Webster, and Didier Brunet. 2010. Discriminating Between Organic Matter in Soil from Grass and Forest by Near-infrared Spectroscopy. *European Journal of Soil Science* 61(2), 207- 216.

Evershed, Richard P. 1993. Biomolecular Archaeology and Lipids. *World Archaeology* 25(1), 74-93.

Evershed, Richard P. 2008. Organic Residue Analysis in Archeology: The Archaeological Biomarker Revolution. *Archaeometry* 50 (6), 895-924.

Foley, William J. 1998. Ecological Applications of Near Infrared Reflectance Spectroscopy - A Tool for Rapid, Cost- effective Prediction of the Composition of Plant and Animal Tissues and Aspects of Animal Performance. *Oecologia* 116(3), 293-305.

Fuller, R. E. 1938. Deuteric Alteration Controlled by the Jointing of Lavas. *American Journal of Science* 35(207), 161- 166.

Hrdlička, Aleš. 1945. *The Aleutian and Commander Islands and Their Inhabitants*. Philadelphia, The Wistar Institute of Anatomy and Biology.

Jochelson, Waldemar. 1925. *Archaeological Investigations in the Aleutian Islands*. Washington, Carnegie Institution of Washington.

Kelemen, P. B., G. M. Yogodzinski, and D. W. Scholl. 2003. Along-strike Variation in Lavas of the Aleutian Island Arc: Implications for the Genesis of High Mg# Andesite and the Continental Crust. In J. Eiler (ed.), *Inside the Subduction Factory*, 293-311. Washington, American Geophysical Union Monograph.

Knecht, Richard A., Richard S. Davis, and Gary A. Carver. 2001. The Margaret Bay Site and Eastern Aleutian Prehistory. In D. E. Dumond (ed.), *Archaeology in the Aleut Zone of Alaska*, 35-69. Eugene, University of Oregon Press.

Knecht, Richard, and Richard Davis. 2004. A*maknak Bridge Site Data Recovery Project Final Report.* South Channel Bridge, Project No. MGS-BR-0310(S)/52930. Report to the Alaska Department of Transportation and Public Facilities, Anchorage. Museum of the Aleutians and Bryn Mawr College.

Maschner, Herbert D. G., and James W. Jordan. 2001. The Russell Creek Manifestation of the Arctic Small Tool Tradition on the Western Alaska Peninsula. In D. E. Dumond (ed.), *Archaeology in the Aleut Zone of Alaska*, 151-172. Eugene, University of Oregon Press.

Richardson, Andrew D., James B. Reeves III, and Timothy G. Gregoire. 2003. Multivariate Analyses of Visible/Near Infrared (VIS/NIR) Absorbance Spectra Reveal Underlying Spectral Differences Among Dried, Ground Conifer Needle Samples from Different Growth Environments. *New Phytologist* 161(1), 291-301.

Rutherford, M. J., and P. M. Hill. 1993. Magma Ascent Rates from Amphibole Breakdown: An Experimental Study Applied to the 1980-1986 Mount St. Helens

Eruptions. *Journal of Geophysical Research* 98(B11), 19667-19685.

Savinetsky, Arkady B., Dixie West, Zhanna A. Antipushina, Bulat F. Khassanov, Nina Kiseleva, Olga Krylovich and Andrei M. Perelodov. 2011. The Reconstruction of Ecosystem History of Adak Island (Aleutian Islands) During the Holocene. Chapter 5 in D. West, V. Hatfield, E. Wilmerding, C. Lefèvre and L. Gualtieri (eds.), *The People Before: The Geology, Paleoecology and Archaeology of Adak Island*, Alaska. Oxford, British Archaeological Reports.

Smith, J. V. 1996. Ductile-brittle Transition Structures in the Basal Shear Zone of a Rhyolite Lava Flow, Eastern Australia. *Journal of Volcanology and Geothermal Research* 72(34), 217-223.

Ui, T., S. Takarada, and M. Yoshimoto. 2000. Debris Avalanches. In H. Sigurdsson (ed.), *Encyclopedia of Volcanoes*, 617-626. San Diego, Academic Press.

Veltre, Douglas, and Allen McCartney. 2001. Ethnohistorical Archaeology at the Reese Bay Site, Unalaska Island. In D. E. Dumond (ed.), *Archaeology in the Aleut Zone of Alaska*, 87-104. Eugene, University of Oregon Press.

Veniaminov, Ivan. 1984. *Notes on the Islands of the Unalaska District*. Lydia Black and R.H. Geoghegan [trans]. Kingston, The Limestone Press.

Waythomas, C. F. 1995. Surficial Geologic Map of Northern Adak Island, Alaska 1:50,000. US Geological Survey Open-File Report 95-128.

Webb, S. L., and D. B. Dingwell. 1990. Non-Newtonian Rheology of Igneous Melts at High Stresses and Strain Rates: Experimental Results for Rhyolite, Andesite, Basalt and Nephelinite. *Journal of Geophysical Research* 95(B10), 15695-15701.

West, Dixie, Debra G. Corbett, and Christine Lefèvre. 1997. NSF Grant Report: The Western Aleutians Archaeological and Paleobiological Project. Washington, DC, National Science Foundation.

West, Dixie, Debra G. Corbett, and Christine Lefèvre. 2001. NSF Grant Report: The Western Aleutians Archaeological and Paleobiological Project. Washington, DC, National Science Foundation.

West, Dixie, Debra G. Corbett, and Christine Lefèvre. 2002. NSF Grant Report: The Western Aleutians Archaeological and Paleobiological Project. Washington, DC, National Science Foundation.

West, Dixie, Zhanna Antipushina, Arkady B. Savinestsky, and Olga Krylovich. 2011. Invertebrate Remains: A Pan-Aleutian Comparison. Chapter 10 in D. West, V. Hatfield, E. Wilmerding, C. Lefèvre and L. Gualtieri (eds.), *The People Before: The Geology, Paleoecology and Archaeology of Adak Island*, Alaska. Oxford, British Archaeological Reports.

Wilmerding, Elizabeth and Virginia Hatfield. 2011. Six Thousand Years of Lithic Technology on Adak in a Broader Aleutian Context. Chapter 12 in D. West, V. Hatfield, E. Wilmerding, C. Lefèvre and L. Gualtieri (eds.), *The People Before: The Geology, Paleoecology and Archaeology of Adak Island, Alaska*. Oxford, British Archaeological Reports.

Yasuhara, Masaya and Kohei Kazahaya. 1999. A Stable Carbon Isotope Study on Direct Precipitation Recharge to the Aquifer in the City of Yamagata, Northern Japan. *Impacts of Urban Growth on Surface Water and Groundwater Quality*. Proceedings of IUGG Symposium. Birmingham, England.

Appendix 14A. Samples used in the study and their $\delta^{13}C$ values.

Type of Material	*Location/Site/Provenance*	$\delta^{13}C$	*%C*
Seafood_Barnacle	Reef below ADK-011	-0.34	2.266
Seafood_Blue mussel	Adak Island;Clam Lagoon	-3.82	2.614
Seafood_Blue mussel (Archeological)	ADK-011; House 1; Unit E; Layer 2; Level 2	-0.71	2.348
Seafood_Blue mussel with barnacles	Adak Island;Clam Lagoon	-3.77	2.695
Seafood_Butter clam shell (Modern)	Adak Island;Clam Lagoon	-1.18	2.292
Seafood_Cod bone (Archeological)	ADK-011; Unit G; Layer 2; Level 2	-16.64	1.901
Seafood_Razor Clam (Modern)	Kuluk Bay	-1.58	2.455
Seafood_Sea Urchin (Modern)	Adak Island; Bering Sea	-4.21	2.176
Grddlestones with carbonized residues	ADK-011; House 1; Unit E; Layer 2; Level 3	-24.08	0.236
Grddlestones with carbonized residues	ADK-011; House 1; Unit E; Layer 2; Level 5	-25.54	0.176
Grddlestones with carbonized residues	ADK-011; House 1; Unit H	-24.83	0.121
Grddlestones with carbonized residues	ADK-011; House 1; Unit H	-25.78	0.285
Grddlestones with carbonized residues	ADK-011; House 1; Unit H	-25.48	0.257
Grddlestones with carbonized residues	ADK-011; House 1; Unit H	-24.62	0.102
Grddlestones with carbonized residues	ADK-011; House 1; Unit G; Layer 2; Level 3	-23.04	0.500
Grddlestones with carbonized residues	ADK-011; House 1; Unit F; Layer 2; Level 2	-25.74	0.296
Platy andesite from quarry	ADK-011; Outcrop-Quarry	-25.90	0.071
Plant_Ascophyllum	Kuluk Bay	-17.01	32.981
Plant_Beach greens	Candlestick Bridge	-26.07	40.859
Plant_Beach greens	Beach below ADK-011	-27.57	36.293
Plant_Beach greens	ADK-0 below ADK-011	-28.49	38.561
Plant_Cotton grass	Path to ADK-011	-25.64	44.395
Plant_Cow parsnip	ADK-011	-27.83	42.318
Plant_Cow parsnip	ADK-011	-28.64	40.768
Plant_Cow parsnip roots	ADK-011	-27.99	42.842
Plant_Cow parsnip seeds	ADK-011	-26.75	50.796
Plant_Cranberry	Clam Lagoon	-26.20	45.137
Plant_Crowberry	Clam Lagoon	-26.82	47.746
Plant_Dock	Finger Bay	-25.58	37.411
Plant_Elymus	Zeto Point	-25.21	44.144
Plant_Elymus	Sand dune site	-25.44	45.288
Plant_Elymus	Sand dune site	-23.47	44.118
Plant_Elymus seeds	ADK-011	-24.38	45.363
Plant_Elymus stem	ADK-011	-26.94	42.832
Plant_Equisetum	Clam Lagoon	-27.13	48.993
Plant_Equisetum	ADK-011	-27.64	38.529
Plant_Fern	Finger Bay	-28.53	43.114
Plant_Fern	ADK-011	-24.67	48.246
Plant_Fern	ADK-011	-25.96	44.218
Plant_Fern	ADK-011	-30.09	44.042
Plant_Fern	ADK-011	-26.27	46.650
Plant_Grass	Finger Bay	-30.00	42.254
Plant_Grass	ADK-011	-25.15	42.490
Plant_Kelp	Kuluk Bay	-17.79	19.439
Plant_Kelp	ADK-011	-16.93	23.320
Plant_Lupine	Finger Bay	-25.68	42.509
Plant_Moss	Clam Lagoon	-26.69	31.473
Plant_Moss	Shagak Bay	-26.32	14.629
Plant_Moss	ADK-011	-27.75	37.073
Plant_Moss	ADK-011	-28.16	39.635
Plant_Purple flowers	ADK-011	-31.40	40.283
Plant_Ulva	Candlestick Bridge	-14.94	29.355

Type of Material	*Location/Site/Provenance*	$\delta^{13}C$	*%C*
Plant_Watermelon Berry	ADK-011	-28.12	30.311
Plant_Watermelon Berry	ADK-011	-29.71	43.888
Plant_Wild rhubarb	Finger Bay	-28.06	39.215
Plant_Wild strawberry	Path to ADK-011	-28.23	37.122
Plant_Yarrow	Clam Lagoon	-28.81	43.797
Plant_Yarrow	ADK-011	-31.23	42.354
Soil	NA	-26.52	5.138
Soil	NA	-26.89	6.571
Soil	NA	-26.86	8.452
Soil	NA	-26.51	7.095
Soil	NA	-27.62	3.660
Soil	ADK-011; Unit H; Layer 1	-26.46	3.098
Soil	ADK-011; Unit G; Layer 2	-26.24	6.350
Soil	ADK-011; Unit G; Floor 1	-26.56	5.806
Soil	ADK-011; Unit H; Layer 3 east; Level 4	-25.63	12.182
Soil	ADK-011; Unit G; Layer 3 east; Level 5	-25.98	5.253
Soil	ADK-011; Unit G; Layer 3	-26.79	4.481
Soil	ADK-011; Unit H; Layer 3 east; Level 5	-26.19	5.721
Soil	ADK-011; Unit H; Layer 3; Level 5	-26.37	2.696
Soil	ADK-011; Unit H; Layer 4; Level 7	-25.31	3.652
Soil	ADK-011; Unit H; Layer 4; Level 8	-25.35	2.315
Soil	ADK-011; Unit H; Layer 4; Level 9	-25.49	5.730
Soil	ADK-011; Unit G; Layer 4; Level 9	-24.60	9.152
Soil	ADK-011; Unit G; Layer 4; Level 9	-24.50	14.067
Soil	ADK-011; Unit H; Layer 4; Level 10	-24.16	1.468
Soil	ADK-011; Unit J	-25.87	4.412
Soil	ADK-011; Unit J; Layer 2; Level 4	-25.65	11.798
Soil	ADK-011; Unit G; West wall	-25.60	5.864
Soil	ADK-011;Unit G, West wall: Layer 2	-26.62	4.525
Soil	ADK-011; Unit G; West wall	-26.24	3.502
Soil	ADK-011; Unit G; West wall	-26.51	3.596
Soil	ADK-011; Unit G; West wall	-26.30	4.814
Soil	ADK-011; Unit G; West wall	-26.04	4.716
Soil	ADK-011; Unit G; West wall	-26.29	4.230
Soil	ADK-011; Unit G; West wall; Layer 3a	-26.40	3.378
Soil	ADK-011; Unit G; West wall; Layer 3b	-26.81	3.123
Soil	ADK-011; Unit G; West wall; Layer 4	-26.60	1.619
Soil	ADK-011; Unit H; East wall	-24.86	4.637
Soil	ADK-011; Unit H; East wall	-26.72	4.407
Soil	ADK-011; Unit H; East wall	-26.49	6.113
Soil	ADK-011; Unit H; East wall	-25.95	6.880
Soil	ADK-011; Unit H; East wall	-26.16	8.350
Soil	ADK-011; Unit F; West wall; Layer 1	-26.07	4.312
Soil	ADK-011; Unit F; West wall; Layer2	-25.70	4.495
Soil	ADK-011; Unit F; West wall	-26.49	3.948
Soil	ADK-011; Unit F; West wall; Layer 3	-27.01	2.803
Soil	ADK-011; Unit F; East wall	-26.77	9.048
Soil	ADK-011; (?Unit F;) East wall; Layer 1	-26.27	5.445
Soil	ADK-011; (?Unit F;) East wall; Layer 2	-25.98	2.945
Soil	ADK-011; (?Unit F;) East wall; Layer 2	-26.42	2.167
Soil	ADK-011; (?Unit F;) East wall; Layer 3	-26.37	4.227
Soil	ADK-011; (?Unit F;) East wall	-13.63	8.811
Soil	ADK-012; Unit 3; Layer4	-27.13	36.488
Soil	ADK-SW; Unit B	-26.26	1.002

Type of Material	*Location/Site/Provenance*	$\delta^{13}C$	*%C*
Soil	ADK-193	-26.29	10.931
Soil	ADK-193	-25.62	15.751
Soil	ADK-SW; Unit E; West wall	-25.72	2.736
Soil	ADK-SW; Unit F; Level3	-26.29	16.121
Soil	ADK-SW; Unit E; West wall	-25.26	4.043
Soil	Island site; Unit 1; Layer3	-26.43	9.203
Soil	ADK-011; House 1; Unit C; Layer 2	-26.06	2.917
Soil	ADK-011; House 1; Unit B; West wall; Layer 2	-25.99	3.486
Soil	ADK-011; House 1; Unit B; West wall; Layer 3	-26.05	4.310
Soil	ADK-011; House 1; Unit B; North wall; Layer 1	-24.90	6.281
Soil	ADK-011; House 1; Unit B; North wall; Layer 2B	-26.04	2.799
Soil	ADK-011; House 1; Unit B; West wall; Layer 2	-25.38	2.834
Soil	ADK-011; House 1; Unit B; Layer 2	-26.06	3.154
Soil	ADK-011; House 1; Unit B; North wall; Layer 2B	-21.49	4.742
Soil	ADK-011; House 1; Unit B; Middle North wall; Layer 2B	-25.85	3.130
Soil	ADK-011; House 1; Unit C; West wall; Layer 2b	-26.78	2.267
Soil	ADK-011; House 1; Unit C; West wall; Layer 2	-26.31	9.339
Soil	ADK-011; House 1; Unit C; West wall	-26.15	2.984
Soil	ADK-011; House 1; Unit C; East wall	-26.25	4.404
Soil	ADK-011; House 1; Unit C; East wall	-26.65	3.539
Soil	ADK-011; House 1; Unit C; East wall	-25.62	2.973
Soil	ADK-011; House 1; Unit C; East wall; Layer 1	-26.53	6.708
Soil	ADK-011; House 1; Unit C; East wall; Layer 3	-26.13	2.310
Soil	ADK-011; House 1; Unit D; East wall	-22.80	7.565
Soil	ADK-011; House 1; Unit D; East wall; Layer 3	-26.57	2.036
Soil	ADK-011; House 1; Unit D; East wall; Layer 1	-26.10	4.542
Soil	ADK-011; House 1; Unit D; East wall; Layer 2	-26.23	3.999
Soil	ADK-011; House 1; Unit E; West wall; Layer 2	-25.65	4.168
Soil	ADK-011; House 1; Unit E; West wall	-25.95	4.493
Soil	ADK-011; House 1; Unit E; West wall; Layer 3	-27.05	2.210
Soil	ADK-011; House 1; Unit E; West wall	-25.51	16.401
Soil	ADK-011; House 1; Unit E; West wall	-25.86	5.504
Soil	ADK-011; House 1; Unit E; West wall	-21.76	0.125
Soil	ADK-011; House 1; Unit F; Layer 1	-26.92	1.394
Soil	ADK-171; Unit H2	-26.63	1.22
Tephra	NA	-28.13	2.204
Tephra	ADK-011; Unit G; Layer 2; Level 2	-26.98	1.295
Tephra	ADK-011; Unit G; West wall	-26.85	0.524
Tephra	ADK-084	-26.83	1.279
Tephra	ADK 084	-26.64	1.898
Tephra	ADK 084	-26.98	1.038
Tephra	ADK 084	-27.22	2.064
Tephra	ADK 012	-27.96	1.512

Chapter 15
Bone Artifacts and Debris from Three Archaeological Sites on Adak Island, Alaska

Dixie West and Virginia Hatfield

Abstract: *Archaeologists with the Central Aleutians Archaeological and Paleobiological Project (CAAPP) recovered bone artifacts and debitage from three archaeological sites, ADK-171, ADK-012 and ADK-011, on Adak Island, Aleutian Islands, Alaska. Although the bone tool assemblages are small, the awls dominate the ADK-171 assemblage. A slightly greater variety of tools including awls, needles and wedges comprise the bone inventory at ADK-012. Bone artifacts recovered from the ADK-011 house feature include personal ornaments, weaponry, and household implements typical in many Aleutian prehistoric village sites. ADK-011 midden artifacts include spent hunting gear transported to the village site inside prey carcasses. Bone debitage in the ADK-011 house and midden represents the roughing out and final shaping stages of bone tool manufacture.*

Keywords: *Adak Island, bone artifacts, bone reduction stages*

Introduction

Along with stone tools, bone implements comprise the characteristic Unangan tool kit. Unlike lithic artifacts, bone tools may or may not survive the Aleutian archaeological record. The percentage of identifiable bone tools (and bones for that matter) in any archaeological site is attributed to a pre- and postdepositional environment conducive to their preservation. CAAPP archaeologists recovered bone artifacts in sites ADK-171, ADK-012, and ADK-011, component 2 on north Adak Island. Bone tools and bone manufacturing debris survived because these sites contain shell that buffers acidic Aleutian soils and promotes bone preservation.

ADK-171 *Clinocardium* Midden

The ADK-171 *Clinocardium* midden is located on a 20m high bluff overlooking the east shores of Clam Lagoon. This is a remnant of a once larger garbage disposal area now largely destroyed. During 1998 USBIA archaeologists first inspected the site and saw large *Clinocardium* shells eroding from a midden. Shells from the site radiocarbon dated approximately 6000 years BP (O'Leary 1998; 2001). In 1999 Russian historical ecologists with the Western Aleutians Archaeological and Paleobiological Project cleared a 3m thick profile at the eroding midden. In 2005 these researchers opened an additional 1m X 1m unit (Pit 3) and a 1m X 0.5m unit (Pit 3 Annex) adjacent to the 1999 excavation (Savinetsky *et al.* Chapter 5 this volume). Historical ecologists excavated two distinct cultural levels (Layers 2 and 4) separated by a thin lens of redeposited Intermediate Ash (Layer 3). Charcoal fragments from layers 2 and 3 dated at 5760±20 BP (UCIAMS-22153) and 5735±30 BP (UCIAMS-22154) respectively and indicate the midden comprises a single component. ADK-171 represents the oldest human occupation in the central Aleutian Islands.

ADK-171 Clinocardium Midden Bone Tools

In 1999 the Western Aleutians Archaeological and Paleobiological Project (WAAPP) recovered six bone awls and a bone needle from Level 1 at the *Clinocardium* midden (Luttrell and Corbett 2000, 47). The 1999 WAAPP Level 1 correlates with 2005 CAAPP Level 2 at the same site. CAAPP bone tool analysis reveals that awls (N=5), made on bird wing bones, continue to dominate the ADK-171 assemblage (Figure 15.1,a-d; Table 15.1). CAAPP archaeologists also recovered a needle preform (Figure 15.1,e) on bird bone. Two worked ribs and a bone scrap probably belong to seal (Figure 15.1, f-g). Archaeologists also recovered a bone point fragment while profile cleaning (Figure 15.1,h). Awls and needles suggest that hide working and clothing manufacture occurred at the occupation affiliated with the *Clinocardium* midden. Archaeologists recovered no bone artifacts in cultural Layer 4.

ADK-012

ADK-012, dated circa 1865 years BP, is a single component village site situated on a barrier beach that separates Clam Lagoon from Sitkin Sound. World War II military bulldozing largely destroyed ADK-012; disturbed site remnants are all that remain. Ted Bank first visited the site in 1949 and USFWS archaeologists relocated and mapped the site in 1999 (Luttrell and Corbett 2000). In 2005 CAAPP archaeologists cleared a profile and excavated a 1m X 1 m unit in a bulldozed knoll at ADK-012. CAAPP recovered animal bones, shell, stone tools, lithic debris, and bone artifacts from this largely destroyed site.

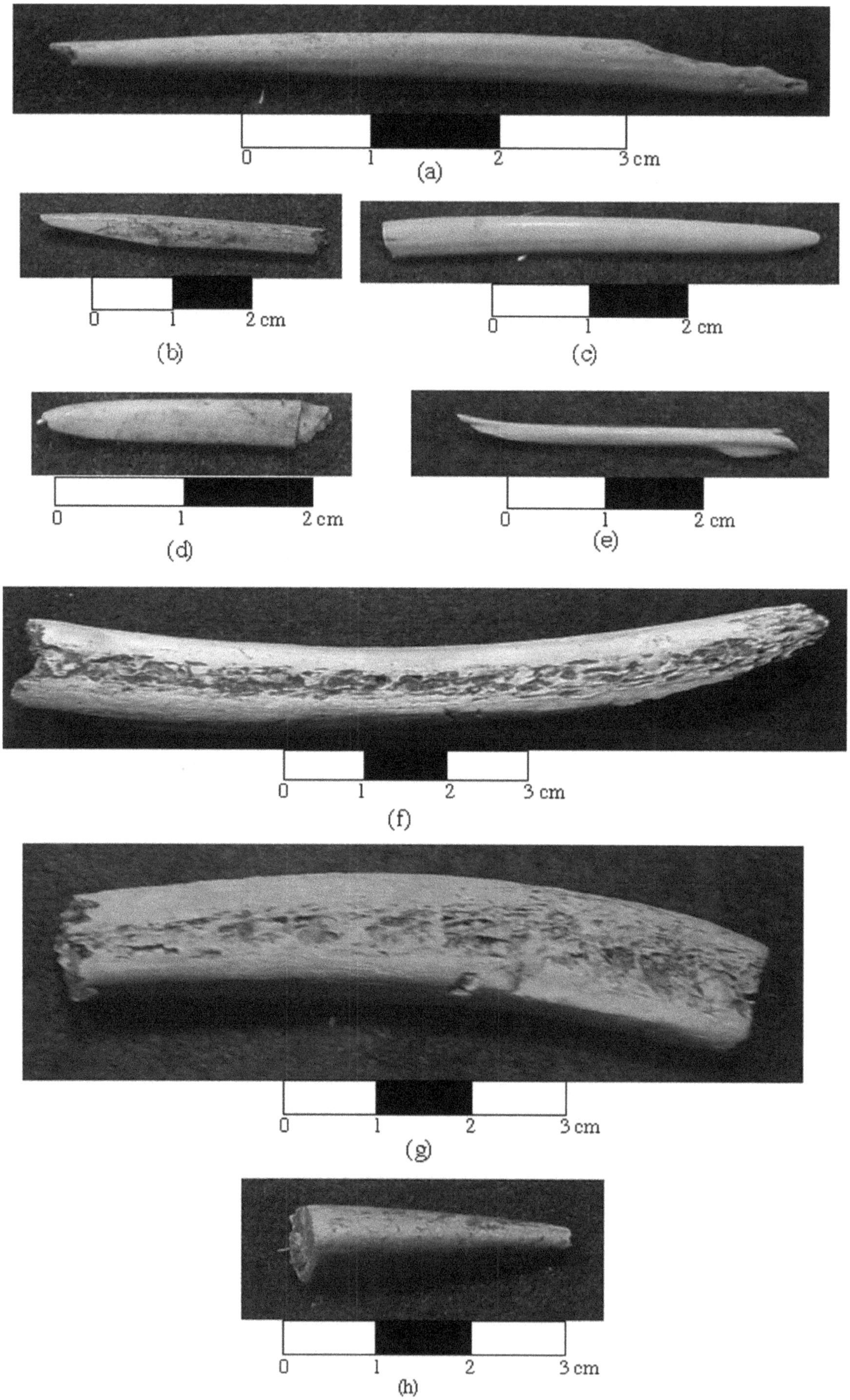

Figure 15.1. ADK-171 artifacts: (a-d) awls; (e) needle preform; (f,g) worked mammal rib; (h) broken point (possibly cut during profile cleaning).

Table 15.1. ADK-171 bone artifacts.

Artifact	Unit/Level/Layer	Bone type	Remarks
Awl	3A/2	Bird	Gull or cormorant radius
Awl	3A/2	Bird	Gull or cormorant radius
Awl	3A/2	Bird	Gull or cormorant radius
Awl	3/2	Bird	Gull or cormorant radius
Awl	3A/2	Bird	Gull or cormorant radius
Worked bone cf. Awl	3/2	Bird	
Needle preform	3A/2	Bird	Shorebird/cf sandpiper
Point tip	3A/profile	Mammal	Rib?
Worked rib	3/2	Mammal	Seal?
Worked rib	3A/2	Mammal	
Worked bone scrap	3/2	Mammal	Seal

Table 15.2. ADK-012 bone artifacts and bone debris.

Artifact	Unit/Layer/Level	Material	Remarks
Awl	3/2	Bird	Gull or cormorant ulna
Awl	2/2	Bird	Cormorant
Awl	2/1	Bird	
Needle	3/1	Bird	Eyeless, nipple end
Needle	2/2/2	Mammal	Eyeless
Fish hook shank	1/3	Mammal	
Bone tube	3/2	Bird	Scored and snapped with burin marks parallel to shaft for needle preform
Bone tube	3/3	Bird	
Wedge	2/2/2	Mammal	
Wedge fragment	3/2	Mammal	
Worked rib	3/1	Mammal	Scoring parallel to shaft
Worked rib	2/1	Mammal	Seal?
Worked bone	3/1	Bird	Awl perform?
Worked bone	1/2	Cf Mammal	

ADK-012 Bone Artifacts

The ADK-012 assemblage consists of three awls (Figure 15.2, e,f), two needles (Figure 15.2,c,d), a fish hook shank (Figure 15.2,g), two cut bird bone tubes (Figure 15.2, a,b), two wedges, two worked mammal rib fragments, and pieces of worked mammal (N=1) and bird bone (N=1) (Table 15.2). ADK-012 needles are eyeless and one has a nipple end. Interestingly, one needle is made of mammal bone, the other of bird. The ADK-012 awls are made on bird wing bones, and the wedges are made of mammal bone. Of interest are the cut bird bone tubes that are scored parallel to the bone shaft (Figure 15.2,a,b). These are not butcher marks that are an incidental part of animal disarticulation, meat extraction and the skinning process. Instead these marks could represent the initial grooves created to snap off needle performs. Once the bird bone is scored, a quick snap to release the preform followed by smoothing produces a new needle.

ADK-011 Zeto Point

Site ADK-011 is a large village site, with over 100 pit features, located on Zeto Point east of Clam Lagoon. This is a multicomponent site with at least two occupations that date 170 to 400 years BP (Component 2) and circa 2500 years BP (Component 1). Previous investigators to this large, well-preserved site include Ted Bank, the USBIA, the USFWS and the Western Aleutians project (Luttrell and Corbett 2000). During 2006 CAAPP archaeologists excavated ten 1m X 1m units at this site. Five excavation units (B-F) transected a garbage disposal area (midden) and five units (G-J) transected a house interior. In ADK-011 Component 2, a large quantity of *Clinocardium* shells buffered soil and preserved bone in the midden and the most recent house occupation.

ADK-011 Component 2 Bone Artifacts

In 1999 WAAPP archaeologists recovered an asymmetrically barbed bone point, a small scored bird bone (possibly a bead), a scored bone fragment, an awl tip, a bone wedge tip, and three worked bone fragments at ADK-011 (Luttrell and Corbett 2000, 24). Although preservation is generally good in ADK-011, Component 2, CAAPP archaeologists found comparatively few bone

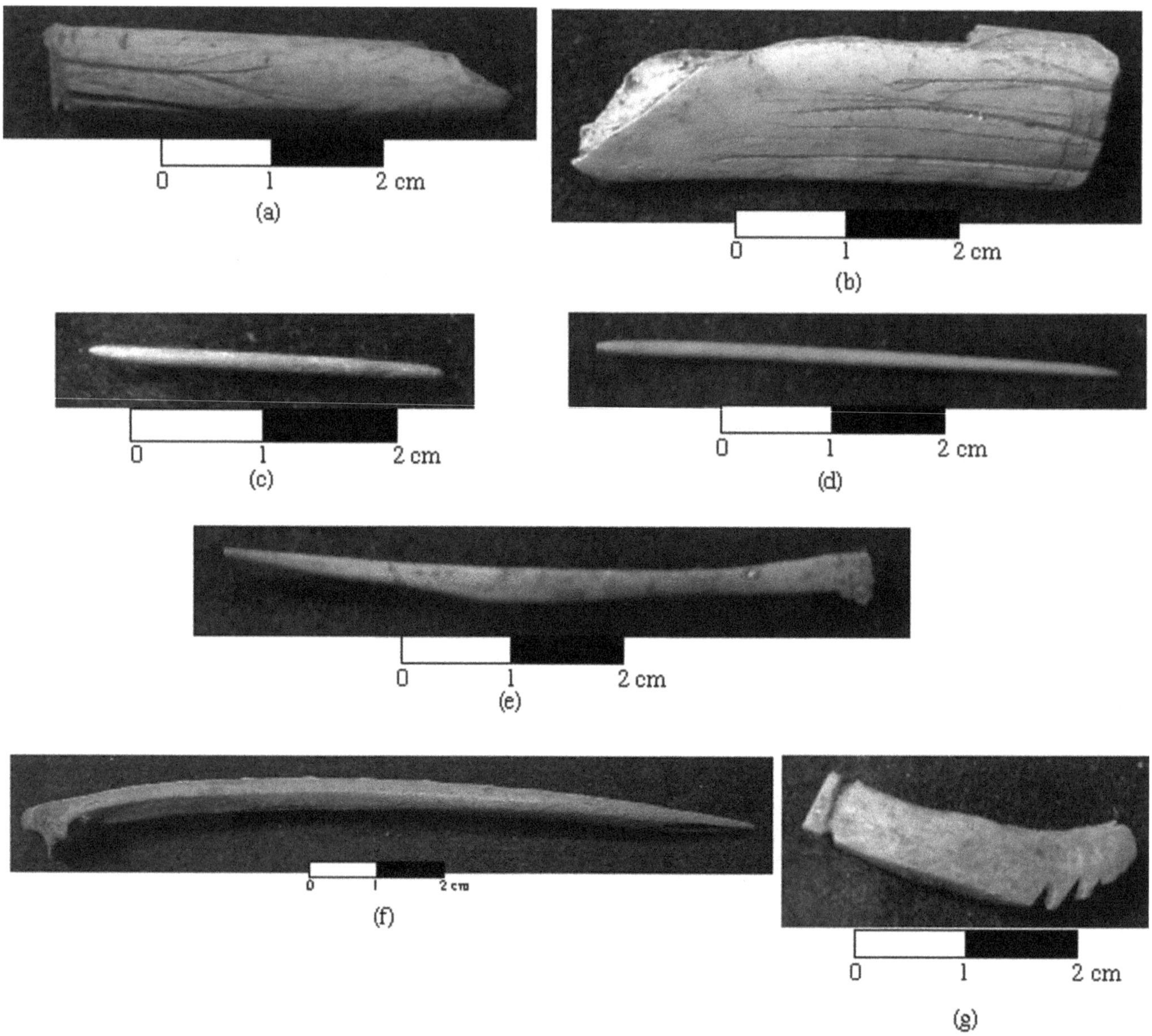

Figure 15.2. ADK-012 artifacts: (a,b) bird bone tubes; (c,d) needles; (e,f) awls; (g) fish hook shank.

tools for the amount of area excavated (Figures 15.3-15.6). Presented in Table 15.3, bone tools are categorized based on where they were found: (a) the ADK-011 house feature and (b) the associated refuse midden.

Archaeologists recovered two finely executed 'pendants' representing a fish and seal in the house feature (Figure 15.3,a,b). A third ornamental piece on ivory or cut tooth, from the midden, could be a pin or some type of fastener (Figure 15.5, b). An unidentified denticulated piece (Figure 15.5,c), recovered from the midden, may be another art object, but so little of the object remains it is difficult to identify. The denticulated piece resembles toothed patterns on a carved ivory fish from Amchitka, Rat Islands (Black 2003; Figure 47). A labret made on mammal bone, recovered from the midden, represents another personal adornment (Figure 15.5,a). A bilateral harpoon shaped object on ivory, recovered from the house, appears more ornamental than utilitarian (Figure 15.3,c).

ADK-011 tools recovered from the house are divided into hunting and household processing tools. In the house, weapons are few and include a fish hook barb, a barb preform and a fish hook shank fragment (Figure 15.3,h,j). All other bone tools are strictly household utilitarian objects. These include complete and fragmented awls made on mammal and bird bone (Figure 15.3,d-g), a needle perform (Figure 15.4,a), and a wedge. Cut bird bone tubes (Figure 15.4,b,c), and cut mammal bone ribs (Figure 15.4,d,e) might represent bone blanks from which tools could be made. A sea mammal epiphysis and two 'tablets' of sea mammal cancellous bone show wear suggesting use as abraders (Figure 15.4,f,g).

Midden utilitarian tools include an extremely weathered, bilaterally worked harpoon, a smaller harpoon form (Figure 15.5,e,f) and a harpoon tip, a needle fragment, and two wedges (e.g., Figure 15.5,h). Interestingly, archaeologists recovered only one awl from the midden area (15.5,g). An

Table 15.3. ADK-011, Component 2 bone artifacts.

Tool type	Number	Material type
House Feature		
Awl	3	mammal
Awl	6	bird
Sharpened rib bone (large awl)	1	mammal
Bone point	1	mammal
Fish hook barb preform	1	
Fish hook shank	1	mammal
Fish hook shank fragment	1	mammal
Needle preform	1	bird
Bilateral harpoon point	1	ivory?
Seal pendant	1	mammal
Fish pendant	1	mammal
Wedge	1	mammal
Cut tooth	1	mammal
Tablet (for smoothing)	2	mammal
Epiphysis smoother	1	mammal
Girdled bone	1	bird
Cut bone	2	bird
Cut bone	6	mammal
Modified bone	1	bird
Modified bone	1	mammal
Modified sea lion ribs	3	mammal
Midden		
Denticulate fragment	1	mammal
Pin or fastener	1	ivory or tooth
Awl	1	bird
Bilateral harpoon (ornamental?)	1	mammal
Bilateral harpoon	1	mammal
Fish hook shank	1	mammal
Fish hook barb	4	mammal
Fish gorge	2	mammal
Harpoon tip	1	mammal
Labret	2	mammal
Needle	1	bird
Wedge	2	mammal
Cut bone	1	mammal
Cut bone	1	bird
Cut tooth	1	mammal
Modified bone	1	mammal

assortment of cut bone could represent bone blanks stored for future tool manufacture (e.g., Figure 15.5,d).

Archaeologists recovered four fish hook barbs, a shank, and two gorges in the garbage disposal area (Figure 15.6). These weapons are affiliated with captured fish carried to the ADK-011 village for processing. Following butchering, broken or hard to retrieve, barbs were discarded with unprocessed carcass parts. A lithic analogue of this scenario is found in Figure 15.7; archeologists found the tip of a stone point embedded in a sea lion rib recovered from the midden.

ADK-011 Bone Debitage

A simple and straightforward *chaîne opératoire* for bone tool manufacture is: bone acquisition> testing and bone preparation> primary trimming> secondary trimming> final smoothing. In this light, the ADK-011 mammal bone debitage is worth noting. West categorized mammal bone debris as (a) chunks, (b) slivers, and (c) 'pinwheels.' Chunks are both large and small, irregularly shaped bone fragments (Figure 15.8,a). These clearly constitute bone refuse from some type of tool manufacture and are analogous to the preparation stage in stone tool manufacture. Bone slivers (primary trimming) constitute the second category of bone debris (Figure 15.8,b). These are comparatively thin bone shavings created when shaping a bone tool once the initial tool form had been roughed out. 'Pinwheels' constitute the final bone *debitage* category (Figure 15.8,c). Pinwheels are tiny bone shavings produced during the final, and more intricate, shaping stages of bone tool manufacture. Once this secondary trimming was complete, final smoothing, if necessary, took place. Of the 733 pieces of bone debris recovered from ADK-011, Component 2, 162 were chunks, 550 were slivers, and 9 were pinwheels.

Bone fractures in a variety of ways but in predictable fashion based on collagen content. Fresh or green bones break in curvilinear fractures. Dry or mineralized bone shatters into straight edged or rectilinear fragments, often with rough surfaces on breaks. The irregularly shaped chunks at ADK-011 were not naturally broken. They bore no spiral fractures, typical signs of marrow extraction, nor did they exhibit dry breaks associated with bones that have lost their collagen. Some showed evidence of cutting. Among the animal bones and fragments recovered from ADK-011, these chunks stand out and represent the discard associated with initial stages of bone tool manufacture where the preform of the bone tool is established. Based on a mental template of what the final bone tool should look like, Unangan bone knappers probably scored the bone with burins and pounded stone flakes into the resulting grooves to lop off bone chunks in order to create a general perform. Chunks ranged from 8.5cm to 0.9cm long, 2.4cm to 0.5cm wide, and 1.4cm to 0.2cm thick.

Once the general outline of the bone tool was established, the knapper sliced off slivers of bone much like one whittles wood. Bone slivers followed the bone 'grain' and were generally longer than they were wide. Sliver measurements varied considerably with lengths ranging between 0.9cm and 8.6cm, widths between 0.3cm and 4.2cm., and thicknesses between 0.1cm to 1.1cm. Larger sliver pieces often contained bone cortex. Some splinters were flat, but others were distinctly curved. If the sliver was removed close to the fresh bone cortex, and if the sliver was suitably thin enough, it tended to curl as it was removed. Jochelson (1925, 79) recovered bent knives of walrus tusk during excavations. He proposed that the implements were once straight but had bent as the dentin and pulp cavity differentially dried, and he proposed that the best way to

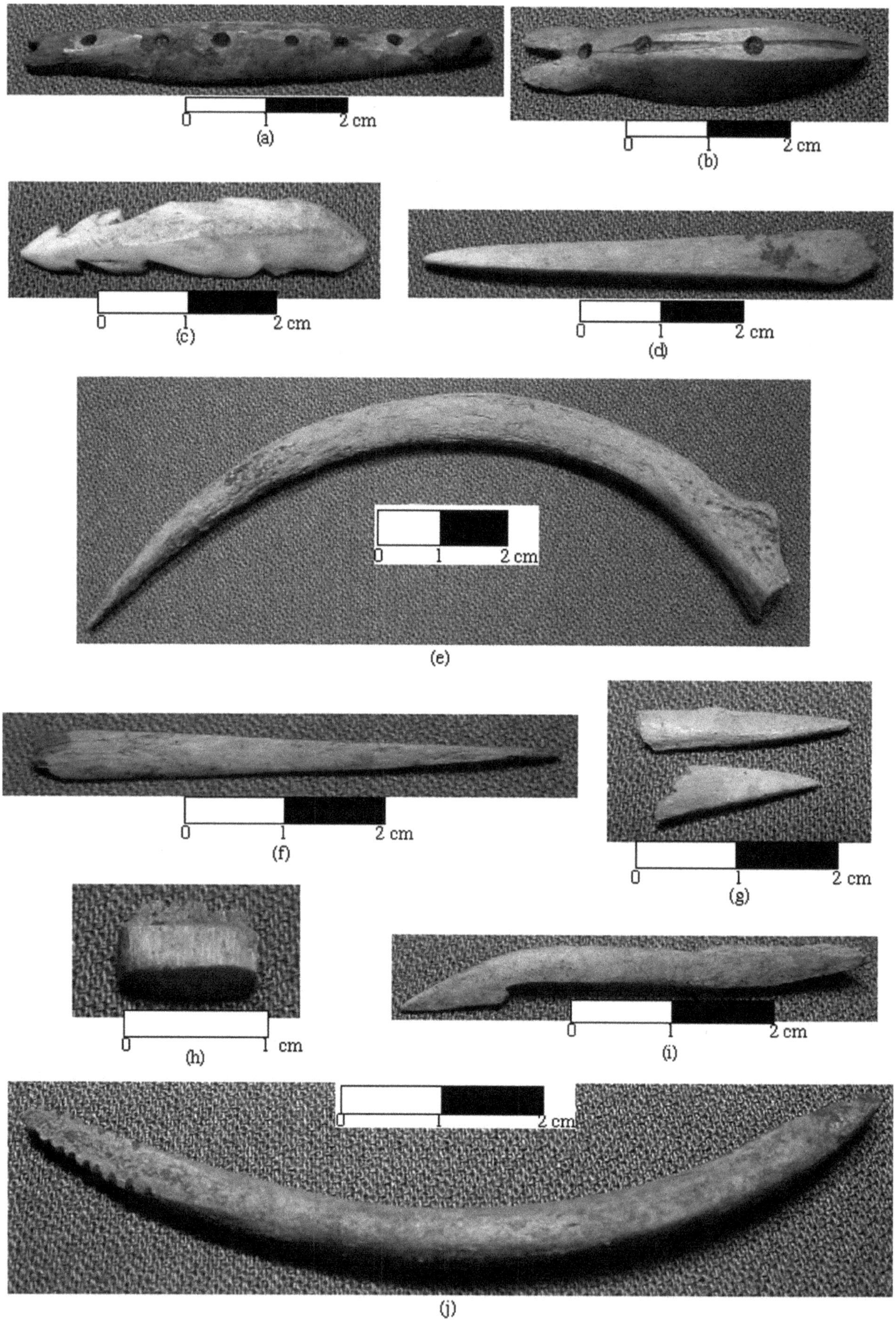

Figure 15.3. ADK-011, Component 2 house feature artifacts: (a) seal pendant of mammal bone; (b) fish pendant of mammal bone; (c) bilaterally worked harpoon point; (d) pointed bone; (e) awl on mammal bone; (f-g) awl tips; (h) fish hook shank fragment; (i) fish hook perform; (j) large fish hook barb.

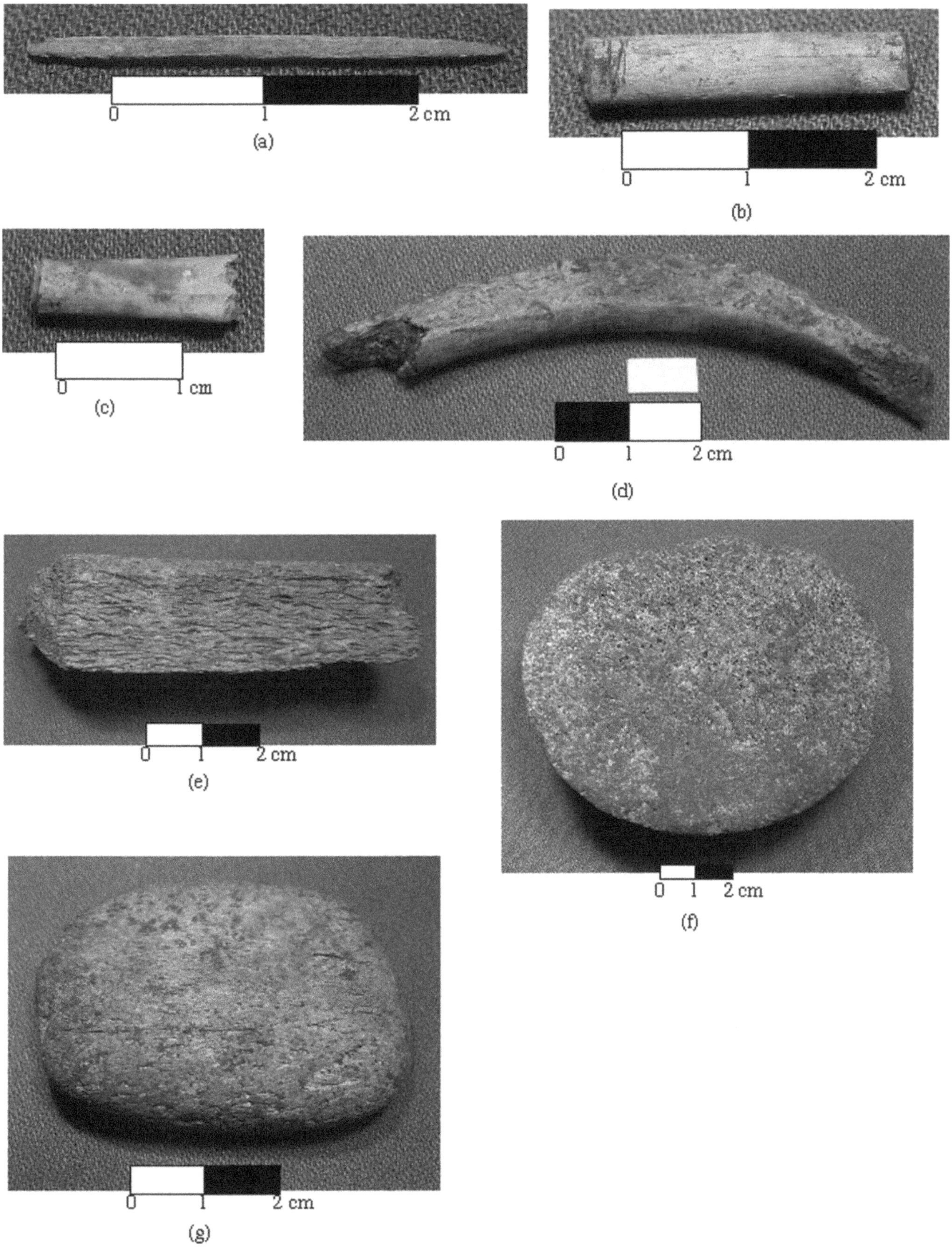

Figure 15.4. ADK-011, Component 2 house feature artifacts: (a) needle perform; (b,c) bird bone tubes; (d) cut mammal rib; (e) cut mammal bone; (f) mammal epiphysis used as a smoother; (g) bone tablet used as a smoother.

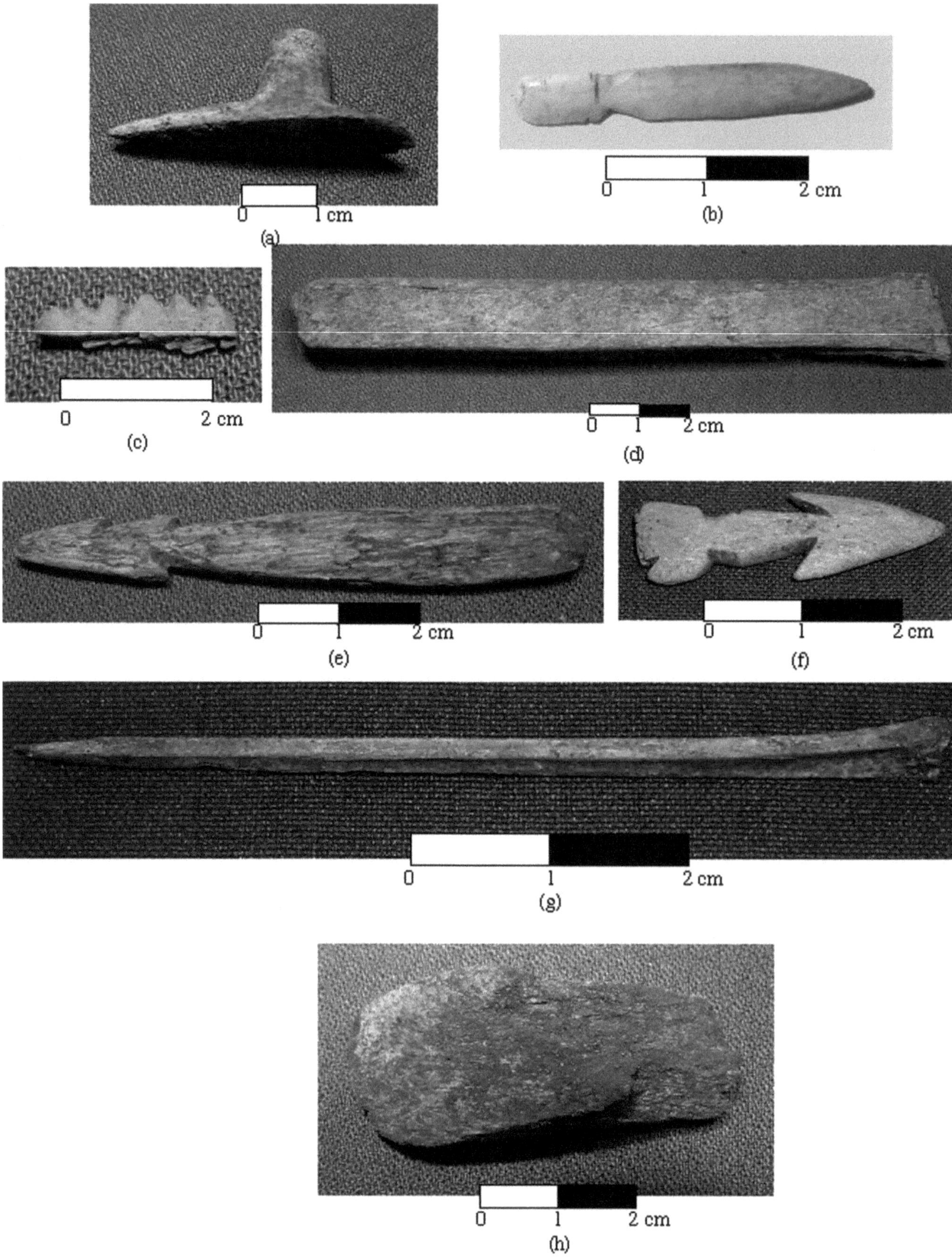

Figure 15.5. ADK-011, Component 2 midden artifacts: (a) labret on mammal bone; (b) pin ornament; (c) denticulated mammal fragment; (d) cut mammal bone blank; (e,f) bilaterally worked harpoon points; (g) awl; (h) bone wedge.

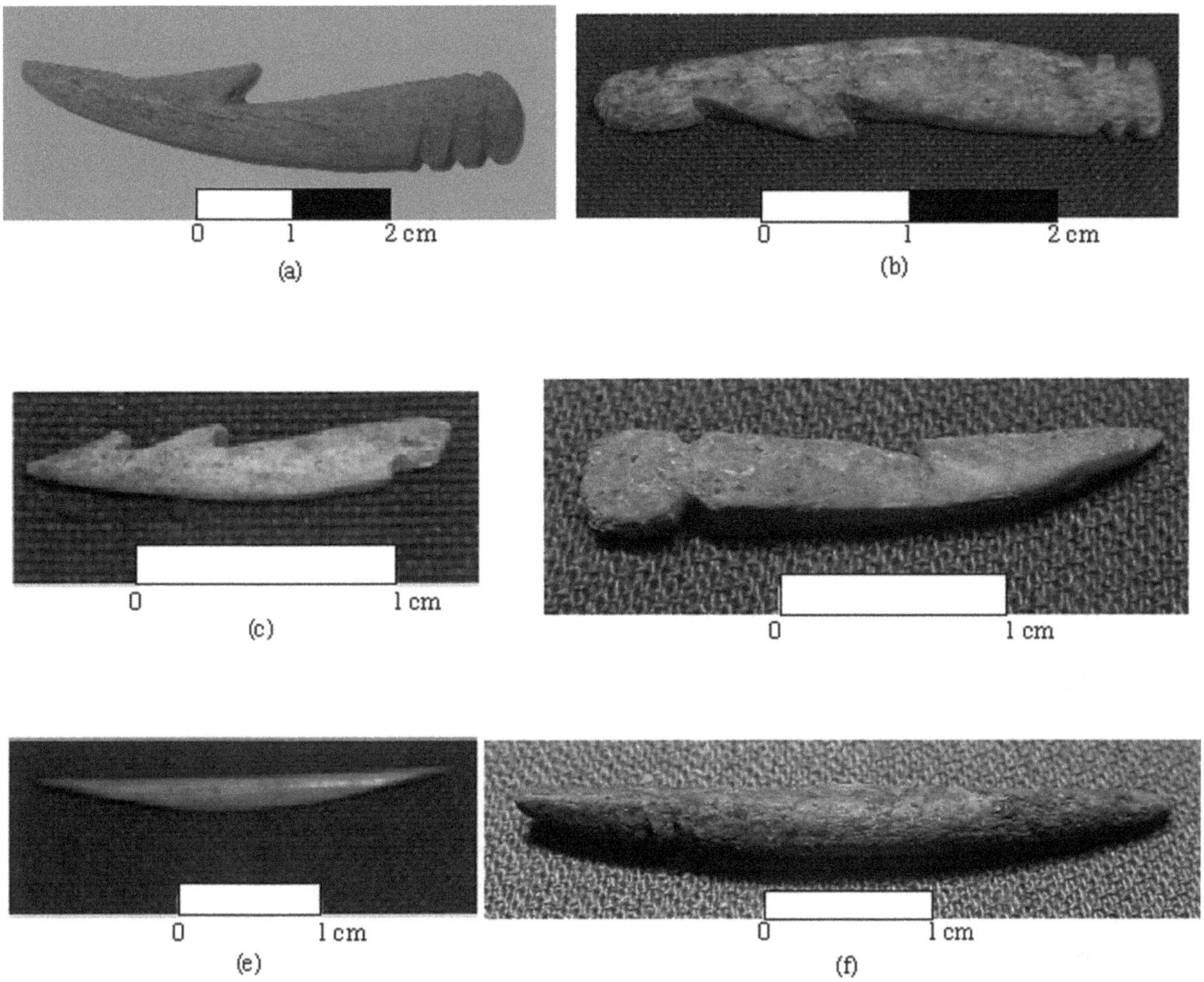

Figure 15.6. ADK-011, Component 2 midden artifacts: (a-d) fish hook barbs; (e,f) fish gorges.

keep a bone tool from bending was to carve it out of a bone of the same density throughout (Jochelson 1925, 81). The same is true for the ADK-011 bone slivers. Slivers curled toward the side of denser substrate, the cortex.

Near the final stages of bone tool manufacture when fine carving was required, bone slivers became extremely thin and slender and curved to form a pinwheel. Archaeologists recovered only nine pinwheels from ADK-011. Because they are small and delicate, and because over time they have lost their collagen, most pinwheels probably did not survive water screening.

Of the 162 chunks, archaeologist recovered 101 in the midden and 61 in the house feature. Archaeologists recovered 313 slivers in the midden and 237 in the house. All nine pinwheels, the final stage of bone tool manufacture, were recovered from the house. The provenience of tool making debris suggests that bone workshops probably occurred around the ADK-011 house with the bulkier bone *debitage* discarded in the midden. Unangan knappers did not bother to remove the smallest bone debris (pinwheels) from the bone workshop.

Also of particular interest are sea lion ribs associated with the house feature. Unangax̂ cut and whittled away the medial side of one rib (Figure 15.9,a). It is unclear if the sea lion rib was intended to be the end tool, or if this is the scrap left behind. Archaeologists also found the medial sides of sea lion ribs at the site (e.g., Figure 15.9,b). No rib fragment could be back fitted to another. One rib had not been cut or split, but was simply cut diagonally (Figure 15.9,c). That the ribs are associated with the house feature, rather than the midden, suggests the Unangax̂ stored these bones for some type of continued or future use.

Conclusions

Archeologists recovered common household objects (awls and needles) in all three Adak prehistoric sites. These ubiquitous implements suggest that hide preparation and clothing construction was a primary focus of people living in the subarctic Bering Sea environment during the last

(a) (b)

Figure 15.7. ADK-011, Component 2 midden: (a) sea lion rib with broken stone point embedded near proximal end; (b) close up of the embedded stone point.

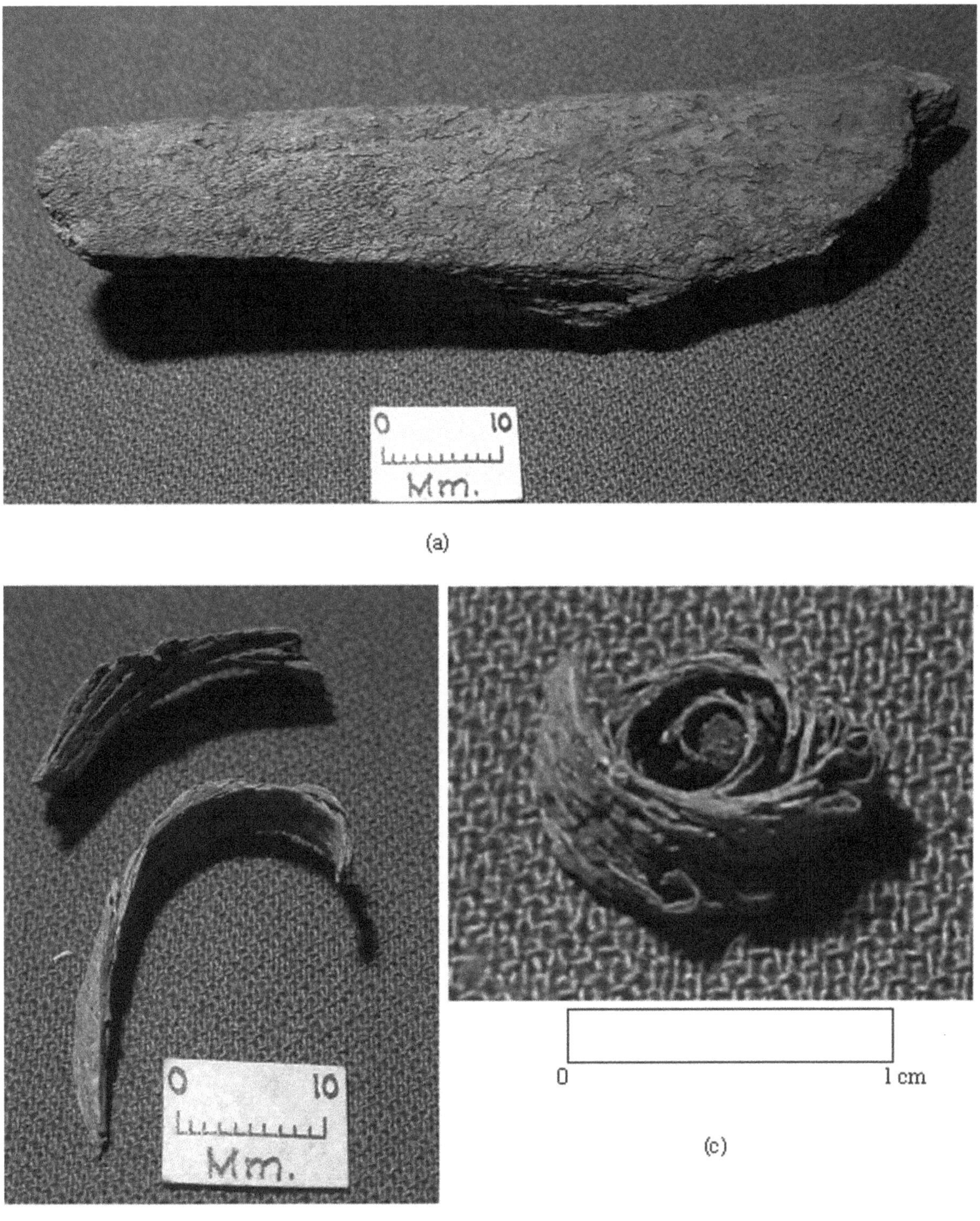

Figure 15.8. ADK-011 Component 2 bone debitage types: (a) bone chunk; (b) bone slivers; (c) bone pinwheel.

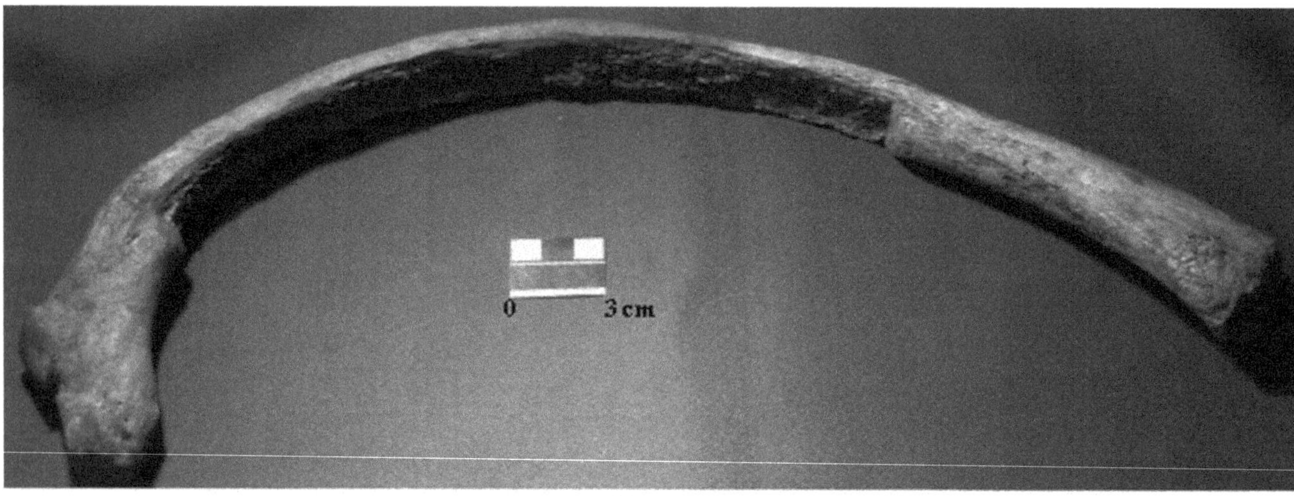

(a)

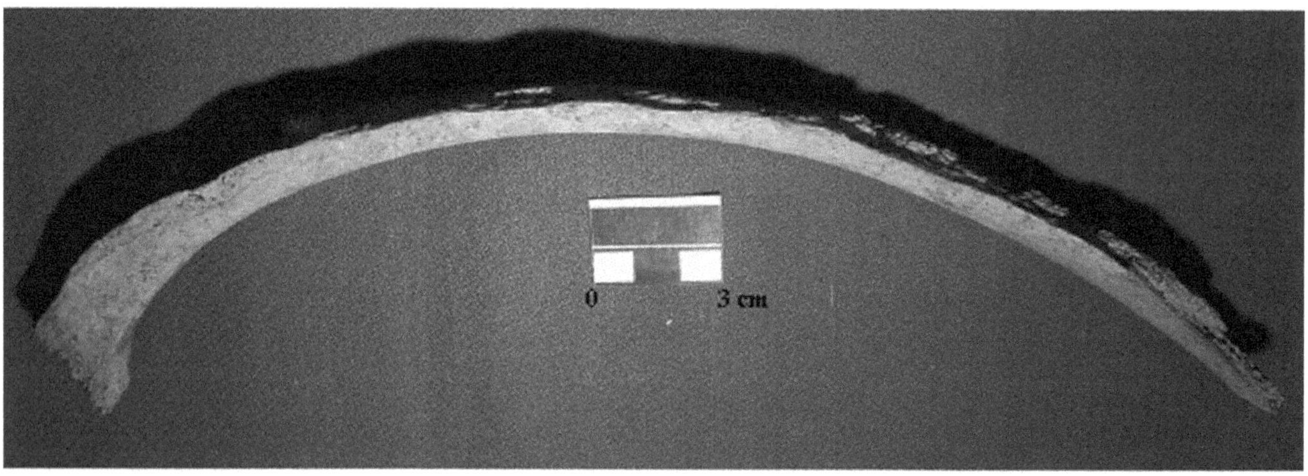

(b)

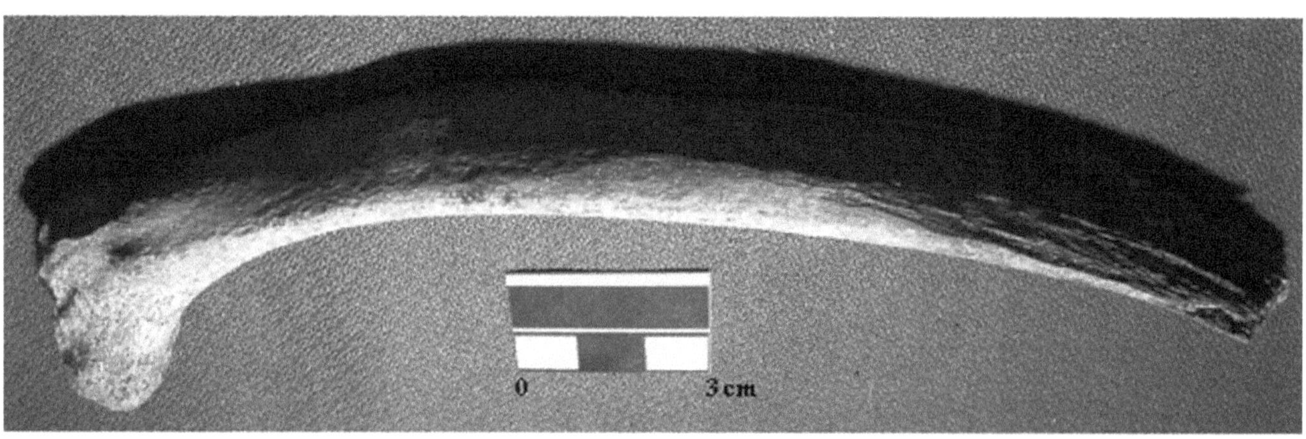

(c)

Figure 15.9. ADK-011, Component 2: (a) sea lion rib with cut marks; (b) medial side of sea lion rib; (c) sea lion rib cut diagonally at distal end.

six thousand years. Bone wedges (ADK-012 and ADK-011 Component 2) suggest woodworking although wood was a rare commodity in Adak archaeological sites. Hooks and gorges (ADK-011, ADK-012) were the weapons of choice to catch large cod and other fish species (Crockford Chapter 6 this volume). Other weapons include bone points; these along with their stone counterparts (Wilmerding and Hatfield Chapter 12 this volume) represent a second category of hunting gear. Archaeologists only recovered personal ornaments (pendants, a labret, and fastener) in the comparatively young house layers at ADK-011. An absence of these vestments in ADK-171 and ADK-012, does not suggest that personal adornments were not used in earlier times on north Adak.

Kay (Chapter 13 this volume) described microplating domes on stone tools from Adak Island. These domes form when stone tools are used against hard contact materials, including bone or ivory. Kay also microscopically analyzed flake tools possessing wear traces consistent with a back and forth movement within a groove or slot. This microscopic evidence, along with a variety of burins found in Adak archaeological sites, suggests that bone working was a common prehistoric task. Wilmerding and Hatfield (Chapter 12 this volume) analyzed pumice and scoria pieces, sometimes with grooves, recovered from all north Adak sites. These abraders also could indicate bone tool manufacture over time on Adak Island.

The Unangax̂ selected cobbles from beaches or targeted locales (Kay Chapter 13 this volume; Jeanotte Chapter 14 this volume) to acquire specific materials for different stone implements. They were also selective in the prey species that became 'quarries' for different bone tool types. The quality of the bone material, as well as the shape and size, significantly constrain the final bone product. Because of their shape and length, cormorant or gulls wing bones (radii and ulnae) were frequently selected for awls. Once removed from the wing, with a quick snap and polish of the green break near the end of the bone, a useful perforating tool was created. Small, delicate needles were nearly always made from dense, smooth bird shaft bones. Bird bone tubes like those found at ADK-012, with grooves parallel to the shaft probably represent the initial stages of needle manufacture. Thicker, heavier mammal bone was required to make larger wedges, fishhooks, and personal ornaments including labrets and pendants. We do not suggest that Central Aleutians people never used bird bone for ornaments; we simply did not find these in our Adak sites.

Acknowledgements

The National Science Foundation's Arctic Social Sciences Program (Award # 0353092) funded this research. Much thanks goes to Ron West for photographing the bone artifacts.

References

Crockford, Susan. 2011. Archaeozoology of Adak Island: 6000 Years of Subsistence History in the Central Aleutians. Chapter 6. In D. West, V. Hatfield, E. Wilmerding, C. Lefèvre, L. Gualtieri (eds), *The People Before: The Geology, Paleoecology and Archaeology of Adak Island, Alaska.* Oxford, British Archaeological Reports.

Jeannotte, Richard, Kirsten Nicolaysen, Taylor Johnson, Dixie West. 2011. Griddlestones: Their Provenance and Associated Organic Molecules, Adak Island, Alaska. Chapter 14 In D. West, V. Hatfield, E. Wilmerding, C. Lefèvre and L. Gualtieri (eds), *The People Before: The Geology, Paleoecology and Archaeology of Adak Island, Alaska.* Oxford, British Archaeological Reports.

Jochelson, Waldemar. 1925. *Archaeological Investigations in the Aleutian Islands.* Carnegie Institute of Washington Publication 367. Washington, DC.

Kay, Marvin. 2011. Clam Lagoon Archaeology and Technology. Chapter 13 in D. West, V. Hatfield, E. Wilmerding, C. Lefèvre and Lyn Gualtieri (eds), *The People Before: The Geology, Paleoecology and Archaeology of Adak Island, Alaska.* Oxford, British Archaeological Reports.

Luttrell, Mark and Debra Corbett. 2000. Archaeological Investigations on Northeast Adak Island, Alaska, July 1999. Report to the US Fish and Wildlife Service, Anchorage, Alaska.

O'Leary, Matthew. 1998. New Site No. 2 ('Tutikoff Site'). Unpublished Report on File at US Department of the Interior, Bureau of Indian Affairs, ANCSA Projects Office, Anchorage.

O'Leary, Matthew. 2001. Volcanic Ash Stratigraphy for Adak Island, Central Aleutian Archipelago. In Don Dumond (ed.), *Archaeology in the Aleut Zone of Alaska, Some Recent Research, University of Oregon Anthropological Papers No. 58.* 215-234. Eugene, University of Oregon Press.

Savinetsky, Arkady B., Dixie L. West, Zhanna A. Antipushina, Bulat F. Khassanov, Nina K. Kiseleva, Olga A. Krylovich and Andrei M. Pereladov. 2011. The Reconstruction of Ecosystems History of Adak Island (Aleutian Islands) During the Holocene. Chapter 5 in D. West, V. Hatfield, E. Wilmerding, C. Lefèvre and L. Gualtieri (eds), *The People Before: The Geology, Paleoecology and Archaeology of Adak Island, Alaska.* Oxford, British Archaeological Reports.

Wilmerding, Elizabeth and Virginia Hatfield. 2011. Six Thousand Years of Aleut Stone Tools and Tool-Making Debris: Adak Island, Alaska. Chapter 12 in D. West, V. Hatfield, E. Wilmerding, C. Lefèvre and L. Gualtieri (eds.), *The People Before: The Geology, Paleoecology and Archaeology of Adak Island, Alaska.* Oxford, British Archaeological Reports.

Chapter 16
A Human Burial from the Andreanof Islands, Alaska

Dixie West, Joan Brenner Coltrain, M. Geoffrey Hayes, Dennis H. O'Rourke, Christine Lefèvre, Debra Corbett, Françoise Bouchet, and Arkady Savinetsky

Abstract: *The lives of children are highly enigmatic to nearly invisible in the archaeological record. It is difficult to determine what activities they may have participated in or how they might have contributed to the overall well being of the social group. The analysis of the Andreanof Island child, a six to seven year old Unangan discovered in the Central Aleutians by the Western Aleutians Archaeological and Paleobiological Project, provides new information about the genetics, general health, diet, and mortuary practices surrounding youngsters in the ancient Aleutian world.*

Keywords: *Aleutian Islands, human burial, DNA, stable isotopes*

Introduction

During 1999, archaeologists with the Western Aleutians Archaeological and Paleobiological Project excavated a minimally disturbed village site in the Andreanof Island group containing approximately 101 house pit depressions (Luttrell 2000; West *et al.* 1999). Some of the village features are large, measuring 14m x 5m although most are approximately 3m x 3m. Archaeological tests produced urchin, fish bone, small amounts of bird and mammal bone, dense beds of shell, charcoal, fire-cracked rock, and evidence of hearths and ash. The excavated feature represents an abandoned house depression, re-used as a dump for human debris. Bone artifacts include: a bone wedge, awls, and one harpoon point; stone tools included: cobble tools and 260 griddle stones and fragments. Radiocarbon dates indicate the pit feature was used between 2490 ± 50 BP. (Beta-132879) (bottom level of unit 2) and 440 ± 40 BP. (level 3 of unit 1) (Beta-132878).

During excavations, a human burial was discovered intruding into a layer of dark brown tephra atop a dense cockle (*Clinocardium* sp.) midden (Figure 16.1). With permission from the Aleut Corporation, archaeologists excavated, exhumed, and analyzed the skeleton in the field laboratory. Culturally modified, although extensively deteriorated, pieces of wood covered the burial. These 'straps' may represent a bentwood bowl or cradle that was inverted over the burial. Bulat Khassanov, paleoethnobotanist at the Russian Academy of Sciences, identified these slat fragments as spruce (*Picea* sp.). Specimens of spruce are very common among driftwood in the Aleutians (Bulat Khassanov, per. comm. 2006). Grave goods were minimal, including a barb and shank of a large fishhook and a multipurpose stone tool.

Selected rib fragments were sent to Beta Analytic for radiocarbon dating. A Standard AMS date using bone collagen extraction indicated a death circa 840 ± 40 BP. (Beta-133745), well before Russian contact. Four additional rib fragments for use in stable isotope (Coltrain) and DNA analyses (Hayes and O'Rourke) were collected with gloved hands and immediately placed in a sealed ziplock bag upon recovery to reduce the risks of contaminating the skeletal samples with modern DNA. Rib fragments collected for radiocarbon dating and DNA analysis were not analyzed in the field and are not included in the counts. Archaeologists also collected soil samples from the abdominal cavity of the burial and the immediate surrounding area for parasitological analysis (Bouchet *et al.* 2001). After documentation was completed, archaeologists reburied the skeleton and associated grave goods in the original burial site.

In-field Analysis

The burial represents a very nearly complete, articulated child, placed on its left side in a fetal position. Knees were drawn up close to the chest and arms were bent with hands positioned just under the chin. Arms, legs, and axial elements were bagged separately to facilitate laboratory identification. Sources used during identification and analysis included: Bass (1971), Krogman (1978), and Steele and Bramblett (1988). The skeleton included the skull and mandible, all vertebrae, a majority of the ribs, phalanges (hand and foot bones), scapulae, pelvis, and all long bones of the right side (Tables 16.1 and 16.2). Bones of the lower left arm and lower left leg were missing, probably due to overburden pressure and intense rootlet acid etching that can readily destroy the soft, largely cartilaginous, bone of young individuals.

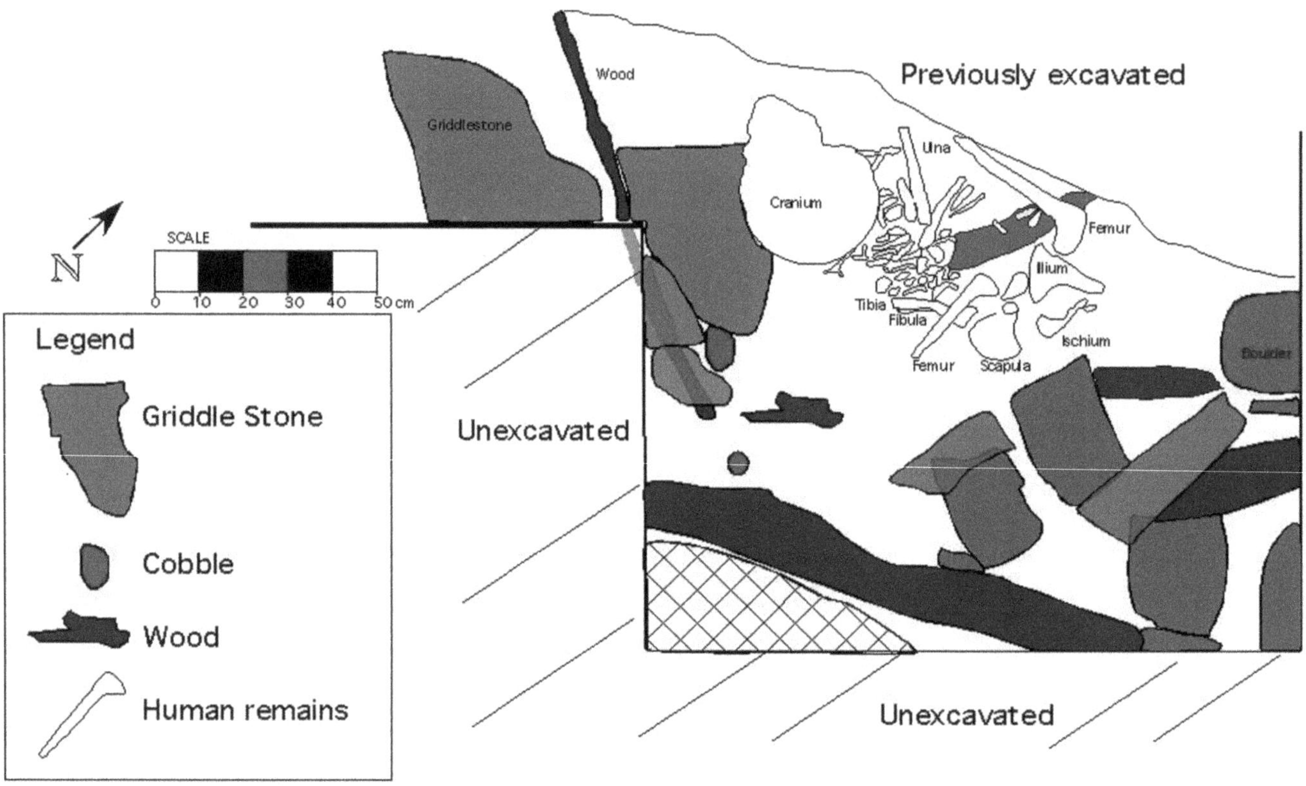

Figure 16.1. Map of the Andreanof Island child grave (drawn by Virginia Hatfield).

Table 16.1. Bones of the Andreanof child burial.

Skeletal Element	***Left***	***Right***	***Axial***
Cranium			+
Mandible			+
Cervical Vertebra			7
Thoracic Vertebra			12
Lumbar Vertebra			5
Sacrum			1
Coccyx			-
Sternum			-
Scapula	+	+	
Clavicle	+	+	
Ribs	8	9	
Humerus	+	+	
Radius	-	+	
Ulna	-	+	
Carpus	1	1	
Innominate	+	+	
Femur	+	+	
Patella	-	-	
Tibia	-	+	
Fibula	-	+	
Tarsus	-	2	
Metacarpus/Metatarsus			13*
Phalanges			21*

+: present; -: absent; *: combined due to difficulties in specific identifications.

Table 16.2. Bone measurements of the Andreanof child burial. (Bones were measured following Bass 1971; bones measured while moist.)

Element	***Measurement in mm.***
Skull:	
interorbital breadth	18.8
zygomatic breadth	103.1
nasal aperture breadth	19.3
greatest breadth of eye orbits	81.3
skull circumference	49.5
Mandible:	
Mandibular length	90.6
Body length	71.3
Ascending ramus breadth	26.7
Greatest breadth of condyles	102.7
Infracrania:	
Right clavicle	80.3
Left clavicle	80.4
Left humerus diaphysis	130.4
Left humerus midshaft breadth	9.3
Right radius midshaft breadth	7.8
Left iliac breadth	83.9
Right tibia proximal epiphysis breadth	49.4
Sacrum: superior segment with alae	56.4
Sacrum: superior segment w/o alae	25.2
Right femur diaphysis + neck length	183.0
Right tibia diaphysis breadth	13.0

Skull

All 29 bones of the skull, including intact nasal conchae, were present (Figure 16.2). Only one tooth, a mandibular right central incisor, was missing from its tooth socket and was not recovered in the screened back dirt. Although the tooth was never recovered, it had not been lost antemortem; no resorption of alveolar bone was observed in the vicinity of the missing tooth socket. The skull and mandible were in excellent condition. No weathering was visible (Behrensmeyer 1978: stage 1) suggesting the corpse, once interred, was never exposed on ground surface. However, the left parietal suffered erosion. This is the surface upon which the skull was resting and is probably associated with the skull placement and subsequent root acid erosion in the soil. This is not a pathological condition of the bone itself, but rather a result of diagenesis (biological, chemical, or physical processes that affect bone after deposition). The skull exhibited no pathologies. A slight concavity of the right parietal just posterior to the fronto/parietal suture could indicate a trauma to the head prior to death that did not have time to heal. However, ground pressure from the overburden covering the skull cannot be ruled out as the cause of the fracture and concavity.

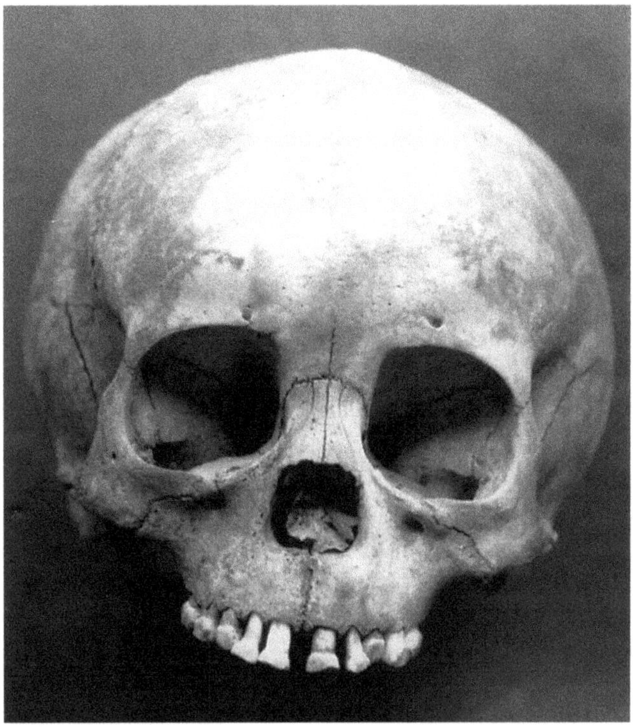

Figure 16.2. Frontal view of the Andreanof Island child skull.

This was clearly a young individual at time of death. The only suture to have commenced fusing was the metopic. This suture closes during the second year of life (Steele and Bramblett 1988, 31). The frontal (metopic) suture was still visible just superior to the naso-frontal (Figure 16.2), but this often remains visible into adulthood. All other cranial sutures were open and the basio-occipital suture was so wide that the basio-occipital could be removed from the base of the skull. This particular suture usually begins to fuse by six years old (Steele and Bramblett 1988, 56).

Teeth

The tooth formula for this individual follows:

$$\frac{i^1, i^2, c, pm^3, pm^4, M^1, M^2}{i_1, i_2, c, pm_3, pm_4, M_1, M_2}$$

The M^1 was just erupting and the M^2 was forming within the alveola. In the mandible the M_1 had erupted, but roots were incomplete; the M_2 was forming within the alveola.

The permanent M^1 erupts in males between 4.8 and 7.9 years and in females between 4.6 and 7.7 years. The I^1 erupts in males between 5.8 and 9.0 years and in females between 5.6 and 8.7 years. The M_1 erupts in males between 4.6 and 7.7 years and in females between 4.3 and 7.5 years. The I_1 erupts in males between 5.0 and 8.0 years and in females between 5.6 and 8.9 years (Steele and Bramblett 1988)). The tooth eruption sequences suggest that this individual was approximately six or seven years old at the time of death (Figure 16.3).

The mandibular incisors, canines, pm_3 and pm_4 showed a minimum amount of wear. On the other hand, the maxillary incisors were distinctly worn (Figure 16.4). This type of wear on the anterior teeth of the youngster might suggest even the youngest members of the group used teeth as tools. Or it might suggest a tooth-grip-and-cut method of eating. The abnormal wear of the left maxillary premolars might indicate: 1) that some object was placed here to 'worry,' 2) use of the teeth as tools or, 3) a habitual chewing quirk unique to this individual.

Although unusually worn, none of the teeth showed decay. The most common cause of dental decay is the introduction of sugars into the diet. Bacteria in the mouth interact with sucrose to form lactic acid. This acid slowly de-mineralizes the tooth enamel allowing further invasion into the center of the tooth. Archaeological evidence (e.g., Crockford Chapter 6 this volume; Lefèvre 2006) shows that the prehistoric Unangan diet in the Andreanof Islands contained little to no refined sugars. The teeth of the Andreanof child showed no signs of enamel hypoplasia, a deviation in normal enamel sometimes associated with vitamin A, C, or D deficiency, infectious disease, premature birth, birth injury, trauma, or local infection. The teeth suggest that this individual did not consume high sugar foods, had an adequate intake of vitamins and minerals, did not suffer any infectious disease, and had incurred no injuries that might influence normal growth and tooth development.

Both maxillary central and lateral incisors exhibited slight shoveling (Figure 16.4). In shoveling the marginal ridges are prominent on the lingual surface of the incisors giving the back of the tooth a distinctive shovel shape. Shoveling is a distinctive inherited characteristic on deciduous and permanent incisors among many Native American groups (Carbonell 1963; Dahlberg 1949; Hrdlička 1920). Unangax̂ have a higher percentage of shovel shaped incisors than Eskimos (Bang and Hasung 1971). There were no signs of dental hypoplasia, crowding, or wear on the interproximal surfaces of teeth.

Vertebral Column

Seven cervical, parts of 12 thoracic, and five lumbar vertebrae were present. A sacrum composed of four unfused segments was observed. No coccygeal vertebrae were recovered during the excavations. Due to the young age of the individual, no epiphyses had developed at time of death. In general, the vertebral column represents 'Stage B' of Steele and Bramblett's (1988, 133) stages of ossification of vertebrae. Neural arches had fused but were often separate from the centra, and no secondary growth centers (epiphyses) had developed. No pathologies were observed on the vertebral column.

Cervical vertebrae:

All neck vertebrae were recovered articulated and in proximity to the occipital condyles of the skull.

Atlas and axis: The arches of the first and second cervicals had fused posteriorly (these fuse at three years of age), but both anterior arches were unfused (these fuse around six years old) (Stewart 1979). The dens was missing from the axis, but this unites only during puberty. The remainder of the axis was present and completely fused.

Cervical 3: The centrum was missing, but the arch was intact. Fusion had not commenced at the time of death.

Cervical 4: The centrum was nearly completely fused with the laminae (arches), but the fusion line was still visible.

Cervicals 5-7: The centra were completely fused with the laminae. The seventh cervical vertebra was cervical in form with a flat low body and two transverse foramina. It possessed no costal facets for ribs.

Thoracic vertebrae:

The arches represented all 12 thoracics, and all 12 centra were recovered during excavations. The arch had united with the body in only one thoracic, and the fusion line was still clearly visible. All arches were completely fused with each other.

Lumbar vertebrae:

The arch had joined the centrum on two of the lumbar vertebrae. Three lumbars still had epiphyses where the arch and centrum were not joined, but two of the centra were missing. According to Steele and Bramblett (1988, 125) in lumbar vertebrae 'the arch joins the body during the sixth year.' This pattern of fusion supports the tooth eruption pattern and suggests the child must have been approximately six to seven years old at time of death.

Sacrum:

The sacrum was composed of four unfused segments. Five sections are usually present in the sacrum, but the most inferior, and smallest, segment of the sacrum was not recovered during excavations and screening. Each segment (body + arch) was fused. The fusion line between the arch and body were still visible on the most superior segment of the sacrum.

Ribs

Heads and necks represented nine right ribs. Eight of these were central ribs and one was a superior rib (probably rib #2). Five head/neck regions represented eight left ribs; three ribs were represented only by shafts. No sternum was recovered.

Clavicle

Both left and right clavicles were recovered.

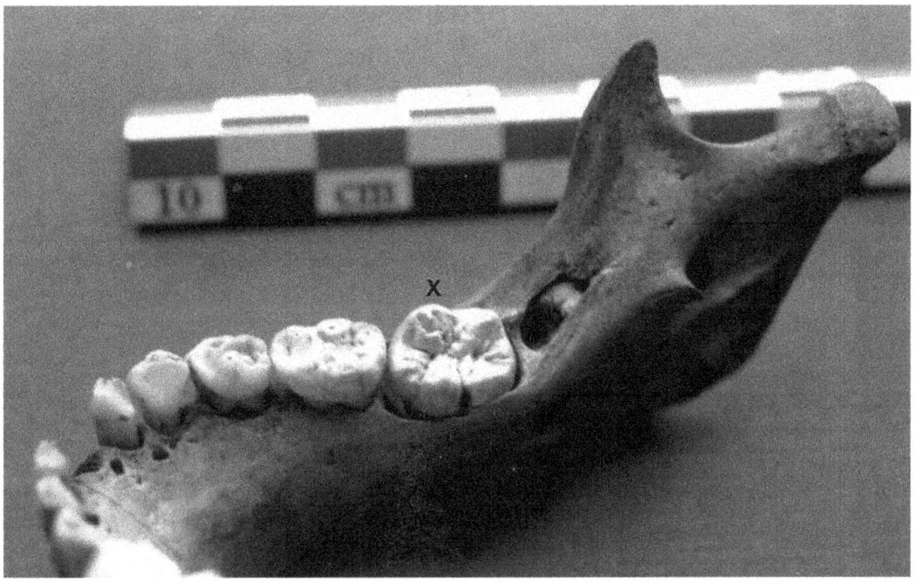

Figure 16.3. Mandible of the Andreanof Island child showing the right permanent first molar (designated with an X) just emerged and the second molar erupting behind it through the alveola. The two teeth just anterior to the first permanent molar are deciduous molars that are shed and replaced by permanent premolars over time.

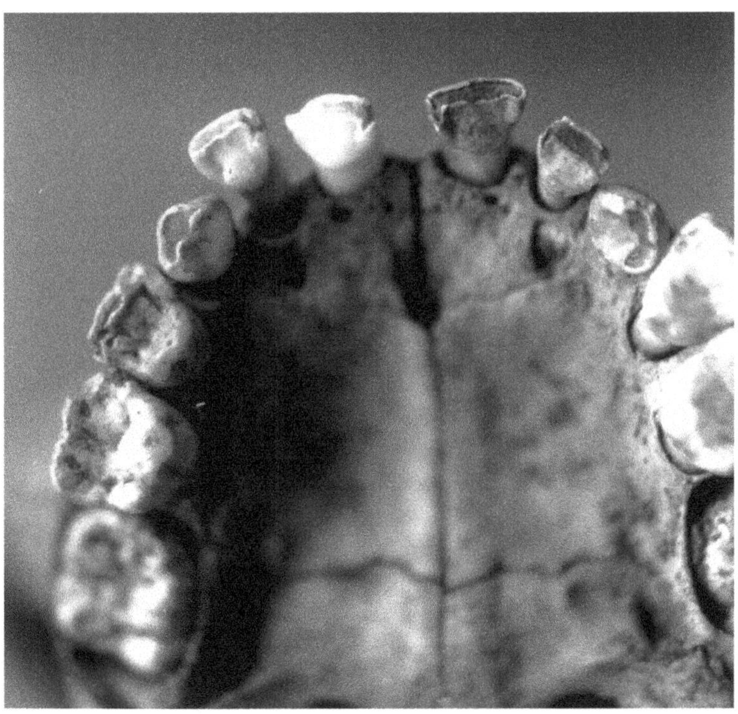

Figure 16.4. Maxillary teeth of the Andreanof child showing wear. Note the shoveling on the central and lateral incisors.

Scapula

The glenoid, coracoid process, infraglenoid tubercle and a portion of the subscapular fossa represented the left scapula. The glenoid and a small fraction of the coracoid process represented the right scapula. The entire body was missing. This is not surprising; the subscapular portion of this bone is thin and highly friable even in adulthood, and does not survive well in an archaeological context.

Arms

The right arm was represented by: a part of the proximal epiphysis and the shaft of the humerus; the diaphysis of the ulna; and the radius with the distal portion of the diaphysis best represented. Root acid etching severely eroded the distal end of the radius although a large area (>80% of bone surface) remained unaffected. Acids had destroyed approximately 50% of the ulna surface, and 70% of the

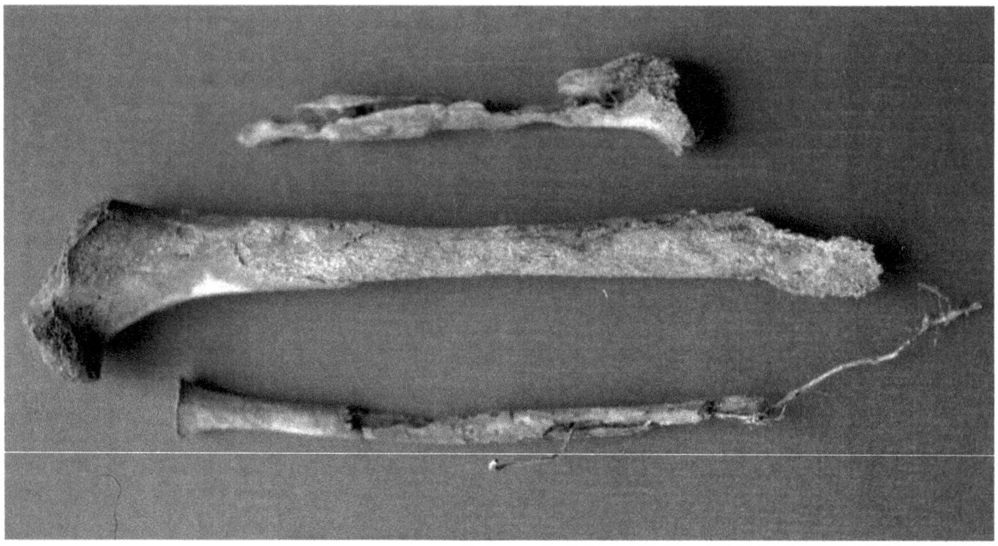

Figure 16.5. Extensive rootlet/acid destruction on bones of the Andreanof Island child.

surface of the neck and shaft of the right humerus had been destroyed by corrosion. Modern roots have left distinctive roadmap tracks on the humerus shaft. The scaphoid of the wrist was also identified.

The shaft of the humerus represented the upper left arm. However, both proximal and distal epiphyses were missing. Acid etching had destroyed approximately 50% of the bone surface of the left humerus. The left ulna and radius were not recovered. Soil acids and/or other diagenetic processes probably destroyed both bones.

Innominate

Left and right pelvic bones were represented by the ilia, ischia, and pubic bones. The three bones of each innominate were unfused; fusion occurs by the twelfth year although this can vary (Bass 1971; Steele and Bramblett 1988). Some erosion had occurred on the superior margin of the ilium. Root etching was visible on the alae of both ilia.

Legs

The right femur was represented by the diaphysis and the proximal epiphysis (head). The posterior portion of the diaphysis in the region of the linea aspera had suffered acid etching. However, approximately 80% of bone table of the anterior surface of this bone was relatively free of root marks, bearing only traces. The epiphysis of the femoral head had suffered some erosion. Diaphyses of the right tibia/fibula were recovered. The complete proximal epiphysis of the right tibia was in excellent condition. The distal third of the tibia shaft on the posterior side suffered extensive acid etching although the corrosion did not eat through the bone table. The distal epiphysis of the right tibia was present and in good condition. A nearly complete shaft represents the right fibula. Both proximal and distal epiphyses were missing.

Roots had attacked the distal three/fourths of the shaft and a modern root was growing inside the medullary cavity of the shaft when it was excavated (Figure 16.5). The right astragalus and calcaneus were also recovered. No root marking or acid etching occurred on these relatively dense bones.

The left femur is represented by the shaft with the best preservation occurring on the proximal end in the region of the neck. This bone has suffered extensive erosion of the bone table on both the anterior and posterior surfaces. Only the area in the vicinity of the medial side of the neck remained unaffected. Both proximal and distal epiphyses of the left femur were missing. However, a portion of a femoral intercondyloid fossa (which could not be assigned to side in the field) was recovered. No lower left limb bones (tibia/fibula) were recovered from the gravesite.

Hands and Feet

Numerous metapodials and phalanges were recovered in the field. Without the aid of a comparative skeleton, and because they possess many epiphyses, these bones could not be readily identified to specific bone. Hand and foot bones included 13 unidentified metapodials. Fingers were represented by eight proximal phalanges, seven middle phalanges, and three distal phalanges. One proximal phalanx and two middle phalanges of the foot were also identified.

Paleoparasitological Analysis

Skeletal preservation allowed archaeologists to precisely locate the abdominal cavity of the buried child and to collect a 150g sample of sediments in this area of the body. A 150g control sample was collected outside, but close to, the burial. Bouchet *et al.* (2001) identified two parasites associated with the abdominal cavity of the child—*Ascaris*

Table 16.3. Native American mtDNA haplogroup definitions and the results from the child burial.

	Haplogroups				Child Burial
Polymorphic Marker	**A**	**B**	**C**	**D**	
*Hae*III np663	+	-	-	-	-
9 bp deletion np8272-8289	-	+	-	-	-
*Hinc*II np13259	+	+	-	+	+
*Alu*I np 5176	+	+	+	-	-

and *Diphyllobothrium*. No parasite egg cases were identified from soil samples surrounding the gravesite.

Ascaris are giant roundworms commonly recovered in ancient human and animal coprolites (Bouchet *et al.* 1999). The specific parasite of man is *A. lumbricoides*. Infestation occurs in the absence of thorough hygiene, including hand washing after defecation, when *Ascaris* eggs enter the body via the mouth. Ingested eggs hatch in the small intestine, and larvae migrate through the gut wall, enter the circulatory system to pass through the heart, liver, and lungs. Larvae break through lung walls and pass to the trachea where they are swallowed and once again enter the intestine where adulthood is reached.

Diphyllobothrium are tapeworms with biological cycles involving marine mammals, fish and humans. Humans can become infected by eating raw or undercooked flesh of sea lions and certain fish. The specific identification for the Andreanof sample is *D. pacificum*, the most common species on the Pacific coast. Tapeworms subsist on blood and can seriously deplete the vitamin B12 of their hosts. In extreme cases, they can provoke a Biemer anemia. No morphological sign, including porotic hyperostosis of the skull—a result of severe anemia sometimes brought on by parasitism, was observed on the skeleton. However, it must be noted that bone is not readily modified due to blood borne anemias (Ortner and Putschar 1985).

If Ascaridiosis is a benign digestive illness, tapeworm infestation is potentially much more debilitating. However, because it is impossible to evaluate the parasite load (density) contained in the intestine of the youngster, it is difficult to evaluate the impact of such parasite infestations and to conclude that these parasites were directly responsible for the death of this individual.

DNA Analysis

O'Rourke and Hayes analyzed rib fragments from the Andreanof child burial. To maximize the potential for DNA extraction and amplification success, as well as comparability with other Beringian human molecular genetic data sets, they restricted their efforts to a low-resolution examination of the mitochondrial DNA (mtDNA) genome. While the bulk of human genetic material is present in the 46 chromosomes of the nuclear genome, the mitochondrion, a cytoplasmic organelle responsible for energy production, also contains genetic material in the form of a small, circular DNA molecule. This mitochondrial genome is a useful tool for studying the genetics of prehistoric individuals due to several unique properties (e.g., Carlyle *et al.* 2000; Hayes *et al.* 2001; Kaestle and Smith, 2001; Kaestle *et al.* 1999; Merriwether *et al.* 1994; O'Rourke *et al.* 1996, 1999, 2000; Parr *et al.* 1996; Stone and Stoneking 1993, 1996, 1998). First, and most important, each cell contains multiple copies of the mitochondrial genome (mtDNA) thereby increasing the likelihood of extracting and amplifying severely degraded ancient DNA (aDNA). Second, the elevated mtDNA mutation rate allows for the discrimination of recently diverged populations, such as the Unangan and other Beringian populations (Forster *et al.* 1996; Shields *et al.* 1993; Saillard *et al.* 2000). And third, mitochondria are exclusively maternally inherited (Giles *et al.*, 1980) which renders an individual haploid for the mtDNA genome (Monnat and Loeb 1985). Hence, sequence differences from generation to generation are due to spontaneous mutation only, not recombination, and degrees of maternal relatedness between individuals can be traced through the mutations in each lineage.

The molecular analysis of the Andreanof burial assessed the presence or absence of three mtDNA restriction sites and a 9bp deletion that define Native American mtDNA haplogroups A through D (see Schurr *et al.* 1990). The relationship between these markers and the haplogroups they define, and their nucleotide pair (np) locations in the mitochondrial genome reference sequence (Anderson *et al.* 1981) are presented in Table 16.3. A '+' denotes the presence of the marker, a '-' denotes its absence.

Three extractions were attempted from the bone fragments (0.50-0.55 g each) which were surface decontaminated with a 15 min 5.25 % sodium hypochlorite (bleach) soak, followed by a 72 hr decalcification in 10 ml of 0.5M EDTA pH 8.5, substituting a fresh change of EDTA every 24 hr (Parr *et al.* 1996). The remaining protein was digested overnight at 56 °C in 1.5 ml of Proteinase K (PK) buffer (50 mM Tris-HCl pH 8, 450 mM EDTA pH 8.5, 0.5 % tween 20, 1 mg/ml PK), and concentrated to 125 µl with centricon C-30s. The resultant DNA-containing solution was added to 625 µl Qiagen PB buffer, the DNA bound to a Qiagen Qiaquick silica-gel matrix column, washed with 750 µl Qiagen PE buffer, and eluted with 35 µl TE buffer [3 mM NaCl, 3 mM Tris pH 8.1, 0.3 mM EDTA] (Yang *et al.* 1998). The aDNA extracts were diluted 1:5 and

Table 16.4. PCR Primers.

Polymorphic Marker	Primer	Sequence (5' - 3')	Anneal Temp (°C)	Reference
*Hae*III	L635	TGAAAATGTTTAGACGGGCTCACATC	63	Handt *et al.*, (1996)
np663	H709	TAGAGGGTGAACTCACTGGAAC		Handt *et al.*, (1996)
9bp	L8215	ACAGTTTCATGCCCATCGTC	58	Wrischnik *et al.*, (1987)
Deletion	H8297	ATGCTAAGTTAGCTTTACAG		Wrischnik *et al.*, 1987
*Hinc*II	L13257	AATCGTAGCCTTCTCCACTTCA	58	Handt *et al.*, (1996)
np13259	H13393	TCCTATTTTTCGAATATCTTGTTC		Ward *et al.*, (1991)
*Alu*I	L5054	TAGGATGAATAATAGCAGCTCTACCG	63	Stone and Stoneking, (1998)
np5176	H5190	GGGTGGATGGAATTAAGGGTGT		Handt *et al.*, (1996)

Table 16.5. Pan-Beringian populations possessing haplogroup D.

Population	Group	N	% D	Reference
Ancient	Unangax̂	30	70.0	Hayes, 2001
Contemporary	Unangax̂	30	67.1	Merriwether *et al.*, 1995; Rubicz and Crawford, 2000
Old Harbor	Yupik Eskimo	115	34.8	Merriwether *et al.*, 1995
Yukagirs	Siberian	27	33.3	Torroni *et al.*, 1993b
Nivkhs	Siberian	57	28.1	Torroni *et al.*, 1993b
Nuu-Chah-Nulth	NW Coast	15	26.7	Torroni *et al.*, 1993a
Gambell	Yupik Eskimo	50	26.0	Merriwether *et al.*, 1995
Bella Coola	NW Coast	36	25.0	Lorenz and Smith, 1996; Torroni *et al.*, 1993a
Siberian	Yupik Eskimo	129	20.2	Torroni *et al.*, 1993b; Starikovskaya *et al.*, 1998
Ouzinkie	Yupik Eskimo	41	14.6	Merriwether *et al.*, 1995
Chukchi	Siberian	90	13.3	Torroni *et al.*, 1993b; Starikovskaya *et al.*, 1998
Haida	NW Coast	29	3.4	Lorenz and Smith, 1996; Torroni *et al.*, 1993a
Inuit/Inupiaq	Inupiaq Eskimo	30	3.3	Lorenz and Smith, 1996
Koryaks	Siberian	201	3.0	Torroni *et al.*, 1993b; Schurr *et al.*, 1999
Savoonga	Yupik Eskimo	49	2.0	Merriwether *et al.*, 1995

1:10 to reduce co-extracting enzymatic inhibitors, thereby decreasing their threat to successful polymerase chain reaction (PCR) amplification (Hummel *et al.* 1992). PCRs were performed in 25 μl with 1x Deep Vent reaction buffer (New England Biolabs), 1x non-acetylated BSA (NEB), 200 mM each dNTP (Stratagene), 2 mM MgSO4 (NEB), 0.15-0.25 μM each primer (Oligos Etc.; see Table 16.4 for sequences), 0.67-1.0 units of Deep Vent (exo-) polymerase (NEB), 5 μl DNA sample and sealed with 1 drop of freshly autoclaved ultrapure mineral oil (Sigma). The reactions were hot-started (Chou *et al.* 1992) at 95 °C for 5 min prior to 40 amplification cycles of 35 sec denaturation (95 °C), 1 min primer anneal (58 or 63 °C; see Table 16.4) and 15 sec extension (72 °C) in a Perkin Elmer 480 thermal cycler. 10 μl of the PCR product were electrophoretically resolved in 3.5 % 3:1 NuSieve agarose gels. If electrophoretic resolution of the PCR product yielded an appropriately sized DNA fragment, potentially containing a Native American haplogroup defining restriction site polymorphism, the remaining 15 μl of PCR product were restriction digested. Digestions were performed overnight at 37 °C according to manufacturer's (NEB) recommendations in 20μl volumes with 10-fold enzyme excess, and again resolved on 3.5 %, 3:1 NuSieve gels.

As with any DNA study, the introduction of modern DNA template during the extraction and amplification process was a constant concern. The protocols used (Carlyle *et al.* 2000; Hayes *et al.* 2005; O'Rourke *et al.* 1996, 2000; Parr *et al.* 1996) were designed to both reduce the likelihood of contamination from spurious modern DNA sources, and detect these events when they do occur. Pre- and post-PCR activities were performed in separate laboratories located on opposite ends of the building, each fully dedicated to aDNA analysis. All pre-PCR procedures were performed in environments (e.g., Airclean 600 enclosure) with continuous HEPA filtered, positively pressured, airflow, and sterilized with bleach, DNA-off (CPG), or DNA Away (MBP), and/or cross-linked with UV light (254nm). Instruments and labware (including aerosol resistant filtered pipettor tips) were freshly autoclaved and UV cross-linked. Reagents were aliquoted into small volumes for single use, freshly autoclaved, and UV-irradiated. Non-autoclavable reagents were 0.22 μm filtered. Lab personnel are masked, coated, sleeved, and gloved. The temp-cycler, located on a separate floor and separate wing of the building, is also irradiated with UV light both prior to and after PCR to further reduce the possibility of cross contamination between samples. Multiple negative extraction controls (all reagents except skeletal sample) and PCR controls (water in place of extracted sample) are used to signal false positives. Results were confirmed from at least two amplifications per marker from the three extractions. These precautions and independent replications meet the criteria for reliable ancient DNA extractions (Cooper and Poinar 2000; Handt *et al.* 1994; O'Rourke *et al.* 1996, 2000; Stoneking 1995).

The results of the mtDNA molecular variation analysis from the Andreanof burial are presented in Table 16.5. The individual lacks the HaeIII np663, 9bp deletion np8272-8289, and AluI np5176 markers and possesses the HincII np13259 marker, signifying the Andreanof child's membership in haplogroup D. This haplogroup is: 1) found at low frequency among US Southwest populations under study in our lab; 2) absent in living Europeans and European-Americans (i.e. all lab and excavation personnel); and 3) the most common haplogroup among prehistoric and contemporary Unangax̂ (Hayes 2001; Merriwether *et al.* 1995; Rubicz *et al.* 2000). This strongly suggests the DNA recovered is not due to modern contamination from lab workers of European descent or cross contamination from other aDNA samples in the lab, but instead represents authentic ancient Unangan DNA. Hence, the data make phylogenetic sense, satisfying an important criterion for authenticating ancient DNA (Handt *et al.* 1994). That a prehistoric Unangan is a member of haplogroup D is not terribly surprising since most ancient and contemporary Unangax̂ are characterized by this haplogroup. In fact, of the pan-Beringian groups for which mtDNA restriction site and length polymorphism data exists, the Unangax̂ have the highest frequency (67.1-70.0 %) of haplogroup D (Table 16.5). This is approximately twice the frequency of the population with the next highest haplogroup D frequency, the Yupik Eskimo from Old Harbor, Kodiak Island (34.8%).

Stable Isotope Analysis

Stable carbon and nitrogen isotope analysis was conducted with a small fragment of rib (1.4g) from the Andreanof child burial. Stable isotope ratios from a single individual are difficult to interpret in the absence of isotope readings from foods likely to be important in the diet of the sampled individual. At present we can compare Andreanof stable isotope signatures to those of an adult burial from the Near Islands and a larger data set from the eastern Aleutians and infer from the range of variation present that the child's diet, like all Unangax̂, was relatively high in marine resources, in this case primarily invertebrates and sea birds.

Among Unangan foragers, skeletal stable carbon isotope ratios ($^{13}C/^{12}C$) covary with reliance on marine foods. These ratios are reported in $\delta\ ^{13}C$ (delta) notation as parts per mil difference from an internationally recognized standard (Craig 1957) and are usually negative because the standard against which carbon isotopes are measured (a Cretaceous marine belemnite) is enriched in ^{13}C relative to naturally occurring Holocene carbons.

The kinetic processes that govern CO_2 uptake and bicarbonate (HCO_3-) formation at the ocean's surface fractionate, or enrich, marine bicarbonates approximately +7 o/oo relative to atmospheric CO_2 (-7.7 o/oo), placing seawater HCO_3- $\delta\ ^{13}C$ values near 0 o/oo (Tauber 1981 1980 in bib; Chisholm *et al.* 1982). Marine plants use a C_3 photosynthetic pathway to metabolize dissolved CO_2 producing tissue $\delta\ ^{13}C$ readings approximately 7-8 o/oo more positive than the tissue of terrestrial C_3 plants taking up atmospheric CO_2 (Ehleringer *et al.* 1997; Fry and Sherr 1988). This difference is passed up the food chain and reflected, with a slight offset (ca. +1 o/oo), in the carbon isotope signature of human populations exploiting marine resources (Ambrose *et al.* 1997; Chisholm *et al.* 1982; McGovern-Wilson and Quinn 1996). Foraging populations heavily dependent on marine animal protein can exhibit $\delta\ ^{13}C$ readings as positive as –12 o/o; whereas foragers exploiting C_3 terrestrial plants and animal protein may have $\delta\ ^{13}C$ values as negative as –21 o/oo (Bender *et al.* 1981; van der Merwe and Vogel 1978).

Stable nitrogen isotope ratios ($^{15}N/^{14}N$) covary with the trophic level of sampled diets. ^{14}N is preferentially excreted at renal membrane boundaries and ^{15}N retained for protein synthesis (Ambrose 1986), systematically enriching the tissue of consumers at each step up the food chain. Unlike $\delta\ ^{13}C$, $\delta\ ^{15}N$ is usually a positive value since atmospheric N_2 is the standard and assigned a value of 0 o/oo. Arctic primary producers such as plankton, kelp and algae exhibit $\delta\ ^{15}N$ readings from ca. +6 o/oo to +10 o/oo, while invertebrates and fish range from ca. +10 o/oo to +16 o/oo and seals and sea lions from +16 o/oo to +18 o/oo (Schell 1999). Polar bear and human foragers who hunt seal and sea lion are at the top of the food chain with $\delta\ ^{15}N$ readings often above +20 o/oo (Coltrain and Hayes 2001; Schell 1999). The trophic level effects of nitrogen metabolism are also well illustrated in nursing infants whose $\delta\ ^{15}N$ values are approximately 3 o/oo more positive than their mothers' (Coltrain1996; Katzenberg 1993).

Adult bone collagen turns over slowly, taking approximately thirty years to replace existing carbon with an equivalent amount of carbon (Libby *et al.* 1964); but, the bone collagen of growing children turns over more rapidly. The isotope signature of bone collagen sampled from growth zones such as long bone epiphyses may reflect a short term or seasonal diet while that sampled from rib should minimally reflect an annual subsistence round, given the long term averaging effects of protein synthesis at a site less subject to rapid remodeling (see Balasse *et al.* 1999).

The Andreanof child exhibited a carbon isotope ratio of –13.9 o/oo and nitrogen isotope reading of +17.0 o/oo (Table 16.6). The rib fragment analyzed was chemically well preserved with a near-modern collagen yield of 25.8% (weight percent collagen in whole bone) and an atomic carbon to nitrogen ratio of 3.0, within the proscribed range for adequately preserved bone collagen. Andreanof and other samples introduced below were soxhlet extracted to

Table 16.6. Stable carbon and nitrogen isotope values for western Aleutian sites and mean values and standard deviations for eastern Aleutian sites.

Site	n=	$\delta^{13}C$ ‰	$\delta^{15}N$ ‰
Andreanof	1	-13.9	+17.0
Near Islands	1	-13.4	+17.8
Kagamil	10	-12.9 ± 0.1	+20.1 ± 0.5
Shiprock	10	-12.4 ± 0.5	+20.4 ± 1.1
Umnak	16	-12.0 ± 0.3	+19.5 ± 0.5

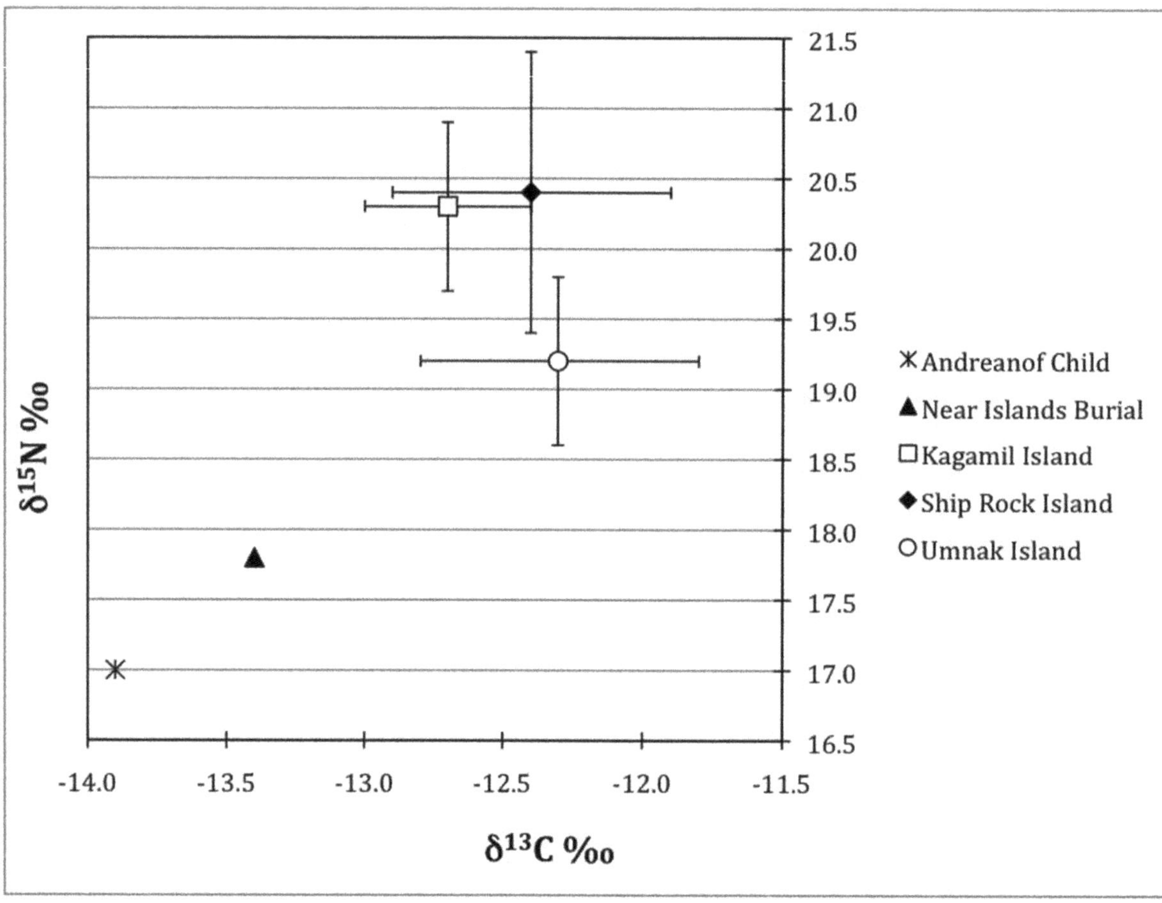

Figure 16.6. Eastern and western Aleutian stable isotope signatures graphed by site.

treat for sea mammal oil contamination then demineralized following Coltrain and Leavitt (2002). Purified collagen was analyzed for stable isotopes on a Finnigan Delta Plus mass spectrometer equipped with a Carlo Erba EA 118 elemental analyzer.

Figure 16.6 illustrates the relationship between the child's isotope values, those of the Near Islands adult burial and isotope signatures from three closely adjacent burial sites in the eastern Aleutians, on Kagamil (N=32), Ship Rock (N=16) and Umnak (N=32) Islands (Coltrain *et al.* 2006; Coltrain 2010). As a group, individuals sampled from the eastern Aleutians subsisted on high trophic level marine diets, regularly exploiting marine mammals and open water fishes. In contrast the Andreanof child appears to be consuming less high trophic level marine mammal tissue and more sea birds and invertebrates, the latter lower on the food chain than those important in eastern Unangan diets. A higher resolution estimate of reliance on marine foods is contingent upon sampling resources likely to have been important in this child's diet.

Conclusions and Discussion

The lives of children are highly enigmatic to nearly invisible in the archaeological record (Bird and Bliege-Bird 2000). It is difficult to determine what activities they may have participated in or how they might have contributed to the overall well being of the social group. The analysis of the Andreanof child provides new information about the diets, eating habits, general health, and mortuary practices surrounding youngsters in the prehistoric Aleutian world.

Centuries ago a six to seven year old Unangan child died and was buried in a shallow grave in an abandoned house midden in the Andreanof Islands, Alaska. The individual was buried on its left side with knees bent and hands placed just below the chin. The youngster was buried with minimal grave goods: a large fishhook and a multipurpose stone tool. A wooden basket made of spruce wood was inverted over the body before it was covered. Morphological analyses indicate that the child did not suffer any traumatic injuries during his/her life. Nor did the bones of the child exhibit any pathological conditions. However, anterior tooth wear on such a young individual suggests that teeth may have been used as tools in either food or hide preparation

Investigations of soils recovered from the abdominal area of the child burial reveal that the youngster suffered from an infestation of both round and tapeworms. This may be expected. The Unangan practice of seasonal sedentism resulted in the build-up of trash middens containing human fecal materials in a moist, cool—but rarely freezing—soil environment which would have favored *Ascaris* egg survival. Disposal of human waste was probably well organized, but parasite eggs could have easily found their way into the mouths of Unangan children as well as older individuals. Hygiene practices, combined with high survival rates of *Ascaris* eggs in refuse areas adjacent to living quarters would have promoted high infection rates. However, seasonal movements, limited band size, and overall small population may have curtailed spread of the parasites among the Unangax̂. Furthermore, Unangax̂, relying on a steady diet of undercooked or raw meat of sea lions and marine fishes, could have easily contracted the *Diphyllobothrium* parasite.

Table 16.7. The importance of commonly exploited resources in Nikolski, Unangan diets (Laughlin 1980, Table 4.2)

Resource Type	***% of Diet***
Marine mammals	30
Fishes	30
Birds and eggs	20
Invertebrates	15
Plants	<5

Stable isotope analysis suggests that the Andreanof child ate a diet high in shellfish. Laughlin (1980, Table 4.2) estimated the average contribution of various resources to the diets of mid-twentieth century Unangax̂ living at Nikolski on Unmak Island in the eastern Aleutians (see Table 16.7). He noted that '[a]ny one of the categories could vary by 5 to 10 percent in successive years, in the same village and between villages, and in different parts of the chain' (Laughlin 1980, 45). As he suggested, minor alterations to his estimates, less intake of marine mammals and more reliance on fish and invertebrates, may account for the slight isotopic depletion evident in Andreanof and Near Island burials relative to eastern Unangan diets. Laughlin (1980, 45) also noted that at Nikolski children and the elderly made a significant contribution to village subsistence. Old women collected terrestrial resources, roots and grass for baskets; and women, the elderly and children foraged along the shoreline for invertebrates. 'The enormous layers of sea urchin shell in the village middens throughout the Aleutian chain testify to the significant contribution of women, children, and old persons' (Laughlin 1980, 45). The latter observation is in keeping with the Andreanof child's diet and suggests that this child may have met a significant portion of his or her own caloric requirements collecting littoral resources.

In the excavated feature where the child had been buried, Lefèvre *et al.* (In press) identified all these taxa, in unequal proportions. Sea mammals were represented by a small number of bones of cetaceans (whales), Steller's sea lion and fur seals. The birds were equally poorly represented: ptarmigans and cormorants were the most abundant, together with various species of alcids (auklets and murres), and some geese and eiders. The species identified here indicate that the people who used this feature hunted terrestrial birds such as ptarmigans and geese, as well as nearshore birds such as cormorants, eiders and alcids. Fish bones were rare, and the most common were cods and rock greenlings. The invertebrates were the most abundant animal remains recovered in this feature. Shells of large heart cockles were particularly abundant, indicating that the inhabitants were exploiting the rich intertidal and subtidal zones.

Although potentially debilitating, there is no evidence to indicate that the child's tapeworm infestation was severe enough to induce starvation and thereby elevate collagen $\delta^{15}N$, nor would starvation be readily detected in bone collagen nitrogen isotope ratios if it were occurring. During periods of extreme caloric shortfall the body metabolizes its own tissues and this may induce near trophic level elevation of $\delta^{15}N$. The trophic level effects of starvation can be monitored incrementally in the collagen of nails or hair strands but is unlikely to be evident in bone since severe caloric shortfall would probably result in death, especially in a juvenile, before the turnover of preexisting proteins. High latitude foragers are also subject to a condition analogous to starvation, induced by high levels of protein consumption in the absence of adequate fat intake, but again the condition would have to be prolonged before detected in bone collagen $\delta^{15}N$ readings. In the absence of food web data or a larger human sample from the central Aleutians, viewing the Andreanof child's $\delta^{15}N$ value as an unbiased indicator of trophic level is the most parsimonious interpretation. Ongoing stable isotope research with bird, fish and sea mammal collections from Shemya Island and invertebrates and terrestrial plant foods from the Andreanof Island group can further clarify these issues.

Although a rib from the child burial dated to 840 ± 40 BP, this date is probably artificially old due to the Marine

Reservoir Effect (Dumond and Griffin 2002). Shellfish extract ancient carbon from their marine environment. When these shells are dated they produce ages much older than when the organism actually lived and died. Because the child relied on a diet high in shellfish, this ancient carbon would have been incorporated into the child's bones. It is probably prudent to slice 500 years off the child's demise, which puts the time of death roughly around 350 years ago.

Acknowledgements

The Aleut Corporation granted permission to exhume and analyze the child burial. Archaeological research was supported by a grant from Office of Polar Programs, National Science Foundation (OPP-9614472) and from the Institut Français pour la Recherche et la Technologie Polaire (Program 301. The U.S. Fish and Wildlife Service provided logistical support. DNA research was supported by the Office of Polar Programs, National Science Foundation, (OPP-9974623, D. H. O'Rourke), as well as support from the Department of Anthropology, College of Behavioral and Social Sciences, and University of Utah. M. G. Hayes was supported by a Natural Sciences and Engineering Research Council of Canada postgraduate fellowship and a University of Utah graduate research fellowship.

References

Ambrose, Stanley H. 1986. Stable Carbon and Nitrogen Isotope Analysis of Human and Animal Diet in Africa. *Journal of Human Evolution* 15, 707-731.

Ambrose, Stanley H., Brian M. Butler, Douglas B. Hanson, Rosalind L. Hunter-Anderson and Harold L. Krueger. 1997. Stable Isotopic Analysis of Human Diet in the Marianas Archipelago, Western Pacific. *American Journal of Physical Anthropology* 104, 343-361.

Anderson, Stephen, Alan T. Bankier, Barclay G. Barrell, Maarten H.L. DeBruijn, Alan R. Coulson, Jacques Drouin, I.C. Eperon, Donald P. Nierlich, Bruce A. Roe, Frederick Sanger, Peter H. Schrier, Andrew J.H. Smith, Roger Staden, and Ian G. Young. 1981. Sequence and Organization of the Human Mitochondrial Genome. *Nature* 290, 457-465.

Balasse, Marie, Herve Bocherens, and Andre Mariotti. 1999. Intra-Bone Variability of Collagen and Apatite Isotopic Compositions Use as Evidence of a Change of Diet. *Journal of Archaeological Science* 26, 593-598.

Bang, Gisele, and A Hasund. 1971. Morphologic Characteristics of the Alaskan Eskimo Dentition. I. Shovel-shape of Incisors. *American Journal of Physical Anthropology* 35(1), 43-47.

Bass, William.M. 1971. *Human Osteology: A Laboratory and Field Manual of the Human Skeleton.* Columbia, Missouri Archeological Society.

Behrensmeyer, Anna K. 1978. Taphonomic and Ecologic Information from Bone Weathering. *Paleobiology* 4, 150-162.

Bender, Margaret M., David A. Baerreis, and Raymond L Steventon. 1981. Further Light on Carbon Isotope and Hopewell Agriculture. *American Antiquity* 46, 346-353.

Bird, Douglas and Rebecca Bliege-Bird. 2000. The Ethnoarchaeology of Juvenile Foragers: Shellfishing Strategies among Meriam Children. *Journal of Anthropological Archaeology* 19, 461-476.

Bouchet, François, Christine Lefèvre, Dixie West, and Debra Corbett. 1999. First Paleoparasitological Analysis of a Midden in the Aleutian Islands (Alaska): Results and Limits. *Journal of Parasitology* 85, 369-372.

Bouchet, François, Dixie West, Christine Lefèvre, and Debra Corbett. 2001. Identification of Parasitoses in a Child Burial from Adak Island (Central Aleutian Islands, Alaska). Compt rendu, *Academy of Sciences* 324, 123-127.

Byers, David A., David Yesner, Jack Broughton, and Joan B. Coltrain. 2011. Stable Isotope Chemistry and Late Prehistoric Subsistence Change in the Aleutian Islands. *Journal of Archaeological Science* 38(1), 183-196.

Carbonell, V. M. 1963. Variations in the Frequency of Shovel-shaped Incisors in Different Populations. In D. R. Brothwell (ed.), *Dental Anthropology*, 211-234. London, Pergamon Press.

Carlyle, Shawn W., Ryan L. Parr, M. Geoffery Hayes, and Dennis H. O'Rourke, 2000. The Context of Maternal Lineages in the Greater Southwest. *American Journal of Physical Anthropology* 113, 85-101.

Chisholm, Brian S., D. Erle Nelson, and Henry P. Schwarcz. 1982. Stable-Carbon Isotope Ratios as a Measure of Marine Versus Terrestrial Protein in Ancient Diets. *Science* 216, 1131-1132.

Chou, Quin, Marion Russell, David E. Birch, Jonathan Raymond, and Will Bloch. 1992. Prevention of Pre-PCR Mis-priming and Primer Dimerization Improves Low-copy Number Amplifications. *Nucleic Acids Research* 20, 1717-1723.

Coltrain, Joan B. 1996. Stable Carbon and Radioisotope Analysis. In R.K. Talbot and L. D. Richins (eds.), Steinaker Gap: An Early Fremont Farmstead. *Brigham Young University Museum of Peoples and Cultures Occasional Papers No. 2.* 115-122. Provo Utah.

Coltrain, Joan B. and M. Geoffery Hayes. 2001. Eastern Arctic Foraging Strategies: A Reconstruction of Coastal Paleodiets from the Bone Chemistry of Dorset, Thule and Proto-historic Burial Assemblages. Paper presented at the 70th Annual Meeting of the American Association of Physical Anthropologists, March, Kansas City, Missouri.

Coltrain, Joan B., and Steven W. Leavitt. 2002. Climate and Diet in Fremont Prehistory: Economic Variability and Abandonment of Maize Agriculture in the Great Salt Lake Basin. *American Antiquity* 67 (3), 453-485.

Coltrain, Joan B., M. Geoffrey Hayes, and Dennis H. O'Rourke. 2006. A Radiometric Evaluation of Hrdlička's Aleutian Population Replacement Hypothesis. *Current Anthropology* 47, 537-548.

Cooper, Alan, and Hendrik N. Poinar. 2000. Ancient DNA: Do It Right or Not at All. *Science* 289, 1139.

Craig, Harmon. 1957. Isotopic Standards for Carbon and Oxygen and Correction Factors for Mass Spectrometer

Analysis of Carbon Dioxide. *Geochimica et Cosmochimica Acta* 12, 133-149.

Dahlberg, Albert. 1949. The Dentition of the American Indian. In William Laughlin (ed.), *The Physical Anthropology of the American Indian*, 138-176. New York, Viking Fund.

Dumond, Don and Dennis Griffin. 2002. Measurements of the Marine Reservoir Effect on Radiocabon Ages in the Eastern Bering Sea. *Arctic* 55 (1), 77-86.

Ehleringer, James R., Thure E. Cerling, and Brent R. Helliker. 1997. C_4 Photosynthesis, Atmospheric CO_2, and Climate. *Oecologia* 122, 285-299.

Forster, Peter, Rosalind Harding, Antonio Torroni, and Hans-Jürgen Bandelt. 1996. Origin and Evolution of Native American mtDNA Variation: A Reappraisal. *American Journal of Human Genetics* 59, 935-945.

Fry, Brian and Evelyn B. Sherr. 1988. ^{13}C Measurements as Indicators of Carbon Flow in Marine and Freshwater Ecosystems. In P.W. Rundel, J.R. Ehleringer and K.A. Nagy (eds.), *Stable Isotopes in Ecological Research*, 196-279. New York, Springer-Verlag.

Giles, Richard E., Hugue Blanc, Howard M. Cann, and Douglas C. Wallace. 1980. Maternal Inheritance of Human Mitochondrial DNA. *Proceedings of the National Academy of Sciences* 77, 6715-6719.

Handt, Olivia, Matthias Höss, Matthias Krings, and Svante Pääbo. 1994. Ancient DNA: Methodological Challenges. *Experientia* 50, 524-529.

Handt, Olivia, Matthias Krings, R.H. Ward, and Svante Pääbo. 1996. The Retrieval of Ancient Human DNA Sequences. *American Journal of Human Genetics* 59, 368-76.

Hayes, M. Geoffery. 2001. Ancestor-Descendant Relationships in North American Arctic Prehistory: Ancient DNA Evidence from the Aleutian Islands and the Eastern Canadian Arctic. *American Journal of Physical Anthropology Supplement 32*, 78.

Hayes, M. G., J. B. Coltrain, and D. H. O'Rourke. 2005. Molecular Archaeology of the Dorset, Thule, and Sadlermiut: Ancestor-descendant Relationships in eastern North American Arctic Prehistory. In P. Sutherland (ed.), *Contributions to the Study of the Dorset Paleo-Eskimos*, 11-32. Mercury Series, Archaeological Paper 167. Hull, Canada, Canadian Museum of Civilization.

Hrdlička, Aleš. 1920. Shovel-shaped Teeth. *American Journal of Physical Anthropology* 3, 429-465.

Hummel, Susanne, Gabrielle Nordsiek, and Bernd Herrmann. 1992. Improved Efficiency in Amplification of Ancient DNA and Its Sequence Analysis. *Naturwissenschaften* 79, 359-360.

Kaestle, Frederika A., Joseph G. Lorenz, and David G. Smith. 1999. Molecular Genetics and the Numic Expansion: A Molecular Investigation of the Prehistoric Inhabitants of Stillwater Marsh. In B. E. Hemphill and C. S. Larsen (eds.), *Prehistoric Lifeways in the Great Basin Wetlands: Bioarchaeological Reconstruction and Interpretation*, 167-183. Salt Lake City, University of Utah Press.

Kaestle, Frederika A. and David G. Smith. 2001. Ancient Mitochondrial DNA Evidence for Prehistoric Population Movement: the Numic Expansion. *American Journal of Physical Anthropology* 115, 1-12.

Katzenberg, Margaret A. 1993. Age Difference and Population Variation in Stable Isotope Values from Ontario, Canada. In J.B. Lambert and G. Grupe, (eds.), *Prehistoric Human Bone: Archaeology at the Molecular Level*, 39-62. Berlin, Springer Verlag.

Krogman, Wilton M. 1978. *The Human Skeleton in Forensic Medicine*. Springfield, Illinois, Charles C. Thomas Publisher.

Laughlin, William S. 1980. *Aleuts: Survivors of the Bering Land Bridge*. Orlando, Florida, Harcourt Brace.

Lefèvre, Christine, Dixie West, Debra G. Corbett. In press. Zooarchaeological Analysis at ADK-911, Adak Island, Central Aleutian Islands, Alaska. In Caroline Funk and Debra Corbett (eds.) Anthropology of the Western Aleutian Islands. *Arctic Anthropology*.

Libby, Willard F., Rainer Berger, James F. Mead, George V. Alexander, and Joseph F. Ross. 1964. Replacement Rates for Human Tissue from Atmospheric Radiocarbon. *Science* 146, 1170-1172.

Lorenz, Joseph G. and David G. Smith. 1996. MtDNA Variation Among Native North Americans: Geographic and Ethnic Distribution of the Four Founding mtDNA Haplogroups. *American Journal of Physical Anthropology* 101, 307-323.

Luttrell, Mark and Debra Corbett. 2000. Archaeological Investigations on Northeast Adak Island, Alaska, July, 1999. Report on file Anchorage, Alaska, US Fish and Wildlife Service.

McGovern-Wilson, Richard and Carol Quinn. 1996. Stable Isotope Analysis of Ten Individuals from Afetna, Saipan, Northern Mariana Islands. *Journal of Archaeological Science* 23, 59-66.

Merriwether, D. Andrew, Francisco Rothhammer, and Robert E. Ferrell. 1994. Genetic Variation in the New World: Ancient Teeth, Bone, and Tissue as Sources of DNA. *Experientia* 50, 592-601.

Merriwether, D. Andrew, Francisco Rothhammer, and Robert E. Ferrell. 1995. Distribution of the Four Founding Haplotypes in Native Americans Suggests a Single Wave of Migration for the New World. *American Journal of Physical Anthropology* 98, 411-430.

Monnat, Raymond J., Jr. and Lawrence A. Loeb. 1985. Nucleotide Sequence Preservation of Human Mitochondrial DNA. *Proceedings of the National Academy of Sciences* 82, 2895-2899.

O'Rourke, Dennis H., Shawn W. Carlyle, and Ryan L. Parr. 1996. Ancient DNA: A Review of Methods, Progress, and Perspectives. *American Journal of Human Biology* 8, 557-571.

O'Rourke, Dennis H., Ryan L. Parr, and Shawn W. Carlyle. 1999. Molecular Genetic Variation in Prehistoric Inhabitants of the Eastern Great Basin. In B. E. Hemphill and C. S. Larsen (eds.) *Prehistoric Lifeways in the Great Basin Wetlands: Bioarchaeological Reconstruction and Interpretation*, 84-102. Salt Lake City, University of Utah Press.

O'Rourke, Dennis H., M. Geoffery Hayes, and Shawn W. Carlyle. 2000. Ancient DNA Studies in Physical Anthropology. *Annual Review of Anthropology* 29, 217-242.

Ortner, Donald and Walter G. J. Putschar. 1985. Identification of Pathological Conditions in Human Skeletal Remains. *Smithsonian Contributions to Anthropology* No. 28. Washington, DC, Smithsonian Institution.

Parr, Ryan L., Shawn W. Carlyle, and Dennis H. O'Rourke. 1996. Ancient DNA Analysis of Fremont Amerindians of the Great Salt Lake Wetlands. *American Journal of Physical Anthropology* 99, 507-518.

Rubicz, Rohina C., Theodore G. Schurr, and Michael H. Crawford. 2000. Mitochondrial DNA Diversity in Modern Aleuts and Their Genetic Relationship to Other Circumarctic Populations. Paper presented at the 8th Annual Arctic Conference, Lawrence, KS, November 3.

Saillard, Juliette, Peter Forster, Niels Lynnerup, Hans-Jürgen Bandelt, and Søren Nørby. 2000. mtDNA Variation Among Greenland Eskimos: The Edge of the Beringian Expansion. *American Journal of Human Genetics* 67, 718-726.

Schell, Donald. 1999. Foodweb and Trophic Level Interactions. Manuscript on file Department of Anthropology, University of Utah, Salt Lake City.

Schurr, Theodore G., Scott W. Ballinger, Yik-Yuen Gan, Judith A. Hodge, D.Andrew Merriwether, Dale N. Lawrence, William C. Knowler, Kenneth M Weiss, and Douglas C. Wallace. 1990. Amerindian Mitochondrial DNAs Have Rare Asian Mutations at High Frequencies, Suggesting They Derived from Four Primary Maternal Lineages. *American Journal of Human Genetics* 46, 613-623.

Schurr, Theodore G., Rem I. Sukernik, Yelena B. Starikovskaya, and Douglas C. Wallace. 1999. Mitochondrial DNA Variation in Koryaks and Itel'men: Population Replacement in the Okhotsk Sea—Bering Sea Region During the Neolithic. *American Journal of Physical Anthropology* 108, 1-39.

Shields, Gerald F., Andrea M. Schmiechen, Barbara L. Frazier, Alan Redd, Mikhail I. Voevoda, Judy K. Reed, and R. H. Ward. 1993. mtDNA Sequences Suggest a Recent Evolutionary Divergence for Beringian and Northern North America Populations. *American Journal of Human Genetics* 53, 549-562.

Starikovskaya, Yelena B, Rem I. Sukernik, Theodore G. Schurr, Andreas M. Kogelnik, and Douglas C. Wallace. 1998. mtDNA Diversity in Chukchi and Siberian Eskimos: Implications for the Genetic History of ancient Beringia and the Peopling of the New World. *American Journal of Human Genetics* 63, 1473-1491.

Steele, D.Gentry and Claud A. Bramblett. 1988. *The Anatomy and Biology of the Human Skeleton*. College Station, Texas A & M University Press.

Stewart, T. Dale. 1979. *Essentials of Forensic Anthropology Especially as Developed in the United States*. Springfield, Illinois, Charles C. Thomas.

Stone, Ann C. and Mark Stoneking. 1993. Ancient DNA from a Pre-Columbian Amerindian Population. *American Journal of Physical Anthropology* 92, 463-471.

Stone, Ann C. and Mark Stoneking. 1996. Genetic Analysis of an 8000-year-old Native American Skeleton. *Ancient Biomolecules* 1, 83-87.

Stone, Ann C. and Mark Stoneking. 1998. mtDNA Analysis of a Prehistoric Oneota Population: Implications for the Peopling of the New World. *American Journal of Human Genetics* 62, 1153-1170.

Stoneking, Mark. 1995. Ancient DNA: How Do You Know When You Have It and What Can You Do With It? American *Journal of Human Genetics* 57, 1259-1262.

Tauber, Hemrik. 1981. ^{13}C Evidence for Dietary Habits of Prehistoric Man in Denmark. *Nature* 292, 332-333.

Torroni, Antonio, Theodore G. Schurr, Margaret F. Cabell, Michael D. Brown, James V. Neel, Merethe Larsen, David G. Smith, Carlos M. Vullo, and Douglas C. Wallace. 1993a. Asian Affinities and Continental Radiation of the Four Founding Native American mtDNAs. *American Journal of Human Genetics* 53, 563-590.

Torroni, Antonio, Rem I. Sukernik, Theodore G. Schurr, Yelena B. Starikovskaya, Margaret F. Cabell, Michael H. Crawford, Anthony G. Comuzzie, and Douglas C. Wallace. 1993b. mtDNA Variation of Aboriginal Siberians Reveals Distinct Genetic Affinities with Native Americans. *American Journal of Human Genetics* 53, 591-608.

Van der Merwe, Nicholas J. and John C. Vogel. 1978. ^{13}C Content of Human Collagen as a Measure of Prehistoric Diet in Woodland North America. *Nature* 276(5690), 815-816.

Ward, R.H., Barbara L. Frazier, Kerry Dew-Jager, and Svante Pääbo. 1991. Extensive Mitochondrial Diversity Within a Single Amerindian Tribe. *Proceedings of the National Academy of Sciences* 88, 8720-8724.

West, Dixie, Debra Corbett, and Christine Lefèvre. 1999. 1999 NSF Grant Report: the Western Aleutians Archaeological and Paleobiological Project. Report on file to the National Science Foundation. Washington DC, National Science Foundation.

Wrischnik, Lisa A., Russell G.Higuchi, Mark Stoneking, Henry A. Erlich, Norman Arnheim, and Allan C. Wilson. 1987. Length Mutations in Human Mitochondrial DNA: Direct Sequencing of Enzymatically Amplified DNA. *Nucleic Acids Research* 25, 529-542.

Yang, Dongya Y., Barry Eng, John Waye, J. Christopher Dudar, and Shelley R. Saunders. 1998. Improving DNA Extraction From Ancient Bones Using Silica-based Spin Columns. *American Journal of Physical Anthropology* 105, 539-543.

Chapter 17
Conclusions

Dixie West and Susan Crockford

'Archaeologists often seem to build more on what their predecessors concluded about the past than on the evidence upon which these conclusions were based.' (Trigger 1989, 16)

'If it were easy, everyone would be out here doing it.' (Debra Corbett 1993, Buldir Island)

Introduction

The Aleutian Islands comprise a distinctive ecosystem that supports diverse and abundant marine life and a human presence that is closely tied to the Bering Sea environment and natural resources. Little is understood about human and ecological interactions in the remote past of the Aleutians and particularly how the central Aleutians fits into any broad geographic patterns for the archipelago as a whole. The Central Aleutians Archaeological and Paleobiological Project (CAAPP) adopted an interdisciplinary approach to Aleutian data analysis by integrating individual research fields across three disciplinary domains: geology, paleobiology, and archaeology. Consequently, ours is the first rigorous assimilation of data to provide an understanding of the fundamental physical, natural, and human context in the Andreanof Islands, Alaska through deep time. When possible, central Aleutians cultural and paleobiological dynamics were evaluated against comparable data from both the eastern and western ends of the Aleutian archipelago. Essentially, this research comprises a pan-Aleutian comparison that, quite frankly, has been impossible until now. Finally, our collaborative efforts provide biologists studying current Bering Sea biocomplexity with meaningful information about ecosystem dynamics in the North Pacific Rim spanning nearly six millennia. As the quote from Trigger above suggests, we remind the reader that our data and analyses are touchstones that future scientists working in the central Aleutians can weigh and test against new data sets.

CAAPP Results: Putting It All Together

Geomorphologic research of the central Aleutians is generally lacking. While geomorphology shapes human occupation of landscapes, it is itself independent of human influence. Similarly, while the relationship between marine deposits and sea level history varies among islands — often in unknown ways — these natural processes strongly influence human occupation. CAAPP integrated geological and archeological investigations to correlate human settlement patterns on north Adak Island with documented changes in earth history. Combined geological and paleobiological data provided a framework for understanding the physical and natural world prior to and during Unangan occupation on Adak. To that end, the project team addressed the following questions. Did specific changes in the physical environment (e.g., glaciation, sea level fluctuations, volcanic eruptions, climate change) influence the initial occupation, continued sustainability or eventual demise of Adak cultures?

Geological evidence suggests that Adak was unglaciated by at least 11,000 years ago (Waythomas 1995) and certainly by the time Unangan peoples settled on Zeto Point approximately 6000 years ago. Sea level change only somewhat influenced Unangan settlement patterns on Adak Island. Gualtieri and colleagues (Chapter 3 this volume) identified fives sets of terraces on Adak and Kanaga Islands. From youngest to oldest, the terraces are (1) 2-3m, (2) 5-6m, (3) 7-10m, (4) 13-15m, and (5) 30-40m above current sea level. Prehistoric sites on Adak exist at different elevations and on different terraces through time. The earliest site ADK-171 (c. 6000 BP) is located 15m above sea level and is on the fourth terrace. ADK-011, with components 1 (c. 2500 BP) and 2 (170-400 BP), is located even higher at 19.8m above current sea level between the fourth and fifth terraces. Low lying site ADK-084 (280-500 BP) was only occupied when the second terrace provided dry real estate for human habitation. Site ADK-012 (c. 1860 BP) is currently situated 5m above sea level; however, the topography may have slightly changed due to military bulldozing. Radiocarbon dated sites indicate that as lower terraces became available for occupation, Unangan people did not necessarily abandon higher ground, but opted to live either at high and low elevations. However, later sites like ADK-084, located on the second terrace, could have only been occupied when the terrace was established well above tide line. Whether high or low, each site probably provided a distinct advantage for human habitation, whether that was protection from the wind and climate, access to valuable food resources, fresh water, a suitable beach for kayak landings, and any number of reasons based on seasonal and/or topographic choices.

The changing coastline associated with sea level change probably did not negatively influence long term Unangan economics and might even have improved it. As noted in West *et al.* (Chapter 10 this volume) when uplift occurred, and if the appropriate substrate was available, invertebrates quickly colonized emerging shorelines by moving vertically. Species displacement created an empty niche that other invertebrate species probably quickly filled. Kelp forests expanded seaward, creating more area for even larger populations of sea otters, harbor seals, kelp-dwelling fishes and perhaps fur seals. The nearshore environment, so critical for year-round human survival in the Aleutians, became larger and more easily accessible. Sea lions, dependent on dry land for rookeries and haul outs, adapted by occupying newly emerged coastlines Similarly, new islets and reef systems created additional roosting places for sea birds while the greater area exposed on islands provided more nesting grounds.

Timing of the formation of the barrier beach that encloses Clam Lagoon is uncertain. An OSL sample from the basal medium-to-coarse sand in a marsh pit associated with the barrier beach yielded a minimal age of 2045 ± 290 years for beach emergence (Gualtieri *et al.* Chapter 3 this volume). The barrier beach that formed the lagoon probably existed much earlier. The formation of the lagoon created an additional human exploitation zone along with two already available, the open ocean and kelp forest surrounding north Adak. *Clinocardium* aggregated in the low energy, sandy mud intertidal zone of the lagoon and became a key invertebrate resource for humans living at ADK-171 and ADK-011 (West *et al.* Chapter 10 this volume). Saffron cod and *Clinocardium* at the earliest site ADK-171 suggests that the lagoon was present when the Unangan people settled Zeto Point by 6000 years ago (Savinetsky *et al.* Chapter 5 this volume; West *et al.* Chapter 10 this volume).

At least five major volcanic events are recorded on north Adak: the Main, Intermediate, Sandwich, YBO, and 40 Year ash occurrences (Black 1976). Volcanism potentially had a significant impact on human habitation. To test previous chronologies of volcanic eruptions, Mitsuru Okuno and colleagues (Chapter 4 this volume) collected charred fragments associated with tephra layers from both archaeological sites and from natural contexts across Adak and Kanaga islands. The ancient volcanic cones of Mt. Moffet and Mt. Adagdak on north Adak produced none of the Holocene aged ash found in either natural or archaeological deposits on north Adak. Apparently, Kanaga volcano produced none of the major, upper ashes on Adak, but varying thicknesses of Adak ash deposits suggest that an unidentified, submerged volcano produced some of the Adak ash. Okuno confirmed that the ash located immediately below site ADK-171 is the Intermediate Ash that fell approximately 7200 years ago. ADK-171 cultural layers located immediately above a poorly developed soil and Intermediate Ash suggests that the volcanic explosion did not prohibit people from settling Zeto Point soon after the ash fell. Nor did the unknown event that redeposited the Intermediate Ash at the ADK-171 *Clinocardium* midden cause the Unangan people to abandon the site. The dates of the Intermediate Ash suggest that radiocarbon dates for the ADK-171 *Clinocardium* midden, ranging between 5405 and 6410 BP, are largely accurate and probably fall closer to the 6500 year range.

Three thousand years later, the Sandwich ash fall, dated circa 4000 BP apparently did not dissuade Unangan people from living at ADK-187, the Dune Site (see West *et al.* Chapter 1 this volume Appendix 1.B). It is currently unknown how quickly island ecosystems recover from catastrophic volcanic explosions and ash falls. Radiocarbon dating and correlation of the geological and archaeological record on Adak are not fine grained enough to observe immediate impacts of volcanic eruptions. Current US Fish and Wildlife research following the 2008 volcanic eruption on Kasatochi Island promises critical information about Aleutian ecosystem devastation and recovery timing.

A currently unexplained, settlement hiatus occurred between 6000 and 4000 BP on north Adak. Diatom analysis at Haven Lake in the sixth zone indicates that climatic conditions were warmer and more stable than the preceding period when ADK-171 was occupied (Savinetsky *et al.* Chapter 5 this volume). Warmer and stable climatic conditions assume less Bering Sea bioproductivity; nevertheless, the small Unangan population probably still had plentiful resources to sustain itself in the north Adak vicinity. No apparent volcanic eruption caused migration from north Adak during this 2000 year time period. The most parsimonious explanations include: (1) Unangan people did not live in the Clam Lagoon vicinity for two millennia and lived on other parts of Adak or adjacent islands or, (2) archaeologists simply have not found deeply buried cultural horizons on higher terraces associated with sites of this time period. ADK-171 cultural horizons are buried 120-140cm deep and do not appear to be affiliated with pits visible on the current ground surface at Zeto Point. Although north Adak may have been uninhabited between 6000 and 4000 BP, we suggest that if archaeological sites from this time do exist, they are buried comparatively deeply on higher terraces and are probably not associated with visible, ground surface pits.

The upper layers of many archaeological sites on north Adak contain 40 Year Ash that fell approximately 400 years ago (Okuno *et al.* Chapter 4 this volume). In the Clam Lagoon vicinity ADK-011 Component 2 is the only known archaeological site where cultural materials dating 170-395 years old occur above this tan ash. O'Leary (2001, 228) noted that ADK-018 at the south entrance of the Bay of Islands and ADK-015 at Shagak Bay contain cultural deposits above 40 Year Ash. New CAAPP volcanic ash dates suggest the sites reported by the USBIA and ADK-011 cultural layers above 40 Year Ash can be no older than 400 years BP (Okuno *et al.* Chapter 4 this volume). CAAPP investigations are in agreement with O'Leary that protohistoric sites are rare on north Adak, and historic sites seem virtually nonexistent. Our geological and environmental data show that no major geological or climatic events forced Unangan people to abandon north Adak in recent prehistory. Historical evidence, translated and summarized by Lydia Black (1984), indicates that Adak was apparently occupied when Russians first explored the archipelago in the 18th century. However, a larger Unangan population was settled on Kanaga Island, west of Adak, when Russians first visited the Andreanof Islands. The occurrence of Atka dacite in Adak lithic tool assemblages (Nicolaysen *et al.* Chapter 11 this volume) suggests that, following initial settlement, these adroit mariners probably moved easily, and possibly frequently, within the Andreanof Island group. Archaeological settlements encompassing the historic time period should be found on Kanaga, other parts of Adak or adjacent islands if the ethnographic record is accurate. Ongoing archaeological work on high elevation sites by Diane Hanson and colleagues (Hanson *et al.* 2008; Hanson and Corbett 2010) should throw new light on Unangan occupations and settlement patterns on Adak.

Although the climate significantly varied from the beginning of Unangan occupation (c. 6000 years ago) to Russian contact (mid 18th century), work by Savinetsky, Crockford, and West (Chapters 5, 6 and 10 this volume) suggests the central Aleutians faunal composition (invertebrates and vertebrates) did not dramatically change throughout prehistory. Sea otters and fur seals are found in all north Adak sites regardless of time period. However, some prey reshuffling occurs over time. Some shifts appear to be environmentally driven; others might have been brought on by human activity.

Saffron cod bones represent 12-13% of identified cod bones in the oldest Adak site ADK-171. Saffron cod, an extreme cold-water species, is not identified in later Adak sites or any prehistoric Aleutian site for that matter and currently does not live in the Aleutian region. The ratio of boreal and north-alpine species of diatoms (fifth zone) at the Haven Lake peat-mineral deposit coupled with the identification of saffron cod at ADK-171, an archaeological site of comparable age to Haven Lake zone 5, indicates humans first settled on Adak during an extremely cold, but highly bioproductive, period. Although humans may have disrupted and ultimately exterminated saffron cod in Clam Lagoon, climatic amelioration accounts for the demise of this cold loving fish in the greater Bering Sea region.

At the ADK-171 midden Unangan people shifted invertebrate collection practices. Species inhabiting the rocky intertidal zone (e.g., *Mytilus*) were replaced by species living within the sandy intertidal zones (*Clinocardium*). This invertebrate shift could be due to decreasing predator pressure of sea otters on infaunal species (cockles) as the earliest Adak immigrants quickly depleted sea otters in Clam Lagoon (Savinetsky *et al.* Chapter 5 this volume). Faunal and isotopic data provided by Crockford and Garong *et al.* (Chapters 6 and 8 this volume) suggest that prehistoric sea otters fed in kelp forests and open ocean, rather than inside the lagoon. The Unangan people probably hunted sea otters in kelp forests surrounding north Adak for most of prehistory. But it is not unreasonable to suppose that sea otters inhabited the lagoon when the first Unangan people colonized the island. Blocked from escape by a narrow inlet to the sea, lagoon otters became easy targets in the enclosed lagoon. Optimal foraging theory suggests that lagoon sea otters would have been the first killed by human predators new on the Aleutian landscape. A rapid depletion of sea otters in the lagoon resulted in a favorable, low energy environment for *Clinocardium* aggregations free from otter predation. *Clinocardium* from late prehistoric levels at ADK-011 quickly grew and had attained a relatively old age by the time they were harvested, suggesting cockles lived in a favorable, nutrient rich environment for maximum growth and longevity (Koike *et al.* Chapter 9 this volume). In contrast, at the ancient ADK-171 *Clinocardium* were harvested younger (on average) but at a larger size (West *et al.* Chapter 10 this volume). An absence of sea otters in the lagoon after the occupation of ADK-171 could account for prehistoric cockle demographics.

Nishida *et al.* (Chapter 7 this volume) genetically analyzed sea otter bones recovered from Adak archaeological sites from oldest to youngest: ADK-171 (c. 6000 BP), ADK-012 (c. 2000 BP), ADK-011 component 2 Layer 2 (285-415 BP), and ADK-011 component 2 Layer 1(285-170 BP). Genetic diversity of mtDNA for the sea otter was generally low through time with only five haplotypes, B, B2, B3, B4 and W3 in Adak samples. Four of these (B2, B3, B4 and W3) are new sea otter haplotypes. Genetic diversity of mtDNA was higher in sea otters from earlier sites ADK-171 and ADK-012 than from those recovered from the late prehistoric site ADK-011. These genetic depletion events apparently occurred either by human predation and/or environmental changes of sea otter habitat. Intensive Unangan sea otter hunting in the immediate area around Clam Lagoon could have resulted in decreased genetic diversity over time. It remains unclear if this genetic depletion is only local and due to Unangan overhunting in a specific area of north Adak or if this pattern encompasses a larger Bering Sea region.

The Aleutians sea otter populations comprise the subspecies *E. lutris kenyoni.* Although contemporary sea otter populations of Amchitka in Aleutian Islands possesses haplotype A (dominant) and B (Larson *et al.* 2002), the ancient Adak sea otter population had no haplotype A. This indicates gene flow was largely restricted between close, neighboring islands in the Aleutian archipelago. Prehistoric intensification of hunting and/or high population densities at one location likely increased the pressure for some individuals to emigrate (Estes 1990). The genetic evidence suggests that for thousands of years Adak hosted a core population of sea otters that was not impacted by immigration from Amchitka but that Amchitka populations were occasionally augmented by sea otters moving from Adak. There appears to be some essential differences in ecological history between Amchitka and Adak that we do not yet understand although disparity in the human component might be an element to explore.

Faunal analyses by Savinetsky *et al.* and Crockford (Chapters 5 and 6 this volume) reveal that fur seals are also found in north Adak archaeological sites. Pups were also identified at ADK-171, ADK-012 and ADK-011. Although fur seals no longer give birth in the Aleutian Islands, Savinetsky's analysis coupled with that of Crockford for later sites, demonstrates that fur seal rookeries existed near Adak Island from 6000 years ago to the 18th century. This evidence fits a larger, well-documented pattern seen across the Eastern North Pacific prehistorically: many locally-breeding fur seal populations existed that were harvested regularly by aboriginal peoples. None of these rookeries are listed in the historic record, suggesting that locally-breeding fur seals became extinct just before European contact or shortly after. In the Aleutians, fur seals may have become extinct in conjunction with the intense commercial hunt for sea otters that began with the Russian occupation of the western-most island groups (Near, Rat and Andreanof) c. 1745. Thus began a period of dramatic change in the Aleutians, for humans and animals, which has become a legacy of the region. Among a litany of other impacts, the quest for sea otter pelts may also have helped drive Aleutian fur seals to extinction. The historic-era collapse of mature kelp forests, which characterized so many Aleutian prehistoric landscapes, appears to link the fates of these two species.

Crockford (Chapter 6 this volume) suggests the intriguing possibility that fur seals used the canopy surface of mature kelp forests as 'rookery' habitat. Although perhaps impossible to prove, the idea nevertheless fits a number of facts we know about these fur seals and about sea otters. If true, it appears that fur seals did not survive the repeated disturbance created by Russian-organized sea otter hunts within their kelp forest habitat or that the reduction in the kelp forests (due to decimated sea otter populations and subsequent consumption of kelp by sea urchins) robbed fur seals of the rookery habitat they required. Whatever the cause, clearly these fur seals did not survive much past the early Russian occupation period in the Aleutians, although they persisted longer further south. What has not yet been explored is the impact that the extinction of these fur seals had on the ecosystem and their closest relatives, Steller sea lions.

Crockford identified Steller sea lion pups in one of the late pre-contact Adak sites. Sea lions of this age are rare in archaeological assemblages throughout their range. This implies that a rookery must have existed close to the site. We know very little about the history of sea lion populations over the last ten thousand years (but see McKechnie and Wigen 2011). The assumption seems to be that sea lions have always been abundant in the North Pacific, but this may not be valid. As Crockford has suggested, competition with locally-breeding fur seals may have kept their numbers low (or out of human range) until fur seals had become extinct.

Overall, the Aleutian pattern of bird harvest is markedly different from one time period to the other and one island or island group to the other. Some researchers attribute such differences across the Aleutians to climate change (temperature and precipitation) over time (Causey *et al.* 2005; Savinetsky, Kiseleva and Khassanov 2004). The archaeological record (Crockford, Chapter 6 this volume; West and Hatfield Chapter 14 this volume) and the ethnographic record (Black 1984; Jochelson 1925) suggest that Unangan hunters targeted specific birds for reasons other than food. These reasons include species preferences for bird skin clothing, bone tools, and feathers for ornamentation. If Unangan hunters did not kill birds in the proportions that they occur in nature (Yesner 1981) then avian taxa in archaeological sites do not necessarily reflect changes in the natural world. They do, however, tell us about changing human tastes, both literally and stylistically. Moreover, specific local environmental conditions including geographically determined suitable nesting sites or access to adequate food sources might account for the different proportions of birds found in sites rather than shifting climate. Corbett *et al.* (2008) reported that Unangan hunters took prey species most intensively within 4km of villages. If this statement is true, then we might expect that the birds found in any archaeological site reflect those species that are most accessible and not the true picture of bird species occurring in any given island group. We propose that climatic change might be better determined by studying geomorphologic change, peat and diatoms in natural contexts, and stable isotopes of appropriate faunal species recovered from archaeological sites (e.g., Savinetsky *et al.* Chapter 5 this volume).

Local Unangan travel within the Andreanof Island group is upstaged by either long-distance voyages, trade, or raiding evidenced by the discovery of obsidian on north Adak. The only currently known sources of Aleutian obsidian are at Okmok and Akutan in the eastern Aleutians. Trace element analysis adds an important dimension to the study of the prehistoric movement of certain goods including, in our case, obsidian. Analysis of the Adak obsidian reveals that it probably originated at Okmok volcano on Umnak (Nicolaysen *et al.* Chapter 11 this volume). Although relatively common in eastern Aleutian sites, the incidence and small size of this volcanic glass 1046km from its source suggests it was coveted to the degree that it was, in some currently unknown way, moved from its source west along the archipelago. The small size of Adak obsidian sharpening flakes implies that cores and tools were curated until they possessed absolutely no remaining use life. If Unangan seafarers risked the 2092km round trip to obtain the parent material for tools we might expect to find larger cores, flakes, and more tools in Adak archaeological sites. It is tempting to suggest that small flake size indicates some type of trade down the chain. People living on Adak at, or near, the end of the trade network received relatively small pieces of obsidian from neighbors immediately to the east. Is it possible that central Aleutian peoples were trading sea mammal pelts for the highly valued obsidian, found only in the eastern Aleutians? This hypothesis may be impossible to test using mitochondrial DNA unless sea mammal pelts were processed with the foot bones attached. Moreover, it is important to remember that the Okmok obsidian occurs naturally in thin seams and may not have initially produced large cores.

This rare resource, along with cherts, occurred most frequently in the earliest Adak site (ADK-171), but was relatively rare at ADK-012 and early and late components at ADK-011. Both obsidian and fine-grained cherts decreased over time as people subsequently relied more heavily on basalts and andesites. This shift from fine to coarse-grained stones over time correlates with a transition from formal to informal tool technology (Wilmerding and Hatfield Chapter 12 this volume). A similar transition is seen in sites within Unalaska Bay in the eastern Aleutians where obsidian and chert comprise a larger percentage of the lithic materials in sites dating to 9000 years BP compared to sites dating after 5000 years BP (Hatfield 2006). This shift away from the use of fine-grained stone and the increasing reliance on coarse-grained lithic materials also corresponds with an increase in informal tools later in time in Unalaska Bay (Hatfield 2006). The decrease in obsidian and chert in the Aleutians may reflect either increasing restriction of certain resources as population densities increased (limiting access to these materials) or decreased importance of tools dependent on these fine-grained materials as other technologies, e.g., bone tools or groundstone tools, increased in importance.

Microscopic use-wear analysis is a valuable addition to any lithic study in that it provides information on how stone artifacts were used that goes beyond the mere shape of the artifact (Kay Chapter 13 this volume). Use-wear analysis indicates a wide array of chipped stone tool function, long use-lives, and recycling on Adak Island. Use wear analysis reveals a number of graving tools used for bone tool manufacture. Although bone preservation can be poor in Aleutian sites (e.g., ADK-SW, ADK-011 Component 1), use wear analysis demonstrates that bone tool manufacture comprised an important technology in all Adak archaeological sites. Although Kay finds the 'usage of basalt for chipped stone tools…curious' Wilmerding and Hatfield (Chapter 12 this volume) note that the shift over time from fine-grained to coarse-grained stones for stone tools is a pattern found in both the eastern and central Aleutians.

CAAPP research also showed the value of petrographic analysis in Aleutian lithic studies. Nicolaysen and Johnson (in Jeannotte Chapter 14 this volume) performed petrographic analysis on flat andesite samples, ethnographically known as griddle stones, from the stratigraphic spread of ADK-011 and two potential source locations on north Adak. Petrographic analysis shows that the source for most of the ADK-011 flat slabs is an andesite outcrop located approximately 5km distance from ADK-011 across Sitkin Sound. Early Aleutian observers (e.g., Jochelson 1925) reported that the Unangan people used flat stones for cooking or warming food. These stones are ubiquitous in most Aleutian sites, and archaeologists currently take it for granted that these were employed for some type of culinary activity. Historical interpretations are subjective, and biases can influence the interpretation of the observer. With this in mind, Richard Jeannotte (Chapter 14 this volume) analyzed the organic residues on flat, andesite stones recovered from archaeological site ADK-011 to determine if these flat stones, many of which are charred, were actually used to heat or cook food. $\delta^{13}C$ stable isotope analysis reveals that the signatures found on the griddle stones could represent four possible sources: (1) the driftwood fuel used to the heat the stones, (2) roots in the archaeological site, (3) other archaeological site matrix, and/or (4) burned organic residues associated with the 40 Year ash. A preliminary investigation using VIS/NIR spectroscopy and principal component analysis reveals that the chemical fingerprint of the residues are different from the archaeological site matrix but share chemical features with burned residues associated with ash and plants ('vegetable material'), but do not clearly show the presence of chemical features associated with potential Unangan food sources. This 'vegetable material' could represent the driftwood used to heat the stones. However, the current biochemical complexity and an apparent absence of evidence does not suggest that the flat, andesite stones were not used for some culinary activity. We simply probably have not teased out the appropriate biochemical signatures for possible food molecules yet.

It's a Question of Time: Comments on Adak Radiocarbon Dates

A precise chronology is essential to testing hypotheses concerning correlations and causes of events in past natural and human systems. North Pacific shell, fishbone, and other materials in the marine food web have radiocarbon ages 600-1000 years older than those for coeval terrestrial material (Dumond and Griffin 2002; Kuzmin *et al.* 2001; Stuiver *et al.* 1986; Yoneda *et al.* 2000) due to the strong upwelling of aged (radiocarbon-depleted) subsurface waters into the North Pacific mixed layer. This offset or reservoir age must be taken into account when evaluating radiocarbon dates on marine materials and terrestrial consumers of marine diets in order to avoid major errors in interpretation. The correction is typically determined by radiocarbon dating shells of known age or other marine materials collected before excess ^{14}C from nuclear testing entered the ocean, or by comparing ^{14}C ages of associated marine and terrestrial material from archaeological sites or paleoshoreline deposits.

The paucity of reservoir age data currently available for the Aleutians is disturbing. Along with required corrections being as high as 600-1000 years (Stuiver *et al.* 1986), oceanographic variability along the Aleutian chain suggests that significant spatial variations may be present (Favorite *et al.* 1976). Observed interannual and decadal shifts in regional climate and oceanographic indices (Stabeno and Reed 1992; Trenberth and Hurrell 1994; Verkhunov and Tkachenko 1992), which could in principle affect upwelling strengths, also raise the possibility of temporal changes, although comparisons of associated wood and shell dates from coastal sites in British Columbia (Southon *et al.* 1990; Southon and Fedje 2003) have thus far shown no evidence for large-scale shifts.

Radiocarbon dates used in this volume are derived from four different sources (Appendix 1.B). CAAPP collaborator, John Southon, Keck Laboratory (UCIAMS), University of California, Irvine, AMS dated archaeological samples of charcoal and bone from ADK-171, ADK-012, ADK-011, ADK-084, and ADK-193. Arkady Savinetsky's laboratory (IEMAE) radiocarbon dated samples from ADK-009 and ADK-171. Prior to the CAAP project US Bureau of Indian Affairs, US Fish and Wildlife, and the Western Aleutians Archaeological and Paleobiological Project used Beta Analytic (Beta) for radiocarbon dates. Additionally, Mitsuru Okuno AMS dated charred materials associated with volcanic tephra layers on north Adak and Kanaga Island.

The CAAP project attempted to address the problems of reservoir age variability and the current lack of data for the central Aleutians by collecting terrestrial organics (wood, charcoal, terrestrial avifauna) for dating comparisons with associated marine materials (shell, fishbone, and marine mammal remains) whenever possible. Additionally, CAAPP coupled AMS dates on bone and charcoal with tephrochronology.

Appendix 1.B demonstrates the need to collect more than one organic sample for dating any Aleutian archaeological site. Some dates simply do not seem to reflect the actual timing of Unangan occupation. An excellent example is the 7760±20 years BP date (UCIAMS-28906) on charcoal from ADK-011 Layer 1. The date of this large chunk of burned wood is clearly anomalous, and it does not correlate with other organic material dates from ADK-011 Component 2 (175-395 years BP). Although Southon assured CAAPP that the sample was not contaminated, the most parsimonious explanation could be that Unangan villagers discovered, collected and burned an extremely old chunk of driftwood during this relatively recent occupation. A second anomalous date comes from the Swetzoff Site on a small island in Clam Lagoon. A stone lamp recovered during 2005 test excavations is located relatively high (50cm below ground surface) in the excavation levels, yet charred residue scraped from it dates to 3835±20 BP (UCIAMS-22149). Charcoal from house floors excavated in 2007, located in 92cm deep strata well below where the lamp was found, are dated to 1735± 25 BP (UCIAMS-42405), 1745 ±20 BP (UCIAMS- 61684) and 1755± 20 BP (UCIAMS-42404). These date disparities could suggest that the lamp was transported from its older, manufacturing location to ADK-283 for re-use as a lamp or for some other purpose. Lamp manufacture would have been a labor-intensive practice, and prehistoric Unangan people possibly opted to recycle labor-intensive tools when they discovered them in older, abandoned sites. If this is the case, such recycled equipment does not provide accurate dates for the occupation of a particular site. On the other hand, if the residue represents sea mammal oil, then the date may be artificially old due to the marine reservoir effect.

At ADK-012, CAAPP selected a variety of organic materials, including rockfish, charcoal and terrestrial bird, for dating. The rockfish dated at 2455±20 BP (UCIAMS-22162) and 2470±15 BP (UCIAMS- 22273) while charcoal dated at 1515±15 BP (UCIAMS-22158) and 1480±20 (UCIAMS-61685). The terrestrial bird leg bone dated at 1865±15 BP (UCIAMS-22161). The rockfish, a marine organic, probably dates too old and the terrestrial charcoal and bird dates probably more reasonably represent the true date of the ADK-012 occupation.

At ADK-011, Susan Crockford was asked to identify bones of terrestrial birds for dating. These terrestrial bone dates were subsequently compared with charcoal dates from ADK-011 deposits. Ptarmigan, bunting, and finch bones place ADK-011 Component 2 layers within the 170-450 BP time range (see Appendix 1.B for greater detail). Wood and charcoal dates fall largely within this time framework.

Joan Coltrain (West *et al.* Chapter 16 this volume) conducted stable carbon and nitrogen isotope analyses on a rib of a Unangan child exhumed in the Andreanof Islands. The Andreanof child's stable isotope ratios (δ^{13}C ‰=-13.9; δ^{15}N ‰ = +17.0) compare favorably with values (δ^{13}C ‰=-13.4; δ^{15}N ‰ = +17.8) of an elderly adult recovered in the Near Islands (West *et al.* 2003; West *et al.* Chapter 16 this volume). Unangan skeletons sampled at Kagamil, Shiprock, and Umnak in the eastern Aleutians (N=36) had δ^{13}C ‰ levels ranging between -12.0±0.3 and -12.9±0.1 and δ^{15}N ‰ levels ranging between +19.5±0.5 and 20.1±0.5 (West *et al.* Chapter 16 this volume). Although sample size is small, preliminary stable isotope analyses seem to indicate that central and western Unangan peoples consumed lower trophic level marine mammal tissue and more sea birds and invertebrates. Although a rib from the child burial dated to 840 ± 40 BP, this date is probably artificially old due to the Marine Reservoir Effect. Shellfish extract ancient carbon from their marine environment. When these shells are dated they produce dates much older than when the organism actually lived and died. Because the child relied on a diet high in shellfish, this ancient carbon was incorporated into the child's bones. In West *et al.* (Chapter 15 this volume) we sliced 500 years off the child's demise, which puts the time of death roughly around 350 years ago.

Final Comments

> 'Archaeological data have been accumulating continuously for several centuries and new data

> are traditionally held to constitute a test of earlier interpretation. Yet what data are collected by what methods are influenced by every archaeologist's sense of what is significant, which in turn reflects his theoretical presuppositions' (Trigger 1989, 15-16).

Several authors in this volume have commented on the small sample sizes collected by the CAAP project. Kay (Chapter 13 this volume) notes: 'Small scale excavation or other recovery programs well suited to highlight environmental conditions and subsistence trends are not always as adept at measuring technological response. It should be clear too that systematic technological analyses require, as a minimum, an adequate sampling of the settlement system. Rarely, if ever, does that system fit conveniently in a 1-meter square or even several of them, no matter how nice it might be if it did.'

Such frustration is understandable and ubiquitous in archaeology. Analysts in subspecialties sometimes under-appreciate the extent to which their needs and objectives must be juggled against the big picture. Had we opted to excavate an entire house at the well preserved ADK-011 Component 2 at the sacrifice of other site exploration we would know *much more* about late prehistoric Unangan dwelling construction, but we would know *much less* about change through time in the central Aleutians. Had we opened a larger area at the ADK-171 *Clinocardium* midden, the site might be totally exhausted for future research. Had we opted to recover as many stone tools as possible, the biological samples might have suffered or analyses would have taken longer to complete. Or, as in some other cases, large quantities of artifacts and ecofacts could sit unanalyzed on museum shelves in memoriam.

Given the amount of time and funds available, we opted to test, excavate, and thoroughly analyze as many materials from as many north Adak sites as possible to gain the maximum amount of geological, environmental and cultural information possible. CAAPP water screened nearly all matrix using 1/8 inch mesh. Scientists and students sorted all water screened matrix recovering small bones, shell fragments, and lithic debitage. This was an arduous and time consuming task under field conditions that were often appallingly difficult. However, had we used any other procedures we would have missed the small obsidian debitage, the fragile long-nose lancetfish bones, and the smallest invertebrates. We have provided the Unangan and scientific communities with a never before possible insight into the prehistoric central Aleutian world because we worked in a region where few previous archaeologists were willing to conduct in depth data collection and analyses. Quite frankly, if it hadn't been difficult, it would already have been done.

Finally, any kind of archaeological research requires gleaning information from publications, dissertations and unpublished reports. Much central Aleutians data are stored at government agencies and are hard to come by. If nothing else, this volume ends with a plea to finance government agencies in order that the massive volumes of gray literature on file can be digitized and be made more easily available upon request for future research endeavors. The Aleutians form the southern boundary for the most biologically and commercially important region of the Pacific—the Bering Sea—and interdisciplinary work is key to understanding this remote region. Such interdisciplinary programs, by their very nature, require pooling data from diverse sources: we cannot build on previous knowledge when original data sources are not readily available.

References

Black, Lydia. 1984. *Atka, An Ethnohistory of the Western Aleutians. Alaska History*. Kingston, Ontario, Canada, The Limestone Press.

Black, Theodore P. 1976. Geology of Umnak Island, Eastern Aleutians, as Related to the Aleuts. *Arctic and Alpine Research* 8(1), 7-35.

Corbett, Debra G., Douglas Causey, Mark Clementz, Paul L. Koch, Angela Doroff, Christine Lefèvre, and Dixie West. 2008. Aleut Hunters, Sea Otters, and Sea Cows, in C. Rick Torben and Jon M. Erlandson (eds.), *Human Impacts on Ancient Marine Ecosystems: A Global Perspective*, 43-75. Berkeley, California, University of California Press.

Crockford, Susan. 2011. Archaeozoology of Adak Island: 6000 Years of Subsistence History in the Central Aleutians. Chapter 6, in D. West, V. Hatfield, E. Wilmerding, C. Lefèvre, L. Gualtieri (eds.), *The People Before: The Geology, Paleoecology and Archaeology of Adak Island, Alaska*. Oxford, British Archaeological Reports.

Dumond, Don E. and Dennis G. Griffin. 2002. Measurements of the Marine Reservoir Effect on Radiocarbon Ages in the Eastern Bering Sea. *Arctic* 55(1), 77-86.

Estes, James A. 1990. Growth and Equilibrium in Sea Otter Populations. *Journal of Animal Ecology* 59, 385-401.

Favorite, F., A. J. Dodimead and K. Nasu. 1976. Oceanography of the Subarctic Pacific Region 1960-71, *International North Pacific Fisheries Commission Bulletin* #33, 187pp.

Garong, Ame, Chiozuru Takashima, Shin Nishida, Akihiro Kano, and Hiroko Koike. 2011. Carbon, Nitrogen, and Oxygen Isotopic Compositions of Bone Collagen and Tooth Enamel Carbonate of the Sea Otter *Enhydra lutris* Archaeological Sites at Adak Island, Aleutian Archipelago, Chapter 8, in D. West, V. Hatfield, E. Wilmerding, C. Lefèvre, L. Gualtieri (eds.), *The People Before: The Geology, Paleoecology and Archaeology of Adak Island, Alaska*. Oxford, British Archaeological Reports.

Gualtieri, Lyn, Brenn Sarata, Mitsuru Okuno, and Dixie West. 2011. Did Holocene Paleoenvironmental Factors Affect Ancient Aleut Occupation and Settlement in the Central Aleutian Islands? Chapter 3, in D. West, V. Hatfield, E. Wilmerding, C. Lefèvre, L. Gualtieri (eds.),

The People Before: The Geology, Paleoecology and Archaeology of Adak Island, Alaska. Oxford, British Archaeological Reports.

Hanson, Diane K., Debra G. Corbett, David P. Staley, and Kimberly Fleming. 2008. Upland Sites in the Aleutian Archipelago. Paper presented at the 73rd Annual Meeting of the Society for American Archaeology, Vancouver, British Columbia.

Hanson, Diane and Debra G. Corbett. 2010. Shifting Ground: Archaeological Surveys of Upland Adak Island, the Aleutian Islands, Alaska and Changing Assumptions of Unangan Land Use Patterns. *Polar Geography* 33, 165-178.

Hatfield, Virginia L. 2006. Historical Continuity from Shemya to Dutch Harbor: An Evolutionary Analysis of Chipped Stone Technology in the Aleutian Islands. Unpublished PhD. Dissertation, Department of Anthropology, University of Kansas, Lawrence.

Jeannotte, Richard, Kirsten Nicolaysen, Taylor Johnson, Dixie West. 2011. Griddlestones from Adak Island, Alaska: Their Provenance and Biological Origins of Their Associated Organic Residues. Chapter 14 in D. West, V. Hatfield, E. Wilmerding, C. Lefèvre, L. Gualtieri (eds.), *The People Before: The Geology, Paleoecology and Archaeology of Adak Island, Alaska.* Oxford, British Archaeological Reports.

Kay, Marvin. 2011. Clam Lagoon Archaeology and Technology. Chapter 13 in D. West, V. Hatfield, E. Wilmerding, C. Lefèvre, L. Gualtieri (eds.), *The People Before: The Geology, Paleoecology and Archaeology of Adak Island, Alaska.* Oxford, British Archaeological Reports.

Koike, Hiroko, Chizuru Takashima and Akihiro Kano. 2011. The Aleut Cockle, *Clinocardium nuttallii* (Conrad, 1837) from ADK-011, Adak Island, Alaska. Chapter 9 in D. West, V. Hatfield, E. Wilmerding, C. Lefèvre, L. Gualtieri (eds.), *The People Before: The Geology, Paleoecology and Archaeology of Adak Island, Alaska.* Oxford, British Archaeological Reports.

Kuzmin, Yaroslav V., G. S. Burr and A. J. Timothy Jull. 2001. Radiocarbon Reservoir Correction Ages in the Peter the Great Gulf, Sea of Japan, and Eastern Coast of the Kunashir, southern Kuriles (northwestern Pacific), *Radiocarbon* 43, 477-481.

Larson, Shawn, Ronald Jameson, James Bodkin, Michelle Staedler and Paul Bentzen. 2002. Microsatellite DNA and Mitochondrial DNA Variation in Remnant and Translocated Sea Otter (*Enhydra lutris*) populations. *Journal of Mammology* 83(3), 893–906.

McKechnie, I. and Wigen, R. J. 2011. Towards a Historical Ecology of Pinniped and Sea Otter Hunting Traditions on the Coast of Southern British Columbia, in T. Braje and T. Rick (eds.), *Human and Marine Ecosystems: Archaeology and Historical Ecology of Northeastern Pacific Seals, Sea Lions, and Sea Otters*, 129-166. Los Angeles, University of California Press.

Nicolaysen, Kirsten, Taylor Johnson, Elizabeth Wilmerding, Virginia Hatfield, Dixie West, and Robert McGimsey. 2011. Provenance of Obsidian Fragments Recovered from Adak Island, Central Aleutian Islands: Evidence for Long-distance Transport of Raw Lithic Material. Chapter 11 in D. West, V. Hatfield, E. Wilmerding, C. Lefèvre, L. Gualtieri (eds.), *The People Before: The Geology, Paleoecology and Archaeology of Adak Island, Alaska.* Oxford, British Archaeological Reports.

Nishida, Shin, Dixie West, Susan Crockford, and Hiroko Koike. 2011. Ancient DNA Analysis for the Sea Otter (*Enhydra lutris*) from Archaeological Sites on Adak, Aleutian Islands. Chapter 7 in D. West, V. Hatfield, E. Wilmerding, C. Lefèvre, L. Gualtieri (eds.), *The People Before: The Geology, Paleoecology and Archaeology of Adak Island, Alaska.* Oxford, British Archaeological Reports.

Okuno, Mitsuru, Keiji Wada, Toshio Nakamura, Lyn Gualtieri, Brenn Sarata, Dixie West, Masayuki Torii. 2011. Holocene Tephra Layers on the Northern Half of Adak Island in the West-central Aleutians, Alaska. Chapter 3 in D. West, V. Hatfield, E. Wilmerding, C. Lefèvre, L. Gualtieri (eds.), *The People Before: The Geology, Paleoecology and Archaeology of Adak Island, Alaska.* Oxford, British Archaeological Reports.

O'Leary, Matthew. 2001. Volcanic Ash Stratigraphy for Adak Island, Central Aleutian Archipelago. In Don Dumond (ed.), *Archaeology in the Aleut Zone of Alaska, Some Recent Research*, 215-234. University of Oregon Anthropological Papers No. 58. Eugene, University of Oregon Press.

Orchard, Trevor J. and Susan J. Crockford. 2010. Size Estimation of Pacific cod (*Gadus macrocephalus*) Remains from Shemya Island. In Debra Corbett, Dixie West and Christine Lefèvre. *The People at the End of the World: The Western Aleutians Project and the Archaeology of Shemya Island*, 93-98. Anchorage, Alaska. Aurora Monograph Series-VIII.

Savinetsky, Arkady B., Nina Kiseleva, and Bulat Khassanov. 2004. Dynamics of Sea Mammal and Bird. Populations of the Bering Sea Region Over the Last Several Millennia. *Palaeogeography, Palaeoclimatology, Palaeoecology* 6, 335-352.

Savinetsky, Arkady B., Nina Kiseleva, and Bulat Khassanov. 2010. Paleoenvironment—Holocene Deposits from Shemya Island. In Debra Corbett, Dixie West and Christine Lefèvre (eds.), *The People at the End of the World: The Western Aleutians Project and the Archaeology of Shemya Island*, 71-82. Anchorage, Aurora Monograph Series-VIII, Alaska Anthropology Association.

Savinetsky, Arkady B., Dixie West, Zhanna A. Antipushina, Bulat F. Khassanov, Nina K. Kiseleva, Olga A Krylovich, and Andrei M. Perelodov. 2011. The Reconstruction of Ecosystems History of Adak Island (Aleutian Islands) During the Holocene. Chapter 5 in D. West, V. Hatfield, E. Wilmerding, C. Lefèvre, L. Gualtieri (eds.), *The People Before: The Geology, Paleoecology and Archaeology of Adak Island, Alaska.* Oxford, British Archaeological Reports.

Southon, John R., D. Erie Nelson and John S. Vogel. 1990. A Record of Past Ocean-Atmosphere

Radiocarbon Differences from the Northeast Pacific, *Paleoceanography* 5, 197-206.

Southon, John R. and Daryl Fedje. 2003. Post-glacial Record of 14C Reservoir Ages for the British Columbia Coast, *Canadian Journal of Archaeology* 27, 95-111.

Stabeno, Phyllis J. and Ronald K. Reed. 1992. A Major Circulation Anomaly in the Western Bering Sea, *Geophysical Research Letters* 19 (16), 1671-1674.

Stuiver, Minze, G. W. Pearson and T. F. Brazuinas. 1986. Radiocarbon Calibration of Marine Samples back to 9000 cal BP, *Radiocarbon* 28, 980-1021.

Trenberth, Kevin E. and J. W. Hurrell. 1994. Decadal Atmosphere-ocean Variations in the Pacific, *Climate Dynamics* 9, 303-319.

Trigger, Bruce G. 1989. *A History of Archaeological Thought*. Cambridge, Cambridge University Press.

Verkhunov, Alexander V. and Y. Y. Tkachenko. 1992. Recent Observations of Variability in the Western Bering Sea Current System, *Journal of Geophysical Research* 97, 14,369-14,376.

Waythomas, Christopher F. 1995. Surficial Geologic Map of Northern Adak Island, Alaska. United States Geological Survey Open-File Report 95-128. Anchorage, Alaska.

West, Dixie, Christine Lefèvre, Elizabeth Wilmerding, Virginia Hatfield, and Lyn Gualtieri. 2011. The Central Aleutians Archaeological and Paleobiological Project. Chapter 1 in D. West, V. Hatfield, E. Wilmerding, C. Lefèvre, L. Gualtieri (eds.), *The People Before: The Geology, Paleoecology and Archaeology of Adak Island, Alaska*. Oxford, British Archaeological Reports.

West, Dixie, Zhanna Antipushina, Arkady B. Savinetsky, and Olga Krylovich. 2011. Invertebrate Remains: A Pan-Aleutian Comparison. Chapter 10 in D. West, V. Hatfield, E. Wilmerding, C. Lefèvre, L. Gualtieri (eds.), *The People Before: The Geology, Paleoecology and Archaeology of Adak Island, Alaska*. Oxford, British Archaeological Reports.

Wilmerding, Elizabeth and Virginia Hatfield. 2011. Six Thousand Years of Lithic Technology on Adak in a Broader Aleutian Context. Chapter 12 in D. West, V. Hatfield, E. Wilmerding, C. Lefèvre, L. Gualtieri (eds.), *The People Before: The Geology, Paleoecology and Archaeology of Adak Island, Alaska*. Oxford, British Archaeological Reports.

Winterhalder, Bruce, and Eric A. Smith. 1981. *Hunter-gatherer Foraging Strategies: Ethnographic and Archaeological Analyses*. Chicago, University of Chicago Press.

Yoneda, Minoru, Hiroyuki Kitagawa, Johannes van der Plicht, Masao Uchida, Atsushi Tanaka, Takashi Uehiro, Yasuyuki Shibata, Masatoshi Morita, and Terufumi Ohno. 2000. Pre-bomb Marine Reservoir Ages in the Western North Pacific: Preliminary Results on Kyoto University Collection, *Nuclear Instruments and Methods* B172,377-381

www.ingramcontent.com/pod-product-compliance
Lightning Source LLC
LaVergne TN
LVHW071203210926
848482LV00015B/717

9781407309057